Advancements and Innovations in Electrochemical Conversion and Energy Storage

Khaoula Alaoui
https://orcid.org/0000-0003-1269-1796
Ibn Tofail University, Morocco

Moussa Ouakki
Ibn Tofaïl University, Morocco

Younes El Kacimi
https://orcid.org/0000-0002-8662-4644
Ibn Tofail University, Morocco

IGI Global
Scientific Publishing
Publishing Tomorrow's Research Today

Vice President of Editorial	Melissa Wagner
Director of Acquisitions	Mikaela Felty
Director of Book Development	Jocelynn Hessler
Production Manager	Mike Brehm
Cover Design	Jose Rosado

Published in the United States of America by
IGI Global Scientific Publishing
701 East Chocolate Avenue
Hershey, PA, 17033, USA
Tel: 717-533-8845 | Fax: 717-533-7115
Website: https://www.igi-global.com E-mail: cust@igi-global.com

Library of Congress Cataloging-in-Publication Data

LCCN: 2025019845 (CIP Data Pending)
ISBN13: 979-8-3373-3156-0
Isbn13Softcover: 979-8-3373-3157-7
EISBN13: 979-8-3373-3158-4
British Cataloguing in Publication Data
A Cataloguing in Publication record for this book is available from the British Library.

Table of Contents

Detailed Table of Contents

Fuel cells offer a transformative solution for clean, sustainable energy across electrochemical energy conversion and storage applications. This chapter presents an in-depth analysis of major fuel cell types, including AFC, SOFC, and PEMFC, highlighting recent advances in materials, system designs, and performance improvements. Developments such as cold start strategies, enhanced catalyst durability, and integration with batteries and electrochemical capacitors are discussed. Applications in electric vehicles, microgrids, hybrid systems, and backup power are explored. The chapter also covers intelligent control strategies using MPPT algorithms combined with fuzzy logic to boost operational efficiency under dynamic conditions. Through a combination of simulation results, academic studies, and real-world case analyses, it offers comprehensive insights into current challenges and future directions. Fuel cells are positioned as vital components in building sustainable, green energy systems, making this chapter valuable for researchers, engineers, and academicians.

Nickel–Metal Hydride (Ni-MH) batteries remain a relevant technology for safe and sustainable energy storage, especially in consumer electronics, renewable energy systems, and hybrid vehicles. The novelty of this chapter lies in its comprehensive discussion of the latest advances in metal hydride alloys for negative electrodes, particularly AB5, AB2, and A2B7 systems, highlighting their impact on hydrogen storage capacity, electrochemical performance, and long-term stability. Special attention is given to how alloy design, microstructural control, and surface modifications address the growing demand for higher energy density and durability. By bridging fundamental electrochemical principles with recent material innovations, this chapter provides new perspectives on the future potential of Ni-MH technology as a competitive and environmentally friendly alternative within the energy storage landscape.

Electrochemical energy storage and conversion (EESC) technologies are important to today's energy systems. They provide sustainable solutions to the growing global demand for efficient and green energy. Among them, metal–air batteries have emerged as promising candidates due to their high theoretical specific energies and energy densities. However, like many electrochemical systems, metal–air batteries are highly susceptible to corrosion and material degradation during operation. These batteries often operate under harsh environmental conditions. Corrosion not only compromises the structural integrity of the battery but also leads to reduced efficiency, safety concerns, and shortened lifespan. Understanding the mechanisms of corrosion in metal–air batteries is therefore essential for improving their durability.

electrical conductivity, durability, and scalability are comprehensively discussed. The most recent strategies in the realm of material hybridization, system integration, and bioinspiration are reviewed. The fast-developing intersection between natural material science and electrochemical engineering keeps driving innovation to the forefront in this interdisciplinary domain towards next-generation sustainable energy technologies.

Calcium phosphates are inorganic materials. They have good electrochemical properties due to their stability, low cost, and compatibility with biological systems. These features make them suitable for many applications. This chapter looks at their role in corrosion protection and explores opportunities for their use in energy storage.it includes an experimental study that compares the anticorrosive effectiveness of different calcium phosphates on carbon steel in harsh neutral and acidic environments. The results show significant inhibition rates, ranging from 87% to 92.5% in a 3% NaCl solution and exceeding 95% in acidic conditions. SEM/EDX analyses confirmed the formation of protective layers on the carbon steel surface, supporting the effectiveness of these inhibitors. Lastly, it addresses some key challenges of calcium phosphates as electrode materials for energy storage and looks at future opportunities to improve performance.

The shift to environmentally friendly mobility is central to slowing down climate change, cutting dependence on fossil fuels, and curbing greenhouse gas emissions, thus speeding up the worldwide use of electric vehicles. Battery systems stand at the centre of this movement as they have direct impacts on EV performance, efficiency, and sustainability. This assessment examines the electrochemical processes, strengths, and weaknesses of different contemporary EV battery types, including lithium-

Foreword

Electrochemical energy storage and conversion involve the transformation of electricity into chemical energy. Crucial technologies in this field include fuel cells, batteries, and electrolyzes, which are vital for a sustainable future. Innovations in electrochemical energy storage and conversion are critically needed to meet the growing demand for renewable energy. However, significant challenges remain in terms of performance, cost, durability, and safety of these technologies.

This book "Advancements and Innovations in Electrochemical Conversion and Energy Storage", provide insights into the latest developments in electrochemical energy storage and conversion technologies, along with their various applications in both academia and industry. This Book is written with a new approach and new philosophy, and it is hoped that it will fulfill their aspirations. This book will aim to provide relevant theoretical frameworks and the latest empirical research findings in the Advancements and Innovations in Electrochemical Conversion and Energy Storage. It will be written to provide a medium for the communication of ideas, best practices, and developments in all aspects of this field and includes both fundamental and practical aspects of novel engineering technologies.

The 1st Section of this Book focuses on electrochemical conversion technologies, and 4 specific chapters are presented.

In the 2nd section, some materials for energy storage and corrosion resistance have carefully been highlighted.

Finaly, section 3 elucidates some engineering of energy storage and applications.

This book is one of indispensable resources for professionals and researchers working in the field of emerging engineering technologies and their industrial applications.

So, the Book «Emerging Engineering Technologies and Industrial Applications» contains four sections, which contains 16 chapters. In fact, the authors presented the recent advances, and future challenges for electrochemical conversion and energy Storage and their practical application, which provides a very good reference with valuable information.

Rachid Touir

Faculty of Science, Advanced Materials and Process Engineering Laboratory, Ibn Tofail University, Morocco & Regional Center of Education and Training Professions (CRMEF), Kenitra, Morocco

Rachid Touir was born in 1977 in Kenitra City, Morocco. He received his Ph.D. in Physics Chemistry, from Ibn Tofail University, Morocco, 2009. He had published more than 100 papers in international journals. His work has been cited 1726 total citations by 976 documents with h-index = 38 and presented several communications in symposia and national/international meetings. He was honored with the award for the best thesis researcher in Ibn Tofail University since the year 2009. He has developed 3 patents and has participated in 6 national research projects. He has supervised 35 Doctoral Theses and 60 Master theses in science and didactics. He is a reviewer at several national and international journals. He is the author of a 3 books. His main research interests are in the field of Physical Chemistry, especially electrochemistry (Corrosion and its inhibition in several media for metals and alloys using various techniques). He is also interested in electroless plating of Ni-P, Ni-Cu-P, Cu-P, Ni-Pd-Cu… and their components in several corrosive medium. He is currently interested in using some quantum chemical methods like density functional theory (DFT) and other semi-empirical methods to study corrosion inhibition efficiencies of organic compounds. He is also interested in the Didactics of physics and chemistry.

Preface

Advancements and Innovations in Electrochemical Conversion and Energy Storage is a comprehensive research book that delves deep into the heat of relentless technological revolution within modern applications of electrochemical energy storage systems, which will be classified into three categories: Batteries, Electrochemical capacitors and fuel Cells. The electrochemical storage system involves the conversion of chemical energy to electrical energy in a chemical reaction involving energy release in the form of an electric current at a specified voltage and time.

Electrochemical batteries consist of electrochemical cells that convert stored chemical energy into electrical energy.

Rechargeable batteries are one of the oldest technologies for electrical energy storage systems, they are extensively used for daily needs and in numerous industrial applications.

Electrochemical batteries are made of electrochemical cells that consist of electrodes made up of different types of metals that are submersed into liquid or solid electrolytes. A flow of electrons travels from the anode to the cathode electrode, until the anode runs out of electrons. In the case of primary batteries, the process ends here. Whereas for secondary batteries that are rechargeable, this reaction can be reversed using a charger. The electrons will flow from the cathode back to the anode, restoring the initial excess of electrons in the anode.

The flow of electrons originates from reduction-oxidation chemical reactions that occur in the electrochemical cell, these are commonly called redox reactions and in the case of an electrochemical cell, one half-reaction occurs in each electrode. Oxidation is the loss of electrons and is the reaction that occurs in the anode, the electrode reacts with the electrolyte producing electrons and we say that the anode is oxidized. Whilst, reduction is the gain of electrons, and is what occurs simultaneously in the cathode, enabling it to accept electrons and we say that the cathode is reduced.

During the discharge cycle, the positive electrode is the cathode, and the negative electrode is the anode, electrons flow from the anode to the cathode generating electricity. Whereas, during the charging cycle the reverse reaction happens, and the positive electrode is the anode, and the negative electrode is the cathode.

Advancements and Innovations in Electrochemical Conversion and Energy Storage book provides emerging research exploring the theoretical and practical aspects of electrochemical energy storage for power generation within a wide span of professional disciplines. Featuring coverage on a broad range of topics such as Batteries, Electrochemical capacitors, and Fuel Cells, this book is ideally designed for engineers, scientist researchers and industrial seeking current research and technological advancements in Electrochemical conversion and energy storage in various professional scopes.

The many academic areas covered in this publication include, but are not limited to:

- Battery energy storage systems
- Rechargeable batteries
- Lead-acid batteries
- Lead-acid batteries
- Sodium-Sulfur batteries
- Flow batteries energy storage
- Fuel cell energy storage
- Electrochemical Capacitors
- Electrochemical Double-Layer capacitors
- Electrochemical energy storage data

This book equips industrial and research engineers, alongside research scientists, with the tools needed to stay relevant and at the forefront of their industries. It offers valuable insights into harnessing these technologies to improve product design development processes and contribute significantly to their respective field's growth.

THE CHALLENGES

The book "Advancements and Innovations in Electrochemical Conversion and Energy Storage" focus on developing higher-performing, sustainable, and cost-effective solutions, particularly for renewable energy integration. Key innovations include new materials, advanced battery architectures, and the use of novel technologies for better management. Topics underscores the urgency to adopt the new technologies, emphasizing that industries or companies which fail to integrate these technologies risk becoming relics of the past. The "Advancements and Innovations in Electrochemical Conversion and Energy Storage" book is written with a new approach and new philosophy, and it is hoped that it will fulfill their aspirations. This book will aim to provide relevant theoretical frameworks and the latest empirical research findings in the emerging engineering technologies. It will be written for provides a medium for the communication of ideas, best practices, developments in all aspects of this field and includes both fundamental and practical aspects of novel engineering technologies.

Target audience of the Advancements and Innovations in Electrochemical Conversion and Energy Storage will be composed of scientists, engineers, technologists, and industrials working in various field and disciplines.

ORGANIZATION OF THE EDITED RESEARCH BOOK

This book is organized into 16 chapters and covers various aspects of Advancements and Innovations in Electrochemical Conversion and Energy Storage; to make it easier to understand, we have chosen to devise it into 3 below sections. Each of these sections contains several chapters, which detail feedback from expert and specialist authors in their fields. The sections are:

Chapter 1: Advances in Fuel Cell Technologies for Efficient Electrochemical Energy Conversion and Storage.

Fuel cells offer a transformative solution for clean, sustainable energy across electrochemical energy conversion and storage applications. This chapter presents an in-depth analysis of major fuel cell types, including AFC, SOFC, and PEMFC, highlighting recent advances in materials, system designs, and performance improvements. Developments such as cold start strategies, enhanced catalyst durability, and integration with batteries and electrochemical capacitors are discussed. Applications in electric vehicles, microgrids, hybrid systems, and backup power are explored. The chapter also covers intelligent control strategies using MPPT algorithms combined with fuzzy logic to boost operational efficiency under dynamic conditions. Through a combination of simulation results, academic studies, and real-world case analyses, it offers comprehensive insights into current challenges and future directions. Fuel cells are positioned as vital components in building sustainable, green energy systems, making this chapter valuable for researchers, engineers, and academicians.

Chapter 2: Nickel-Metal Hydride Batteries a Reliable Energy Storage Solution for Consumer Electronics and Renewable Energy Storage

Nickel–Metal Hydride (Ni-MH) batteries remain a relevant technology for safe and sustainable energy storage, especially in consumer electronics, renewable energy systems, and hybrid vehicles. The novelty of this chapter lies in its comprehensive discussion of the latest advances in metal hydride alloys for negative electrodes, particularly AB5, AB2, and A2B7 systems, highlighting their impact on hydrogen storage capacity, electrochemical performance, and long-term stability. Special attention is given to how alloy design, microstructural control, and surface modifications address the growing demand for higher energy density and durability. By bridging fundamental electrochemical principles with recent material innovations, this chapter provides new perspectives on the future potential of Ni-MH technology as a competitive and environmentally friendly alternative within the energy storage landscape.

Chapter 3: Understanding Corrosion and Materials Deterioration in Metal–Air Batteries

Electrochemical energy storage and conversion (EESC) technologies are important to today's energy systems. They provide sustainable solutions to the growing global demand for efficient and green energy. Among them, metal–air batteries have emerged as promising candidates due to their high theoretical specific energies and energy densities. However, like many electrochemical systems, metal–air batteries are highly susceptible to corrosion and material degradation during operation. These batteries often operate under harsh environmental conditions. Corrosion not only compromises the structural integrity of the battery but also leads to reduced efficiency, safety concerns, and shortened lifespan. Understanding the mechanisms of corrosion in metal–air batteries is therefore essential for improving their durability.

Chapter 4: Effects of Corrosion on the Durability of Battery

The term "corrosion" refers to the degradation of metallic materials (metals and their alloys) caused by a reaction with environmental factors. This phenomenon causes serious financial and health problems. Corrosion is one of the electrochemical phenomena that manifests itself as the formation of clusters on the surface of metals, where one electrode, known as the "anode", consumes the other, known as the "cathode", which remains dormant. In reality, these clusters form as soon as the metal surface becomes heterogeneous, forming the anode and cathode domains where the two primary electrochemical reactions (oxidation and reduction) take place. For example, dissolution or passivation of active electrode materials and dissolution, oxidation or passivation of current collectors are the main causes of corrosion in batteries. An in-depth analysis of battery corrosion is necessary due to the rapid development of battery research. We begin with an overview of recent developments in corrosion and electrode protection for different battery types.

Chapter 5: Recent Advances in Supercapacitor Materials: A Review on Performance Enhancement, Flexibility, and Scalability for Next-Generation Energy Storage

Supercapacitors are attracting attention as high-performance energy storage devices, especially for flexible electronics. They offer high power density, long cycle life, and fast charge/discharge, making them strong alternatives to batteries in applications requiring quick energy delivery. This study reviews recent advances in supercapacitor materials, focusing on enhancing electrode performance through nanostructures and conductive composites. These innovations have significantly improved specific capacitance and electrochemical stability. The development of heat-resistant separators and highly conductive electrolytes has further improved performance under harsh conditions. Research is also advancing flexible supercapacitors to meet demand for wearable and portable electronics. Their improved mechanical flexibility and solid-state performance enable new tech integrations. However, large-scale production faces challenges such as high manufacturing costs and complex material synthesis. Continued efforts are needed to reduce costs and enable commercial use across energy storage sectors

Chapter 6: Design and Optimisation of Metal-Air Batteries: A Critical Review of Mechanisms and Performance

Metal-air batteries (Li-air, Zn-air, Na-air, and Al-air) represent a promising alternative for energy storage due to their high energy density and potential for sustainable applications. However, their performance is severely limited by challenges associated with metallic anodes, including corrosion, passivation, and uncontrolled hydrogen evolution, which reduce cycle life and faradaic efficiency. This critical review examines the electrochemical mechanisms associated with metallic anodes in metal-air batteries, comparing the performance of Li-air systems (3500 Wh/kg), Zn-air (300 cycle), Na-air (1000 Wh/kg), and Al-air (2790 Wh/kg). Recent solutions, such as the use of alloyed anodes, surface modifications, and the incorporation of inhibitor additives in alkaline electrolytes, have significantly improved the performance and durability of metal-air batteries. The findings from this analysis highlight recent advancements and persistent challenges in the optimisation of metallic anodes, paving the way for more efficient and durable metal-air batteries for large-scale applications.

Chapter 7: Advances in Plant-Based Materials for Sustainable Electrochemical Energy Devices

Transitions to circular and sustainable energy systems require the design of innovative material platforms that are safe for the environment and highly resource-use efficient. In this respect, plant-derived biomaterials are increasingly recognized as highly promising alternatives to conventional components in batteries, supercapacitors, and fuel cells. This chapter underlines the role that plant-based materials play in the development of greener and more sustainable energy infrastructures by presenting their integration within the above-mentioned energy systems. Major issues such as electrical conductivity, durability, and scalability are comprehensively discussed. The most recent strategies in the realm of material hybridization, system integration, and bioinspiration are reviewed. The fast-developing intersection between natural material science and electrochemical engineering keeps driving innovation to the forefront in this interdisciplinary domain towards next-generation sustainable energy technologies.

Chapter 8: Exploring the Electrochemical Potential of Calcium Phosphates: From Corrosion Inhibition to Energy Storage

Calcium phosphates are inorganic materials. They have good electrochemical properties due to their stability, low cost, and compatibility with biological systems. These features make them suitable for many applications. This chapter looks at their role in corrosion protection and explores opportunities for their use in energy storage. it includes an experimental study that compares the anticorrosive effectiveness of different calcium phosphates on carbon steel in harsh neutral and acidic environments. The results show significant inhibition rates, ranging from 87% to 92.5% in a 3% NaCl solution and exceeding 95% in acidic conditions. SEM/EDX analyses confirmed the formation of protective layers on the carbon steel surface, supporting the effectiveness of these inhibitors. Lastly, it addresses some key challenges of calcium phosphates as electrode materials for energy storage and looks at future opportunities to improve performance.

Chapter 9: A Comprehensive Review of State-of-the-Art Battery Energy Systems for Green Transportation

The shift to environmentally friendly mobility is central to slowing down climate change, cutting dependence on fossil fuels, and curbing greenhouse gas emissions, thus speeding up the worldwide use of electric vehicles. Battery systems stand at the centre of this movement as they have direct impacts on EV performance, efficiency, and sustainability. This assessment examines the electrochemical processes, strengths, and weaknesses of different contemporary EV battery types, including lithium-ion, lead-acid, nickel-based, sodium-based, and those employed in hybrid, plug-in hybrid, and fuel cell vehicles. Lithium-ion batteries are notable for their ability to store a lot of energy per unit mass, have long lifetimes, and are efficient, yet they are the most widely used battery type in most EVs. Though cheaper, lead-acid and others are generally employed in HEVs and PHEVs because of lesser performance. A comparison highlights the role of each type in EV evolution. The significance of Battery Management Systems for safety, lifespan, and performance is also highlighted.

Chapter 10: Breakthroughs in Electrochemical Innovations into Scalable Applications

Electrochemical technologies are emerging as pivotal solutions for the global transition to sustainable and scalable energy systems. This chapter bridges recent laboratory breakthroughs with pathways for industrial deployment, emphasizing innovations that extend beyond conventional proof-of-concept studies. Advances in nanostructured electrodes, solid-state electrolytes, and high-performance catalysts have redefined benchmarks of efficiency, stability, and safety. Novel manufacturing strategies including additive manufacturing and roll-to-roll processing are presented as transformative enablers of cost-effective scale-up, while AI-driven optimization accelerates the design-to-application cycle. The chapter also highlights electrochemical routes for green hydrogen production, CO2 reduction, and nitrogen fixation, which are reshaping sustainable chemical manufacturing. Distinctive case studies in battery storage and hydrogen transport illustrate how scientific progress translates into real-world applications.

Khaoula Alaoui
Ibn Tofail University, Morocco

Moussa Ouakki
Ibn Tofail University, Morocco

Younes El Kacimi
Ibn Tofail University, Morocco

Acknowledgment

The editor's would like to acknowledge the help of all the people involved in this project and, more specifically, to the authors and reviewers that took part in the review process. Without their support, this book would not have become a reality.

First, the editor's would like to thank each one of the authors for their contributions. Our sincere gratitude goes to the chapter's authors who contributed their time and expertise to this book.

Second, the editor's wishes to acknowledge the valuable contributions of the reviewers regarding the improvement of quality, coherence, and content presentation of chapters. Several authors also served as referees, we highly appreciate their double task.

Finally, we would like to thank all the IGI publishing team for professional assistance and support.

Editorial Advisory Board

Chapter 1

Advances in Fuel Cell Technologies for Efficient Electrochemical Energy Conversion and Storage.

Anusha Gaddala
https://orcid.org/0009-0006-5664-6689
SR University, Ananthasagar, India

Ram Ragotham Deshmukh
SR University, Ananthasagar, India

A. V. V. Sudhakar
SR University, Ananthasagar, India

ABSTRACT

Fuel cells offer a transformative solution for clean, sustainable energy across electrochemical energy conversion and storage applications. This chapter presents an in-depth analysis of major fuel cell types, including AFC, SOFC, and PEMFC, highlighting recent advances in materials, system designs, and performance improvements. Developments such as cold start strategies, enhanced catalyst durability, and integration with batteries and electrochemical capacitors are discussed. Applications in electric vehicles, microgrids, hybrid systems, and backup power are explored. The chapter also covers intelligent control strategies using MPPT algorithms combined with fuzzy logic to boost operational efficiency under dynamic conditions. Through a combination of simulation results, academic studies, and real-world case analy-

DOI: 10.4018/979-8-3373-3156-0.ch001

ses, it offers comprehensive insights into current challenges and future directions. Fuel cells are positioned as vital components in building sustainable, green energy systems, making this chapter valuable for researchers, engineers, and academicians.

1. INTRODUCTION

The worldwide energy sector experiences deep changes because of two concurrent reasons: first, we need to decrease greenhouse gas emissions to fight global warming, and second, we need to shift to sustainable renewable power generation. The electrical energy creation process through redox reactions represents a core element in this energy conversion transformation because it performs direct chemical-to-electric conversion without the thermal conversion methods that typically result in performance losses and emissions (Zheng et al., 2021). The direct energy conversion method enhances power plant performance while producing significantly lower pollution during electricity generation. Modern society depends heavily on electrochemical energy conversion systems to sustain its transformation towards cleaner sources of energy that replace fossil fuels. Electrochemical power systems consist mainly of batteries, supercapacitors and fuel cells, which find critical uses throughout various applications, including consumer electronics, grid storage and transportation and industrial power systems (Bagotsky et al., 2015). Fuel cells stand out as an important technology because they continuously produce electricity when given fuel together with an oxidant. The continuous operation of this technology diverges from standard batteries since they hold finite amounts of energy while needing regular recharging cycles(Goodenough, 2013).

Figure 1. Diagram of electrochemical energy conversion process (battery vs. fuel cell vs. supercapacitor) (Hou et al., 2011).

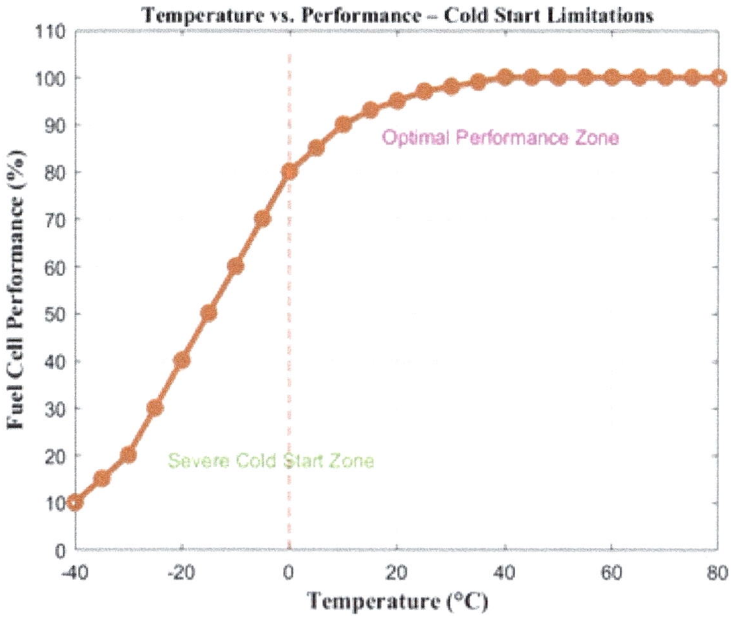

The exceptional characteristics of fuel cells are their clean operational nature and high efficiency, combined with flexibility. The use of hydrogen fuel cells presents itself as an encouraging solution for obtaining zero-emission energy generation since their sole manufacturing outputs consist of water alongside heat (Daud et al., 2011). The conversion efficiency range for these systems reaches 40% to 60% and increases further when implemented as Combined Heat and Power (CHP) systems. Fuel cells can accept multiple fuel types such as hydrogen and methanol, and additional hydrocarbon ingredients, making them versatile for different energy systems and transition approaches. Through their modular design, fuel cells can adjust power capacity from small portable units to large stationary electricity generators(Ellis et al., 2001). The modular nature of fuel cells allows them to serve as fundamental elements for decentralized power grids because they enable effective decentralization over traditional centralized power systems. FCEVs demonstrate how fuel cells help build eco-friendly transportation systems while providing an alternate driving system to internal combustion engines and helping battery electric vehicles reach their purpose(Soleimani et al., 2024).

Figure 2. Global market share or growth trends in electrochemical storage technologies(Antipov et al., 2019).

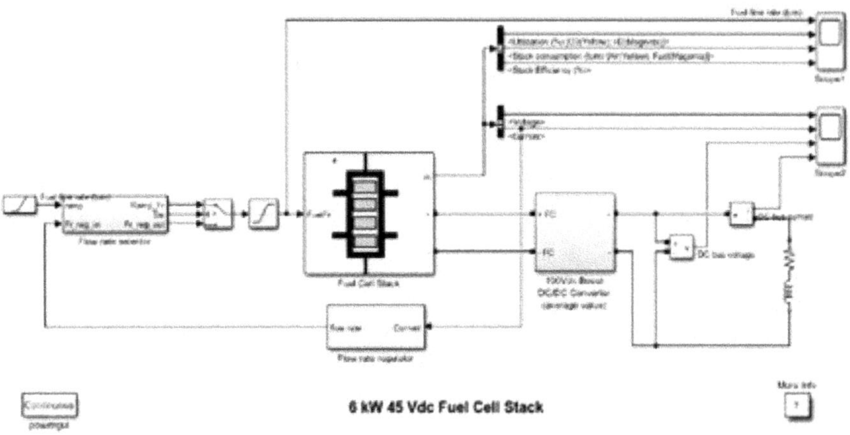

Table 1. Comparison of Energy Storage Technologies(Smallbone et al., 2017).

Parameter	Batteries	Fuel Cells	Electrochemical Capacitors
Energy Conversion Efficiency	80–90%	40–60% (up to 65% with CHP)	85–95%
Energy Density	100–265 Wh/kg (Li-ion)	800–3000 Wh/kg (with fuel supply)	1–10 Wh/kg
Power Density	250–1000 W/kg	100–2000 W/kg	1000–10,000 W/kg
Cycle Life	1000–5000 cycles	>5000 hours (depends on system)	>1,000,000 cycles
Response Time	Seconds	Seconds to minutes	Milliseconds
Start-up Time	Fast	Moderate to slow (esp. cold start)	Instantaneous
Cost (per kWh)	$150–300 (Li-ion)	$4000–10,000+ (initial setup)	$1000–10,000
Maintenance	Moderate	High (due to catalysts & system)	Low
Environmental Impact	Moderate (recycling issues)	Low (if hydrogen is green)	Low
Typical Applications	Portable electronics, EVs	Backup power, EVs, grid support	Power smoothing, regenerative braking, UPS

Fuel cells play a crucial role in renewable energy management due to their ability to store intermittent power from solar and wind before allowing its distribution back to the electricity grid (Ehteshami & Chan, 2014a). They function as essential components in building smart grids and energy systems that are both resilient against disruptions and low-carbon emission-based. Fuel cells, together with electrochemical energy systems constitute essential foundations for energy technology advancement in power utilization and production(Sharaf & Orhan, 2014). Moving towards sustainable practices requires essential fuel cell integration across clean energy systems, while smart transportation and Internet of Vehicles (Ivo) architectures demand this technology strongly. These power systems offer dependable, efficient, clean power services to different application areas, thereby establishing their significance in future energy technology development (Luo et al., 2015).

Table 2. Different types of fuel cells (Carrette et al., 2000).

Type	Electrolyte	Operating Temperature	Key Features	Applications
PEMFC (Proton Exchange Membrane Fuel Cell)	Solid polymer membrane	60–100°C	High power density, fast startup	Transportation (FCEVs), backup power
SOFC (Solid Oxide Fuel Cell)	Solid ceramic (zirconia-based)	600–1000°C	High efficiency, internal reforming	Stationary power, industrial CHP
DMFC (Direct Methanol Fuel Cell)	Solid polymer membrane	50–120°C	Uses liquid methanol directly	Portable electronics, military
AFC (Alkaline Fuel Cell)	Potassium hydroxide in water	60–250°C	High efficiency with pure hydrogen	Space applications, niche markets
MCFC (Molten Carbonate Fuel Cell)	Molten carbonate salt	~650°C	High efficiency, CO_2 tolerance	Large-scale stationary power
PAFC (Phosphoric Acid Fuel Cell)	Phosphoric acid	~200°C	Good durability, moderate efficiency	Stationary power, commercial buildings

In summary, the fundamental motivation behind fuel cell research lies in achieving efficient, low-emission electrochemical conversion. To better understand how these systems function at the cell level, the following section explains the working principles and reaction mechanisms that define fuel cell operation.

2. WORKING PRINCIPLE OF FUEL CELLS

Every fuel cell separates itself from ordinary energy production systems by using basic electrochemical reactions. The conversion of electrical energy from chemical energy takes place within fuel cells through reduction-oxidation (redox) reactions, unlike conventional thermal power plants and internal combustion engines that convert fuel heat to mechanical or electrical energy(Albatayneh et al., 2020). Direct energy conversion through this process produces better efficiency results without generating numerous safety hazards that come from combustion. The anode of a typical hydrogen fuel cell receives the most investigated H_2 hydrogen gas, which triggers its catalytic decomposition process(Yin et al., 2004). Platinum acts as a catalyst during the reaction that divides hydrogen molecules into H^+ protons along with electrons. Hydrogen dissociates into protons and electrons at the anode, followed by proton transfer through a membrane barrier while electrons produce an electric current for use in powering electronics as well as powering vehicles or transmitting energy to an electrical grid (Abu-Rub et al., 2014).

Figure 3. Schematic of a basic fuel cell (PEMFC preferred) showing anode, cathode, electrolyte, and catalyst (Khorshidian & Sedighi, 2019).

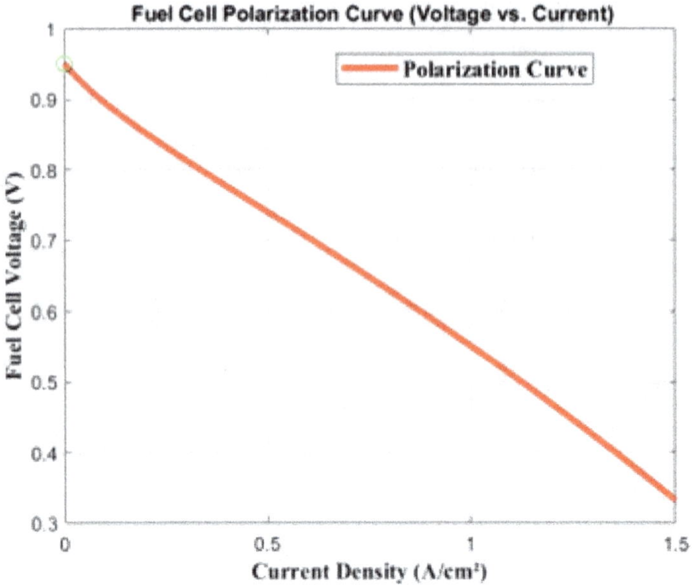

Oxygen (O_2) from ambient air enters the process at the cathode simultaneously. The electrolyte ambits allow protons to reform with external circuit electrons and oxygen molecules into water molecules (H_2O) (Dai et al., 2019). The clean byproduct formed by fuel cells depends on temperature and design and emerges either as gas vapours or liquid substances. The complete process operates with high efficiency, yielding an emissions-free outcome when using pure hydrogen as fuel. A fuel cell relies on the partnership between its four vital elements: anode, cathode, electrolyte and catalyst to power up an electrochemical reaction that produces electricity. The negative anode functions as the electrode, which enables the oxidation reactions to occur at this location. To create power in hydrogen fuel cells, the anode accepts hydrogen gas, which gets divided through catalysis into protons (H^+) and electrons (e^-) (Sorensen & Spazzafumo, 2018a). To support efficient service along with operational longevity, the component must show electrical conductivity and chemical resistance under relevant fuel-cell operating parameters. The positive electrode located at the cells opposite end operates as the cathode during reduction events. The electrolyte enables proton migration and oxygen molecule interaction, so electrons can flow through circuits while resulting in water formation as a product(Chang et al., 2004). A strong electrocatalysis together with operational stability under oxidising conditions represents essential requirements for cathode materials. The electrolyte acts as a fundamental ionic conductor which directs H^+ ions specifically through PEMFCs allowing only electron flow from the anode toward the cathode (Dickinson & Smith, 2020).

Figure 4. Electron and ion flow in a fuel cell with hydrogen and oxygen reactions(-Virkar, 2012).

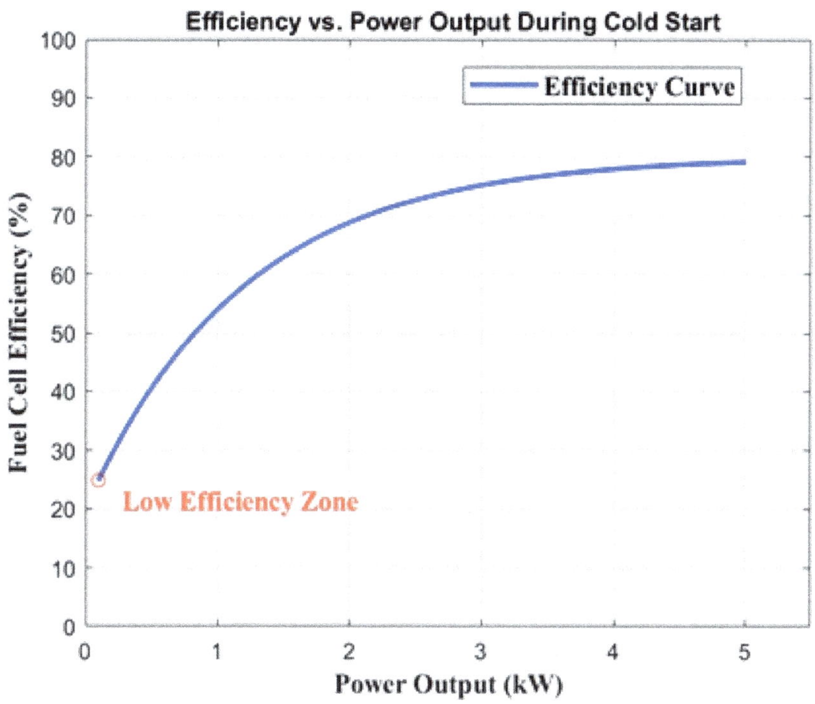

The electrons migrate through an external power circuit, which leads to power generation. The electrolyte functions as an electrical insulator, thus, it stops the occurrence of internal short circuits. At both electrodes, platinum and platinum-group metals serve as catalyst agents that enhance electrochemical reaction speeds without disappearing during operation(Hughes et al., 2021). The commercial success of fuel cells faces barriers due to both the high expense and low material availability of catalysts, which are fundamental to efficient operation. These components separately work to determine how efficiently fuel cells function and how durable they are alongside their economic viability (J. Wang, 2017). Fuel cells exhibit superior performance to classical combustion-generated power systems, with capabilities of achieving high-efficiency levels along with exceptional power density. Most of the fuel chemical energy (20-30%) produced by traditional internal combustion engines escapes into heat, while power generation reaches only 20-30%. Different types of fuel cells, along with their operational environments, enable electrical efficiencies ranging from 40% to 60%. The total system efficiency surpasses 80% when waste heat collection enables Combined Heat and Power (CHP) applications (Al-Sayyab et al., 2023). Fuel cells demonstrate their value in applications requiring the highest possible energy utilization due to their operating efficiency between 40% to 60%. The high mass-based energy density of hydrogen fuel cells enables their ideal application in the aerospace and automotive industries, along with maritime sectors, since these sectors are limited by their primary restriction. Fuel cell-powered vehicles can operate as long or longer than gasoline vehicles thanks to their high specific energy level(Pramuanjaroenkij & Kakaç, 2023).

Table 3. Theoretical vs. Actual Voltage and Efficiency of Different Fuel Cell Types(-Doddathimmaiah & Andrews, 2009)

Fuel Cell Type	Operating Temp. (°C)	Electrolyte	Theoretical Voltage (V)	Typical Actual Voltage (V)	Theoretical Efficiency (%)	Typical Efficiency (%)
PEMFC (Proton Exchange Membrane)	60–100	Proton exchange membrane	1.23	0.6–0.8	~83	40–60
SOFC (Solid Oxide)	600–1000	Solid oxide ceramic (e.g., YSZ)	1.10	0.7–1.0	~65	45–60 (up to 85 with CHP)
AFC (Alkaline)	60–90	Aqueous KOH	1.23	0.8–1.0	~83	60–70

continued on following page

Table 3. Continued

Fuel Cell Type	Operating Temp. (°C)	Electrolyte	Theoretical Voltage (V)	Typical Actual Voltage (V)	Theoretical Efficiency (%)	Typical Efficiency (%)
MCFC (Molten Carbonate)	600–700	Molten carbonate salt mixture	1.10	0.7–0.9	~65	45–55 (up to 70 with CHP)
PAFC (Phosphoric Acid)	150–220	Phosphoric acid	1.14	0.7–0.8	~70	40–50
DMFC (Direct Methanol)	50–120	Proton exchange membrane	1.21	0.3–0.6	~90	30–40

The main obstacle exists in the volumetric energy density measurement (energy per unit volume) due to gaseous hydrogen having less capacity than liquid fuels. Precise storage solutions represented by high-pressure tanks, together with cryogenic storage and solid-state storage materials, become necessary to preserve hydrogen in restricted space applications (Abdalla et al., 2018). Fuel cells demonstrate better functionality than batteries because they offer multiple operational benefits. Fuel cells provide continuous operation if they receive fuel alongside an oxidizing agent, whereas batteries store limited electrical power, which needs recharging after depletion (Peng & Chen, 2009). The ability to operate continuously and fuel efficiently makes fuel cells better choices when applications require long-duration performance and speedy recharging, along with consistent reliability, thus suitable for backup power systems and military usage and fuel cell electric vehicles (FCEVS)(Abdin, 2024a). Fuel cells surpass most battery systems by lacking limitations regarding scalability and modular design. The ability to stack these power cells according to need allows them to provide from milliwatt-sized electronics power up to megawatt-class industrial power generation (Das et al., 2017).The electrochemical principles discussed above form the basis for various fuel cell architectures developed to meet different application requirements. The next section categorizes these major fuel cell types, highlighting their unique materials, temperature ranges, and operational features.

3.TYPES OF FUEL CELLS

Figure 5. Individual illustrations for each type: PEMFC, SOFC, AFC, MCFC, PAFC (each with working temperature, electrolyte, and reaction)(Mehr et al., 2021a).

✓ Proton Exchange Membrane Fuel Cells (PEMFC)

PEMFCS, which stands for Proton Exchange Membrane Fuel Cells and Polymer Electrolyte Membrane Fuel Cells, represent the most researched and commercially available fuel cell type because they manage to be efficient and versatile while maintaining compact dimensions (G. Zhang et al., 2022). An electrochemical reaction in PEMFC depends on a solid polymer electrolyte membrane to conduct protons (H^+ ions) between the anode and cathode ends. PEMFCS function efficiently within an operating temperature range of 60°C to 100°C because this allows them to start up quickly, therefore making them good for motorized vehicles and portable devices(Rosli et al., 2017). The PEMFC design contains features that enable it to generate high power densities by delivering considerable amounts of power from a minimal size and weight structure. PEMFCS provide excellent transportation applications in fuel-cell electric vehicles (FCEVS) because of their small footprint and weight, which is essential for vehicles. PEMFC operates quietly while they scale and produce clean energy, making them suitable for stationary backup use, residential power units and portable electronic devices(Qasem & Abdulrahman, 2024a).

The operation of PEMFC demands static high-purity hydrogen gas as anode fuel alongside oxygen or ambient air at the cathode. Platinum-based catalysts enable redox reactions between these gases that are located on both electrode surfaces (Huang et al., 2021). The anode stage produces hydrogen molecule decomposition into protons and electrons. The electrolyte membrane allows hydrogen protons to travel toward the cathode as the electrons use an external circuit to create electrical power (Badwal et al., 2006). The PEMFC energy conversion technology produces only water (H_2O) emissions through water formation when combining protons and electrons with oxygen at the cathode. The technical and economic obstacles alongside PEMFC advantages, which include zero emissions, high efficiency and fast dynamic response, but also modular design, benefit the technology (Qi & Wu, 2011). The successful operation of PEMFC cells depends on three main limitations: the high costs of platinum catalysts and their limited availability, and the strict standards for water management and temperature maintenance and hydrogen purity because catalyst degradation and shorter life occur from impurities in hydrogen(Becker et al., 2023). The performance and commercial potential of PEMFCS receive ongoing improvement through continued advancements in materials science, as well as membrane technology and hydrogen infrastructure development. PEMFCS establish a central position within sustainable low-carbon energy systems because they thrive in green transportation fields together with decentralised energy applications(Abdin, 2024b).

Anode Reaction: $H_2 \rightarrow 2H^+ + 2e^-$

Cathode Reaction: $\frac{1}{2}O_2 + 2H^+ + 2e^- \rightarrow H_2O$

Overall, Cell Reaction: $H_2 + \frac{1}{2}O_2 \rightarrow H_2O$

✓ **Solid Oxide Fuel Cells (SOFC)**

Solid Oxide Fuel Cells (SOFCS) operate as high-temperature fuel cells through their solid ceramic electrolyte, combined with heat-operating temperatures between 600°C to 1000°C. The standard electrolyte of SOFCS employs yttria-stabilised zirconia (YSZ) that behaves as an excellent oxygen ion (O^{2-}) conductor between the cathode and anode (Brett et al., 2008). The electromagnetic reactions that power electricity generation proceed without liquid electrolytes through ion movement that benefits the system's mechanical stability. The SOFC technology benefits from its high operating temperature through multiple distinctive advantages. The major benefit of these cells emerges from their capability to transform hydrocarbon fuels, including natural gas and biogas and additional light hydrocarbons directly inside the fuel cell array (Sorensen & Spazzafumo, 2018b). Inside the cell, the stack operates its reforming procedure, which removes requirements for external reformers, thus streamlining the total system design and enhancing fuel adaptability. SOFCS provide the advantage of operating with various fuel types instead of requiring pure hydrogen because of their internal hydrocarbon fuel reforming ability(Boder & Dittmeyer, 2006).

The primary benefit of using SOFC technology involves its outstanding electrical efficiency capacity that extends between 45% and 60% while operating independently for power generation. CHP systems with integrated waste heat recovery enable power production to increase beyond 85% efficiency when operating at high levels (Corigliano et al., 2022a). SOFCS demonstrate outstanding efficiency, which makes them ideal for a stand-alone power supply in commercial facilities alongside industrial installations and utility sites, with their controlled environments suitable for high-temperature operation. SOFCS serve distributed power generation needs because they generate dependable electrical power with simultaneous heat output for industrial facilities, as well as data centres, hospitals, and residential communities(Corigliano et al., 2022b) .

The system's flexible design allows manufacturers to use it in industrial cogeneration systems for cost-efficient heat and power generation operations. SOFCS undergo research to embrace renewable energy systems where they offer persistent dispatchable power output during wind or solar periods of absence(STRANIERO, 2016). Approximately one major obstacle for SOFCS arises from high operating temperature requirements that create difficulties with thermal control and material reliability, together with slower system start-up functions. Advanced ceramic materials, along with high temperature sealing technologies, increase the initial system costs and make the system more complex(Bartnitzek et al., 2016). Progress in materials science, together with fabrication techniques and system design, actively solves current limitations to improve the market prospects of SOFCS(Zakaria et al., 2020).

Anode Reaction: $H_2 + O^{2-} \rightarrow H_2O + 2e^-$

Cathode Reaction: $\frac{1}{2}O_2 + 2e^- \rightarrow O^{2-}$

Overall, Cell Reaction: $H_2 + \frac{1}{2}O_2 \rightarrow H_2O$

✓ **Alkaline Fuel Cells (AFC)**

Alkaline Fuel Cells (AFCS) represent among the first generation and most developed kinds of fuel cell technology. Aqueous solutions of potassium hydroxide (KOH) serve as the electrolyte, which transmits hydroxide ions (OH^-) from the cathode to the anode through their operations(Hull et al., 1970). The operating temperature range of these fuel cells extends from 60°C to 250°C, which enables high electrical efficiency above 60% and fast start-up capabilities for quick power delivery. The primary benefit of AFCS includes their capability to function with catalyst materials such as nickel instead of platinum, which would typically be more expensive

(Ozoemena, 2016). Such a characteristic helps minimize the expense of fuel cell systems through reduced material costs, mainly when operating under controlled conditions where fuel Decay is not required. These cells operate more efficiently due to the alkaline environment that promotes rapid electrochemical reaction kinetics, particularly for the oxygen reduction reaction (ORR) (Xie et al., 2022a).

The anode receives hydrogen gas for a chemical reaction, which produces water and releases electrons through the process. The electron movement through the external circuit generates electrical power (Xie et al., 2022b). The cathode process combines oxygen from air elements with water and returned circuit electrons to form hydroxide ions that migrate through the electrolyte to the anode for the cycle conclusion. The reaction generates water as the main output during an efficient process. The main drawback of AFCS arises from their reaction sensitivity to carbon dioxide (CO_2) (Alonso-Moreno & García-Yuste, 2016). The presence of CO_2 in reactant gases causes potassium hydroxide electrolyte to produce potassium carbonate (K_2CO_3), that reduces cell efficiency and ionic conductivity. The cells need protection against carbon dioxide contamination because this compound reduces cell performance and efficiency and prevents the use of AFCS with natural air supplies without dedicated purification equipment. Higher purity supplies of hydrogen and oxygen make AFCS, which proves the most efficient technology for these applications(Du et al., 2021).

NASA established Alkaline Fuel Cells as important through their utilisation in the Apollo and Space Shuttle missions in the past. The space missions implemented AFCS to supply power generation and water production via decay products, allowing their invaluable function in spacecraft systems (Reid et al., 2013). The present-day application of AFCS remains under study for military and aerospace functions because these sectors require high-purity fuel combined with dependable performance, swift reaction times, and maximum efficiency (Jha, 2012). Submarine propulsion systems, underwater vehicles, and remote sensing stations are under review for use of AFC technology because they require silent operation combined with high efficiency in controlled environments (Miyazaki et al., 2003).

Anode Reaction:$H_2 + 2OH^- \rightarrow 2H_2O + 2e^-$
Cathode Reaction:$\frac{1}{2}O_2 + H_2O + 2e^- \rightarrow 2OH^-$
Overall, Cell Reaction:$H_2 + \frac{1}{2}O_2 \rightarrow H_2O$

✓ **Molten Carbonate Fuel Cells (MCFC)**

People classify Molten Carbonate Fuel Cells (MCFCS) as advanced high-temperature fuel cells that function with a molten carbonate salt blend made up primarily of lithium, potassium and sodium carbonates. Operating at 650°C delivers multiple advantages to MCFCS, including modified fuel versatility, speeded-up electrochemical reactions, and usable waste heat production, which enables their adoption in extended-scale power generation(Tomczyk, 2006). The MCFC conducts carbonate ions (CO_3^{2-}) through a process from the cathode to the anode as its fundamental operational principle. CO_2, along with O_2, enters the cathode where they produce carbonate ions by reacting with external circuit electrons. The molten electrolyte conducts carbonate ions to the anode, so these ions react with fuel sources, including hydrogen, carbon monoxide and hydrocarbons, to produce water (H_2O), and carbon dioxide, along with electrons (Bar-Nun & Chang, 1983). Electrons travel from the cathode through an external circuit before returning to the cathode, thus producing electricity. The closed-loop operation of MCFCS permits direct fuel usage that eliminates the requirement for external reformers in most cases(Nouman, 2018).

The high operating temperature of MCFCS facilitates the internal reforming of hydrocarbon fuels, such as natural gas and biogas, within the fuel cell itself. The system design becomes simpler, and fuel processing achieves higher efficiency through this approach(Mehr et al., 2021b). MCFC technology runs at high temperatures that speed up electrochemical reactions, thus achieving 45–55% electrical efficiency that can reach 70% when operating in combined heat and power or combined cycle setups. The output combination of electricity and heat makes MCFCS an ideal choice for industrial power generation in cogeneration systems as well as utility-scale power plants (Hsu & Nathanson, 1995). The main operational advantage of MCFCS consists of their fuel flexibility for both hydrogen and synthesis gas (syngas) and biogas, besides other fossil fuels. The ability to operate on various fuel types makes MCFCS essential for waste-to-energy facilities and landfill gas collection operations, as well as sewage treatment facilities and industrial plants. Wastewater treatment facilities benefit from MCFCS converting their methane-rich biogas into electricity and heat, which provides self-generated power and cuts down carbon emissions (Torvelainen, 2023).

The technological challenges exist despite the various advantages MCFCS possess. The highly corrosive molten carbonate electrolyte causes the degradation of materials, including electrodes and interconnects, throughout operation time (P. Wang et al., 2023). A substantial engineering challenge exists to achieve stability of cell components regarding heat and chemical exposure, particularly in extended operating periods. Special materials and complex thermal management systems are needed due to MCFC operating temperatures at high levels, which leads to system complexity and higher costs(Mehr et al., 2021c). MCFCS provide a conversion process which produces a low environmental impact. The production of carbon dioxide occurs as both an input reactant and reaction product in MCFC operation, but the system accepts carbon capture methods to minimize total emissions(Balafkandeh et al., 2022a). Running MCFCS on renewable biogas and waste-derived fuels makes them vital components for establishing sustainable energy infrastructure. MCFC technology finds practical use as a power source in industrial combined heat and power systems, as well as utility-scale power plants and municipal infrastructure networks throughout the world. Their ability to operate persistently and scale up alongside their dual-fuel efficiency allows MCFCS to function as promising power generation systems for decentralized networks alongside grid stabilization operations in areas pursuing energy transition with a reliable power supply (Balafkandeh et al., 2022b).

Anode Reaction: $H_2 + CO_3^{2-} \rightarrow H_2O + CO_2 + 2e^-$

Cathode Reaction: $\frac{1}{2}O_2 + CO_2 + 2e^- \rightarrow CO_3^{2-}$

Overall, Cell Reaction: $H_2 + \frac{1}{2}O_2 + CO_2 \rightarrow H_2O + CO_2$

✓ **Phosphoric Acid Fuel Cells (PAFC)**

Phosphoric Acid Fuel Cells (PAFCS) introduced the market to mature fuel cell technology with shown reliability for generating stationary power. A fuel cell uses phosphoric acid (H_3PO_4) in concentrated form as its electrolyte, which is embedded within a silicon carbide matrix or an alternative porous substance. A PAFC operates at temperatures between 150°C and 220°C due to an optimal temperature range between PEMFC operation temperature and SOFC or MCFC operation temperature(Sammes et al., 2004). The anode section of a PAFC splits hydrogen gas particles into proton (H^+) molecules and electron (e^-) particles. Electrons stream via an external circuit while protons move through the phosphoric acid electrolyte so that they reach the cathode, where electricity is generated. The cathode enables oxygen in the air to unite with protons, coupled with electrons to create water. The hydrogen fuel cell operation proceeds identically to regular cells using phosphoric acid electrolytes, which introduce distinct thermal behaviour and electrochemical properties (Saslow, 1999).

Phosphoric Acid Fuel Cells (PAFCS) introduced the market to mature fuel cell technology with shown reliability for generating stationary power. A fuel cell uses phosphoric acid (H_3PO_4) in concentrated form as its electrolyte, which is embedded within a silicon carbide matrix or an alternative porous substance(R.-H. Song et al., 2002). A PAFC operates at temperatures between 150°C and 220°C due to an optimal temperature range between PEMFC operation temperature and SOFC or MCFC operation temperature. The anode section of a PAFC splits hydrogen gas particles into proton (H^+) molecules and electron (e^-) particles. Electrons stream via an external circuit while protons move through the phosphoric acid electrolyte so that they reach the cathode, where electricity is generated(Vilčiauskas et al., 2012). The cathode enables oxygen in the air to unite with protons, coupled with electrons to create water. The hydrogen fuel cell operation proceeds identically to regular cells using phosphoric acid electrolytes, which introduce distinct thermal behaviour and electrochemical properties(Tian et al., 2017).

The catalysts used in PAFCS are typically platinum-based, which adds to the overall cost, although the more forgiving fuel requirements somewhat offset this expense. The corrosive nature of phosphoric acid, especially at elevated temperatures, requires a careful selection of materials for cell components to ensure chemical and thermal stability(Haile, 2003). In terms of commercial deployment, PAFCS have been successfully implemented in a range of distributed power generation projects. For instance, they have powered office buildings, telecommunications centres, and manufacturing facilities, providing both base-load power and backup capabilities(Ahmed et al., 2018). The quiet operation, low emissions, and ability to provide both electricity and heat make PAFCS a viable option for urban and suburban settings where clean, efficient, and reliable power is needed. In conclusion, Phosphoric Acid Fuel Cells occupy an important niche in the fuel cell landscape. With their proven track record, moderate operating conditions, and ability to integrate into CHP systems, they continue to be a reliable and effective technology for stationary power generation, especially in environments that benefit from the dual outputs of electricity and heat. Their robustness and impurity tolerance make them an attractive option for long-term energy solutions in commercial and institutional settings (Gür, 2018).

Anode Reaction: $H_2 \rightarrow 2H^+ + 2e^-$

Cathode Reaction: $\frac{1}{2}O_2 + 2H^+ + 2e^- \rightarrow H_2O$

Overall, Cell Reaction: $H_2 + \frac{1}{2}O_2 \rightarrow H_2O$

Table 4. Comparative Chart of Fuel Cell Types(Mekhilef et al., 2012a)

Fuel Cell Type	Operating Temperature (°C)	Efficiency (%)	Fuel Type	Typic4al Applications
PEMFC	60–100	40–60	Pure hydrogen	Transportation (FCEVS), portable power, backup systems
SOFC	600–1000	45–60 (up to 85 CHP)	Hydrogen, CO, hydrocarbons	Stationary power, auxiliary power units (APUS), micro-CHP
AFC	60–90	60–70	Hydrogen	Spacecraft (NASA), military, niche portable applications
MCFC	600–700	45–55 (up to 70 CHP)	Hydrogen, CO, natural gas	Large-scale stationary power, industrial cogeneration
PAFC	150–220	40–50	Hydrogen, reformate gases	Commercial power, small-scale cogeneration
DMFC	50–120	30–40	Methanol	Portable electronics, military, and limited mobility systems

While the preceding section outlined conventional fuel cell types, ongoing research continues to enhance their performance and practicality. The following section focuses on recent advancements in materials, catalysts, and control strategies that have reshaped modern fuel cell technology.

4. RECENT ADVANCEMENTS

Fuel cell technology has undergone rapid development in recent years, driven by the demand for cleaner energy sources, improved efficiency, and enhanced reliability. The following recent advancements are transforming the landscape of fuel cell systems, especially in transportation and stationary applications(Abdin, 2024c).Advances in material science have significantly enhanced the performance, efficiency, and durability of fuel cells, particularly through innovations in membrane and electrode design. One major development is the incorporation of inorganic fillers such as zirconium phosphate, silica, titania, and graphene oxide into traditional Nafion membranes to create composite membranes (Lee et al., 2014). These enhancements improve proton conductivity and thermal stability, especially under high-temperature and low-humidity conditions, which are common in automotive and stationary applications(Wong et al., 2019). Another important trend is the emergence of high-temperature proton exchange membranes (HT-PEMS), particularly those based on phosphoric acid-doped polybenzimidazole (PBI). These membranes enable fuel cell operation at temperatures above 120°C, offering greater carbon monoxide (CO) tolerance and simplifying water management due to reduced humidification requirements (Z. Zhang et al., 2024). Furthermore, advanced electrode materials have been developed to improve the catalytic activity and durability of fuel cells. The use of nanostructured carbon supports, metal-organic frameworks (MOFS), and porous carbon networks has significantly increased the electrochemical surface area, leading to enhanced catalyst dispersion, better reactant accessibility, and accelerated reaction kinetics. These material innovations collectively contribute to the improved overall performance and commercial viability of modern fuel cell systems(Hall & Kerr, 2003).

Figure 6. Material innovations – nano-structured electrodes, membrane upgrades(Mangaiyarkkarasi, 2025).

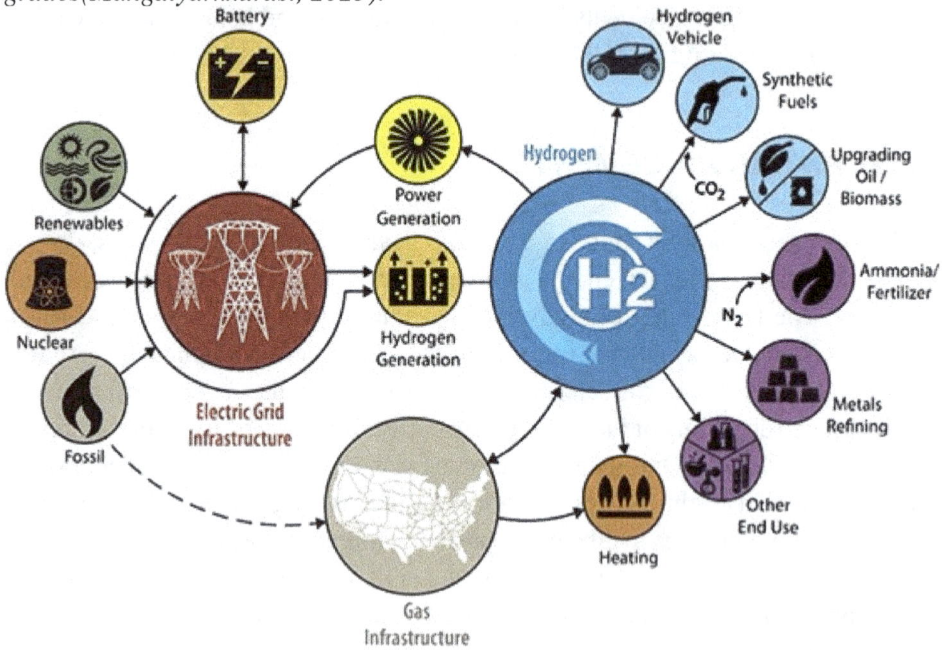

Catalyst innovation plays a pivotal role in reducing the overall cost of fuel cells while enhancing their performance and long-term sustainability. One of the most significant advancements in this area is the development of platinum group metal-free (PGM-free) catalysts (C. Song, 2002). These catalysts, particularly those based on transition metal-nitrogen-carbon (M–N–C) structures using elements such as iron and cobalt, have shown promising activity in the oxygen reduction reaction (ORR), which is critical to fuel cell efficiency(L. Zhang et al., 2018). Additionally, alloying platinum with more abundant metals like nickel or copper has proven effective in improving catalytic performance while significantly lowering the required platinum content. Another breakthrough involves the design of core-shell nanostructures, where a non-precious metal core is coated with a thin layer of platinum, thereby maintaining high catalytic activity with reduced reliance on costly materials (X. Wang et al., 2014). To further improve durability, researchers have introduced protective coatings on catalyst particles, refined dispersion methods for better active site exposure, and employed corrosion-resistant support materials. These collective strategies not only extend the operational life of fuel cells but also bring them closer to widespread commercial adoption by addressing key challenges in cost and durability(Abdin, 2024d).

Figure 7. Performance curve – catalyst performance vs. load cycles or temperature (Ramanathan & Oh, 2014).

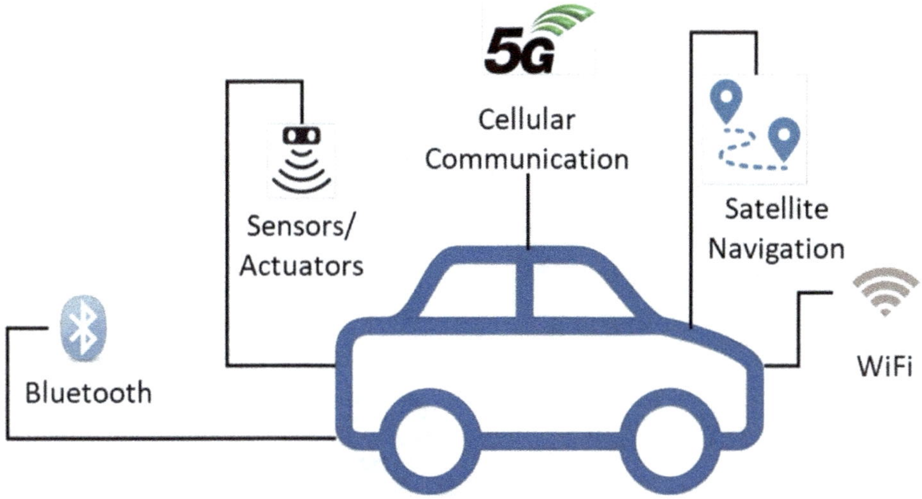

Cold start remains one of the most critical challenges in fuel cell technology, especially in sub-zero environments where the presence of water within the system can lead to freezing, causing irreversible damage to fuel cell components(Yan et al., 2006). To address this issue, recent advancements have focused on improving thermal management, control systems, and power electronics. Enhanced thermal management systems now incorporate integrated heaters, phase change materials, and active insulation to maintain optimal temperatures during startup, reducing the risk of ice formation(Oró et al., 2012). In parallel, optimized control algorithms are being developed to intelligently adjust startup sequences based on real-time inputs such as ambient temperature, humidity levels, and voltage feedback, ensuring smoother and safer transitions from inactive to active states(Bradshaw, 2010).

Figure 8. Cold start temperature vs. start-up time(Deng et al., 2018a)

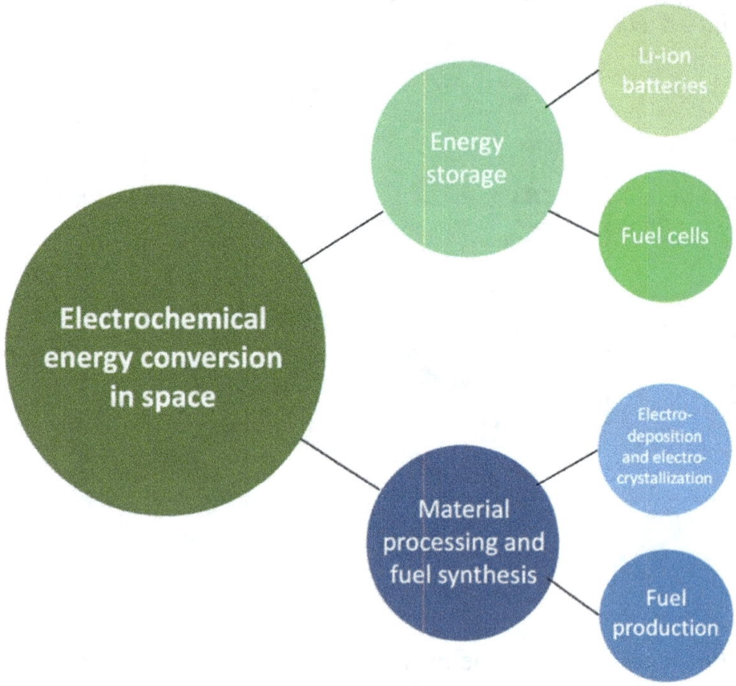

Furthermore, the introduction of high-efficiency, high-gain DC-DC converters allows for rapid pre-heating and faster system initialization without placing excess load on the fuel cell stack, thereby preserving its integrity(Kim et al., 2022). Complementing these strategies are innovative water management techniques, including controlled gas humidification and precise purging protocols, which help regulate internal moisture levels and prevent ice accumulation. Together, these advancements significantly enhance the cold start capabilities of fuel cell systems, improving their reliability and viability in colder climates(Amamou et al., 2016).

Figure 9. Cold start mechanism diagram (your system, if possible)(Yuan & Hernandez, 2023).

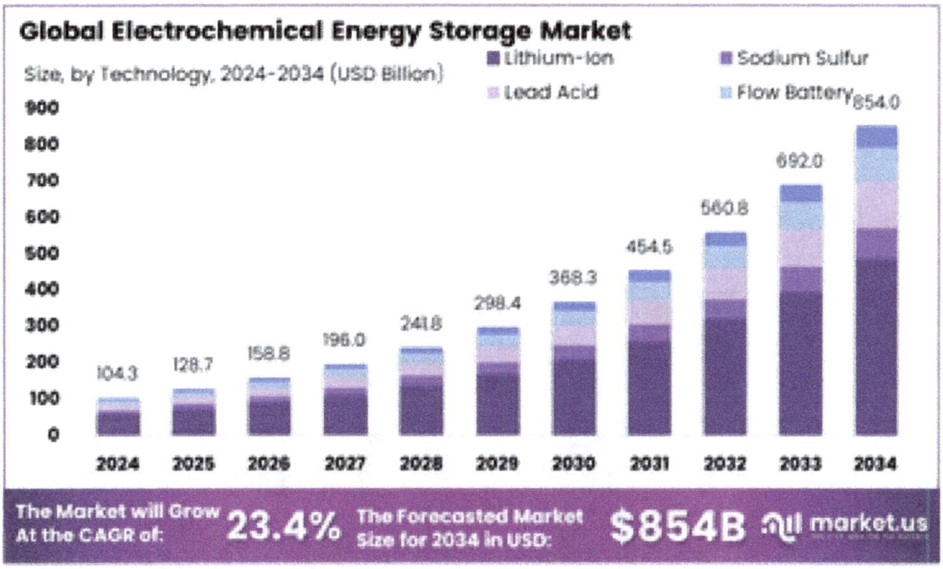

Hybrid systems that integrate fuel cells with secondary energy storage devices such as batteries and supercapacitors have emerged as a promising approach to enhance both the energy density and dynamic power responsiveness of powertrains. In battery–fuel cell hybrid configurations, lithium-ion batteries play a critical role by supplying high power during acceleration and capturing energy during regenerative braking, thereby reducing the instantaneous load on the fuel cell and improving overall system efficiency(C. Zhang et al., 2021). Similarly, fuel cell–cell-supercapacitor systems leverage the ability of supercapacitors to deliver bursts of power almost instantaneously, which is particularly beneficial during sudden load changes. This setup minimizes stress on the fuel cell stack and enhances its longevity. These hybrid architectures are further enabled by the development of advanced bidirectional DC-DC converters and intelligent energy management algorithms, which coordinate power flow seamlessly between the fuel cell and the auxiliary storage components based on real-time demands and operating conditions(Deng et al., 2009). Moreover, the scalability and modular integration of these systems make them adaptable across a wide range of applications, from compact vehicles and drones to heavy-duty transport and stationary backup systems, making them highly versatile and attractive for next-generation energy solutions(Gupta et al., 2024).

Figure 10. Block diagram of hybrid FC-Battery/Supercapacitor system (Fathabadi, 2018).

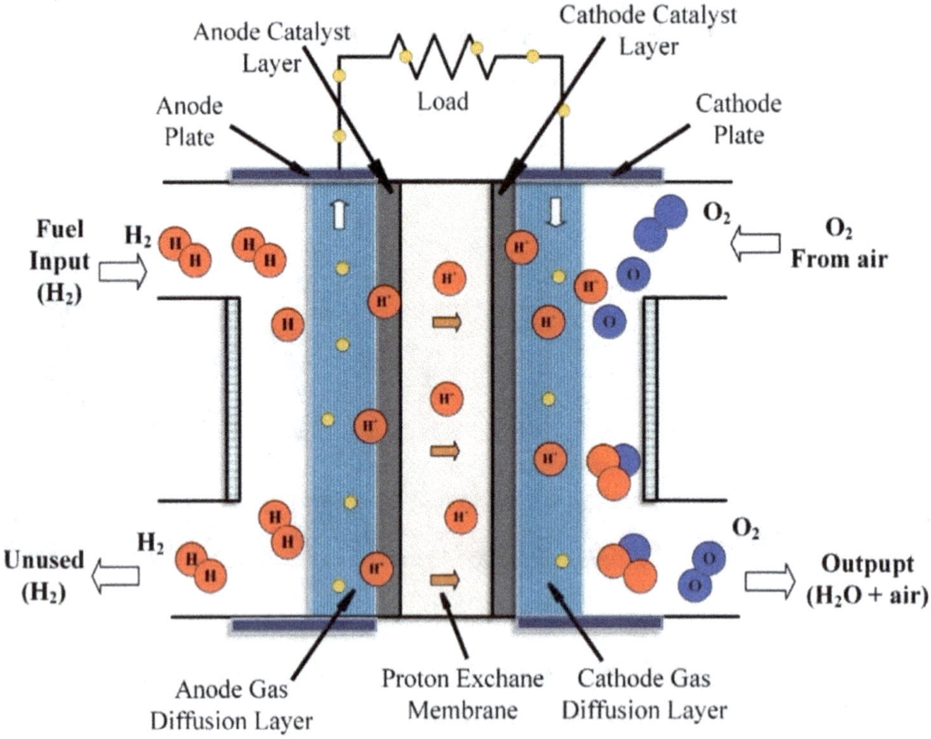

Artificial Intelligence (AI) and Machine Learning (ML) are revolutionizing the operation, diagnostics, and optimization of fuel cell systems, offering intelligent solutions to enhance performance, reliability, and longevity. One of the key applications of ML is in predictive maintenance, where models trained on historical and real-time operational data can accurately forecast component failures, thereby minimizing unplanned downtime and extending the fuel cell's service life(Dirkes et al., 2023). AI-driven control systems also play a crucial role in optimising real-time efficiency by dynamically adjusting parameters such as fuel and air supply rates, stack temperature, and load distribution in response to varying operational conditions. In addition, advanced deep learning and data-driven approaches are being employed for State-of-the-art (SoH) estimation, providing continuous insight into the degradation level of the fuel cell and enabling proactive interventions(Abdin, 2024e). Furthermore, AI techniques such as fuzzy logic, artificial neural networks, and reinforcement learning are being increasingly utilized to implement Maximum Power Point Tracking (MPPT) algorithms, enabling fuel cells to operate at their optimal power output under fluctuating environmental conditions. These intelligent systems not only boost the overall performance of fuel cells but also contribute to their wider adoption by addressing key operational challenges (Hassan et al., 2023).

Figure 11. Smart energy management using AI/ML flowchart or neural network diagram(Chander & Gopalakrishnan, 2024).

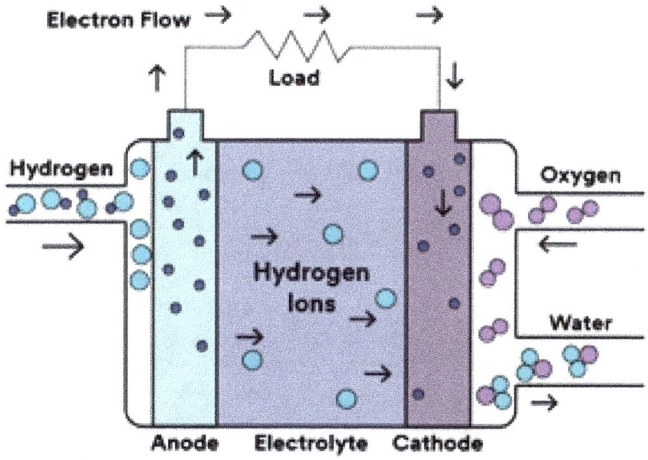

Recent innovations have not only improved fuel cell efficiency but have also expanded their role in broader energy systems. Consequently, the next section explores how fuel cells integrate into energy storage architectures, enabling stable, renewable power supply and hybrid configurations.

5. FUEL CELL SYSTEMS FOR ENERGY STORAGE

Fuel cell systems are emerging as a promising solution for energy storage, particularly in applications where clean, efficient, and scalable power solutions are required. Their ability to store energy in the form of hydrogen and convert it back to electricity on demand makes them highly suitable for integration with renewable energy sources. The following developments highlight the key aspects of fuel cell systems in energy storage applications (Belmonte et al., 2017). The intermittent nature of renewable energy sources such as solar and wind presents a major challenge for grid stability and energy reliability. Fuel cell systems, when integrated with electrolysers, offer an efficient method for storing excess energy generated during peak production periods(Keshavarzzadeh et al., 2019). During times of surplus electricity, for instance, sunny or windy periods, electrolysers use this energy to split water into hydrogen and oxygen. The hydrogen is then stored and later fed into fuel cells to generate electricity when renewable generation is low or demand is high (Ehteshami & Chan, 2014b). This process, often referred to as Power-to-Hydrogen-to-Power (P2H2P), creates a closed-loop system that enhances the viability of renewables. The integration not only enables long-duration energy storage but also contributes to grid balancing, frequency regulation, and peak shaving. Furthermore, decentralized fuel cell systems allow energy to be stored and utilized locally, reducing transmission losses and enhancing the overall efficiency of the power system(Agüero, 2012).

Figure 12. Integration of a fuel cell with solar/wind in a hybrid system(Talaat et al., 2021).

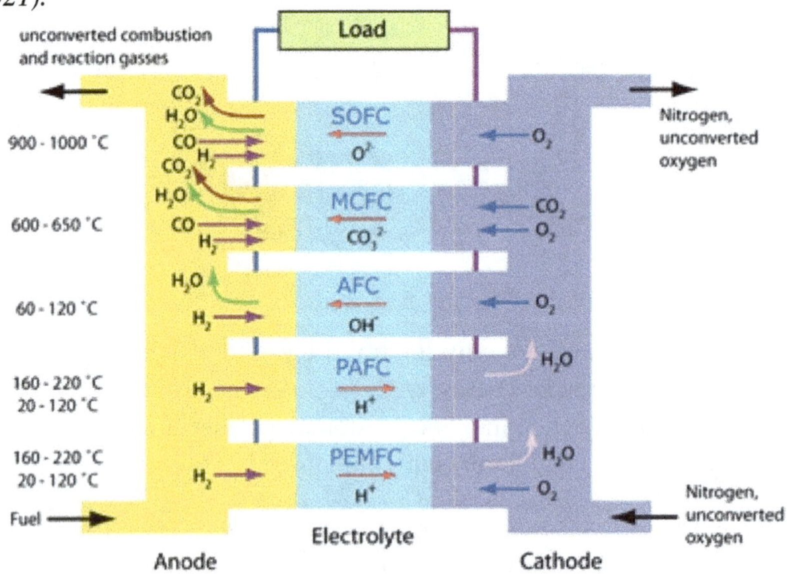

At the core of the energy conversion process in a fuel cell system lies the fuel cell stack, which consists of multiple individual cells connected in series to generate the required voltage and power output. These stacks efficiently convert hydrogen fuel into electrical energy through electrochemical reactions, producing only water and heat as by-products, making them environmentally friendly(Stamenkovic et al., 2017). Recent advancements in stack design have focused on improving power density, optimizing thermal management, and enhancing durability to meet the demands of various applications. Complementing these developments are innovations in hydrogen storage techniques, which play a crucial role in determining the overall system performance and efficiency (C. Zhang et al., 2022). The most widely adopted method is compressed hydrogen storage, where hydrogen is stored at high pressures (typically 350–700 bar) in specially designed tanks, offering simplicity and rapid refuelling. Another approach is liquid hydrogen storage, which involves cooling hydrogen to cryogenic temperatures (−253°C) to achieve higher energy density, albeit with the need for sophisticated insulation and handling systems (Vevere et al., 2024). Additionally, solid-state storage methods utilize metal hydrides or other chemical carriers to store hydrogen safely at a high volumetric density, though this comes with trade-offs in terms of system weight and slower charging speeds. By optimizing both stack architecture and hydrogen storage methods, fuel cell systems can achieve higher efficiency, improved safety, and greater scalability, making them suitable for a wide range of mobile and stationary energy storage applications (Kebede et al., 2022).

Figure 13. Bidirectional power flow control in FC-battery/supercapacitor systems (Thounthong et al., 2009).

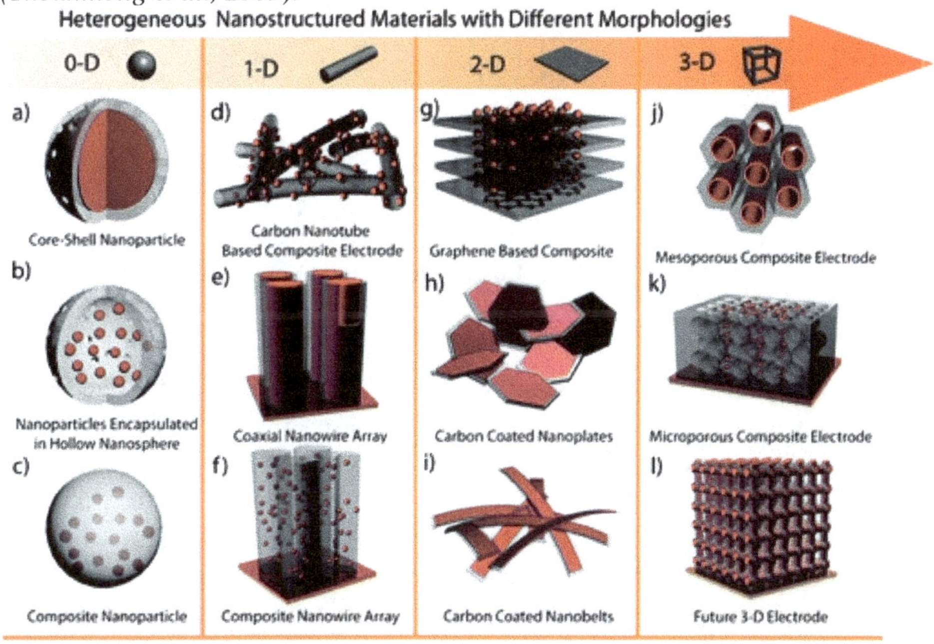

Fuel cell–battery hybrid systems are gaining traction in advanced energy storage due to their ability to support bidirectional power flow. In these systems, fuel cells provide a steady base power, while batteries manage rapid power fluctuations and store excess energy. Bidirectional DC-DC converters play a critical role in enabling power to flow in both directions, charging the battery when there is surplus energy and discharging it when there is a peak in load demand (Ero lu et al., 2021).

Figure 14. System efficiency vs. power demand in standalone vs. hybrid FC systems(Behzadi & Niasati, 2015).

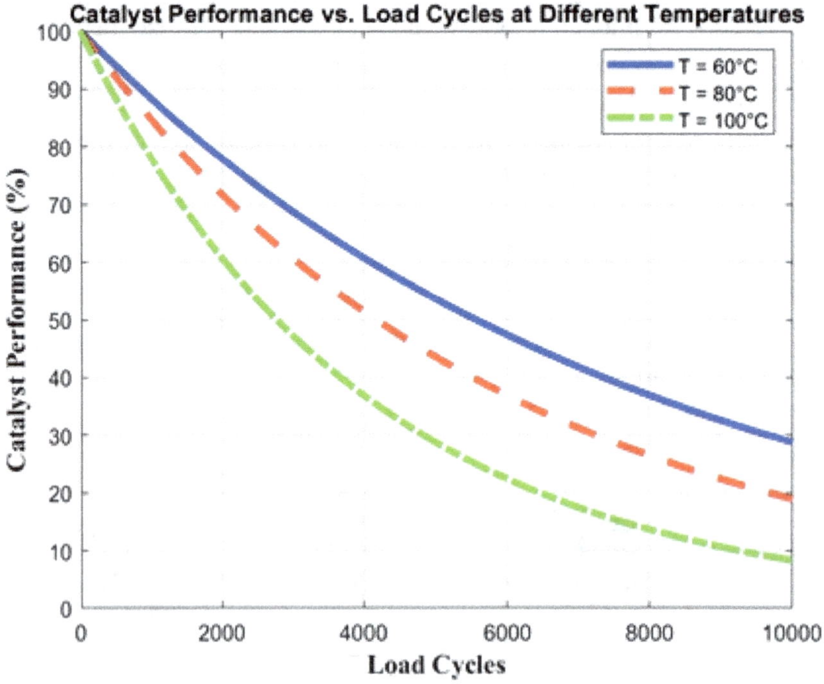

This bidirectional capability enhances the load-following characteristics of the system, allowing it to respond effectively to dynamic energy demands. It also supports regenerative braking in electric vehicles and grid services such as peak load shaving and voltage stabilization (Yang et al., 2024). Moreover, intelligent energy management algorithms coordinate the interaction between the fuel cell and battery, optimizing energy usage, prolonging battery life, and maintaining fuel cell efficiency. These systems are not only effective for mobile applications like fuel cell electric vehicles (FCEVS) but are also ideal for stationary applications such as microgrids and off-grid power systems(Waseem et al., 2023).Although fuel cell–based storage systems offer promising performance, several technical and economic challenges still hinder large-scale deployment. The following section critically examines these limitations to identify areas requiring further development and optimization.

6. CHALLENGES AND LIMITATIONS

Despite significant advancements, fuel cell technologies still face several technical and economic challenges that hinder their widespread commercialization. Understanding these limitations is essential to guide future research and improve the feasibility of fuel cell-based energy systems. One of the most persistent barriers to the large-scale adoption of fuel cell systems is their high cost. The reliance on expensive materials, particularly platinum group metals (PGMS) used in catalysts, significantly increases production costs. Moreover, manufacturing processes for components like membrane electrode assemblies (MEA), bipolar plates, and gas diffusion layers are complex and not yet fully optimized for mass production (Huang & Qi, 2022). Scaling up fuel cell systems to meet industrial or grid-level demands also poses logistical challenges in terms of system integration, hydrogen storage infrastructure, and supply chain readiness. To overcome these issues, current efforts are focused on developing PGM-free catalysts, cost-effective composite membranes, and modular system designs that allow for easier assembly and scalability (Martinaiou & Daletou, 2024).

Table 5. Summary of Key Limitations by Fuel Cell Type(Mekhilef et al., 2012b)

Fuel Cell Type	Cost	Degradation Rate	Start-Up Time	Operating Complexity	Key Limitations
PEMFC	High (Pt catalyst)	Moderate	Short (30–60 sec)	Low to Medium	Sensitive to CO poisoning, humidification required, expensive catalyst
SOFC	High (ceramic materials)	Low to Moderate	Long (minutes to hours)	High (thermal cycling issues)	Slow start-up, brittle materials, high operating temp (600–1000°C)
AFC	Moderate	High	Short	High	CO_2 contamination, electrolyte management
MCFC	High	Moderate	Long (10–60 min)	High	Corrosive electrolyte, CO_2 management, thermal management
PAFC	Moderate to High	Moderate	Moderate (5–10 min)	Medium	Acid corrosion, bulky, moderate efficiency
DMFC	High	High	Short (under 1 min)	Low to Medium	Low power density, methanol crossover, slow kinetics

Durability remains a critical issue, especially in applications requiring continuous and long-term operation, such as in vehicles or stationary power systems. Fuel cells are subject to chemical, thermal, and mechanical degradation mechanisms that affect performance over time. Common degradation pathways include catalyst sintering or dissolution, membrane thinning or cracking, and corrosion of carbon-based support structures (Zhao & Li, 2019). These effects lead to reduced power output and lower efficiency, requiring more frequent maintenance or replacement. Enhancing the resilience of materials, incorporating protective coatings, and optimizing operating conditions are some of the strategies being explored to extend fuel cell lifespan and ensure stable performance throughout their service cycle.

Figure 15. Degradation mechanism over time (ageing, catalyst poisoning, membrane failure) (LaConti et al., 2003).

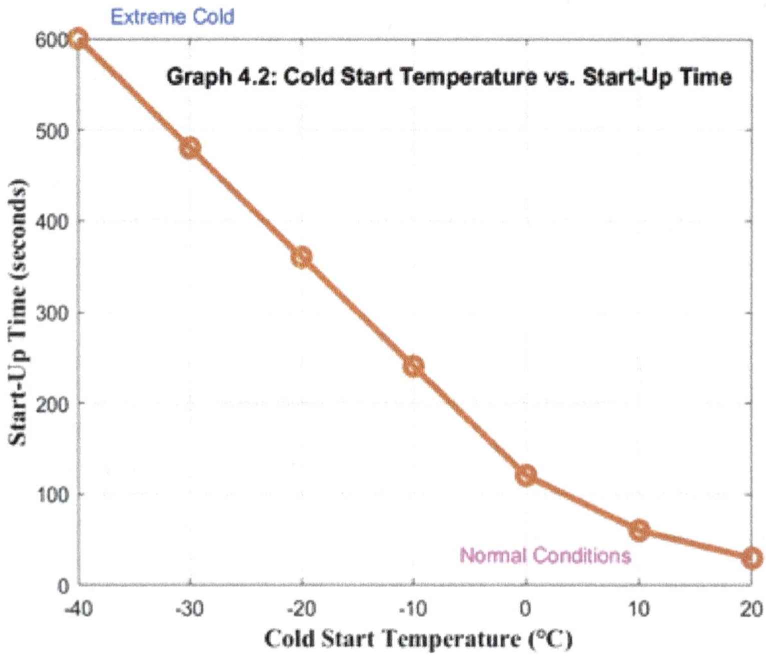

Start-up behaviour is particularly important in automotive and portable applications, where immediate power availability is often required. Fuel cells, particularly proton exchange membrane fuel cells (PEMFCS), generally require preheating to reach their optimal operating temperature range, which can result in delays during cold starts(Deng et al., 2018b). In freezing conditions, the water produced during operation can solidify within the stack, causing mechanical stress and performance loss. To address this, researchers have developed enhanced thermal management systems, including integrated heaters, phase change materials, and optimized flow field designs. Additionally, improved membrane materials and control algorithms are being used to accelerate warm-up times and maintain performance in low-temperature environments, ensuring that fuel cell systems remain viable across a wide range of climatic conditions (Yu et al., 2022).

Figure 16. Temperature-performance curve showing cold start limitations(L. Li et al., 2019).

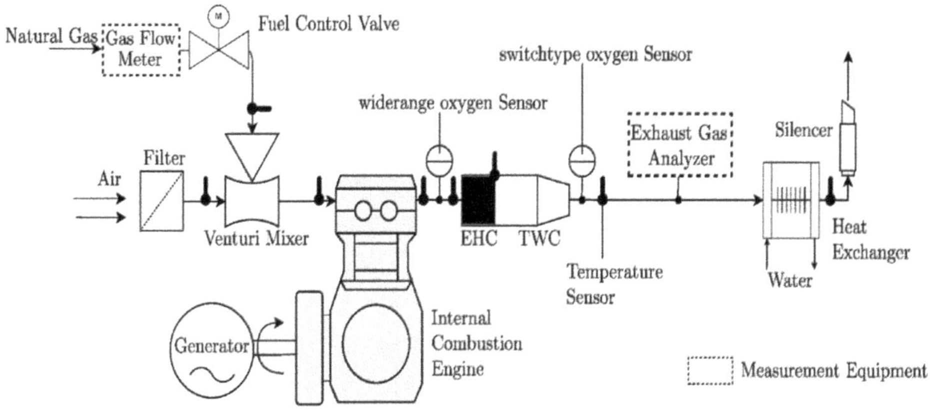

To address these challenges, simulation tools have become essential for analyzing and improving real-world performance. The next section demonstrates model-based design and simulation studies, particularly using MATLAB/Simulink, to validate system behavior and control strategies.

7. CASE STUDIES OR SIMULATION MODELS

Model-based design using simulation tools has become indispensable in the development of modern fuel cell systems. Among these, MATLAB/Simulink stands out as a widely adopted platform that provides an extensive environment for designing, simulating, and optimizing fuel cell-based energy systems(Gallo et al., 2020). Its graphical modelling capabilities, control system toolboxes, and integration with embedded platforms make it highly suitable for research and development purposes. In this work, a comprehensive Proton Exchange Membrane Fuel Cell (PEMFC) system was modelled and simulated using MATLAB/Simulink. The model included key components such as the fuel cell stack, high-gain DC-DC converter, hybrid energy storage unit, and intelligent MPPT (Maximum Power Point Tracking) controller. This simulation aimed to analyse the system's dynamic performance, control strategy, and efficiency, particularly under variable loading and cold start conditions(Deng et al., 2018c).

Figure 17. MATLAB/Simulink model architecture of a fuel cell system (Bayrak & Cebeci, 2014)

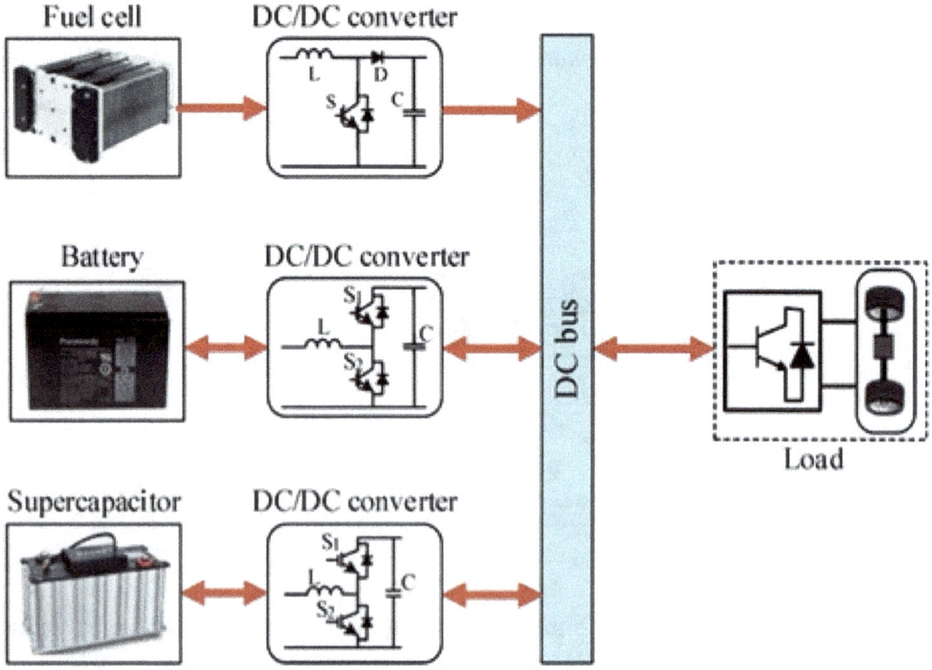

Figure 18. Voltage vs. current (polarization curve) from simulation(Arif et al., 2020).

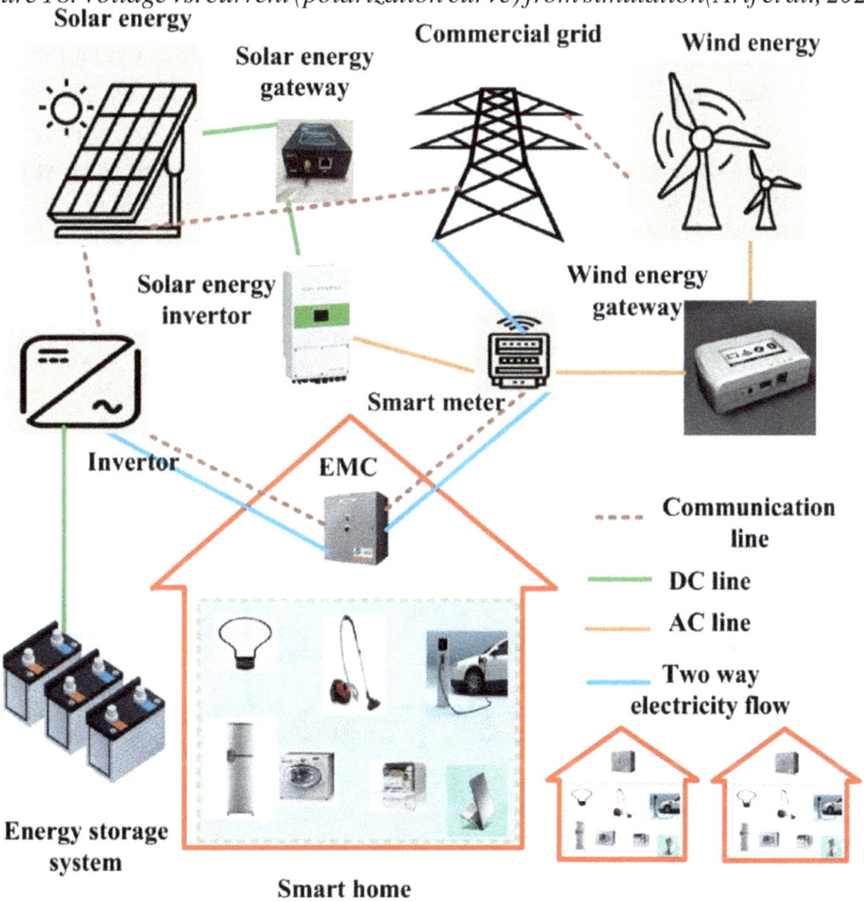

The simulation model developed in MATLAB/Simulink integrates several critical subsystems to emulate the behaviour of a complete Proton Exchange Membrane Fuel Cell (PEMFC) system under realistic operating conditions. At its core, the fuel cell stack was modelled using empirical equations that accurately represent activation, ohmic, and concentration losses. This allowed for dynamic assessment of the stack's voltage-current characteristics in response to varying load, temperature, and pressure conditions (Y. Li et al., 2015). To efficiently manage the inherently low voltage output of the PEMFC, a high-gain interleaved boost converter was designed. This topology not only elevated the voltage to desired levels but also reduced current ripple and improved thermal distribution, ensuring higher conversion efficiency during transient and high-load scenarios. A fuzzy logic-based Maximum Power Point Tracking (MPPT) controller was implemented to continuously optimize power extraction from the stack. Unlike traditional Perturb & Observe or Incremental

Conductance techniques, the fuzzy logic controller offered faster response and higher accuracy under dynamic environmental changes (Tang & Ahmad, 2024). Additionally, a hybrid energy storage system, comprising a lithium-ion battery and/or a supercapacitor, was integrated via a bidirectional DC-DC converter. This hybrid configuration enabled seamless power balancing, voltage stabilization, and effective energy recovery during regenerative braking, making the system particularly suitable for fuel cell electric vehicle applications (Yildiz & Özel, 2021).

Figure 19. Efficiency vs. power output during cold start (Deng et al., 2018d).

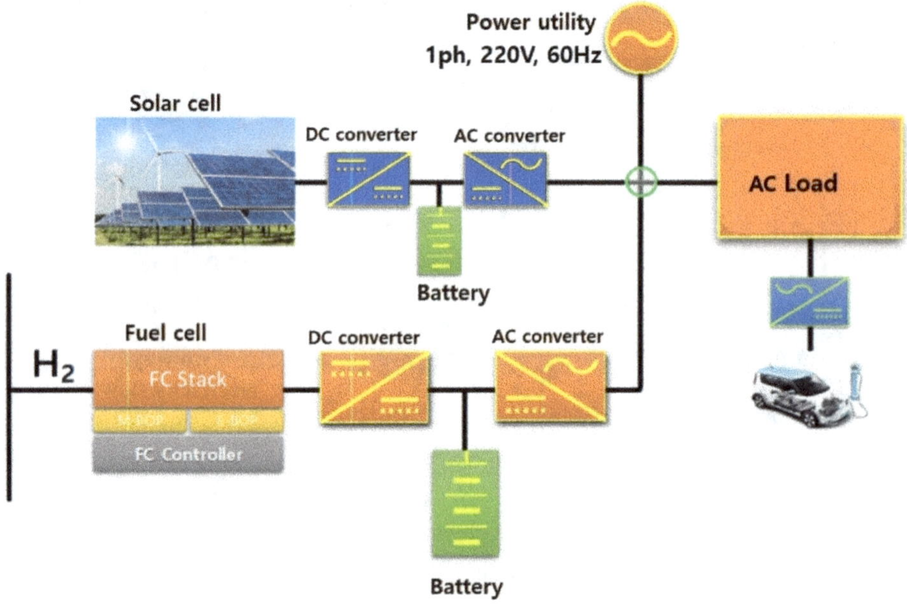

Figure 20. Controller model for cold start(Deng et al., 2018e) .

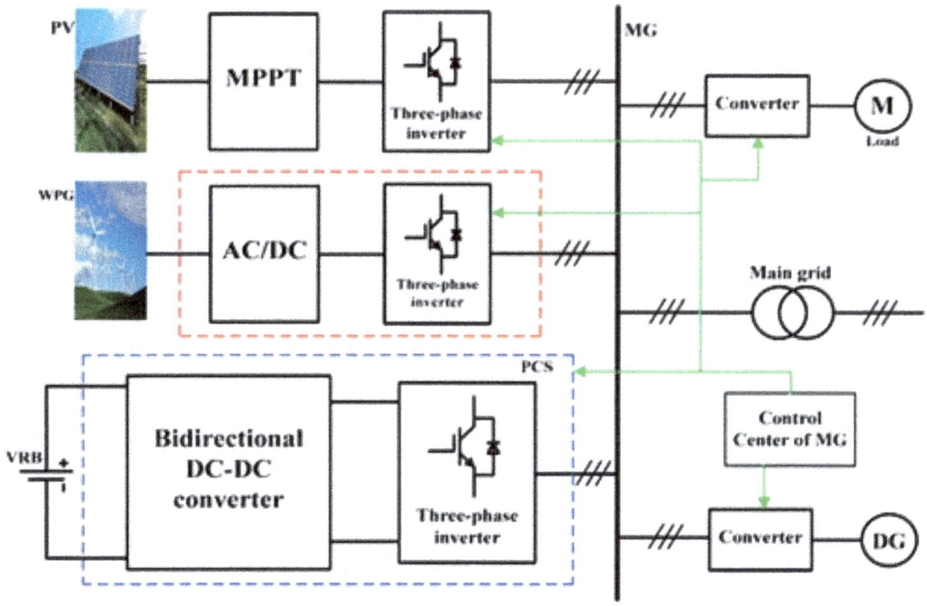

The simulation results indicated significant improvements in overall system performance. The fuzzy MPPT controller enhanced energy extraction by maintaining fuel cell operation near its maximum power point, leading to an 8–12% increase in efficiency under dynamic loading. The converter regulated output voltage effectively while maintaining low ripple and thermal stability(Redl & Sun, 2009). During simulated cold start scenarios, the high-gain converter rapidly brought the system to operational conditions without stressing the stack components. Voltage response, current stability, state of charge (SOC) management, and temperature effects were all monitored and optimized through the simulation. These results validate the robustness and flexibility of the modelled architecture and highlight the effectiveness of intelligent control strategies(Liu et al., 2023).The simulation analysis confirms the feasibility of integrating advanced control and converter topologies within fuel cell systems. Building on these findings, the following section discusses prospects and emerging global trends, including hydrogen infrastructure, green mobility, and digital integration.

8. PROSPECTS

The hydrogen economy represents a transformative pathway toward a clean and sustainable energy future. At its core, it revolves around the generation, storage, transportation, and utilization of hydrogen as a primary energy carrier. Recent initiatives across the globe have focused on expanding the hydrogen infrastructure, including large-scale green hydrogen production through water electrolysis powered by renewable sources such as solar and wind. This approach eliminates carbon emissions associated with traditional hydrogen production methods like steam methane reforming. Additionally, governments are investing heavily in the deployment of hydrogen refuelling stations, especially in regions like Europe, Japan, South Korea, and parts of North America, to support fuel cell electric vehicles (FCEVs) and industrial applications (Samsun et al., 2022). Technological advancements in hydrogen storage, including high-pressure tanks, cryogenic liquefaction, and solid-state hydride systems, are improving safety and efficiency, making hydrogen a more viable and scalable energy solution. With regulatory frameworks, international collaboration, and investment incentives in place, the hydrogen economy is gradually becoming a cornerstone of global decarbonization efforts (Asif & Schmidt, 2021).

Figure 21. Future hydrogen economy ecosystem (production, storage, FC usage) (Dou et al., 2023).

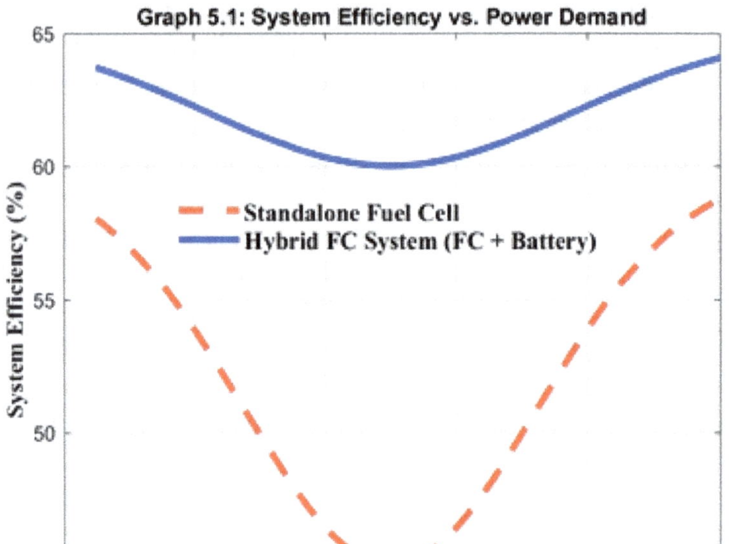

Fuel cell technology has seen a notable expansion into both portable and automotive domains due to its high energy density, rapid refuelling, and zero-emission profile. In portable applications, Proton Exchange Membrane Fuel Cells (PEMFCS) are being used to power backup systems, communication devices, medical equipment, military electronics, and off-grid sensors (Qasem & Abdulrahman, 2024b). These systems offer silent operation and long runtime, making them ideal for remote or mission-critical environments. On the automotive front, fuel cell electric vehicles are gaining traction as a cleaner alternative to internal combustion engine vehicles, particularly for medium- and heavy-duty transportation such as trucks, buses, and trains. Companies like Toyota, Hyundai, and Honda have already introduced commercial FCEV models, supported by expanding hydrogen refuelling infrastructure. The advantage of quick refuelling and extended driving ranges gives fuel cell vehicles a competitive edge in applications where battery-electric solutions face limitations (Togun et al., 2024). Moreover, ongoing research into lightweight stack materials, compact high-gain converters, and intelligent hybridization is helping improve vehicle performance and overall fuel cell system reliability.

Figure 22. Green Internet of Vehicles (GIoV) architecture with fuel cell integration(J. Wang et al., 2021).

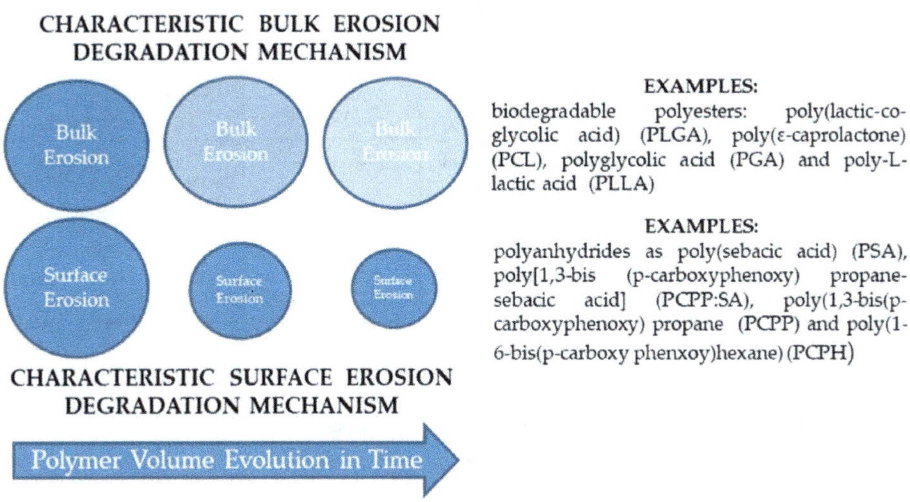

The Green Internet of Vehicles (GIoV) is an emerging paradigm that integrates environmentally friendly vehicular technologies with advanced communication and data processing systems. It represents a convergence of sustainable transportation, digital connectivity, and artificial intelligence(Chen et al., 2019). In the context of hydrogen-powered vehicles, GIoV enables real-time monitoring, energy optimization, predictive diagnostics, and intelligent routing based on traffic and environmental conditions. The use of AI and machine learning allows for smart energy management within fuel cell hybrid systems, ensuring efficient power sharing between fuel cells, batteries, and supercapacitors. Additionally, vehicles can communicate with infrastructure (V2I), other vehicles (V2V), and the grid (V2G) to optimize energy usage and reduce emissions. Through integration with renewable energy sources and decentralized energy storage networks, GIoV also supports the concept of Vehicle-to-Everything (V2X), playing a critical role in future smart cities. As the hydrogen economy matures, GIoV offers an intelligent, low-carbon transportation model that enhances efficiency, safety, and environmental sustainability (Yazdi et al., 2023).

Table 6. Comparison of Research Focus Areas and TRL for Each Fuel Cell Type (Gulotta et al., 2022)

Fuel Cell Type	Current TRL	Primary Research Focus Areas	Commercial Maturity	Prospects
PEMFC	8–9	Catalyst cost reduction, cold start strategies, and durability	High (automotive & portable)	The main candidate for automotive and Iov integration
SOFC	6–8	High-temp materials, thermal cycling, and hybrid integration	Moderate (stationary, CHP)	Promising grid support and hydrogen infrastructure
AFC	5–6	CO_2 tolerance, electrolyte stability	Low to moderate	Niche applications in aerospace, backup power
MCFC	6–7	CO_2 management, corrosion resistance, system scale-up	Moderate (industrial CHP)	Suitable for large-scale CO_2-utilizing systems
PAFC	7–8	System miniaturization, acid corrosion mitigation	Moderate	Commercial in some CHP systems, but slowly advancing
DMFC	5–6	Methanol crossover control, improved membrane performance	Low (portable electronics)	Portable systems with simplified fuel logistics

8.1 Environmental and Economic Analysis of Hydrogen Technologies

The environmental and economic implications of hydrogen-based energy systems are increasingly significant in shaping future energy policies. From an environmental perspective, hydrogen fuel cells produce only water and heat during operation, eliminating greenhouse gas emissions associated with fossil fuel combustion. When hydrogen is generated from renewable sources, the entire energy cycle becomes nearly carbon neutral. Compared with battery systems, hydrogen storage offers higher gravimetric energy density and longer operational lifespans, contributing to lower lifecycle emissions. Economically, the cost of hydrogen production remains a major challenge, with green hydrogen currently estimated between $3–6 per kg, though technological advances and global scaling are expected to reduce this cost below $2 per kg by 2030. The increasing adoption of large-scale electrolyzers, government subsidies, and carbon pricing mechanisms are accelerating cost parity with traditional fuels. Moreover, hydrogen infrastructure, comprising pipelines, refuelling stations, and storage facilities, represents a long-term investment opportunity that supports energy independence and grid resilience. Several international policies, such as the European Union's "Hydrogen Strategy for a Climate-Neutral Europe" and India's "National Green Hydrogen Mission," emphasize the dual role of hydrogen in decarbonization and economic development. Thus, integrating hydrogen technologies not only promotes environmental sustainability but also drives industrial innovation, energy security, and employment growth, making it a cornerstone of the future clean energy economy.

Table 7. Comparative Environmental and Economic Analysis of Hydrogen, Battery, and Fossil Fuel Technologies

Parameter	Hydrogen Fuel Cells	Battery Systems	Fossil Fuels
CO_2 Emissions (g/kWh)	0 (with green H_2)	100–150 (from manufacturing and recycling)	800–1000
Energy Density (Wh/kg)	800–3000	100–250	12,000 (but with high emissions)
Lifecycle Cost (USD/kWh)	0.12–0.18	0.10–0.15	0.08–0.12
Policy Support Level	High (EU, Japan, India, USA)	Moderate	Declining
Environmental Impact	Near-zero emissions; recyclable materials	Moderate (limited recyclability)	High pollution, CO_2-intensive
Economic Outlook (2030)	Cost parity expected with fossil fuels via green H_2 scaling	Stable with recycling innovation	Gradual phase-out due to carbon taxes

The global momentum toward hydrogen adoption and intelligent vehicle integration marks a pivotal shift in the energy transition. The final section synthesizes these insights, presenting key conclusions and research directions that can drive future innovations in fuel cell technology.

9. CONCLUSION

In recent years, fuel cell technology has witnessed transformative innovations across multiple domains, enhancing its performance, reliability, and applicability. From the development of novel electrode and electrolyte materials to platinum-free catalyst advancements and high-temperature membrane technologies, material science has played a key role in improving efficiency and durability. Cold start strategies, hybridization with batteries and supercapacitors, and intelligent power management using AI/ML algorithms have addressed critical operational challenges, particularly in automotive and portable systems. Simulation-based modelling, including high-gain DC-DC converters and fuzzy logic MPPT control, has further contributed to optimizing fuel cell systems for real-world performance. Moreover, integration with renewable energy sources and the emergence of the hydrogen economy underscore the growing role of fuel cells in global energy transition efforts. Concepts such as the Green Internet of Vehicles (GIoV) signify a paradigm shift toward digitally connected, zero-emission transportation networks.

Looking forward, potential research directions include the development of low-cost, earth-abundant catalysts with enhanced durability, scalable solid-state hydrogen storage methods, and advanced thermal management systems for all-climate operability. There is also a need to explore multi-objective control algorithms that combine AI with real-time sensor feedback for adaptive energy management in hybrid systems. The integration of fuel cells into smart grids, microgrids, and decentralized energy networks opens new avenues for research on grid-responsive control and energy exchange protocols. Finally, further investigation into cyber-physical security, interoperability, and AI ethics will be essential as fuel cell systems become increasingly connected within smart transportation ecosystems. Continued interdisciplinary efforts will be crucial to unlocking the full potential of fuel cells in building a sustainable, intelligent, and resilient energy future.

REFERENCES

Abdalla, A. M., Hossain, S., Nisfindy, O. B., Azad, A. T., Dawood, M., & Azad, A. K. (2018). Hydrogen production, storage, transportation and key challenges with applications: A review. *Energy Conversion and Management*, *165*, 602–627. DOI: 10.1016/j.enconman.2018.03.088

Abdin, Z. (2024a). Shaping the stationary energy storage landscape with reversible fuel cells. *Journal of Energy Storage*, *86*, 111354. DOI: 10.1016/j.est.2024.111354

Abdin, Z. (2024b). Shaping the stationary energy storage landscape with reversible fuel cells. *Journal of Energy Storage*, *86*, 111354. DOI: 10.1016/j.est.2024.111354

Abdin, Z. (2024c). Shaping the stationary energy storage landscape with reversible fuel cells. *Journal of Energy Storage*, *86*, 111354. DOI: 10.1016/j.est.2024.111354

Abdin, Z. (2024d). Shaping the stationary energy storage landscape with reversible fuel cells. *Journal of Energy Storage*, *86*, 111354. DOI: 10.1016/j.est.2024.111354

Abdin, Z. (2024e). Shaping the stationary energy storage landscape with reversible fuel cells. *Journal of Energy Storage*, *86*, 111354. DOI: 10.1016/j.est.2024.111354

Abu-Rub, H., Malinowski, M., & Al-Haddad, K. (2014). *Power electronics for renewable energy systems, transportation and industrial applications*. John Wiley & Sons. DOI: 10.1002/9781118755525

Agüero, J. R. (2012). Improving the efficiency of power distribution systems through technical and non-technical losses reduction. *PES T&D*, *2012*, 1–8.

Ahmed, F., Naeem, M., Ejaz, W., Iqbal, M., & Anpalagan, A. (2018). Resource management in cellular base stations powered by renewable energy sources. *Journal of Network and Computer Applications*, *112*, 1–17. DOI: 10.1016/j.jnca.2018.03.021

Al-Sayyab, A. K. S., Mota-Babiloni, A., & Navarro-Esbrí, J. (2023). Renewable and waste heat applications for heating, cooling, and power generation based on advanced configurations. *Energy Conversion and Management*, *291*, 117253. DOI: 10.1016/j.enconman.2023.117253

Albatayneh, A., Assaf, M. N., Alterman, D., & Jaradat, M. (2020). Comparison of the overall energy efficiency for internal combustion engine vehicles and electric vehicles. *Rigas Tehniskas Universitates Zinatniskie Raksti*, *24*(1), 669–680. DOI: 10.2478/rtuect-2020-0041

Alonso-Moreno, C., & García-Yuste, S. (2016). Environmental potential of the use of CO2 from alcoholic fermentation processes. The CO2-AFP strategy. *The Science of the Total Environment*, *568*, 319–326. DOI: 10.1016/j.scitotenv.2016.05.220

Amamou, A. A., Kelouwani, S., Boulon, L., & Agbossou, K. (2016). A comprehensive review of solutions and strategies for cold start of automotive proton exchange membrane fuel cells. *IEEE Access : Practical Innovations, Open Solutions*, *4*, 4989–5002. DOI: 10.1109/ACCESS.2016.2597058

Antipov, E. V., Abakumov, A. M., Drozhzhin, O. A., & Pogozhev, D. V. (2019). Lithium-ion electrochemical energy storage: The current state, problems, and development trends in Russia. *Thermal Engineering*, *66*(4), 219–224. DOI: 10.1134/S0040601519040013

Arif, M., Cheung, S. C. P., & Andrews, J. (2020). A systematic approach for matching simulated and experimental polarization curves for a PEM fuel cell. *International Journal of Hydrogen Energy*, *45*(3), 2206–2223. DOI: 10.1016/j.ijhydene.2019.11.057

Asif, U., & Schmidt, K. (2021). Fuel cell electric vehicles (FCEV): Policy advances to enhance commercial success. *Sustainability (Basel)*, *13*(9), 5149. DOI: 10.3390/su13095149

Badwal, S. P. S., Giddey, S., & Ciacchi, F. T. (2006). Hydrogen and oxygen generation with polymer electrolyte membrane (PEM)-based electrolytic technology. *Ionics*, *12*(1), 7–14. DOI: 10.1007/s11581-006-0002-x

Bagotsky, V. S., Skundin, A. M., & Volfkovich, Y. M. (2015). *Electrochemical power sources: batteries, fuel cells, and supercapacitors*. John Wiley & Sons.

Balafkandeh, S., Mahmoudi, S. M. S., & Gholamian, E. (2022a). Design and tri-criteria optimization of an MCFC based energy system with hydrogen production and injection: An effort to minimize the carbon emission. *Process Safety and Environmental Protection*, *166*, 299–309. DOI: 10.1016/j.psep.2022.08.020

Balafkandeh, S., Mahmoudi, S. M. S., & Gholamian, E. (2022b). Design and tri-criteria optimization of an MCFC based energy system with hydrogen production and injection: An effort to minimize the carbon emission. *Process Safety and Environmental Protection*, *166*, 299–309. DOI: 10.1016/j.psep.2022.08.020

Bar-Nun, A., & Chang, S. (1983). Photochemical reactions of water and carbon monoxide in Earth's primitive atmosphere. *Journal of Geophysical Research*, *88*(C11), 6662–6672. DOI: 10.1029/JC088iC11p06662

Bartnitzek, T., Thelemann, T., Apel, S., & Suphan, K.-H. (2016). Advantages and limitations of ceramic packaging technologies in harsh applications. *International Symposium on Microelectronics,* 2016(1), 581–585. DOI: 10.4071/isom-2016-THP23

Bayrak, G., & Cebeci, M. (2014). Grid connected fuel cell and PV hybrid power generating system design with Matlab Simulink. *International Journal of Hydrogen Energy, 39*(16), 8803–8812. DOI: 10.1016/j.ijhydene.2013.12.029

Becker, H., Murawski, J., Shinde, D. V., Stephens, I. E. L., Hinds, G., & Smith, G. (2023). Impact of impurities on water electrolysis: A review. *Sustainable Energy & Fuels, 7*(7), 1565–1603. DOI: 10.1039/D2SE01517J

Behzadi, M. S., & Niasati, M. (2015). Comparative performance analysis of a hybrid PV/FC/battery stand-alone system using different power management strategies and sizing approaches. *International Journal of Hydrogen Energy, 40*(1), 538–548. DOI: 10.1016/j.ijhydene.2014.10.097

Belmonte, N., Luetto, C., Staulo, S., Rizzi, P., & Baricco, M. (2017). Case studies of energy storage with fuel cells and batteries for stationary and mobile applications. *Challenges, 8*(1), 9. DOI: 10.3390/challe8010009

Boder, M., & Dittmeyer, R. (2006). Catalytic modification of conventional SOFC anodes with a view to reducing their activity for direct internal reforming of natural gas. *Journal of Power Sources, 155*(1), 13–22. DOI: 10.1016/j.jpowsour.2004.11.075

Bradshaw, V. (2010). *The building environment: Active and passive control systems.* John Wiley & Sons.

Brett, D. J. L., Atkinson, A., Brandon, N. P., & Skinner, S. J. (2008). Intermediate temperature solid oxide fuel cells. *Chemical Society Reviews, 37*(8), 1568–1578. DOI: 10.1039/b612060c

Carrette, L., Friedrich, K. A., & Stimming, U. (2000). Fuel cells: Principles, types, fuels, and applications. *ChemPhysChem, 1*(4), 162–193. DOI: 10.1002/1439-7641(20001215)1:4<162::AID-CPHC162>3.0.CO;2-Z

Chander, B., & Gopalakrishnan, K. (2024). Integration of artificial intelligence techniques for energy management. *Sustainable Management of Electronic Waste*, 1–46.

Chang, C. J., Chang, M. C. Y., Damrauer, N. H., & Nocera, D. G. (2004). Proton-coupled electron transfer: A unifying mechanism for biological charge transport, amino acid radical initiation and propagation, and bond making/breaking reactions of water and oxygen. *Biochimica et Biophysica Acta (BBA)-. Biochimica et Biophysica Acta. Bioenergetics, 1655*, 13–28. DOI: 10.1016/j.bbabio.2003.08.010

Chen, H., Zhao, T., Li, C., & Guo, Y. (2019). Green internet of vehicles: Architecture, enabling technologies, and applications. *IEEE Access : Practical Innovations, Open Solutions*, *7*, 179185–179198. DOI: 10.1109/ACCESS.2019.2958175

Corigliano, O., Pagnotta, L., & Fragiacomo, P. (2022a). On the technology of solid oxide fuel cell (SOFC) energy systems for stationary power generation: A review. *Sustainability (Basel)*, *14*(22), 15276. DOI: 10.3390/su142215276

Corigliano, O., Pagnotta, L., & Fragiacomo, P. (2022b). On the technology of solid oxide fuel cell (SOFC) energy systems for stationary power generation: A review. *Sustainability (Basel)*, *14*(22), 15276. DOI: 10.3390/su142215276

Dai, A., Li, Q., Liu, T., Amine, K., & Lu, J. (2019). Fundamental Understanding of Water-Induced Mechanisms in Li–O2 Batteries: Recent Developments and Perspectives. *Advanced Materials*, *31*(31), 1805602. DOI: 10.1002/adma.201805602

Das, V., Padmanaban, S., Venkitusamy, K., Selvamuthukumaran, R., Blaabjerg, F., & Siano, P. (2017). Recent advances and challenges of fuel cell based power system architectures and control–A review. *Renewable & Sustainable Energy Reviews*, *73*, 10–18. DOI: 10.1016/j.rser.2017.01.148

Daud, W. R. W., Najafpour, G., & Rahimnejad, M. (2011). Clean energy for tomorrow: towards zero emission and carbon free future: a review. *Iranica Journal of Energy and Environment*, *2*(3), 262–273.

Deng, Y., Foo, S., & Li, H. (2009). Real time simulation of power flow control strategies for fuel cell vehicle with energy storage by using real time digital simulator (RTDS). *2009 IEEE 6th International Power Electronics and Motion Control Conference*, 2323–2327.

Deng, Y., Liu, H., Zhao, X., & Chen, J. (2018a). Effects of cold start control strategy on cold start performance of the diesel engine based on a comprehensive preheat diesel engine model. *Applied Energy*, *210*, 279–287. DOI: 10.1016/j.apenergy.2017.10.093

Deng, Y., Liu, H., Zhao, X., & Chen, J. (2018b). Effects of cold start control strategy on cold start performance of the diesel engine based on a comprehensive preheat diesel engine model. *Applied Energy*, *210*, 279–287. DOI: 10.1016/j.apenergy.2017.10.093

Deng, Y., Liu, H., Zhao, X., & Chen, J. (2018c). Effects of cold start control strategy on cold start performance of the diesel engine based on a comprehensive preheat diesel engine model. *Applied Energy*, *210*, 279–287. DOI: 10.1016/j.apenergy.2017.10.093

Deng, Y., Liu, H., Zhao, X., & Chen, J. (2018d). Effects of cold start control strategy on cold start performance of the diesel engine based on a comprehensive preheat diesel engine model. *Applied Energy*, *210*, 279–287. DOI: 10.1016/j.apenergy.2017.10.093

Deng, Y., Liu, H., Zhao, X., & Chen, J. (2018e). Effects of cold start control strategy on cold start performance of the diesel engine based on a comprehensive preheat diesel engine model. *Applied Energy*, *210*, 279–287. DOI: 10.1016/j.apenergy.2017.10.093

Dickinson, E. J. F., & Smith, G. (2020). Modelling the proton-conductive membrane in practical polymer electrolyte membrane fuel cell (PEMFC) simulation: A review. *Membranes (Basel)*, *10*(11), 310. DOI: 10.3390/membranes10110310

Dirkes, S., Leidig, J., Fisch, P., & Pischinger, S. (2023). Prescriptive Lifetime Management for PEM fuel cell systems in transportation applications, Part I: State of the art and conceptual design. *Energy Conversion and Management*, *277*, 116598. DOI: 10.1016/j.enconman.2022.116598

Doddathimmaiah, A., & Andrews, J. (2009). Theory, modelling and performance measurement of unitised regenerative fuel cells. *International Journal of Hydrogen Energy*, *34*(19), 8157–8170. DOI: 10.1016/j.ijhydene.2009.07.116

Dou, Y., Sun, L., Ren, J., & Dong, L. (2023). Opportunities and future challenges in hydrogen economy for sustainable development. In *Hydrogen economy* (pp. 537–569). Elsevier. DOI: 10.1016/B978-0-323-99514-6.00008-X

Du, Z., Liu, C., Zhai, J., Guo, X., Xiong, Y., Su, W., & He, G. (2021). A review of hydrogen purification technologies for fuel cell vehicles. *Catalysts*, *11*(3), 393. DOI: 10.3390/catal11030393

Ehteshami, S. M. M., & Chan, S. H. (2014a). The role of hydrogen and fuel cells to store renewable energy in the future energy network–potentials and challenges. *Energy Policy*, *73*, 103–109. DOI: 10.1016/j.enpol.2014.04.046

Ehteshami, S. M. M., & Chan, S. H. (2014b). The role of hydrogen and fuel cells to store renewable energy in the future energy network–potentials and challenges. *Energy Policy*, *73*, 103–109. DOI: 10.1016/j.enpol.2014.04.046

Ellis, M. W., Von Spakovsky, M. R., & Nelson, D. J. (2001). Fuel cell systems: Efficient, flexible energy conversion for the 21st century. *Proceedings of the IEEE*, *89*(12), 1808–1818. DOI: 10.1109/5.975914

Ero lu, F., Kurto lu, M., & Vural, A. M. (2021). Bidirectional DC–DC converter based multilevel battery storage systems for electric vehicle and large-scale grid applications: A critical review considering different topologies, state-of-charge balancing and future trends. *IET Renewable Power Generation*, *15*(5), 915–938. DOI: 10.1049/rpg2.12042

Fathabadi, H. (2018). Novel fuel cell/battery/supercapacitor hybrid power source for fuel cell hybrid electric vehicles. *Energy*, *143*, 467–477. DOI: 10.1016/j.energy.2017.10.107

Gallo, M., Costabile, C., Sorrentino, M., Polverino, P., & Pianese, C. (2020). Development and application of a comprehensive model-based methodology for fault mitigation of fuel cell powered systems. *Applied Energy*, *279*, 115698. DOI: 10.1016/j.apenergy.2020.115698

Goodenough, J. B. (2013). Evolution of strategies for modern rechargeable batteries. *Accounts of Chemical Research*, *46*(5), 1053–1061. DOI: 10.1021/ar2002705

Gulotta, T. M., Salomone, R., Lanuzza, F., Saija, G., Mondello, G., & Ioppolo, G. (2022). Life Cycle Assessment and Life Cycle Costing of unitized regenerative fuel cell: A systematic review. *Environmental Impact Assessment Review*, *92*, 106698. DOI: 10.1016/j.eiar.2021.106698

Gupta, P., Toksha, B., & Rahaman, M. (2024). A critical review on hydrogen based fuel cell technology and applications. *Chemical Record (New York, N.Y.)*, *24*(1), e202300295. DOI: 10.1002/tcr.202300295

Gür, T. M. (2018). Review of electrical energy storage technologies, materials and systems: Challenges and prospects for large-scale grid storage. *Energy & Environmental Science*, *11*(10), 2696–2767. DOI: 10.1039/C8EE01419A

Haile, S. M. (2003). Fuel cell materials and components. *Acta Materialia*, *51*(19), 5981–6000. DOI: 10.1016/j.actamat.2003.08.004

Hall, J., & Kerr, R. (2003). Innovation dynamics and environmental technologies: The emergence of fuel cell technology. *Journal of Cleaner Production*, *11*(4), 459–471. DOI: 10.1016/S0959-6526(02)00067-7

Hassan, Q., Azzawi, I. D. J., Sameen, A. Z., & Salman, H. M. (2023). Hydrogen fuel cell vehicles: Opportunities and challenges. *Sustainability (Basel)*, *15*(15), 11501. DOI: 10.3390/su151511501

Hou, J., Shao, Y., Ellis, M. W., Moore, R. B., & Yi, B. (2011). Graphene-based electrochemical energy conversion and storage: Fuel cells, supercapacitors and lithium ion batteries. *Physical Chemistry Chemical Physics*, *13*(34), 15384–15402. DOI: 10.1039/c1cp21915d

Hsu, M., & Nathanson, D. (1995). *Zteks ultra high efficiency fuel cell/gas turbine combination*. USDOE Morgantown Energy Technology Center (METC), WV (United States).

Huang, L., & Qi, R. (2022). A comprehensive review on assembly design strategies on proton exchange membrane applications. *International Journal of Hydrogen Energy*, *47*(80), 33903–33918. DOI: 10.1016/j.ijhydene.2022.07.271

Huang, L., Zaman, S., Tian, X., Wang, Z., Fang, W., & Xia, B. Y. (2021). Advanced platinum-based oxygen reduction electrocatalysts for fuel cells. *Accounts of Chemical Research*, *54*(2), 311–322. DOI: 10.1021/acs.accounts.0c00488

Hughes, A. E., Haque, N., Northey, S. A., & Giddey, S. (2021). Platinum group metals: A review of resources, production and usage with a focus on catalysts. *Resources*, *10*(9), 93. DOI: 10.3390/resources10090093

Hull, M. N., Ellison, J. E., & Toni, J. E. (1970). The anodic behavior of zinc electrodes in potassium hydroxide electrolytes. *Journal of the Electrochemical Society*, *117*(2), 192. DOI: 10.1149/1.2407463

Jha, A. R. (2012). *Next-generation batteries and fuel cells for commercial, military, and space applications*. CRC Press Boca Raton, FL, USA: Kebede, A. A., Kalogiannis, T., Van Mierlo, J., & Berecibar, M. (2022). A comprehensive review of stationary energy storage devices for large scale renewable energy sources grid integration. *Renewable & Sustainable Energy Reviews*, *159*, 112213.

Keshavarzzadeh, A. H., Ahmadi, P., & Safaei, M. R. (2019). Assessment and optimization of an integrated energy system with electrolysis and fuel cells for electricity, cooling and hydrogen production using various optimization techniques. *International Journal of Hydrogen Energy*, *44*(39), 21379–21396. DOI: 10.1016/j.ijhydene.2019.06.127

Khorshidian, M., & Sedighi, M. (2019). A review of the main mechanisms of catalyst layer degradation in polymer electrolyte membrane fuel cell (PEMFC) and different performance recovery methods. *Hydrogen, Fuel Cell &. Energy Storage*, *6*(2), 91–115.

Kim, T.-H., Ha, C.-S., Jang, S.-R., Kim, J.-S., & Han, S.-T. (2022). Suppression of a long-term instability of a commercial magnetron with low-ripple DC power supply for heater. *IEEE Access : Practical Innovations, Open Solutions*, *10*, 32235–32243. DOI: 10.1109/ACCESS.2022.3158020

LaConti, A. B., Hamdan, M., & McDonald, R. C. (2003). Mechanisms of membrane degradation. *Handbook of Fuel Cells*, *3*, 647–662.

Lee, D. C., Yang, H. N., Park, S. H., & Kim, W. J. (2014). Nafion/graphene oxide composite membranes for low humidifying polymer electrolyte membrane fuel cell. *Journal of Membrane Science*, *452*, 20–28. DOI: 10.1016/j.memsci.2013.10.018

Li, L., Wang, S., Yue, L., & Wang, G. (2019). Cold-start method for proton-exchange membrane fuel cells based on locally heating the cathode. *Applied Energy*, *254*, 113716. DOI: 10.1016/j.apenergy.2019.113716

Li, Y., Zhao, X., Liu, Z., Li, Y., Chen, W., & Li, Q. (2015). Experimental study on the voltage uniformity for dynamic loading of a PEM fuel cell stack. *International Journal of Hydrogen Energy*, *40*(23), 7361–7369. DOI: 10.1016/j.ijhydene.2015.04.058

Liu, Z., Zhang, X., Sun, Y., & Zhou, Y. (2023). Advanced controls on energy reliability, flexibility and occupant-centric control for smart and energy-efficient buildings. *Energy and Building*, *297*, 113436. DOI: 10.1016/j.enbuild.2023.113436

Luo, X., Wang, J., Dooner, M., & Clarke, J. (2015). Overview of current development in electrical energy storage technologies and the application potential in power system operation. *Applied Energy*, *137*, 511–536. DOI: 10.1016/j.apenergy.2014.09.081

Mangaiyarkkarasi, J. (2025). Nanostructural Innovations: Powering the Future of Energy Storage. In *Design, Fabrication, and Significance of Advanced Nanostructured Materials* (pp. 51–74). IGI Global.

Martinaiou, I., & Daletou, M. K. (2024). Enhancing Electrode Efficiency in Proton Exchange Membrane Fuel Cells with PGM-Free Catalysts: A Mini Review. *Energies*, *17*(14), 3443. DOI: 10.3390/en17143443

Mehr, A. S., Lanzini, A., Santarelli, M., & Rosen, M. A. (2021a). Polygeneration systems based on high temperature fuel cell (MCFC and SOFC) technology: System design, fuel types, modeling and analysis approaches. *Energy*, *228*, 120613. DOI: 10.1016/j.energy.2021.120613

Mehr, A. S., Lanzini, A., Santarelli, M., & Rosen, M. A. (2021b). Polygeneration systems based on high temperature fuel cell (MCFC and SOFC) technology: System design, fuel types, modeling and analysis approaches. *Energy*, *228*, 120613. DOI: 10.1016/j.energy.2021.120613

Mehr, A. S., Lanzini, A., Santarelli, M., & Rosen, M. A. (2021c). Polygeneration systems based on high temperature fuel cell (MCFC and SOFC) technology: System design, fuel types, modeling and analysis approaches. *Energy*, *228*, 120613. DOI: 10.1016/j.energy.2021.120613

Mekhilef, S., Saidur, R., & Safari, A. (2012a). Comparative study of different fuel cell technologies. *Renewable & Sustainable Energy Reviews*, *16*(1), 981–989. DOI: 10.1016/j.rser.2011.09.020

Mekhilef, S., Saidur, R., & Safari, A. (2012b). Comparative study of different fuel cell technologies. *Renewable & Sustainable Energy Reviews, 16*(1), 981–989. DOI: 10.1016/j.rser.2011.09.020

Miyazaki, H., Watanabe, T., Kishinaga, S., & Kawakami, F. (2003). Active field control (AFC)-reverberation enhancement system using acoustical feedback control. *Audio Engineering Society Convention 115.*

Nouman, J. (2018). *Development of a carbon capture and storage system in oil refineries using molten carbonate fuel cells.*

Oró, E., De Gracia, A., Castell, A., Farid, M. M., & Cabeza, L. F. (2012). Review on phase change materials (PCMs) for cold thermal energy storage applications. *Applied Energy, 99*, 513–533. DOI: 10.1016/j.apenergy.2012.03.058

Ozoemena, K. I. (2016). Nanostructured platinum-free electrocatalysts in alkaline direct alcohol fuel cells: Catalyst design, principles and applications. *RSC Advances, 6*(92), 89523–89550. DOI: 10.1039/C6RA15057H

Peng, B., & Chen, J. (2009). Functional materials with high-efficiency energy storage and conversion for batteries and fuel cells. *Coordination Chemistry Reviews, 253*(23–24), 2805–2813. DOI: 10.1016/j.ccr.2009.04.008

Pramuanjaroenkij, A., & Kakaç, S. (2023). The fuel cell electric vehicles: The highlight review. *International Journal of Hydrogen Energy, 48*(25), 9401–9425. DOI: 10.1016/j.ijhydene.2022.11.103

Qasem, N. A. A., & Abdulrahman, G. A. Q. (2024a). A recent comprehensive review of fuel cells: History, types, and applications. *International Journal of Energy Research, 2024*(1), 7271748. DOI: 10.1155/2024/7271748

Qasem, N. A. A., & Abdulrahman, G. A. Q. (2024b). A recent comprehensive review of fuel cells: History, types, and applications. *International Journal of Energy Research, 2024*(1), 7271748. DOI: 10.1155/2024/7271748

Qi, Y., & Wu, X. (2011). Low-carbon technologies integrated innovation strategy based on modular design. *Energy Procedia, 5*, 2509–2515. DOI: 10.1016/j.egypro.2011.03.431

Ramanathan, K., & Oh, S. H. (2014). Modeling and analysis of rapid catalyst aging cycles. *Chemical Engineering Research & Design, 92*(2), 350–361. DOI: 10.1016/j.cherd.2013.06.020

Redl, R., & Sun, J. (2009). Ripple-based control of switching regulators—An overview. *IEEE Transactions on Power Electronics*, *24*(12), 2669–2680. DOI: 10.1109/TPEL.2009.2032657

Reid, C. M., Miller, T. B., Hoberecht, M. A., Loyselle, P. L., Taylor, L. M., Farmer, S. C., & Jansen, R. H. (2013). History of electrochemical and energy storage technology development at NASA Glenn Research Center. *Journal of Aerospace Engineering*, *26*(2), 361–371. DOI: 10.1061/(ASCE)AS.1943-5525.0000323

Rosli, R. E., Sulong, A. B., Daud, W. R. W., Zulkifley, M. A., Husaini, T., Rosli, M. I., Majlan, E. H., & Haque, M. A. (2017). A review of high-temperature proton exchange membrane fuel cell (HT-PEMFC) system. *International Journal of Hydrogen Energy*, *42*(14), 9293–9314. DOI: 10.1016/j.ijhydene.2016.06.211

Sammes, N., Bove, R., & Stahl, K. (2004). Phosphoric acid fuel cells: Fundamentals and applications. *Current Opinion in Solid State and Materials Science*, *8*(5), 372–378. DOI: 10.1016/j.cossms.2005.01.001

Samsun, R. C., Rex, M., Antoni, L., & Stolten, D. (2022). Deployment of fuel cell vehicles and hydrogen refueling station infrastructure: A global overview and perspectives. *Energies*, *15*(14), 4975. DOI: 10.3390/en15144975

Saslow, W. M. (1999). Voltaic cells for physicists: Two surface pumps and an internal resistance. *American Journal of Physics*, *67*(7), 574–583. DOI: 10.1119/1.19327

Sharaf, O. Z., & Orhan, M. F. (2014). An overview of fuel cell technology: Fundamentals and applications. *Renewable & Sustainable Energy Reviews*, *32*, 810–853. DOI: 10.1016/j.rser.2014.01.012

Smallbone, A., Jülch, V., Wardle, R., & Roskilly, A. P. (2017). Levelised Cost of Storage for Pumped Heat Energy Storage in comparison with other energy storage technologies. *Energy Conversion and Management*, *152*, 221–228. DOI: 10.1016/j.enconman.2017.09.047

Soleimani, A., Hosseini Dolatabadi, S. H., Heidari, M., Pinnarelli, A., Mehdizadeh Khorrami, B., Luo, Y., Vizza, P., & Brusco, G. (2024). Progress in hydrogen fuel cell vehicles and up-and-coming technologies for eco-friendly transportation: An international assessment. *Multiscale and Multidisciplinary Modeling. Experiments and Design*, *7*(4), 3153–3172.

Song, C. (2002). Fuel processing for low-temperature and high-temperature fuel cells: Challenges, and opportunities for sustainable development in the 21st century. *Catalysis Today*, *77*(1–2), 17–49. DOI: 10.1016/S0920-5861(02)00231-6

Song, R.-H., Dheenadayalan, S., & Shin, D.-R. (2002). Effect of silicon carbide particle size in the electrolyte matrix on the performance of a phosphoric acid fuel cell. *Journal of Power Sources, 106*(1–2), 167–172. DOI: 10.1016/S0378-7753(01)01055-2

Sorensen, B., & Spazzafumo, G. (2018a). *Hydrogen and fuel cells: emerging technologies and applications.*

Sorensen, B., & Spazzafumo, G. (2018b). *Hydrogen and fuel cells: emerging technologies and applications.*

Stamenkovic, V. R., Strmcnik, D., Lopes, P. P., & Markovic, N. M. (2017). Energy and fuels from electrochemical interfaces. *Nature Materials, 16*(1), 57–69. DOI: 10.1038/nmat4738

STRANIERO, D. (2016). *Regenerative solid oxide cells (RSOCs) for energy storage applications.*

Talaat, M., Elgarhy, A., Elkholy, M. H., & Farahat, M. A. (2021). Integration of fuel cells into an off-grid hybrid system using wave and solar energy. *International Journal of Electrical Power & Energy Systems, 130*, 106939. DOI: 10.1016/j.ijepes.2021.106939

Tang, H. H., & Ahmad, N. S. (2024). Fuzzy logic approach for controlling uncertain and nonlinear systems: A comprehensive review of applications and advances. *Systems Science & Control Engineering, 12*(1), 2394429. DOI: 10.1080/21642583.2024.2394429

Thounthong, P., Rael, S., & Davat, B. (2009). Energy management of fuel cell/battery/supercapacitor hybrid power source for vehicle applications. *Journal of Power Sources, 193*(1), 376–385. DOI: 10.1016/j.jpowsour.2008.12.120

Tian, C., Lin, F., & Doeff, M. M. (2017). Electrochemical characteristics of layered transition metal oxide cathode materials for lithium ion batteries: Surface, bulk behavior, and thermal properties. *Accounts of Chemical Research, 51*(1), 89–96. DOI: 10.1021/acs.accounts.7b00520

Togun, H., Aljibori, H. S. S., Abed, A. M., Biswas, N., Alshamkhani, M. T., Niyas, H., Mohammed, H. I., Rashid, F. L., & Paul, D. (2024). A review on recent advances on improving fuel economy and performance of a fuel cell hybrid electric vehicle. *International Journal of Hydrogen Energy, 89*, 22–47. DOI: 10.1016/j.ijhydene.2024.09.298

Tomczyk, P. (2006). MCFC versus other fuel cells—Characteristics, technologies and prospects. *Journal of Power Sources, 160*(2), 858–862. DOI: 10.1016/j.jpowsour.2006.04.071

Torvelainen, J. (2023). *Viability of self-generated energy in paper and board mills: study on manufacturing costs.*

Vevere, L., Yakushin, V., Sture-Skela, B., Andersons, J., & Cabulis, U. (2024). Cryogenic Insulation—Towards Environmentally Friendly Polyurethane Foams. *Polymers, 16*(17), 2406. DOI: 10.3390/polym16172406

Vilčiauskas, L., Tuckerman, M. E., Bester, G., Paddison, S. J., & Kreuer, K.-D. (2012). The mechanism of proton conduction in phosphoric acid. *Nature Chemistry, 4*(6), 461–466. DOI: 10.1038/nchem.1329

Virkar, A. V. (2012). Transport through mixed proton, oxygen ion and electron/hole conductors: Analysis of fuel cells and electrolyzer cells using Onsager equations. *International Journal of Hydrogen Energy, 37*(17), 12609–12628. DOI: 10.1016/j.ijhydene.2012.05.118

Wang, J. (2017). System integration, durability and reliability of fuel cells: Challenges and solutions. *Applied Energy, 189*, 460–479. DOI: 10.1016/j.apenergy.2016.12.083

Wang, J., Zhu, K., & Hossain, E. (2021). Green Internet of Vehicles (IoV) in the 6G era: Toward sustainable vehicular communications and networking. *IEEE Transactions on Green Communications and Networking, 6*(1), 391–423. DOI: 10.1109/TGCN.2021.3127923

Wang, P., Du, K., Yin, H., & Wang, D. (2023). Corrosion and protection of metallic materials in molten carbonates for concentrating solar power and molten carbonate electrolysis applications. *Corrosion Communications, 11*, 58–71. DOI: 10.1016/j.corcom.2023.01.003

Wang, X., He, B., Hu, Z., Zeng, Z., & Han, S. (2014). Current advances in precious metal core–shell catalyst design. *Science and Technology of Advanced Materials, 15*(4), 043502. DOI: 10.1088/1468-6996/15/4/043502

Waseem, M., Amir, M., Lakshmi, G. S., Harivardhagini, S., & Ahmad, M. (2023). Fuel cell-based hybrid electric vehicles: An integrated review of current status, key challenges, recommended policies, and future prospects. *Green Energy and Intelligent Transportation, 2*(6), 100121. DOI: 10.1016/j.geits.2023.100121

Wong, C. Y., Wong, W. Y., Ramya, K., Khalid, M., Loh, K. S., Daud, W. R. W., Lim, K. L., Walvekar, R., & Kadhum, A. A. H. (2019). Additives in proton exchange membranes for low-and high-temperature fuel cell applications: A review. *International Journal of Hydrogen Energy, 44*(12), 6116–6135. DOI: 10.1016/j.ijhydene.2019.01.084

Xie, X., Du, L., Yan, L., Park, S., Qiu, Y., Sokolowski, J., Wang, W., & Shao, Y. (2022a). Oxygen evolution reaction in alkaline environment: Material challenges and solutions. *Advanced Functional Materials*, *32*(21), 2110036. DOI: 10.1002/adfm.202110036

Xie, X., Du, L., Yan, L., Park, S., Qiu, Y., Sokolowski, J., Wang, W., & Shao, Y. (2022b). Oxygen evolution reaction in alkaline environment: Material challenges and solutions. *Advanced Functional Materials*, *32*(21), 2110036. DOI: 10.1002/adfm.202110036

Yan, Q., Toghiani, H., Lee, Y.-W., Liang, K., & Causey, H. (2006). Effect of sub-freezing temperatures on a PEM fuel cell performance, startup and fuel cell components. *Journal of Power Sources*, *160*(2), 1242–1250. DOI: 10.1016/j.jpowsour.2006.02.075

Yang, C., Sun, T., Wang, W., Li, Y., Zhang, Y., & Zha, M. (2024). Regenerative braking system development and perspectives for electric vehicles: An overview. *Renewable & Sustainable Energy Reviews*, *198*, 114389. DOI: 10.1016/j.rser.2024.114389

Yazdi, M., Moradi, R., Pirbalouti, R. G., Zarei, E., & Li, H. (2023). Enabling safe and sustainable hydrogen mobility: Circular economy-driven management of hydrogen vehicle safety. *Processes (Basel, Switzerland)*, *11*(9), 2730. DOI: 10.3390/pr11092730

Yildiz, A., & Özel, M. A. (2021). A comparative study of energy consumption and recovery of autonomous fuel-cell hydrogen–electric vehicles using different powertrains based on regenerative braking and electronic stability control system. *Applied Sciences (Basel, Switzerland)*, *11*(6), 2515. DOI: 10.3390/app11062515

Yin, S. F., Xu, B. Q., Zhou, X. P., & Au, C. T. (2004). A mini-review on ammonia decomposition catalysts for on-site generation of hydrogen for fuel cell applications. *Applied Catalysis A, General*, *277*(1–2), 1–9. DOI: 10.1016/j.apcata.2004.09.020

Yu, X., Chang, H., Zhao, J., Tu, Z., & Chan, S. H. (2022). Application of self-adaptive temperature recognition in cold-start of an air-cooled proton exchange membrane fuel cell stack. *Energy and AI*, *9*, 100155. DOI: 10.1016/j.egyai.2022.100155

Yuan, H., & Hernandez, A. A. (2023). User cold start problem in recommendation systems: A systematic review. *IEEE Access : Practical Innovations, Open Solutions*, *11*, 136958–136977. DOI: 10.1109/ACCESS.2023.3338705

Zakaria, Z., Awang Mat, Z., Abu Hassan, S. H., & Boon Kar, Y. (2020). A review of solid oxide fuel cell component fabrication methods toward lowering temperature. *International Journal of Energy Research*, *44*(2), 594–611. DOI: 10.1002/er.4907

Zhang, C., Chen, R., Wang, S., Dui, H., & Zhang, Y. (2022). Resilience efficiency importance measure for the selection of a component maintenance strategy to improve system performance recovery. *Reliability Engineering & System Safety*, *217*, 108070. DOI: 10.1016/j.ress.2021.108070

Zhang, C., Zeng, T., Wu, Q., Deng, C., Chan, S. H., & Liu, Z. (2021). Improved efficiency maximization strategy for vehicular dual-stack fuel cell system considering load state of sub-stacks through predictive soft-loading. *Renewable Energy*, *179*, 929–944. DOI: 10.1016/j.renene.2021.07.090

Zhang, G., Qu, Z., Tao, W.-Q., Wang, X., Wu, L., Wu, S., Xie, X., Tongsh, C., Huo, W., Bao, Z., Jiao, K., & Wang, Y. (2022). Porous flow field for next-generation proton exchange membrane fuel cells: Materials, characterization, design, and challenges. *Chemical Reviews*, *123*(3), 989–1039. DOI: 10.1021/acs.chemrev.2c00539

Zhang, L., Wilkinson, D. P., Liu, Y., & Zhang, J. (2018). Progress in nanostructured (Fe or Co)/N/C non-noble metal electrocatalysts for fuel cell oxygen reduction reaction. *Electrochimica Acta*, *262*, 326–336. DOI: 10.1016/j.electacta.2018.01.046

Zhang, Z., Mao, J., & Liu, Z. (2024). Advancements and insights in thermal and water management of proton exchange membrane fuel cells: Challenges and prospects. *International Communications in Heat and Mass Transfer*, *153*, 107376. DOI: 10.1016/j.icheatmasstransfer.2024.107376

Zhao, J., & Li, X. (2019). A review of polymer electrolyte membrane fuel cell durability for vehicular applications: Degradation modes and experimental techniques. *Energy Conversion and Management*, *199*, 112022. DOI: 10.1016/j.enconman.2019.112022

Zheng, Z., Liu, T., Liu, Q., Lei, J., & Fang, J. (2021). A distributed energy system integrating SOFC-MGT with mid-and-low temperature solar thermochemical hydrogen fuel production. *International Journal of Hydrogen Energy*, *46*(38), 19846–19860. DOI: 10.1016/j.ijhydene.2021.03.137

Chapter 2
Nickel–Metal Hydride Batteries:
A Reliable Energy Storage Solution for Consumer Electronics and Renewable Energy Storage

Oscar Sotelo
https://orcid.org/0000-0002-3664-1286
Instituto de Ciencias Fisicas, Mexico

John Henao
https://orcid.org/0000-0002-8954-6039
SECIHTI-CIATEQ, Mexico

Nestor Starlin Flores-Garcia
https://orcid.org/0000-0002-9352-9688
Instituto de Ciencias Fisicas, Mexico

Victor Zezatti
https://orcid.org/0000-0001-6470-6772
CIICAp, Mexico

Carlos A. Poblano-Salas
CIATEQ, Mexico

Horacio Martinez-Valencia
Instituto de Ciencias Fisicas, Mexico

ABSTRACT

Nickel–Metal Hydride (Ni-MH) batteries remain a relevant technology for safe and sustainable energy storage, especially in consumer electronics, renewable energy systems, and hybrid vehicles. The novelty of this chapter lies in its comprehensive discussion of the latest advances in metal hydride alloys for negative electrodes, particularly AB5, AB2, and A2B7 systems, highlighting their impact on hydrogen storage capacity, electrochemical performance, and long-term stability. Special attention is given to how alloy design, microstructural control, and surface modifications address the growing demand for higher energy density and durability. By bridging fundamental electrochemical principles with recent material innovations,

DOI: 10.4018/979-8-3373-3156-0.ch002

this chapter provides new perspectives on the future potential of Ni-MH technology as a competitive and environmentally friendly alternative within the energy storage landscape.

INTRODUCTION

Batteries have played a pivotal role in the technological advancements of modern society, from the invention of the Voltaic cell to the development of contemporary lithium-ion batteries. The ongoing pursuit of high-energy, high-power-density, low-cost, and environmentally friendly rechargeable batteries is driven by the increasing energy demands of applications ranging from spacecraft and smartphones to electric vehicles. These compact energy sources have significantly transformed modern life and continue to shape technological progress (Notten & Latroche, 2009).

The history of batteries dates to the late 18th century, when Italian physicist Alessandro Volta made a groundbreaking discovery. In 1800, he developed the first electric battery, the Voltaic cell, comprising alternating discs of copper and zinc, separated by pieces of cardboard soaked in saltwater. Although rudimentary, this invention marked the first step toward electrochemical energy storage (Notten & Latroche, 2009).

Over the following century, battery technology evolved gradually. The first commercially available batteries, such as zinc–carbon and lead–acid types, emerged in the mid-19th century. Despite their size and weight, these batteries provided reliable energy storage for early technologies like telegraphs and electric lighting. A breakthrough occurred in the late 20th century with the introduction of cylindrical cell formats, which remain widely used today in various applications (Ying et al., 2006).

The development of rechargeable batteries started in the 1890s, when the growing demand for portable power sources led to the fabrication of the first nickel-cadmium (NiCd) batteries. These were invented by Swedish engineer Waldemar Jungner in 1899. Jungner patented the battery and conducted experiments with various electrode materials, primarily nickel and cadmium. The NiCd battery represented the first rechargeable alternative to lead-acid batteries, offering several advantages, although the materials were costly and progress in commercialization was gradual (Dehghani-Sanij et al., 2019).

By 1910, NiCd batteries were introduced to the Swedish market, but significant improvements in electrode technology, such as the deposition of active materials within a porous nickel-plated electrode, were not achieved until 1932. The technology reached the United States in 1946.

NiCd cells possess unique physical and electrical properties. The active materials in the electrodes undergo oxidation state changes without physical degradation.

Unlike many other battery systems, the charge and discharge reactions in NiCd batteries do not involve the transfer of material between electrodes. Consequently, electrodes exhibit long service life, as their active materials are neither consumed during operation nor affected by storage (Omar et al., 2014).

The advantages of NiCd batteries include long service life, durability, good charge retention, excellent performance during long-term storage, low maintenance requirements, and uniform discharge characteristics. However, their main disadvantages are relatively low energy density, higher cost compared to lead-acid (Pb-A) batteries, and a pronounced memory effect. Additionally, cadmium is a highly toxic heavy metal that requires proper disposal; discarded NiCd batteries are a significant source of cadmium contamination in municipal solid waste (Ovshinsky et al., 1993).

In 1947, further advancements were achieved by incorporating mechanisms to absorb gases generated during charging, leading to the development of the modern sealed NiCd battery. For many years, NiCd batteries were the preferred power source for two-way radios, emergency medical equipment, professional video cameras, and power tools. In the late 1980s, the introduction of Ultra High Capacity NiCd batteries marked a significant milestone, offering capacities 60% greater than standard NiCd cells. This improvement was made possible by increasing the amount of active material within the cell; however, it was accompanied by higher internal resistance and a reduced number of charge–discharge cycles. It was later discovered that certain types of rechargeable batteries exhibited a memory effect. This phenomenon occurs when batteries are not fully discharged before recharging, leading to a reduction in their effective storage capacity over time. Despite this limitation, rechargeable batteries represented a breakthrough in energy storage technology (Young & Yasuoka, 2016).

By the end of the 20th century, NiMH batteries emerged as a prominent type of rechargeable battery widely used across numerous devices. The origins of NiMH technology trace back to 1899 with the invention of the NiCd battery. This technology evolved until the development of the NiMH battery in 1970, which distinguished itself as a more environmentally friendly alternative. NiMH batteries were developed by Stanford Ovshinsky, who later founded the Ovonic Battery Company in 1982 (Ovshinsky et al., 1993).

The NiMH battery consists of three main components. The anode is composed of a hydrogen-absorbing alloy, which releases hydrogen ions during the discharge process, whereas the cathode is made of nickel hydroxide and the electrolyte is a potassium-sodium hydroxide aqueous solution. The development of the positive electrode was led by Masahiko Oshitani, who pioneered high-energy paste electrode technology. The combination of this high-energy positive electrode with advanced hydride alloys for the negative electrode, discovered by Philips Laboratories and the French Centre National de la Recherche Scientifique (CNRS) in the 1970s, resulted in the creation of a new, environmentally friendly, NiMH battery. The first NiMH batteries were introduced to the market in the late 1980s (Young & Yasuoka, 2016).

The NiMH battery is a type of rechargeable battery like the NiCd battery but replaces the cadmium negative electrode with a hydrogen-absorbing alloy. The positive electrode, as in NiCd batteries, consists of nickel oxyhydroxide (NiOOH). NiMH batteries typically offer two to three times the capacity of equivalently sized NiCd batteries. NiMH batteries offer several advantages, including lower toxicity and greater environmental friendliness, recyclability, reduced memory effect, and higher capacity compared to NiCd batteries. However, NiMH batteries also have some disadvantages, such as sensitivity to deep discharge, shortened lifespan, and heat generation during rapid charging and high-load discharging. They exhibit higher self-discharge rates than many other battery types and require careful maintenance, including frequent full discharges to prevent crystal formation. Furthermore, NiMH batteries are generally more expensive than NiCd batteries (Young & Yasuoka, 2016).

A hybrid variant of the nickel-metal hydride battery with low self-discharge, introduced in 2005 by Sanyo, is the LSD NiMH (Low Self-Discharge NiMH) battery that features a significantly lower self-discharge rate. This means it retains a charge longer when not in use, making it ideal for devices that are used intermittently or for long-term storage (Dehghani-Sanij et al., 2019). Currently, NiMH batteries are positioned as an alternative to lithium-ion batteries in certain technological applications. The primary distinction between the two technologies lies in their chemical composition; NiMH batteries are considered significantly more environmentally friendly. As a result, they are commonly used in electronic devices and hybrid electric vehicles. However, compared to lithium-ion batteries, NiMH cells exhibit lower volumetric energy density and higher self-discharge rates. Despite these limitations, NiMH batteries offer enhanced safety, as they are more tolerant to overcharging and overdischarging conditions.

The global NiMH battery market size was valued at USD 2,785 million in 2024 and is projected to grow up to USD 3,766.9 million by 2031, exhibiting a compound annual growth rate (CAGR) of 4.41% over the forecast period. The expansion of the NiMH battery market is driven by the rising demand for green energy solutions, advancements in battery technology, the increasing adoption of hybrid and electric vehicles, and strict environmental regulations that promote the implementation of sustainable practices. Figure 1 presents the classification of batteries into primary and secondary types, along with the approximate time periods when each was introduced into the market (Quintero V. et al., 2021).

Figure 1. Timeline of first market introduction for different types of batteries (Adapted from "Lithium ion batteries: features and applications" (p.15), Quintero, V. et al., 2021. Revista de I+D Tecnológico, 17(1),14-22).

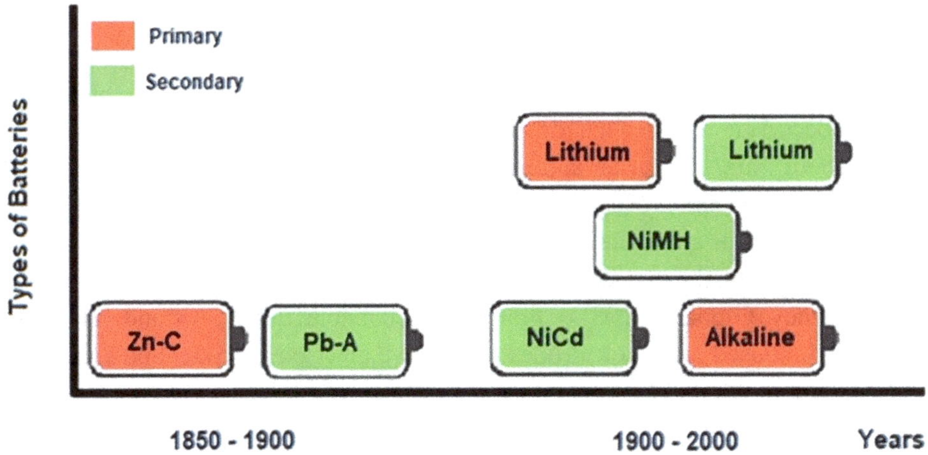

The NiMH battery market is segmented by battery type, application, and component. This segmentation facilitates the analysis of growth opportunities, the formulation of market strategies, and the identification of key application areas and distinctions among target markets. Based on type, the NiMH battery market is segmented into small-sized NiMH batteries for consumer electronics and large-sized NiMH batteries for hybrid electric vehicles (HEVs). Based on applications, the NiMH battery market is segmented into automotive, cordless phones, dust collectors, personal care, lighting tools, and power tools. Finally, the NiMH battery market is segmented by component into positive electrode, negative electrode, and others.

Furthermore, NiMH batteries are widely preferred due to their relatively lower cost compared to lithium-ion alternatives, particularly in applications requiring moderate energy density and reliable performance. Leading companies are investing in research and development. Their goals are to enhance energy density, extend battery lifespan, and improve electrode materials and cell design. These efforts address key market demands (Notten & Van Beek, 2000).

PRINCIPLES OF OPERATION OF NICKEL-METAL HYDRIDE BATTERIES

Nickel-metal hydride (NiMH) batteries are rechargeable electrochemical energy storage devices that are widely used in hybrid electric vehicles, portable electronic devices, and, increasingly, in renewable energy systems (Gibbard, 1994) (Yi et al., 2024) (Young et al., 2011). Their operation relies on a series of reversible redox reactions occurring between a hydrogen-absorbing alloy anode and a nickel-based cathode, both immersed in an alkaline electrolyte. The chemistry of NiMH batteries offers high safety, excellent reversibility, and efficient energy conversion, establishing them as a reliable alternative to other secondary battery technologies. Compared with NiCd batteries, NiMH cells exhibit a 25–40% improvement in energy efficiency (Gibbard, 1993) (Ambrosio & Ticianelli, 2001). A standard NiMH battery is composed of two electrodes, a negative electrode (anode) and a positive electrode (cathode), that are separated by a porous membrane and submerged in a concentrated alkaline electrolyte, typically a 6 molar potassium hydroxide (KOH) solution. The separator allows ionic transport while preventing electrical contact between electrodes, thereby avoiding short circuits (Figure 2).

To enable the formation of reversible metal hydrides (MH), the anode is composed of hydrogen-absorbing alloys (Notten & Latroche, 2009). Lanthanide-nickel ($LaNi_5$), titanium, zirconium, vanadium-nickel, and chromium are among the most frequently used alloys. Mischmetal-based alloys ($MmNi_5$) are also common. Mischmetal is a naturally occurring mixture of rare earth elements, including lanthanum (La), cerium (Ce), neodymium (Nd), and praseodymium (Pr). During charging, the hydrogen-absorbing alloy forms MH by incorporating hydrogen. During discharge, the stored hydrogen is released in the form of protons and electrons, contributing to the electrochemical reaction. The cathode is typically composed of nickel hydroxide [$Ni(OH)_2$], which transforms nickel oxyhydroxide (NiOOH) during the charging process (Dansui et al., 2000). This redox pair is like that used in NiCd batteries but avoids the environmental hazards associated with cadmium toxicity. The alkaline electrolyte, typically a 6 M KOH solution, ensures high ionic conductivity while maintaining chemical stability throughout charge and discharge cycles. Although it actively participates in ion transport, primarily through hydroxide ions (OH^-), the electrolyte itself is not consumed in electrochemical reactions.

Figure 2. Schematic structure of a Ni-MH battery adapted from electricity-magnetism, built with GeneratePress, 2025 (https://www.electricity-magnetism.org/electric-battery/nimh-battery/))

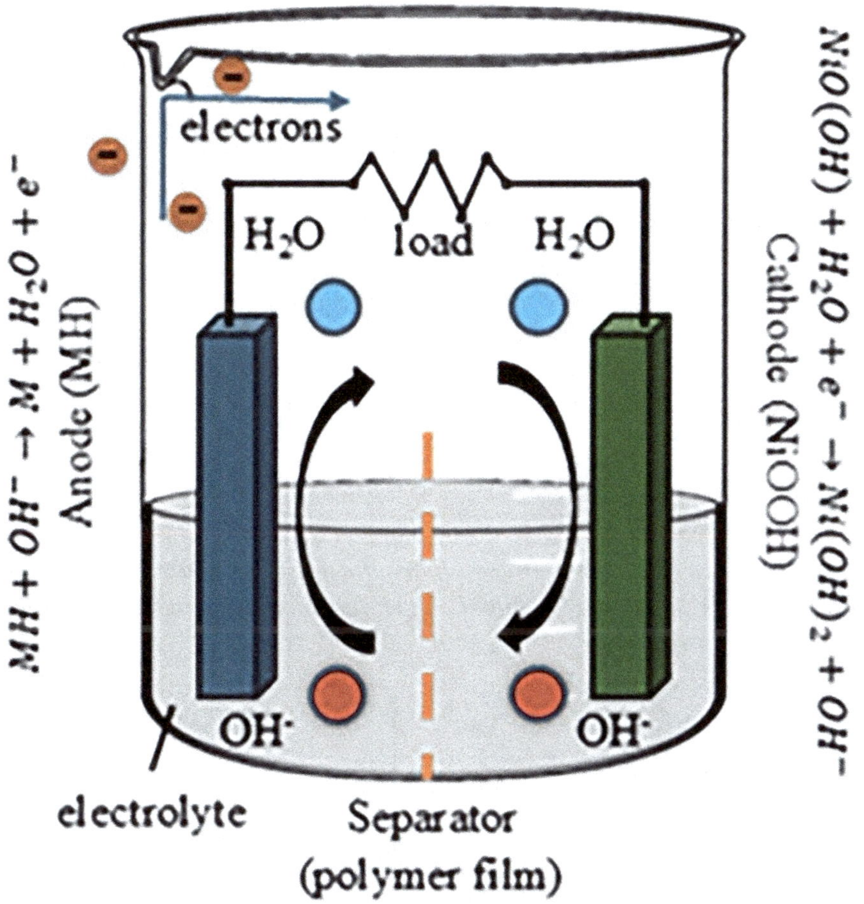

The materials and alloys used in the electrodes and electrolyte indirectly influence the overall efficiency of the battery. Factors such as hydrogen bonding and diffusion behavior, along with the microstructure of the alloy, play a critical role in determining hydrogen storage capacity. The internal resistance of the electrode is affected by the distribution of hydrogen within the alloy, which subsequently impacts the charge/discharge rate and contributes to electrochemical aging (Kitayama et al., 2002).

In hydrogen storage, the negative electrode (anode) is of paramount importance. Hydrogen ions (H^+) and hydroxide ions (OH^-) are produced from the dissociation of water molecules at the electrode–electrolyte interface during the charging process (Notten & Van Beek, 2000). The H^+ ions penetrate the crystal lattice of the alloy, forming a metal hydride (MH), while electrons from the external circuit participate in the reduction process, as shown in Figure 2. The half-reaction at the anode during charging is:

$$M + H_2O + e^- \rightarrow MH + OH^- \qquad (1)$$

During discharge, this reaction is reversed. The hydrogen stored in the metal hydride is oxidized, releasing H^+ ions that combine with OH^- in the electrolyte to form water. In contrast, the electrons are released to the external circuit:

$$MH + OH^- \rightarrow M + H_2O + e^- \qquad (2)$$

This reversible hydrogen absorption–desorption process enables the battery to be cycled multiple times with minimal active material loss, provided that the alloy´s structural integrity is preserved over time.

Figure 3. Mechanism of hydrogen absorption and desorption in the metal hydride alloy adapted from Martínez, C. (8 de septiembre de 2022). Guantes de protección contra productos químicos. Tecnitool. https://tecnitool.es/guantes-de-proteccion-contra-productos-quimicos

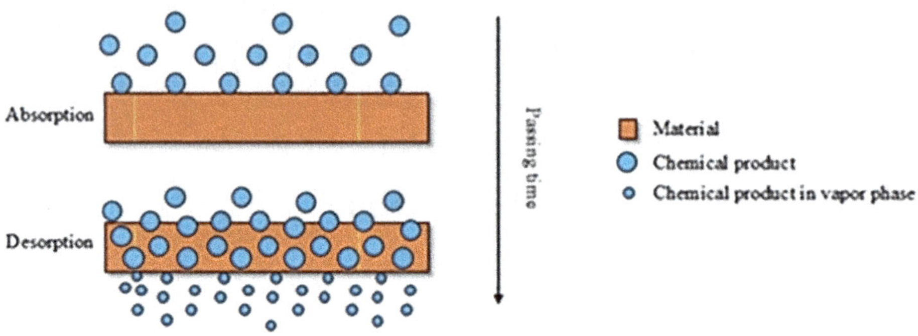

At the positive electrode (cathode), the redox reactions involve nickel-based compounds. During charging, nickel hydroxide ($Ni(OH)_2$) is oxidized to nickel oxy-hydroxide (NiOOH), releasing water and an electron (Borthomieu & Bernard, 2009):

$$Ni(OH)_2 + OH^- \rightarrow NiOOH + H_2O + e^- \qquad (3)$$

Conversely, during discharge, NiOOH is reduced back to $Ni(OH)_2$:

$$NiOOH + H_2O + e^- \rightarrow Ni(OH)_2 + OH^- \qquad (4)$$

These reactions are reversible and maintain the electrochemical balance of the cell. The hydroxide ions (OH^-) generated or consumed are transported through the electrolyte to sustain charge neutrality.

The electrolyte, typically a potassium hydroxide (KOH) solution, is chemically stable and remains unchanged throughout battery operation. Its primary function is to serve as a conductive medium for ion migration, primarily hydroxide ions (OH^-), thereby maintaining the continuity of the internal circuit. One of the key physicochemical properties of the electrolyte is its typical concentration of 6 mol/L, although concentrations ranging from 3 to 9 M are also commonly employed. However, it is important to avoid exceeding concentrations above 10 M, as the increased viscosity of the electrolyte reduces ion mobility and negatively impacts battery performance. Another key property is its high ionic conductivity, up to 0.6 S/cm at room temperature (Figure 4), and a wide operating temperature range, from −20 °C to +60 °C.

Figure 4. KOH ionic conductivity vs. concentration at 25°C (author's own elaboration).

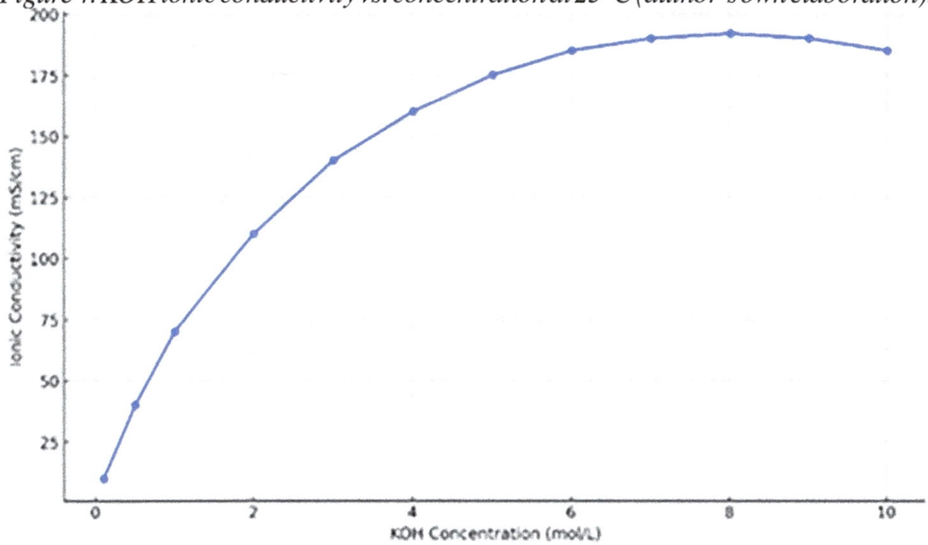

The alkalinity of the electrolyte is determined by the amount of acid that a solution can absorb without significantly altering its pH or becoming neutralized. The alkaline pH of KOH (around 14) helps stabilize both NiOOH and $Ni(OH)_2$, preventing the formation of inactive compounds. Conversely, the absorption of atmospheric CO_2, which results in the formation of carbonates (K_2CO_3), is a phenomenon known as carbonation, in addition to alkalinity, that may occur. This process may reduce the electrolyte's conductivity and, consequently, shorten the cell's lifespan. High KOH concentrations may also accelerate corrosion of internal metallic components. Furthermore, at low temperatures, if the electrolyte becomes dehydrated, crystallization may occur (Yan et al., 2020).

By combining the half-cell reactions from both electrodes, the global redox reaction of a Ni-MH cell can be expressed as:

$$MH + NiOOH \rightleftharpoons M + Ni(OH)_2 \qquad (5)$$

This reaction demonstrates the intercalation-deintercalation of hydrogen into the metal alloy (anode) and the oxidation–reduction of nickel compounds at the cathode. Electrons flow through the external circuit, and hydroxide ions move internally through the electrolyte, enabling efficient charge transfer and energy conversion.

The service life of Ni-MH batteries can be significantly impacted by their operational parameters, either positively or negatively. Current density is one of the most critical parameters, as the degradation of each electrode can be accelerated by high current densities in response to the accumulation of secondary reaction products and heat generation. Ultimately, energy efficiency is diminished because of increased internal polarization and ohmic resistance that are induced by elevated temperatures. At temperatures above 45°C, degradation of active materials and the separator is accelerated. In contrast, at low temperatures (below 0°C), the available capacity decreases due to reduced ionic mobility in the electrolyte and slower hydrogen diffusion. The optimal operating temperature typically ranges between 10°C and 35°C (Young & Yasuoka, 2016).

Additionally, the depth of discharge (DoD) is a critical parameter that quantifies the extent to which the battery's stored energy has been depleted. Repeated overcharge conditions can lead to nickel dendrite growth, internal pressure buildup due to gas accumulation (H_2, O_2), and a significant loss of capacity. Regular full discharges at 100% DoD can reduce cycle life to fewer than 500 cycles, while partial discharges (between 20–40% DoD) can extend cycle life to over 2,000 cycles (Figure 5) (Chu et al., 2020).

Figure 5. Life cycles vs. depth of discharge (DoD) (author's own elaboration).

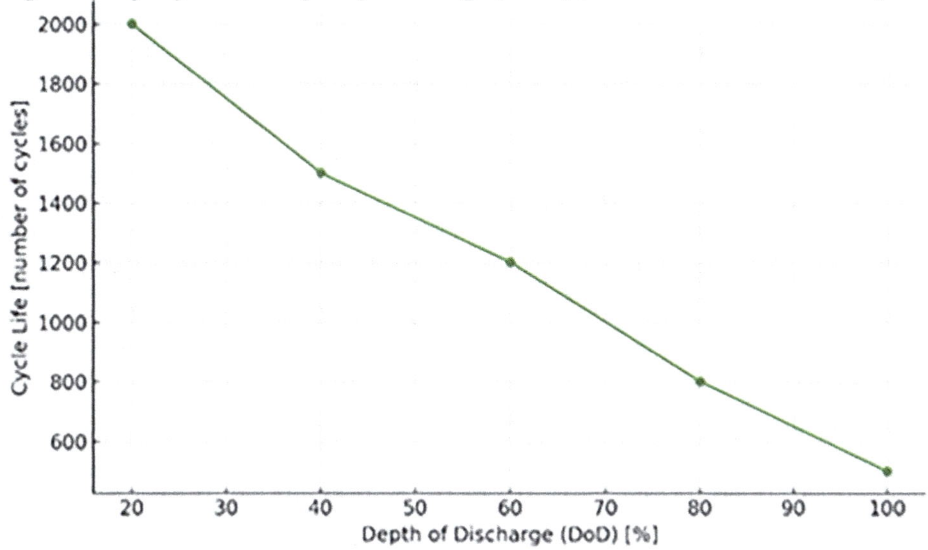

The lifespan of a battery is limited by the number of charge-discharge cycles it can withstand. This limitation is primarily caused by the degradation of the metal alloy due to the repeated absorption and desorption of hydrogen. Surface oxidation of the alloy is an additional degradation mechanism that results in the formation of passive layers of metal oxides or hydroxides that impede hydrogen transport. Impurity accumulation is also common, often resulting from the presence of contaminating metal ions in the electrolyte. These issues may arise from manufacturing residues or internal corrosion, both of which interfere with redox reactions (Monnier et al., 2024).

Strategies must be implemented at both the material and system levels to mitigate these degradation mechanisms. Progressive anode corrosion is one of the primary degradation pathways. Optimizing the alloy composition to enhance corrosion resistance may help to mitigate these issues. Elements such as aluminum (Al), manganese (Mn), cobalt (Co), and tin (Sn) are commonly added as alloying elements to inhibit corrosion. Beyond modifying the alloy's composition, refining its microstructure at the micrometer or nanometer scale may reduce internal stresses caused by the expansion and contraction of the metal hydride during cycling, thereby improving structural stability (Karwowska et al., 2018).

The KOH electrolyte may experience evaporation or ion migration during charge and discharge cycles, particularly at elevated temperatures. To mitigate this issue, stabilizing additives, such as silicates or polymers, or gel-based electrolytes may be implemented to enhance ionic conductivity and minimize water loss. Leakage can also lead to moisture loss, which can be prevented by implementing more robust sealing methods using materials resistant to internal pressure buildup (Karwowska et al., 2020).

One of the most critical degradation factors is temperature. Corrosion, electrolyte loss, and electrode imbalance are all caused by operating the battery outside the recommended temperature range. It is imperative to maintain thermal stability by incorporating thermal management systems, whether they are passive (e.g., heat sinks or ventilation) or active (e.g., temperature sensors, controllers, or cooling fans). These systems aid in temperature monitoring and regulation, thereby reducing internal thermal stresses and premature aging of the cell. Batteries are also susceptible to external heat exposure during improper storage, even when not in use. Battery storage is advised in a dry environment, away from external heat sources, at moderate temperatures (15°C to 25°C) (Yang et al., 2023).

NiMH battery degradation is the result of multiple interacting factors. Therefore, to optimize battery performance and extend battery life, an effective strategy must integrate environmental, electronic, structural, and material strategies (Figure 6).

Figure 6. Strategies to avoid degradation in cells (author's own elaboration).

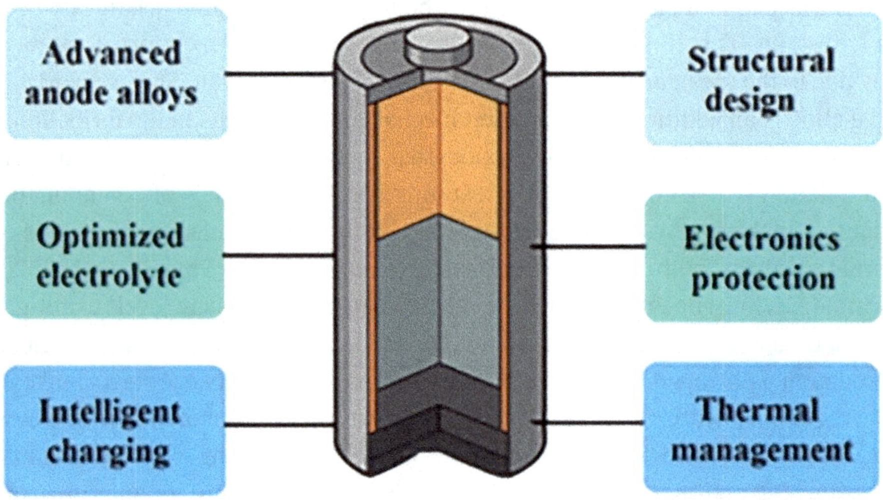

MATERIALS USED IN HYDRIDE ALLOYS

It is worth noting that there are specific prerequisites for hydrogen storage electrode alloys to be considered suitable candidates for use in NiMH batteries. For example,

(1) high reversible hydrogen storage capacity, (2) good electrochemical catalyst for hydrogen charge (reduction/absorption) and discharge (desorption/oxidation), (3) easy activation and excellent corrosion resistance in alkaline electrolyte, (4) a hydrogen equilibrium pressure not higher than 5 atm at 25 °C, (5) good charge/absorption and discharge/desorption kinetics for efficient operation, and (6) long cycle life and low cost (Liu et al., 2011). As already mentioned, the performance of NiMH batteries depends strongly on the intrinsic properties of the negative electrode active materials (hydrogen storage alloys).

Since 1990, the development of NiMH batteries has experienced significant growth. Alloys based on classical AB compounds, such as AB_2 (which contains a Laves phase) and AB_5 (which has a $CaCu_5$-type structure), have demonstrated excellent performance as negative active materials due to their composition of elements that form stable metal hydrides. The electrochemical capacity of currently advanced AB_5-type alloys has reached 310-330 mAh/g at a 0.2C rate and room temperature (Lota et al., 2020). In fact, AB_5 alloys are the most promising materials for hydrogen storage due to high hydrogen storages capacities, reversibility, kinetics, cycle life and corrosion resistance. It is important to mention that the partial substitution of A or B elements is desirable to improve hydrogen diffusion and extend life cycle (Lota et al., 2020). The A component is one or more rare earth elements, and the mainly used B elements are transition metals such as Al, Co and Mn. It has been observed that the partial substitution with Al improves the cycling performance, activation and stability of the alloy. Cobalt enhances the cycle life and contributes to antielectron-oxidation properties, whereas Mn facilitates the formation of metal hydride (by reducing the formation pressure). Such formation leads to the development of both α-phase (in which some hydrogen atoms are absorbed), and β-phase (where hydrides are formed). It should be noted that the formation of metal-hydride may deteriorate due to exposure to impurities. One way to increase the performance of the metal alloy is the introduction of Platinum Group Metals (PGM). It has been observed that electrochemical activation is not required with a Pd-modified AB_5 alloy (Hubkowska et al., 2019). Besides PGM metals, Mg increases electrochemical capacity because it prevents the pulverization of hydrides (Dymek et al., 2019).

Figure 7. Crystal structure of $LaNi_5$ alloy (a) and the tetrahedral sites (b) and octahedral sites (c) for hydrogen occupancy (Adapted from "Advanced hydrogen storage alloys for Ni/MH rechargeable batteries" (p.4748), Liu et al. 2011, Journal of Materials Chemistry, 21(13), 4743-4755).

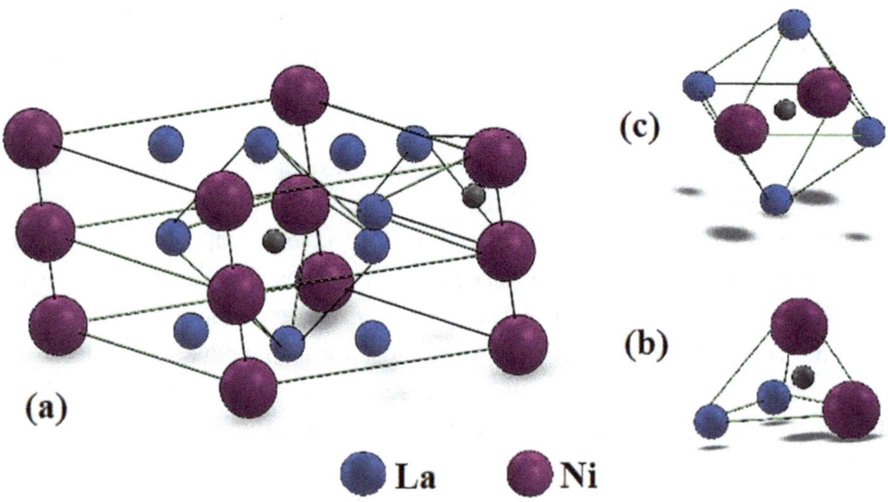

On the other hand, RNi_5 (AB_5, where R denotes a rare earth element) intermetallic compounds form hydrides with high hydrogen storage capacity (RNi_5H_6) and exhibit nearly complete reversible characteristics (Mani & Ramaprabhu, 2004). Figure 7 illustrates three octahedral and three tetragonal sites within the $CaCu_5$-type structure, where hydrogen preferentially occupies tetrahedral sites containing A_2B_2, AB_3, and B_4 configurations. One $LaNi_5$ unit admits up to six hydrogen atoms to form $LaNi_5H_6$ hydride, reaching a theoretical electrochemical capacity of approximately 372 mAh/g (Kang Hu, W. 2000) (Liu et al., 2011). The rare earth-based AB_5-type alloys have been investigated extensively as electrode materials for rechargeable NiMH batteries. However, the high cost of lanthanum in $LaNi_5$-based alloys and a decrease of discharge capacity with the increment of charge/discharge cycles due to alloy pulverization and La oxidation into $La(OH)_3$ in the alkali solution on cycling have been strong reasons for substituting La by a mischmetal system (cerium-rich mischmetal Mm or lanthanum-rich mischmetal ML) (Mungole et al., 1999) (Tliha et al., 2007). Mm is a mixture of rare-earth elements; its composition corresponds to that of natural ores. It contains 50-55% cerium, 18-28% lanthanum, 12-18% neodymium, 4-6% praseodymium and other rare earth elements in small quantities, as well as metallic impurities. A typical composition of a presently used AB_5-type hydrogen storage alloy is $MmNi_{3.5}Co_{0.7}Mn_{0.4}Al_{0.3}$ (Ruetschi et al., 1995). Mm(Ni,

Mn, Co, Al)$_5$-based alloys are promising candidate materials for hydrogen storage applications due to their favorable solution thermodynamics, hydriding behavior, hysteresis characteristics, magnetic properties, and hydrogen desorption kinetics. Additionally, factors such as cost, cycle life, and storage capacity are also taken into consideration (Tliha et al., 2007).

A study conducted by M. N. Mungole et al. employed a mischmetal (Mm) alloy composed of 43 wt% Ce, 23 wt% La, 18 wt% Nd, 5 wt% Pr, 3 wt% Sm, and 8 wt% Fe. This was combined with high-purity (99.99%) Ni, Al, Mn, and Sn in the required stoichiometric ratios to produce the compositions $MnNi_{5-X}Al_X$ (x = 0.4 and 0.8), $MmNi_{5-X}Mn_X$ (x = 0.4 and 0.8), and $MnNi_{4\cdot6}Sn_{0\cdot4}$. The alloys were synthesized via multiple arc melting in an argon atmosphere. The resulting button samples were cast and subsequently annealed at 1273 K for 24 hours under a pressure of 1.33×10^{-6} MPa. The microstructure of all studied alloys was formed by the metallic matrix as the major constituent decorated with second phase precipitates. Partial substitution of Ni in $MmNi_5$ with Al, Mn, and Sn resulted in distinct phase morphologies. The hydrogen storage capacity of these alloys correlated strongly with the volume fraction of the matrix phase. For instance, increasing Al substitution led to a decrease in the matrix phase volume fraction, which corresponded to a reduction in hydrogen storage capacity. Conversely, higher Mn substitution increased the matrix phase fraction, resulting in enhanced capacity. The Sn-substituted alloy showed a lower hydrogen storage capacity, consistent with its reduced matrix volume fraction. In conclusion, a clear correlation is established between the volume fraction of the matrix phase, which contains the strong hydride-forming elements, and hydrogen storage capacity. Therefore, minimizing the fraction of the precipitate phase is crucial for optimizing performance (Mungole et al., 1999).

On the other hand, as mentioned above, AB_2-based alloys have emerged as promising hydrogen storage materials (Hariprakash et al., 2024). In this formula, "A" represents the hydride-forming element, typically zirconium or titanium, which are the most important metals for hydrogen absorption in AB_2 alloys, while "B" denotes other metals such as nickel, cobalt, vanadium, manganese, aluminum, or chromium (Ruetschi et al., 1995). As mentioned for AB_5 alloys, the composition of B-elements in AB_2 alloys strongly influences the performance of the negative electrode. For example, cobalt (Co) facilitates easy activation, vanadium (V) increases capacity, nickel (Ni) improves cycle life, and chromium (Cr) enhances charge retention (Cuevas et al., 2001) (Young & Nei, 2013). As mentioned above, AB_2 represents the intermetallic compounds with a Laves phase structure corresponding to a compact stacking for atomic radius ratio $R_A/R_B = 1.225$. There are three types of Laves phase structures corresponding to the hexagonal C14 ($MgZn_2$, *P63/mmc*), the cubic C15 ($MgCu_2$, *Fd-3m*) and the hexagonal C36 ($MgNi_2$, *P6/mmc*). The C14 and C15 Laves phases are good hydrogen absorbers while the C36 phase shows limited ability to absorb hydrogen. Hydrogen is stored in tetrahedral interstices of the C14 and C15 Laves phases (see Figure 8). The initial research on these alloys was focused on binary crystalline and glassy systems; however, interest soon turned to multi-element pseudobinary intermetallics in applications for Ni-MH batteries (Liu et al., 2011). The AB_2-type alloys reach capacity ranges from 370 to 450 mAh/g, which is higher than that of AB_5-type alloys. It should be noted that elements as Al, Mn and Cr promote a larger fraction of C14 phase, while Fe and Co favor the C15 structure. An example of a commercial AB_2-type alloy is $Ti_{0.51}Zr_{0.49}V_{0.70}Ni_{1.18}Cr_{0.12}$. However, alloys AB_2 as $Ti_{0.55}Zr_{0.45}V_{0.36}Ni_{1.03}Cr_{0.3}Mn_{0.3}$ and $Zr_{0.57}Ti_{0.43}V_{0.36}Ni_{1.03}Cr_{0.3}Mn_{0.3}$ show discharge capacities around 400 mAh/g at 0.25C rate. Also, the Zr0. 8Ti0.2Ni0.8Mn0.8V0.2Co0.15Al0.05 alloy with C14 structure has a reversible electrochemical capacity around 380 mAh/g at a 0.2 C rate. However, compared to rare-earth AB_5-type electrode alloys, AB_2-type electrode alloys suffer from slower activation and lower rate capabilities (Liu et al., 2011).

Figure 8. Crystal structure of C14 (a) and C15 (b) Leaves phases and their possible interstitial sites for hydrogen occupancy (Adapted from "Advanced hydrogen storage alloys for Ni/MH rechargeable batteries" (p.4749), Liu et al. 2011, Journal of Materials Chemistry, 21(13), 4743-4755) .

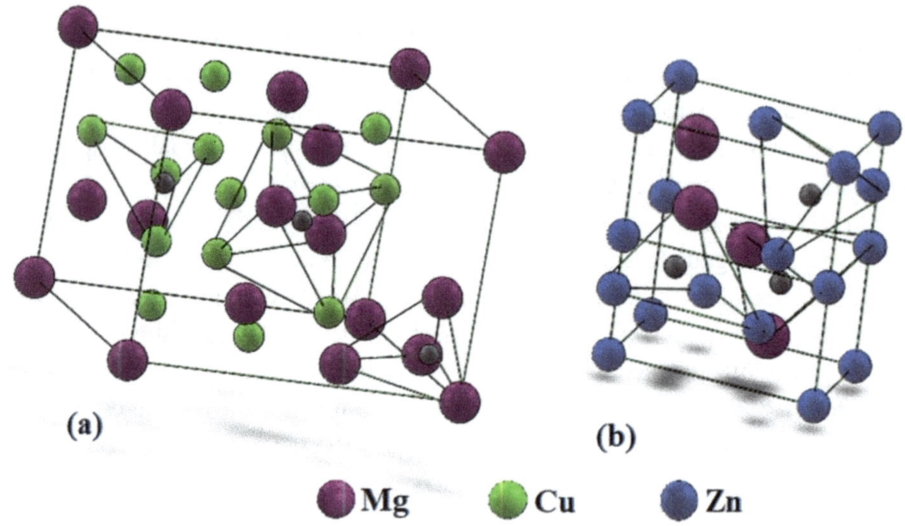

Table 1 presents the approximate weight compositions of AB_5 and AB_2 alloys used in Ni-MH batteries, depending on the battery format.

Table 1. Chemical composition of Ni-MH batteries (wt%) (Ruetschi et al., 1995).

Element	Button cells AB_5 alloys	Cylindrical cells AB_5 alloys	Cylindrical cells AB_2 alloys	Prismatic large cells AB_5 alloys
Ni	29-39	36-42	37-39	38-40
Fe	31-47	22-25	23-25	6-9
Co	2-3	3-4	1-2	2-3
La, Ce, Nd, Pr	6-8	8-10		7-8
Zr, Ti, V, Cr			13-14	
H and O[a]	8-10	15-17	15-17	16-18
C[b]	2-3	0-1		0-1
Plastics[C]	1-2	3-4	3-4	16-19
K	1-2	1-2	1-2	2-3
Others	2-3	2-3	1-2	3-4

[a]In water and oxides.
[b]Graphite
[c]Polypropilene, polyamide, etc.
[d]With plastic casing.

It is well known that there are other hydrogen storages alloys, such as AB_3, A_2B_7, A_5B_{19}, AB, Mg and V-based alloys. For instance, Mg-based alloys are considered good candidates as hydrogen storage systems in NiMH batteries because of their high electrochemical capacity, low cost, and abundant resources. The theoretical capacity of Mg is as high as 2200 mAh/g when forming MgH_2. The Mg_2Ni phase achieves a capacity of approximately 1080 mAh/g through the formation of Mg_2NiH_4. However, a major drawback of Mg-based alloys is their poor hydriding and dehydriding kinetics at room temperature, along with rapid degradation in alkaline solutions. A effective strategy to improve the activation of Mg-based alloy electrodes is the formation of amorphous and nanocrystalline alloys. For example, the $Mg_{50}Ni_{50}$ amorphous alloy exhibits an initial capacity ~ 500 mAh/g; nevertheless, some studies have reported that the discharge capacity of this alloy can be increased to over 1000 mAh/g by having a nano-/amorphous microstructure, which enhances hydrogen diffusion and charge transfer reactions (Liu et al., 2011). The performance of Mg-based alloy electrodes can also be significantly enhanced by alloying with other elements such as scandium (Sc), titanium (Ti), vanadium (V), and chromium (Cr); these alloys demonstrate very high reversible hydrogen storage capacities, up to six times greater than those of commercial AB_5-type materials. Specifically, $Mg_{80}Sc_{20}$, $Mg_{80}Ti_{20}$, $Mg_{80}V_{20}$, and $Mg_{80}Cr_{20}$ alloys exhibit capacity values of 1790 mAh/g, 1750 mAh/g, 1700 mAh/g, and 1270 mAh/g, respectively. This remarkable performance is attributed to the formation of ternary hydrides with a CaF_2-type fluorite crystal structure, which can accommodate more hydrogen atoms compared to the rutile structure of pure MgH_2. The cycle life in charge/discharge test can be increased in Mg_2Ni alloys by the partial substitution of Ti, Al, Co, Ce, Zr, Cr, Sn, Mn, Ca or Y for Mg, and Co, Mn, Cu, Fe, W, Al, Pd or C for Ni. These elements improve the corrosion resistance of the alloy in alkaline solutions by promoting the formation of a protective oxide layer on the alloy surface. This oxide layer slows down further oxidation of magnesium and results in a nickel-enriched layer on the surface, enhancing the alloy's overall stability.

On the other hand, new intermetallic compounds, such as rare earth–Mg–Ni-based alloys, have been explored for electrochemical applications in NiMH hydrogen storage. Specifically, the La–Mg–Ni system enables the synthesis of La_{1-} Mg Ni_X (3 < x < 4, 0 < y < 1) ternary compounds. The incorporation of magnesium reduces the molar mass, significantly enhancing the electrochemical storage capacity of these materials, reaching values up to 400 mAh/g [Li et al., 2023]. Finally, compounds with a composite structure AB_3, A_2B_7, A_5B_{19}, with A being a rare earth and B a transition element, so called superlattice compounds, are used in NiMH batteries. It is known that A_2B_7 intermetallic compounds in the La-Ni system crystallize in either a rhombohedral Gd_2Co_7-type structure or a hexagonal Ce_2Ni_7-type structure (Guzik et al., 2012) (Zhang et al., 2024). It has been mentioned that compounds as Gd_2Co_7 and Ce_2Ni_7 belong to the family of compounds built up with composition between AB_5 ($CaCu_5$-structure type, Haucke phases) and AB_2 slabs ($MgZn_2$- and $MgCu_2$-type structure), C-14 and C-15 Laves phases (see Figure 9) in the hexagonal and rhombohedral crystal system, respectively (Parthe & Lemaire, 1975).

Figure 9. The C15 unit cell characteristics of Laves phases (adapted from. "Intermetallics: Crystal structures Encyclopedia of materials" (p.3), Varin, R. A. 2000, Science and Technology, 1-4).

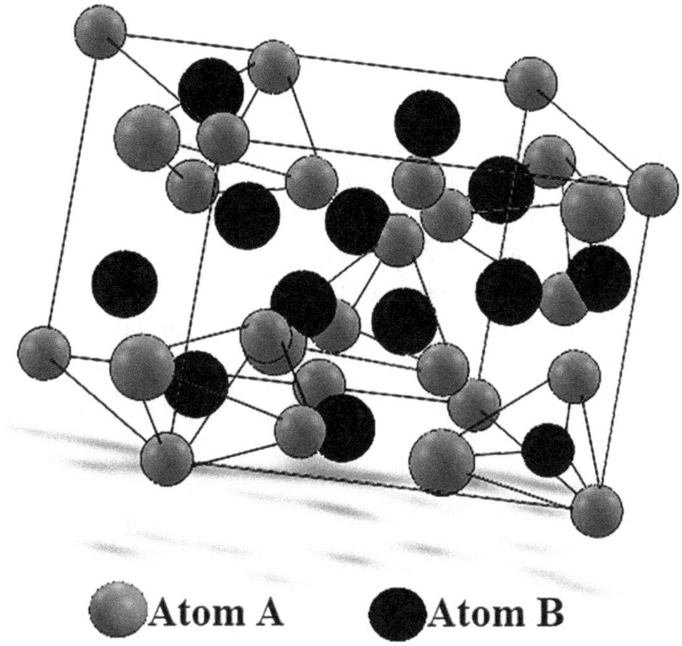

In the development of advanced NiMH batteries, amorphous metals have emerged as promising candidates for electrode materials. These materials are characterized by their disordered atomic structure, lacking the long-range order typical of crystalline alloys. Their amorphous nature provides significant advantages, such as enhanced hydrogen diffusion, a higher density of active sites for hydrogen absorption, and improved electrochemical kinetics. For instance, metallic glasses exhibit excellent corrosion resistance in alkaline environments, contributing to longer life cycle and stable capacity retention. Alloys based on Mg, Ti, and Fe, have been successfully fabricated in amorphous form via rapid solidification techniques (Eyövge et al., 2018) (Sotelo et al., 2024) (Li et al., 2024). The exploration of amorphous metal electrodes represents a promising direction for improving the performance and durability of NiMH battery technologies. These alloys are under evaluation and there are still many questions about their performance in a battery cell.

PERFORMANCE CONSIDERATIONS OF NIMH BATTERIES COMPARED TO OTHER TECHNOLOGIES

While advances in electrodes highlight the potential for improving the intrinsic performance of NiMH batteries, it is equally important to situate these developments within the broader landscape of battery technologies and their market positioning. Understanding how NiMH systems compare with alternative chemistries, such as NiCd and Li-ion, provides critical context for assessing their competitiveness, adoption trends, and areas where future innovations may generate the greatest impact. This transition from material-level innovations to performance and market considerations allows for a more comprehensive evaluation of NiMH technology.

Among the most widely used battery types today are NiCd, NiMH, and Li-ion batteries. As already mentioned, NiCd batteries have largely been replaced by NiMH systems due to the presence of cadmium, a highly toxic heavy metal. In NiMH cells, the negative electrode is based on hydrogen-absorbing materials rather than cadmium. These materials form metal hydrides with certain metals and alloys, enabling a safer and environmentally friendly electrochemical storage (Lota et al., 2020). An added advantage is that hydrogen stored in the form of metal hydrides occupies significantly less volume than compressed gaseous hydrogen, allowing for denser packing and improved thermal management within the battery system (Anani et al., 1994). Furthermore, the use of nickel/metal hydride batteries instead of nickel/hydrogen systems reduces overall weight and increases specific energy. However, the primary motivation behind the development of NiMH technology has been the implementation of environmental regulations aimed at limiting the production of NiCd batteries due to cadmium's high toxicity (Anani et al., 1994).

Since the early 2000´s, significant efforts have been made by battery manufacturers to increase the energy density across various battery technologies, including lead-acid (Pb-acid), nickel-cadmium (Ni-Cd), nickel-metal hydride (NiMH), lithium-ion (Li-ion), and lithium-ion polymer batteries. Table 2 presents a comparison of energy density values for these systems, covering both primary and secondary battery types.

Table 2. Open circuit potentials and energy density values for primary and secondary batteries (Notten & Van Beek, 2000).

	System	Open-circuit Potential [V]	Energy density*	
			Wh/kg	Wh/l
Primary Batteries	$Zn\text{-}MnO_2$	1.6	100-140	300-400
	Zn-air	1.6	200-290	275-1000
Secondary Batteries	SLA	2.0	30-40	70-80
	NiCd	1.3	35-65	100-200
	NiMH	1.3	40-90	160-310
	Li-ion	3.8	100-160	200-390
	Li-Polymer	3.8	100-175	200-300

*Strongly dependent on discharge rate

NiMH batteries differ from NiCd batteries in terms of their charging behavior. NiMH batteries are less tolerant to overcharging and require more precise charging control. This is primarily due to the differences in the electrochemical reactions at the negative electrodes: while the cadmium electrode reaction is endothermic, the metal hydride electrode reaction is exothermic. As a result, the temperature of NiCd batteries remains relatively stable during charging, provided that oxygen evolution is minimal, whereas NiMH batteries tend to heat up earlier in the charging process. Kinetic factors may also contribute to these differences in thermal behavior (Notten & Van Beek, 2000).

On the other hand, a comparison of various battery types in terms of gravimetric and volumetric energy density is shown in Figure 10. Higher gravimetric and volumetric energy densities correspond to greater energy delivery with reduced weight and volume, respectively. For instance, lithium-ion batteries exhibit energy densities of approximately 180 Wh/kg. However, their cost and safety issues remain significant challenges for applications such as electric vehicles (EVs), hybrid electric vehicles (HEVs), and plug-in hybrid electric vehicles (PHEVs). In this context, NiMH batteries represent a promising alternative for use in portable power tools, EVs, HEVs, and PHEVs. Key advantages of NiMH batteries include: 1) High specific power (power density), 2) Flexible cell capacities ranging from 30 mAh to 250 Ah, 3) Safe operation at high voltages (above 320 V), 4) Easy integration into series and series/parallel configurations, 5) Tolerance to overcharge and overdischarge, 6) Maintenance-free operation, 7) Good thermal stability, 8) Simple and cost-effective charging and control electronics, and 9) Environmental compatibility and use of recyclable materials (Liu et al., 2011).

Figure 10. Comparison among different battery types in terms of gravimetric and volumetric energy (Adapted from "Advanced hydrogen storage alloys for Ni/MH rechargeable batteries" (p.4744), Liu et al. 2011, Journal of Materials Chemistry, 21(13), 4743-4755)).

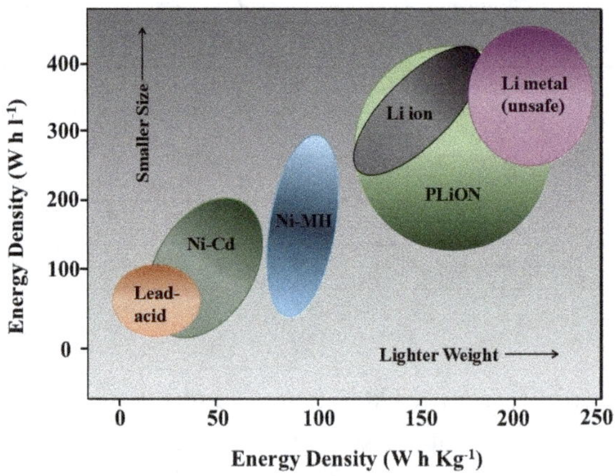

Although NiMH batteries offer a higher specific energy than lead-acid and nickel-cadmium systems, they are still significantly outperformed by modern rechargeable lithium-ion (Li-ion) batteries. However, NiMH batteries are less expensive than lithium-ion batteries due to the higher manufacturing costs associated with Li-ion cell production. It has been estimated that the approximate cost per watt-hour for NiCd, NiMH, and Li-ion batteries follows a ratio of 1:1.35:2–3, respectively. Additionally, lithium-ion batteries require individual electronic control for each cell to ensure safe operation, making them suitable primarily for applications where maximizing energy content is critical.

NiMH batteries are generally heavier and bulkier than their lithium-ion counterparts. However, in terms of energy storage capacity, both technologies are comparable. Li-ion batteries have the advantage of faster charge and discharge rates and exhibit little to no memory effect, which is more pronounced in nickel-based systems. While the memory effect is most severe in NiCd batteries, it also occurs to a lesser extent in NiMH batteries.

The "memory effect" refers to a phenomenon where batteries are repeatedly recharged before being fully discharged. This results in a temporary loss of capacity, often perceived as a voltage drop during discharge. The cause is associated with physical changes in the electrode material, particularly in areas that have remained in a charged state for extended periods without cycling. These areas develop increased internal resistance, leading to higher overvoltage and a more pronounced voltage drop during use. Importantly, this loss of capacity is reversible and can be mitigated by performing several full charge-discharge cycles, which restore the battery's full capacity (Notten & Van Beek, 2000).

It is important to mention that both NiMH and lithium-ion batteries are long-lasting and have been widely used in various devices for many years. However, NiMH batteries have an advantage in terms of durability. In particular, some lithium-ion batteries may experience a reduced lifespan under severe temperature conditions, especially in hot climates (Arun et al., 2022). Table 3 presents a comparison of the specifications between Li-ion and NiMH batteries.

Table 3. Comparison of Li-ion vs NiMH (Arun et al., 2022).

No.	Specifications	Li-ion	Ni-MH
1	Energy	160 Wh/kg	90 Wh/kg
2	Voltage	3.6 V	1.2 V
3	Size and capacities	Customized and 2000-6000 mAh	3XAA and 2000 mAh
4	Self-discharge rate per month	2-8%	30-50%
5	Price range (depends on battery type)	30 to 500$	20-150$
6	Maintenance	No	Low

Specifically, lithium-ion batteries are the most common choice in electric vehicles (EVs) due to their high power-to-weight ratio and excellent high-temperature performance. In EVs, weight is a critical factor because lighter vehicles can travel longer distances on a single charge. Additionally, lithium-ion batteries can store a large amount of energy relative to their cost and tend to retain their charge longer than conventional batteries thanks to their low self-discharge rate. On the other hand, NiMH batteries are more commonly used in hybrid electric vehicles (HEVs) than in all-electric cars, although both types of vehicles may use them (Arun et al., 2022). For example, cylindrical NiMH batteries were used in early HEVs such as the first-generation Toyota Prius in 1997, which was the world's first mass-produced hybrid. This battery achieved a specific power of 600 W/kg and demonstrated a lifetime exceeding 100,000 km of driving. The second-generation Prius, released in 2000, improved the specific power to 1000 W/kg (Taniguchi et al., 2001).

Currently, NiMH portable battery cells have a specific energy of around 80 Wh/kg, which is attractive for HEV applications. In contrast, lithium-ion batteries offer higher specific energy, around 120-130 Wh/kg, consistent with the values shown in Table 3. Regarding volumetric energy density, lithium-ion and NiMH batteries are similar, with cell-specific values between 200-300 Wh/L (Köhler et al., 2002). However, lithium-ion batteries generally have a shorter service life compared with

NiMH batteries, which is one of their weaknesses. The service life of lithium-ion cells today is primarily limited by calendar life.

TECHNOLOGICAL ADVANCEMENTS AND FUTURE PERSPECTIVES OF NICKEL-METAL HYDRIDE BATTERIES

Nickel-metal hydride (NiMH) batteries have played a significant role in the evolution of rechargeable battery technology since their commercial introduction in the early 1990s (Feng et al., 2001). They served as a critical bridge between older nickel-cadmium (NiCd) systems and the more energy-dense lithium-ion (Li-ion) chemistries that dominate today's portable and vehicular energy storage applications. Despite the rise of Li-ion systems, NiMH batteries have retained a solid position in specific markets due to their intrinsic safety, robust cycling performance, and competitive cost profile (Almadani et al., 2024).

As reviewed in this chapter, in NiMH batteries, its specific materials configuration offers multiple advantages: relatively high energy density, excellent thermal stability, and environmental benefits due to the absence of toxic cadmium (Chang et al., 2016). However, it also introduces challenges related to the sourcing and recycling of rare earth elements and transition metals, which has become a growing focus of global battery research in recent times (Pradhan et al., 2022). NiMH batteries offer a specific energy range of 60–120 Wh/kg and a volumetric energy density of 140–300 Wh/L. While these numbers are lower than those of modern Li-ion chemistries, NiMH cells offer other performance benefits such as high energy efficiency (60–92%), long cycle life (typically 1800–2000 cycles), and operational safety under a wide range of environmental conditions (Quintero-Pulido et al., 2025). One of the critical advantages of NiMH batteries lies in their inherent thermal and chemical stability. Unlike Li-ion systems, NiMH batteries are far less susceptible to thermal runaway, significantly reducing the risk of fire or explosion. This safety feature has made them the preferred choice for applications in consumer electronics and energy storage devices where stringent safety requirements exist (Almadani et al., 2024) (Guarnieri, 2022).

Furthermore, NiMH batteries are available globally and can be manufactured at a lower cost than high-performance Li-ion systems, especially when second-life materials or recycled components are used (Arun et al., 2022). In 2025, the cost of a NiMH storage system ranged from EUR 400/kWh to EUR 700/kWh with a life duration between 10 and 15 years, a similar system made of Li-Ion technology is currently ranging from EUR 1000/kWh to EUR 10000/kWh. Although these costs have fluctuated due to raw material prices and supply chain issues, NiMH technology remains an economically viable solution for many medium-demand applications (https://ease-storage.eu/wp-content/uploads/2016/03/EASE_TD_NiMH.pdf).

Much of the recent progress in NiMH battery development has focused on improving the electrode materials. Efforts have been made to enhance the hydrogen storage capacity of the anode alloys by optimizing their microstructure, composition, and surface treatment. Innovations in using AB_2- and AB_5-type alloys, where "A" represents rare earth or alkali metals and "B" includes transition metals, have significantly improved electrochemical kinetics and resistance to corrosion in alkaline electrolytes (Kalita et al., 2025). On the cathode side, modifications to the nickel hydroxide particles, such as doping with cobalt or manganese, have led to better conductivity, structural stability, and improved utilization efficiency. These developments directly contribute to better cycle performance and energy retention over prolonged use (Darıcıoğlu & Öztürk, 2024).

Researchers have also explored the use of modified electrolytes to enhance ionic conductivity, reduce gas generation, and mitigate self-discharge. The addition of sodium hydroxide (NaOH) to the potassium hydroxide (KOH) electrolyte has been shown to stabilize the pH and extend the effective lifespan of the electrolyte, contributing to longer battery life (Xu & Mulder, 2021). Advancements in polymer science have led to the development of improved separators that exhibit higher porosity, better wettability, and greater chemical resistance. These new materials allow for enhanced ion transport while maintaining mechanical integrity and chemical compatibility with the aggressive alkaline environment inside the battery (May, 2024). Integration of thermal sensors and more intelligent battery management system (BMS) units has improved charge/discharge efficiency and safety in large-format NiMH systems. These technologies help monitor battery temperature, prevent overcharging, and extend service life, particularly in hybrid electric vehicle (HEV) applications (Dharaneesh et al., 2025).

NiMH batteries have long been the workhorse for a wide array of applications, from portable electronic devices to large-scale energy storage systems. They were the dominant battery technology in hybrid vehicles before the widespread adoption of Li-ion batteries. The Toyota Prius, for instance, relied heavily on NiMH batteries for more than a decade due to their safety and longevity (Martínez-Sánchez et al., 2024).

Even today, NiMH systems are employed in industrial applications, power backup systems, and some hybrid and plug-in hybrid electric vehicles (PHEVs). Their robust cycling ability and resilience to abuse make them ideal for public safety systems, emergency lighting, medical devices, and tools requiring reliable, maintenance-free power (Islam et al., 2023) (Nawaz et al., 2022).

According to market estimates, the NiMH battery market is projected to reach USD 2.7 billion by 2032 from an estimated USD 1.96 billion in 2023, at a CAGR of 3.62%. Increasing adoption of electric vehicles, growing demand for portable electronics, and rising popularity of power tools are key drivers of market growth (https://www.marketresearchfuture.com/reports/nickel-metal-hydride-battery-market-22315). This sustained market interest can be also attributed to rising environmental concerns, regulatory pressures for recycling, and the need for safer and cost-effective alternatives in specific sectors. In fact, the recycling and reuse of NiMH batteries have become central topics in the context of circular economy. These batteries contain valuable and finite resources such as rare earth metals and transition metals, which can be recovered and reintroduced into the supply chain (Tülin & Fırat, 2025).

Significant progress has been made in scaling up recycling processes, including hydrometallurgical and pyrometallurgical techniques that allow for efficient recovery of lanthanum, cerium, nickel, and cobalt. European regulatory frameworks and international guidelines (e.g., Battery Directive 2006/66/EC and the forthcoming Battery Regulation) have encouraged manufacturers and recyclers to invest in recovery technologies. Furthermore, second-life applications, where NiMH batteries are repurposed for less demanding energy storage roles after their first life, are gaining momentum. These include residential energy storage systems, backup power for telecommunications, and grid stabilization units. Such strategies not only extend the functional lifespan of the batteries but also reduce environmental impact and material waste (Dong et al., 2025).

Given the widespread use of NiMH batteries, a robust framework of international and regional standards has been developed to ensure quality, safety, and interoperability. Standards such as the IEC 61951-2:2017+AMD1:2022 CSV outline the requirements for design, performance, and testing of secondary sealed cells and batteries. Additional standards address aspects such as design and manufacturing controls, performance benchmarking, safety validation, environmental testing and compliance. Organizations such as the Joint Research Centre (JRC) have actively contributed to the analysis and implementation of these standards. Their efforts have helped harmonize battery testing methodologies and facilitate the integration of NiMH technologies into broader energy systems (https://publications.jrc.ec.europa.eu/repository/handle/JRC140136).

Despite the increasing dominance of Li-ion and emerging solid-state technologies, NiMH batteries are expected to remain relevant in niche and mid-range markets for the foreseeable future. According to projections, NiMH systems will continue to serve as a reliable energy storage option for portable devices and hybrid and plug-in hybrid vehicles until at least 2032 (https://www.coherentmarketinsights.com/industry-reports/nickel-metal-hydride-battery-market).

Several future developments are anticipated to further improve NiMH battery performance and sustainability:

1. **Development of Cobalt-Free Alloys:** As cobalt prices fluctuate and ethical sourcing becomes a greater concern, there is growing interest in developing cobalt-free electrode materials without compromising capacity and cycle life (Li et al., 2025).
2. **Advanced Manufacturing Techniques:** Additive manufacturing and nanoscale material engineering may enable more compact and efficient battery structures, reduce material usage and increasing energy density (Wang et al., 2025).
3. **Integration with Renewable Energy Systems:** NiMH batteries, particularly in their second-life form, are expected to play a role in off-grid and microgrid applications where low maintenance and long-term reliability are essential (https://www.coherentmarketinsights.com/industry-reports/nickel-metal-hydride-battery-market).
4. **Digitalization and Predictive Maintenance:** Embedding sensors and leveraging IoT technologies for real-time monitoring and predictive diagnostics will enhance safety, performance tracking, and lifecycle management (Njoku et al., 2025).

Nickel-metal hydride batteries remain a vital technology in the energy storage landscape due to their safety, durability, and adaptability. Ongoing advancements in materials science, recycling technologies, and system integration are enhancing their competitiveness. Although Li-ion and newer battery chemistries may dominate the high-performance market, NiMH batteries are likely to continue providing valuable solutions in specific applications, especially where cost-effectiveness, safety, and sustainability are priorities. The next decade will likely witness continued investment and innovation in NiMH systems, driven by both regulatory imperatives and practical market needs. As the world transitions toward more sustainable energy systems, NiMH batteries are poised to play a supportive but crucial role in achieving global decarbonization and circular economy goals.

CONCLUSIONS

Nickel-metal hydride (NiMH) batteries remain a pivotal and versatile energy storage technology. Their historical development, robust electrochemistry, and adaptability have sustained their relevance in diverse applications despite competition from lithium-ion and emerging solid-state batteries. The key conclusions and future perspectives can be summarized as follows:

- **Market resilience**: NiMH batteries continue to grow in hybrid vehicles, portable electronics, and grid storage, supported by their safety, cost-effectiveness, and recyclability.
- **Environmental and regulatory alignment**: The high recovery value of lanthanum, nickel, and cobalt reinforces NiMH batteries as central to circular economy models. Advances in hydrometallurgical and pyrometallurgical recycling processes, coupled with international regulations (e.g., EU Battery Directive), are accelerating sustainable practices.
- **Extended life cycle**: Second-life applications, such as residential energy storage, telecommunications backup, and grid stabilization, expand the usefulness of NiMH systems beyond their primary operational phase.
- **Standardization and safety**: Development of international standards has ensured interoperability, quality control, and safe deployment across different sectors.
- **Innovation trends**: Future research directions include cobalt-free alloy development, advanced manufacturing methods, digitalization, predictive maintenance, and integration with renewable energy systems. These innovations will sustain the competitiveness of NiMH batteries in medium-performance niches.
- **Strategic role in sustainability**: NiMH technology supports global decarbonization efforts, fosters circular economy practices, and facilitates democratized access to reliable and affordable energy storage.

Overall, while NiMH batteries may not dominate high-performance markets, their robustness, recyclability, and safety ensure their long-term relevance. They are poised to remain an indispensable contributor to sustainable energy transitions and a valuable complement to other emerging storage technologies in the coming decades.

REFERENCES

Almadani, M., Oni, O. E., Longe, O. M., & Olatomiwa, L. (2024). Comparison of battery chemistries for electric vehicle applications. In *5th African International Conference on Industrial Engineering and Operations Management*, pp. 23-25. DOI: 10.46254/AF05.20240159

Ambrosio, R. C., & Ticianelli, E. A. (2001). Nickel-metal hydride batteries, an alternative for the nickel-cadmium batteries. *Quimica Nova*, *24*(2), 243–246. DOI: 10.1590/S0100-40422001000200015

Anani, A., Visintin, A., Petrov, K., Srinivasan, S., Reilly, J. J., Johnson, J. R., Schwarz, R. B., & Desch, P. B. (1994). Alloys for hydrogen storage in nickel/hydrogen and nickel/metal hydride batteries. *Journal of Power Sources*, *47*(3), 261–275. DOI: 10.1016/0378-7753(94)87005-5

Arun, V., Kannan, R., Ramesh, S., Vijayakumar, M., Raghavendran, P. S., Ramkumar, M. S., Anbarasu, P., & Sundramurthy, V. P. (2022). Review on Li-Ion Battery vs Nickel Metal Hydride Battery in EV. *Advances in Materials Science and Engineering*, *2022*(1), 1–7. DOI: 10.1155/2022/7910072

Borthomieu, Y., & Bernard, P. (2009). SECONDARY BATTERIES – NICKEL SYSTEMS | Nickel–Hydrogen. *Elsevier eBooks* (pp. 482-493).

Chang, S., Young, K. H., Nei, J., & Fierro, C. (2016). Reviews on the US Patents regarding nickel/metal hydride batteries. *Batteries*, *2*(2), 10.

Chu, A. H., Chen, X., Yuan, Y. N., Zhang, T., Yohei, H., & Xu, G. C. (2020). Research on Life Assessment of Ni-MH Battery through Orthogonal Experiment under Multiple Aging Factors. [IOP Publishing.]. *IOP Conference Series. Earth and Environmental Science*, *508*(1), 012192. DOI: 10.1088/1755-1315/508/1/012192

Cuevas, F., Joubert, J. M., Latroche, M., & Percheron-Guegan, A. (2001). Intermetallic compounds as negative electrodes of Ni/MH batteries. *Applied Physics. A, Materials Science & Processing*, *72*(2), 225–238. DOI: 10.1007/s003390100775

Dansui, Y., Kato, F., Suzuki, K., & Yuasa, K. (2000). *U.S. Patent No. 6,074,785.* Washington, DC: U.S. Patent and Trademark Office.

Darıcıoğlu, N. Ö., & Öztürk, T. (2024). Search for carbon additives in Co-free Ni (OH) 2 cathodes for NiMH batteries. *Journal of Energy Storage*, *90*, 111850. DOI: 10.1016/j.est.2024.111850

Dehghani-Sanij, A. R., Tharumalingam, E., Dusseault, M. B., & Fraser, R. (2019). Study of energy storage systems and environmental challenges of batteries. *Renewable & Sustainable Energy Reviews, 104*, 192–208. DOI: 10.1016/j.rser.2019.01.023

Dharaneesh, V. M., Aruna, S., Hariprasath, E., & Somasundaram, P. L. (2025, March). Electric Vehicle Battery Parameters Monitoring System using IoT. In *2025 International Conference on Visual Analytics and Data Visualization (ICVADV)* (pp. 386-391). IEEE.

Dong, Z., Tao, Y., Lai, S., Wang, T., & Zhang, Z. (2025). Powering Future Advancements and Applications of Battery Energy Storage Systems Across Different Scales. *Energy Storage and Applications, 2*(1), 1. DOI: 10.3390/esa2010001

Dymek, M., Nowak, M., Jurczyk, M., & Bala, H. (2019). Electrochemical characterization of nanocrystalline hydrogen storage La1.5Mg0.5Ni6.5Co0.5 alloy covered with amorphous nickel. *Journal of Alloys and Compounds, 780*, 697–704. DOI: 10.1016/j.jallcom.2018.11.196

Eyövge, C., Şahin, E. O., & Öztürk, T. (2018). "Surface Modified Metal Hydrides as Negative Electrodes in NiMH Batteries", 1, https://hdl.handle.net/11511/84895

Feng, F., Geng, M., & Northwood, D. O. (2001). Electrochemical behaviour of intermetallic-based metal hydrides used in Ni/metal hydride (MH) batteries: A review. *International Journal of Hydrogen Energy, 26*(7), 725–734. DOI: 10.1016/S0360-3199(00)00127-0

Gibbard, H. F. (1993). Nickel metal hydride battery technology. In *Proceedings of WESCON'93 IEEE,* 215–219. DOI: 10.1109/WESCON.1993.488437

Gibbard, H. F. (1994). Nickel metal-hydride battery applications. *Annual Battery Conference on Applications and Advances*, 127. DOI: 10.1109/BCAA.1994.283603

Guarnieri, M. (2022). Secondary Batteries for Mobile Applications: From Lead to Lithium [Historical]. *IEEE Industrial Electronics Magazine, 16*(4), 60–68. DOI: 10.1109/MIE.2022.3212242

Guzik, M. N., Hauback, B. C., & Yvon, K. (2012). Hydrogen atom distribution and hydrogen induced site depopulation for the $La_{2-x}Mg_xNi_7$–H system. *Journal of Solid State Chemistry, 186*, 9–16. DOI: 10.1016/j.jssc.2011.11.026

Hariprakash, B., Shukla, A., & Venugoplan, S. (2024). Batteries – Battery types – Nickel Batteries | Nickel-Metal Hydride: Overview. *Elsevier eBooks* (pp. 674-681).

Hubkowska, K., Soszko, M., Krajewski, M., & Czerwiński, A. (2019). Enhanced kinetics of hydrogen electrosorption in AB_5 hydrogen storage alloy decorated with Pd nanoparticles. *Electrochemistry Communications*, *100*, 100–103. https://ease-storage.eu/wp-content/uploads/2016/03/EASE_TD_NiMH.pdf. DOI: 10.1016/j.elecom.2019.02.007

Islam, J., Anwar, R., Shareef, M., Zabed, H. M., Sahu, J. N., Qi, X., & Chowdhury, F. I. (2023). Rechargeable metal-metal alkaline batteries: Recent advances, current issues and future research strategies. *Journal of Power Sources*, *563*, 232777. DOI: 10.1016/j.jpowsour.2023.232777

Kalita, G., Otsuka, R., Endo, T., & Furukawa, S. (2025). Recent advances in substitutional doping of AB5 and AB2 type hydrogen storage metal alloys for Ni-MH battery applications. *Journal of Alloys and Compounds*, *1020*, 179352. DOI: 10.1016/j.jallcom.2025.179352

Kang Hu, W. (2000). Improvement in capacity of cobalt-free Mm-based hydrogen storage alloys with good cycling stability. *Journal of Alloys and Compounds*, *297*(1-2), 206–210. DOI: 10.1016/S0925-8388(99)00567-8

Karwowska, M., Fijalkowski, K. J., & Czerwiński, A. A. (2018). Corrosion of Hydrogen Storage Metal Alloy LaMm-$Ni_{4.1}Al_{0.3}Mn_{0.4}Co_{0.45}$ in the Aqueous Solutions of Alkali Metal Hydroxides. *Materials (Basel)*, *11*(12), 2423. DOI: 10.3390/ma11122423 PMID: 30513583

Karwowska, M., Jaron, T., Fijalkowski, K. J., Leszczynski, P. J., Rogulski, Z., & Czerwinski, A. (2020). Influence of electrolyte composition and temperature on behavior of AB_5 hydrogen storage alloy used as negative electrode in Ni-MH batteries. *Journal of Power Sources*, *263*, 304–309. DOI: 10.1016/j.jpowsour.2014.04.050

Kitayama, H., Hayashida, H., Yamamoto, M., Bando, N., Miyamoto, K., & Suzuki, H. (2002). *U.S. Patent No. 6, 399, 247*. Washington, DC: U.S. Patent and Trademark Office.

Köhler, U., Kümpers, J., & Ullrich, M. (2002). High performance nickel-metal hydride and lithium-ion batteries. *Journal of Power Sources*, *105*(2), 139–144. DOI: 10.1016/S0378-7753(01)00932-6

Li, J., Cai, Y., Huang, J., Zhao, S., & Cheng, D. (2024). Enhanced Cycling Stability of Amorphous MgNi-Based Alloy Electrodes through Corrosion Prevention by Incorporating $Al_2(SO_4)_3{\cdot}18H_2O$ into the Electrolyte. *Metals*, *14*(10), 1142. DOI: 10.3390/met14101142

Li, J., Li, H., Jiao, J., & Xu, Y. (2025). Which policy can effectively alleviate the impact of raw material price fluctuations on battery recycling? *Environmental Impact Assessment Review*, *115*, 107996. DOI: 10.1016/j.eiar.2025.107996

Li, R., Lu, H., Pan, X., Zhao, J., Wang, W., Li, Y., Xi, N., Peng, Q., Han, S., & Zhang, L. (2023). Improvement on cyclic stability of AB4-type La–Mg–Ni-based hydrogen storage alloys via merging Y element for nickel-metal hydride batteries. *International Journal of Hydrogen Energy*, *48*(84), 32849–32859. DOI: 10.1016/j.ijhydene.2023.05.011

Liu, Y., Pan, H., Gao, M., & Wang, Q. (2011). Advanced Hydrogen Storage Alloys for Ni/MH Rechargeable Batteries. *Journal of Materials Chemistry*, *21*(13), 4743–4755. DOI: 10.1039/C0JM01921F

Lota, K., Swoboda, P., Acznik, I., Sierczyńska, A., Mańczak, R., Kolanowski, L., & Lota, G. (2020). Electrochemical properties of modified negative electrode for Ni-MH cell. *Current Applied Physics*, *20*(1), 106–113. DOI: 10.1016/j.cap.2019.10.014

Mani, N., & Ramaprabhu, S. (2004). Effect of substitutional elements on hydrogen absorption properties in Mm-based AB_5 alloys. *Journal of Alloys and Compounds*, *363*(1-2), 275–291. DOI: 10.1016/S0925-8388(03)00487-0

Martínez-Sánchez, R., Molina-García, A., Mateo-Aroca, A., & Ramallo-González, A. P. (2024). Sustainable Energy Innovation: Patent Analysis and Assessment for NiMH Battery Regeneration.

May, G. J. (2024). Separators: An essential barrier between electrodes. In *Advanced Materials for Battery Separators* (pp. 35-55). Elsevier.

Monnier, J., Zhang, J., Paul-Boncour, V., & Cuevas, F. (2024). Chapter: Corrosion in Nickel-Metal Hydride (Ni-*M*H) Batteries—Recent Developments. In Saji, V. S. (Ed.), *Corrosion and Degradation in Fuel Cells, Supercapacitors and Batteries* (pp. 219–249). Springer. DOI: 10.1007/978-3-031-57012-4_10

Mungole, M. N., Balasubramaniam, R., & Rai, K. N. (1999). Correlation between microstructure and hydrogen storage capacity in $MmNi_5$ alloys with Al, Mn and Sn substitutions. *International Journal of Hydrogen Energy*, *24*(5), 467–471. DOI: 10.1016/S0360-3199(98)00091-3

Nawaz, M., Ahmed, J., & Abbas, G. (2022). Energy-efficient battery management system for healthcare devices. *Journal of Energy Storage*, *51*, 104358. DOI: 10.1016/j.est.2022.104358

Njoku, J. N., Nkoro, E. C., Medina, R. M., Nwakanma, C. I., Lee, J. M., & Kim, D. S. (2025). Leveraging Digital Twin Technology for Battery Management: A Case Study Review. *IEEE Access : Practical Innovations, Open Solutions, 13*, 21382–21412. DOI: 10.1109/ACCESS.2025.3531833

Notten, P. H. L., & Latroche, M. (2009). Secondary batteries-nickel systems: Nickel–metal hydride: Metal hydrides: Metal Hydrides. In *Encyclopedia of electrochemical power sources* (pp. 502–521). Elsevier. DOI: 10.1016/B978-044452745-5.00164-7

Notten, P. H. L., & Van Beek, J. R. G. (2000). Nickel-Metal Hydride Batteries: From Concept to Characteristics. *Chemistry & Industry, 54*(3), 102–115.

Omar, N., Firouz, Y., Monem, M. A., Samba, A., Gualous, H., Coosemans, T. C., & Van Mierlo, J. (2014). *Analysis of nickel-based battery technologies for hybrid and electric vehicles. Reference module in chemistry, molecular sciences and chemical engineering*. Elsevier.

Ovshinsky, S. R., Fetcenko, M. A., & Ross, J. (1993). A nickel metal hydride battery for electric vehicles. *Science, 260*(5105), 176–181. DOI: 10.1126/science.260.5105.176 PMID: 17807176

Parthe, E., & Lemaire, R. (1975). Structure block stacking in intermetallic compounds. I. The rhombohedral-hexagonal Mn+1X5n-1 and the monoclinic-hexagonal-trigonal-orthorhombic Mn+1X5n+2 structure series. *Acta Crystallographica. Section B, Structural Crystallography and Crystal Chemistry, 31*(7), 1879–1889. DOI: 10.1107/S0567740875006413

Pradhan, S., Nayak, R., & Mishra, S. (2022). A review on the recovery of metal values from spent nickel metal hydride and lithium-ion batteries. *International Journal of Environmental Science and Technology, 19*(5), 4537–4554. DOI: 10.1007/s13762-021-03356-5

Quintero V., Che O., Ching E., Auciello O., & De Obaldía E. (2021). Lithium ion batteries: features and applications. *Revista de I+D Tecnológico, 17*(1), 14-22.

Quintero-Pulido, D. F., Covrig, C. F., & Bruchhausen, M. (2025). On the Performance of Portable NiMH Batteries of General Use. *Batteries, 11*(1), 30. DOI: 10.3390/batteries11010030

Ruetschi, P., Meli, F., & Desilvestro, J. (1995). Nickel-metal hydride batteries. The preferred batteries of the future? *Journal of Power Sources, 1-2*(57), 85–91. DOI: 10.1016/0378-7753(95)02248-1

Sotelo, O., Henao, J., Poblano, C., Campillo, B., Castañeda, E., Flores, N., Molina, A., & Martínez, H. (2024). Evaluating a Fe-Based Metallic Glass Powder as a Novel Negative Electrode Material for Applications in Ni-MH Batteries. *Batteries, 10*(9), 312. DOI: 10.3390/batteries10090312

Taniguchi, A., Fujioka, N., Ikoma, M., & Ohta, A. (2001). Development of nickel/metal-hydride batteries for EVs and HEVs. *Journal of Power Sources, 100*(1-2), 117–124. DOI: 10.1016/S0378-7753(01)00889-8

Tliha, M., Mathlouthi, H., Lamloumi, J., & Percheron-Guegan, A. (2007). AB_5-type hydrogen storage alloy used as anodic materials in Ni-MH batteries. *Journal of Alloys and Compounds, 436*(1-2), 221–225. DOI: 10.1016/j.jallcom.2006.07.012

Tülin, U., & Fırat, B. (2025). The recent developments in EV batteries and REEs recovery processes from spent NiMH batteries. *Review Article Published: 2024-10-10* .DOI: 10.20935/AcadEng7355

Varin, R. A. (2000). Intermetallics: Crystal structures Encyclopedia of materials. *Science and Technology*, 1-4.

Wang, L., Su, Z., Wang, R., Liang, H., Fang, B., & Mo, R. (2025). Recent status, key strategies and challenging perspectives of smart batteries for next-generation batteries. *Journal of Materials Chemistry. A, Materials for Energy and Sustainability, 13*(27), 21116–21171. DOI: 10.1039/D5TA01989C

Xu, Y., & Mulder, F. M. (2021). Non-alloy Mg anode for Ni-MH batteries: Multiple approaches towards a stable cycling performance. *international journal of hydrogen energy, 46*(37), 19542-19553.

Yan, S., Li, P., Nei, J., Young, K.-H., & Ng, S. (2020). Effects of Salt Additives to the KOH Electrolyte Used in Ni/MH Batteries. *Battery, 1*(1), 54–73. DOI: 10.3390/batteries1010054

Yang, C. Z., Lou, Y., Zhang, J., Xie, X., & Xia, B. (2023). Chapter, Mechanism Research of Cycle Process for MH/Ni Battery. In book, Materials and Working Mechanisms of Secondary Batteries, 297-313.

Yi, X., Kirk, C., & Robertson, N. (2024). Nickel hydroxide-based energy storage devices: Nickel-metal hydride batteries vs. nickel hydroxide hybrid supercapacitors. *Carbon Neutrality, 3*(39), 1–17. DOI: 10.1007/s43979-024-00114-7

Ying, T. K., Gao, X. P., Hu, W. K., Wu, F., & Noréus, D. (2006). Studies on rechargeable NiMH batteries. *International Journal of Hydrogen Energy, 31*(4), 525–530. DOI: 10.1016/j.ijhydene.2005.04.018

Young, K., Fierro, C., & Fetcenko, M. A. (2011). Status of Ni/MH battery research and industry. *Power and Energy Society General Meeting*, 1–3.

Young, K. H., & Nei, J. (2013). The current status of hydrogen storage alloy development for electrochemical applications. *Materials (Basel)*, *6*(10), 4574–4608. DOI: 10.3390/ma6104574 PMID: 28788349

Young, K. H., & Yasuoka, S. (2016). Capacity degradation mechanisms in nickel/metal hydride batteries. *Batteries*, *2*(1), 3. DOI: 10.3390/batteries2010003

Zhang, Y., Wu, G., Gu, J., Kang, H., Li, Y., Zhou, D., Wang, W., Zhang, L., & Han, S. (2024). A2B7-type La–Mg–Ni alloys prepared by Mg thermal diffusion for improved hydrogen storage performance. *Rare Metals*, *43*(7), 3260–3272. DOI: 10.1007/s12598-024-02627-7

Chapter 3
Understanding Corrosion and Materials Deterioration in Metal–Air Batteries

Malak Rehioui
https://orcid.org/0000-0001-7833-6426
Faculty of Sicences, Ibn Tofail University, Morocco

Ouigua Rochdi
https://orcid.org/0009-0001-1755-6318
Faculty of Sciences, Ibn Tofaïl University, Morocco

Hamid Erramli
Faculty of sciences, Ibn Tofail University, Morocco

ABSTRACT

Electrochemical energy storage and conversion (EESC) technologies are important to today's energy systems. They provide sustainable solutions to the growing global demand for efficient and green energy. Among them, metal–air batteries have emerged as promising candidates due to their high theoretical specific energies and energy densities. However, like many electrochemical systems, metal–air batteries are highly susceptible to corrosion and material degradation during operation. These batteries often operate under harsh environmental conditions. Corrosion not only compromises the structural integrity of the battery but also leads to reduced efficiency, safety concerns, and shortened lifespan. Understanding the mechanisms of corrosion in metal–air batteries is therefore essential for improving their durability.

DOI: 10.4018/979-8-3373-3156-0.ch003

1. INTRODUCTION

Electrochemical energy storage and conversion (EESC) systems play a critical role in the transition to cleaner and more sustainable energy. The increasing demand for advanced energy storage technologies is underscored by the impact of climate change and the decline of finite fossil fuel resources. Renewable energy sources like wind, solar and hydro are inherently intermittent and require effective storage solutions (Sharmoukh, 2025). Additionally, as global strategies to lower carbon emissions intensify, the requirement for efficient, reliable, and environmentally friendly energy storage solutions continues to expand. EESC devices encompass several categories, including primary batteries (single-use), secondary batteries (rechargeable), fuel cells that convert chemical energy into electricity without storing it, and supercapacitors that store energy via charge separation (Bhattacharjee et al., 2023; Saji, 2024). Each devices operate differently (Detka & Górecki, 2023), and they are broadly adopted in electric and hybrid vehicles, portable electrical devices (laptops, mobile phones, wearable devices, etc.) (Fang et al., 2022), and power stations for space and stationary applications. Supercapacitors and fuel cells, offering unique advantages, are limited by low energy density or complex system requirements (Dissanayake & Kularatna-Abeywardana, 2024; L. Gao et al., 2025; Sebbani et al., 2025). Lithium-ion batteries, once considered the benchmark technology with widespread adoption in hybrid and electric vehicles (HEVs/EVs) and energy storage systems (Fang et al., 2022; Hasan et al., 2025), now face growing concerns related to resource limitations and safety. Limited lithium reserves and accelerated consumption present significant challenges, especially for large-scale energy storage applications (C. Liu et al., 2020). Aluminum-ion batteries have emerged as one of the most promising alternatives, offering ultrahigh volumetric capacity, enhanced safety, and cost-effectiveness, thanks to the abundance of aluminum (T.-T. Wei et al., 2021). Additionally, metal–air batteries are gaining attention as a new generation of devices capable of bridging the gap between high capacity and environmental sustainability, offering exceptionally high theoretical energy densities, lightweight design, and the possibility of using earth-abundant and low-toxicity metals (Chen et al., 2022; Chung et al., 2025; Salado & Lizundia, 2022). Despite huge advances in developing these systems, corrosion and material degradation remain major challenges as they reduce the performance, safety, and lifespan of EESC devices (Du et al., 2023). While Researches (Dilshad et al., 2025; Shabeer et al., 2025) underline their potential for transportation and grid-level storage, yet also emphasize that their commercialization depends on overcoming severe corrosion processes occurring at the anode-electrolyte interface.

In fact, corrosion is an irreversible phenomenon influenced by multiple factors, including material properties, environmental conditions, surface coatings, microbial activity, and mechanical stress. Most commonly associated with metals, corrosion involves electrochemical reactions typically the oxidation of metals into ions, accompanied by electron release. Corrosion can also affect other materials and components within EESC devices (Queiroz et al., 2024; Serebrennikova et al., 2009). Several recent publications can be found on corrosion issues with EESC devices, viz. bipolar plate corrosion, carbon corrosion, current collector corrosion in supercapacitors and metal-ion batteries (Brandstätter et al., 2016; L. Wu et al., 2023a; S. S. Zhang & Jow, 2002), and anode corrosion in metal-air batteries and various other associated degradation issues and corrosion types (Sousa et al., 2025). Corrosion may occur externally, such as damage to the casing, or/and internally, such as corrosion of the electrodes or current collectors. Most EESC devices contain two electrodes (anode and cathode) and an electrolyte. In batteries and fuel cells, energy is generated through chemical reactions at the electrodes. In supercapacitors, energy is stored through charge separation or redox reactions at the surface. Therefore, understanding corrosion processes and developing strategies to prevent them remain vital research priorities in the pursuit of safer, more durable energy technologies.

Another compelling solution lies in metal–air batteries (MABs), featuring electrolytes and bifunctional air electrodes that utilize oxygen from the atmosphere (Shabeer et al., 2025). Known as "air batteries," these systems consist of a metal anode, an air-exposed cathode, and an electrolyte. During discharge, oxidation occurs at the anode while oxygen is reduced at the cathode, generating energy. Metal–air batteries (MABs) have a simple design, lightweight nature and particularly promising owing to their substantial theoretical energy densities and environmental benefits. Despite their potential, metal–air batteries face significant technical hurdles, including anode corrosion, low practical energy density, limited cycle life, and parasitic side reactions. Unlike the relatively well-documented degradation pathways in lithium-ion batteries (Guo et al., 2023a; Mohammadipour & Gharahcheshmeh, 2025; L. Wu et al., 2023b; Zhou et al., 2025), corrosion in metal–air systems involves complex electrochemical and chemical interactions with both the electrolyte and atmospheric oxygen. Addressing corrosion challenges is crucial for their widespread deployment. Ultimately, understanding and mitigating corrosion processes is essential to advancing EESC technologies. A comprehensive grasp of material behavior and degradation mechanisms will support the development of more durable, efficient, and sustainable energy systems. This chapter presents the recent information on the various corrosion problems in metal air batteries and their mitigation strategies.

2. LITTERATURE REVIEW

2.1 Metal-air batteries and corrosion challenges

Metal-air batteries (MaBs) operate on a principle similar to conventional lithium-ion batteries (LIBs), substituting the traditional closed system with an open-cell configuration that allows ambient oxygen to serve as the cathodic reactant (Olabi et al., 2021a). Using air as the cathode in metal-air batteries significantly reduces both cost and weight. These batteries have a metal anode (such as Al, Mg, Li, Na, or K), while oxygen from the air is activated at the cathode, often with the help of electrocatalysts. The battery also includes an electrolyte (either aqueous or non-aqueous) and a separator placed between the anode and cathode, as shown in figure 1. The selection of the electrolyte mainly depends on the metal used for the anode (Barman et al., 2025). The working mechanism of metal air batteries is based on the redox activity of a metal anode and a porous air-breathing cathode, separated by an electrolyte and a physical separator (Y. Wang et al., 2023).

Figure 1. Metal-air battery.

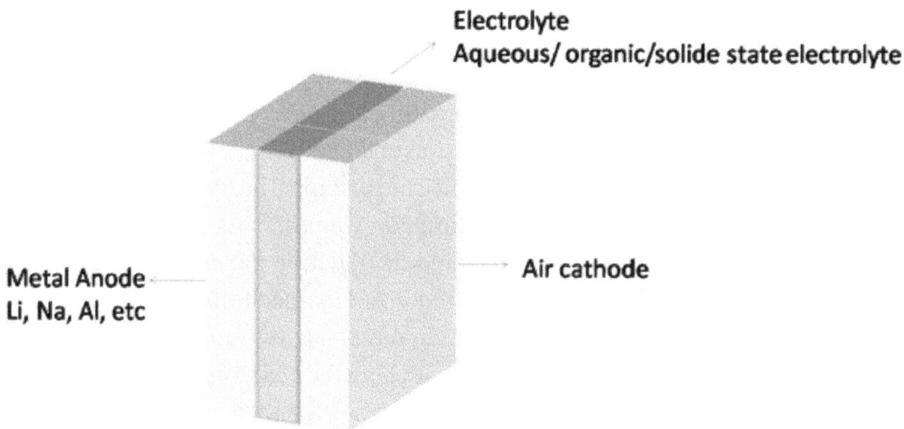

The core redox reactions in metal air batteries (MABs involve the oxidation of a metal anode and the reduction of oxygen at the air cathode [30]. Specifically, the oxygen reduction reaction (ORR), including four-electron pathway:

$O_2 + 2\,H_2O + 4\,e^- \rightarrow 4\,OH^-$

and two-plus-two-electron pathway:

$O_2 + H_2O + 2\,e^- \rightarrow HO_2^- + OH^-$

The oxygen evolution reaction (OER) of:

$4\,OH^- \rightarrow O_2 + 2\,H_2O + 4\,e^-$

Both oxygen reduction reaction (ORR) and oxygen evolution reaction (OER) processes are essential in the discharge and charge cycles, and play a pivotal role in enhancing metal–air batteries performance (Rebrov & Gao, 2023). In aqueous environments, the anode releases electrons through oxidation while oxygen molecules are electrochemically reduced at the cathode, typically forming hydroxide ions. In non-aqueous systems, reduced oxygen species such as superoxides or peroxides may be generated instead. These redox interactions facilitate electron flow through an external load, producing usable electrical energy. A major advantage of MABs lies in their exceptional theoretical energy densities, often exceeding those of traditional lithium-ion batteries. Because atmospheric oxygen serves as the oxidant, there's no need for storing oxygen within the battery, leading to a significant reduction in system weight. This makes metal–air batteries especially appealing for weight-sensitive and high-capacity applications like electric vehicles EVs). Additionally, due to the potential use of earth-abundant, low-toxicity materials, they are often viewed as a sustainable and environmentally responsible energy solution (Guo et al., 2023b). Metal anodes such as Li, Zn, Al, and Mg differ significantly in performance due to differences in reactivity, valency, and compatibility with electrolyte media (Q. Liu et al., 2020). While Li-air batteries offer exceptionally high theoretical energy densities (~3458 $Wh{\cdot}kg^{-1}$), they suffer from instability in aqueous media due to lithium's high reactivity. Zn-air systems, in contrast, provide lower energy densities (~1084 $Wh{\cdot}kg^{-1}$) but exhibit better safety and cost profiles (Salado & Lizundia, 2022). Al-air batteries present a promising compromise with high theoretical energy density and raw material abundance, though they remain mechanically rechargeable due to the irreversibility of $Al^{3+} \rightarrow Al^{0}$ in aqueous systems. A major limitation in multivalent systems (e.g., Ca^{2+}, Mg^{2+}, Al^{3+}) is the sluggish ionic mobility and poor reversibility of the metal plating/stripping process (Salado & Lizundia, 2022). This results primarily from the absence of suitable electrolytes capable of stabilizing multivalent ions.

The selection of electrolyte plays a pivotal role in determining the overall performance, safety, and rechargeability of metal-air batteries (MABs). Depending on their composition and properties, electrolytes can be categorized into several types, each influencing key factors such as efficiency, stability, and the longevity of the battery. Electrolytes can be classified as (J. Lee et al., 2011):

- Aqueous electrolytes: Common in Mg-air, Zn-air, Fe-air, and Al-air batteries. They are cost-effective and eco-friendly but susceptible to carbonate formation in alkaline media, particularly due to CO_2 ingress from ambient air. Neutral or acidic electrolytes (e.g., Leclanché-type) are alternatives that mitigate carbonate formation but may compromise cell integrity or ZnO precipitation in Zn-air systems.

- Non-aqueous electrolytes (aprotic solvent): Used predominantly in Li-air, Na-air, and K-air batteries. They offer broader electrochemical windows and are compatible with high-voltage operation but pose flammability and stability risks.
- Solid-state electrolytes: These provide mechanical stability and avoid leakage or ignition issues associated with liquid systems, while allowing wider compatibility with diverse metal anodes.
- Hybrid electrolytes: Combining aqueous and polymeric or solid components to balance ionic conductivity, electrochemical stability, and structural integrity.

Electrolyte decomposition, passivation layer formation, and oxygen side reactions (e.g., generation of superoxides) pose significant degradation risks (Phogat et al., 2025). In aqueous systems, a solid electrolyte interphase (SEI) can isolate Li metal, reduce self-discharge and allow use of acidic/neutral media.

Separators, typically fabricated from polyethylene (PE), polypropylene (PP), or glass fiber mats, prevent short-circuits while maintaining ionic conductivity. Recent innovations include renewable polymer-based separators and functionalized commercial membranes. For instance, the use of graphene nanosheets and redox mediator-doped separators (e.g., 5,10-dihydro-5,10-dimethylphenazine) has shown improved dendrite suppression and high round-trip efficiency (90% in Li–O_2 systems) (S. H. Lee et al., 2017).

The porous air cathode facilitates ORR and OER reactions using ambient oxygen as an electron acceptor. However, exposure to humidity and CO_2 accelerates cathode degradation through carbonate precipitation and electrode fouling. Optimizing the cathode requires bifunctional catalysts for reversible ORR and OER, high surface area structures for improved mass transport and corrosion resistance in reactive oxygen environments. Transition-metal-based catalysts (e.g., Fe_2N/Fe_3C nanoparticles) and noble-metal-free alternatives such as carbon-based materials have demonstrated promising ORR/OER activity with high stability in acidic media (e.g., half-wave potential of 0.764 V vs. RHE). Electrospun nanofiber cathodes and metal-organic framework (MOF)-derived catalysts (e.g., Mn/Fe-HIB MOFs) exhibit enhanced bifunctional electrocatalytic activity and durability.

Covalent Organic Frameworks (COFs), formed via lightweight organic monomers with defined periodicity and pore structures, offer a complementary strategy. Porphyrin-based COFs enable uniform metal dispersion and stable channels for O_2 transport, achieving extended cycling (180 cycles at 300 $mA{\cdot}g^{-1}$, 1000 $mAh.g^{-1}$ fixed capacity). Super-assembled carbon nanofibers with ZIF-67 decoration have also yielded specific capacities as high as 1635 $mAh.g^{-1}$ at 20 $mA.cm^{-2}$. Despite promising theoretical metrics (e.g., Zn-air: 1353 $Wh.kg^{-1}$), practical devices typically achieve 350–500 Wh.kg^{-1}, limited by air cathode inefficiencies, anode degradation, and electrolyte instability. Commercial Zn-air batteries are mostly primary cells used in low-power devices like hearing aids. Rechargeable configurations face energy efficiencies below 60% due to morphological and chemical degradation pathways.

However, the practical deployment of metal–air batteries faces several significant challenges, with corrosion being one of the most critical. Aqueous-based systems are particularly prone to severe anode corrosion, which adversely affects both the performance and safety of the battery. Corrosion mechanisms include electrode passivation, where oxidation products form insoluble films on the anode, reducing its active surface area and impeding further reactions. Additionally, uneven metal dissolution can cause surface degradation, increasing the risk of short circuits. These processes also hinder the reversibility of the oxidation reaction, contributing to inefficient rechargeability. Moreover, unwanted by-products from corrosion degrade battery materials, reducing the overall cycle life. Although switching to organic electrolytes can help mitigate corrosion, this introduces new concerns, such as flammability and toxicity. Furthermore, non-aqueous systems, particularly lithium–oxygen batteries, are susceptible to parasitic side reactions involving reactive oxygen species formed during discharge and recharge. These include superoxide (O_2^-), a radical species with strong nucleophilic and basic properties that can degrade water and electrolyte solvents; peroxide (O_2^{2-}), which contributes to electrolyte decomposition despite being less reactive than superoxides; and singlet oxygen ($^1O_2$) (Wandt et al., 2016), an excited, highly reactive form of oxygen capable of oxidizing electrolytes and carbon-based cathodes.

Anode corrosion in metal-air batteries is a serious threat. Most of the studies in this category focused on Li-air, Al-air, Mg-air and Zn-air batteries, each offering distinct advantages and limitations. The choice of anode material plays a crucial role in determining the battery's energy density and vulnerability to corrosion. These anode materials, often metals or metal alloys, are exposed to harsh chemical and electrochemical environments during repeated charge and discharge cycles. Over time, this exposure leads to corrosion, manifesting as the degradation of the anode, reduction in active surface area, loss of efficiency, capacity fade, and eventual battery failure.

2.2 Corrosion mechanisms in specific metal-air batteries

2.2.1 Li-air batteries

Lithium-air (Li–air) batteries are regarded as a promising next-generation energy storage system because of their exceptionally high theoretical energy density, which significantly exceeds that of standard lithium-ion batteries (Kwon et al.,

2023; Rani et al., 2024). Central to their architecture is the lithium metal anode, which is favored for its low negative electrochemical potential (−3.045 V vs. SHE) (M. Gao et al., 2021), low density (0.534 $g.cm^{-3}$ at 20°C) (Kurzweil, 2015), and attracted considerable theoretical specific capacity (3842 mAh g^{-1}) (Balaish et al., 2014). The unique operation of Li–O_2 batteries involves complex electrochemical reactions at a dynamic three-phase interface of lithium, electrolyte, and oxygen. Upon discharge, oxygen reduction (ORR) occurs at the cathode where molecular oxygen is first turned into superoxide (O_2^-), followed by the formation of lithium superoxide (LiO_2) and its subsequent transformation into lithium peroxide (Li_2O_2) (Liang et al., 2022). Two principal pathways have been proposed for Li_2O_2 formation:

(i) A disproportionation process: $2\ LiO_2 \rightarrow Li_2O_2 + O_2$
(ii) A direct surface-mediated mechanism: $LiO_2 + Li^+ + e^- \rightarrow Li_2O_2$

The electrolyte's donor number and solvent–ion interactions influence whether Li_2O_2 is formed via a surface-growth or a solution-growth mechanism (H. Wang et al., 2020). During charging, the reverse oxygen evolution reaction (OER) takes place. Here, two main mechanisms are debated: a one-step two-electron process ($Li_2O_2 \rightarrow 2Li^+ + O_2 + 2e^-$) and a two-step one-electron route involving the LiO_2 intermediate. Recent in situ analyses (e.g., Raman spectroscopy and differential electrochemical mass spectroscopy) suggest both mechanisms may coexist, depending on factors like electrolyte composition, catalyst properties, and cathode porosity. The addition of redox mediators (RMs) further complicates the picture but offers significant advantages by lowering overpotentials and suppressing side reactions. RMs operate through reversible redox cycles that chemically oxidize Li_2O_2, offering enhanced energy efficiency but introducing risks like the "shuttle effect" and increased anode corrosion (S. H. Lee et al., 2017; H. Wang et al., 2020).

One of the most formidable barriers to the commercialization of Li–O_2 batteries is the severe corrosion of the lithium metal anode. Figure 2 illustrate the corrosion mechanism in Li-air batteries. Because of the high reactivity of lithium, it readily reacts with electrolyte solvents, oxygen species, water vapor, and impurities (e.g., CO_2) from ambient air that may diffuse through the porous cathode and separator. These parasitic reactions lead to the formation of a thick, heterogeneous, and electrically insulating solid electrolyte interphase (SEI), the growth of inactive "dead Li," and ultimately, a sharp decline in coulombic efficiency, power output, and cycle life. Moreover, during repeated cycling, lithium dendrites tend to form at the anode surface due to non-uniform current density and ion flux, posing safety risks such as internal short-circuiting (Abdollahifar & Paolella, 2025). The combination of lithium corrosion and dendrite growth significantly hinders the structural and electrochemical stability of Li–O_2 batteries. The corrosion problem is exacerbated in the presence of soluble redox mediators. While RMs are critical for promoting efficient Li_2O_2 decomposition, their diffusion toward the lithium anode can result in direct redox reactions with lithium metal, inducing self-discharge and accelerating anode degradation (H. Wang et al., 2020).

Researches have focused on engineering protective layers and electrolyte systems to stabilize the lithium anode. One such strategy involves the use of a durable, hydrophobic composite polymer electrolyte film, which can act as an effective barrier against moisture and oxygen infiltration. These flexible films have demonstrated the ability to significantly reduce lithium corrosion while preserving the fundamental Li–O_2 electrochemistry, improving cycling stability from 24 to 95 cycles (T. Liu et al., 2019). Another innovative approach employs a superhydrophobic and hydrogel-derived porous carbon layer to encapsulate the lithium anode. This design effectively suppresses dendrite formation and corrosion, enabling more stable operation in oxygen-rich and ambient air environments (C. Li et al., 2020).

In brief, while Li–O_2 batteries hold enormous potential as high-energy-density storage systems, their widespread implementation is currently limited by severe lithium corrosion, dendritic growth, and interface instability. The intricate interplay between electrochemical reactions, redox mediators, and the ambient environment necessitates a multi-faceted approach to stabilization. Protective coatings, hydrophobic films, novel electrolyte formulations, and intelligent electrode architectures have emerged as promising solutions. Continued research, particularly with in situ diagnostic techniques and materials innovation, is essential to unravel the complex degradation pathways and to pave the way toward practical and durable Li–O_2 battery systems.

Figure 2. Corrosion mechanism in Li-air batteries.

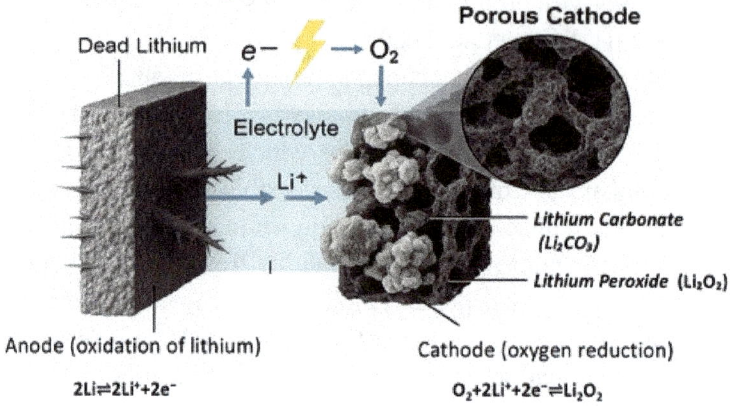

2.2.2 Zn-air batteries

Zinc has long been regarded as an important negative electrode material in the battery industry. Its theoretical capacity (~820 mAh/g), relative abundance, and low toxicity make it attractive for grid-scale and portable energy storage. Zn-based batteries are considered viable competitors to lithium-ion batteries in terms of cost, safety and energy density (Lu et al., 2023). However, the electrochemical instability of zinc in aqueous media is influenced by pH, especially in highly alkaline electrolytes such as 7 M KOH (pH > 14), which severely limits its long-term performance (Lei et al., 2020).

The corrosion mechanism in Zn-based batteries is depicted in figure 5. It arises through two primary pathways: The first is chemical oxidation of Zn, which occurs when Zn interact with aqueous electrolyte, affecting the Zn whether the battery is at rest or in operation. The second process is electrochemical hydrogen evolution corrosion (HER) during charge-discharge cycling of the battery (Cai et al., 2023a). The former dominates in resting states, while the latter is exacerbated during cycling. The HER occurs due to the favorable thermodynamic potential of hydrogen evolution compared to zinc oxidation.

In acidic environments: $2H^+ + 2e^- \rightarrow H_2$

In neutral or alkaline environments: $2H_2O + 2e^- \rightarrow H_2 + 2OH^-$.

While Pourbaix diagrams provide thermodynamic insights into Zn stability, real-world conditions also depend on kinetic factors such as exchange current (i_0) and Tafel slope (b), both of which influence the overpotential required for HER (Lei et al., 2020). In alkaline environments, Zn reacts with OH^- to form soluble species like $Zn(OH)_4^{2-}$, which eventually convert to ZnO. This dense ZnO layer can act as a passivation barrier. However, in neutral or mildly acidic electrolytes (e.g., $ZnSO_4$), Zn tends to form loose, porous compounds such as zinc hydroxysulfate (ZHS), leading to continued corrosion and hydrogen evolution. This difference in corrosion product morphology significantly influences the stability and performance of Zn anodes (Cai et al., 2023b).

To combat corrosion, several engineering strategies have been developed. Alloying Zinc with elements like Al, Bi, In, Sn, and CuO have shown significant improvements in corrosion resistance. For instance, Zn-10Sn alloy electrodes demonstrated low voltage hysteresis and resistance to dendrite formation over prolonged cycling. Similarly, Zn-Bi alloys achieved discharge capacity retention >90%, compared to 74.4% for pure Zn. Protective surface coatings, including Al_2O_3 (via sol-gel methods), polyaniline, and carbon treatments (e.g., Zn/C with glucose-derived carbon) have proven effective in reducing HER and corrosion. Coated electrodes exhibited superior self-discharge behavior and higher capacity retention over 24-hour storage tests. Additionally, the incorporation of electrolyte additives like branched polyethylenimine can modify the local electrolyte environment, suppress Zn corrosion and alter the nucleation kinetics of Zn deposition (Cai et al., 2023b).

Figure 3. Corrosion mechanism in Zn-air batteries.

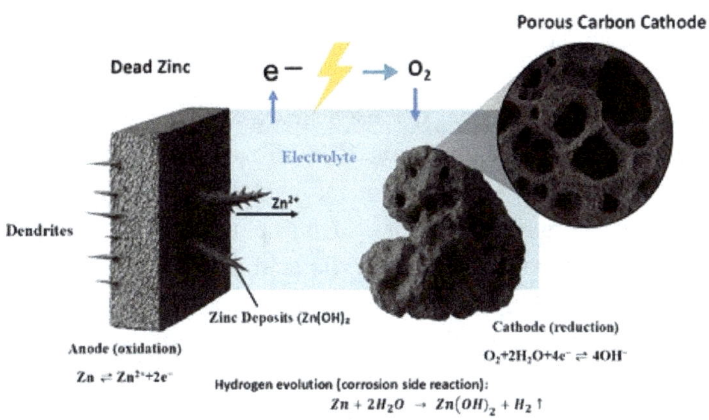

2.2.3 Al-air batteries

Aluminum (Al), the most prevalent metallic element in the Earth's crust, constitutes approximately 8 wt.% of its total composition (Xu et al., 2022). Its unique combination of low cost, light weight, high safety, and favorable electrochemical properties has rendered it an attractive candidate for energy storage systems. Notably, aluminum can exchange three electrons during electrochemical reactions, resulting in an impressive theoretical specific capacity of 2,980 $mAh.g^{-1}$ and a volumetric capacity of 8,040 $mAh.cm^{-3}$ (T.-T. Wei et al., 2021). These characteristics place aluminum-air (Al-air) batteries among the top promising substitutes for conventional lithium-ion batteries, with a theoretical energy density of up to 8,100 $mWh.g^{-1}$ second only to lithium-based systems. Aluminum exhibits several intrinsic advantages relevant to battery technologies. It has a low density (2.7×10^3 $kg.m^{-3}$), good specific strength, and excellent thermal and electrical conductivity. While its conductivity is about 60% that of copper by volume, on a mass basis, aluminum actually surpasses copper. Furthermore, its strong resistance to corrosion stems from the spontaneous formation of a protective oxide film (Al_2O_3) upon exposure to air, a property that also makes it ideal for use as current collectors in lithium-ion batteries (Xu et al., 2022). Despite its many advantages, the deployment of aluminum anodes in battery systems is hindered by corrosion, as illustrated in figure 3. In aqueous electrolytes, aluminum is susceptible to various forms of localized corrosion, including pitting, especially in near-neutral environments such as seawater and humid air. The self-corrosion of aluminum anodes during discharge leads to parasitic hydrogen evolution reactions (HER), which reduce battery efficiency, compromise safety, and contribute to capacity fade (Wen et al., 2020). Some researchers demonstrated that Al-air batteries exhibit performance degradation at elevated temperatures. At 60 °C, the corrosion rate increases significantly due to accelerated electrochemical reactions with the acidic electrolyte, leading to more rapid anode deterioration and capacity fade (Li et al., 2020).

To address these limitations, extensive research has focused on suppressing the HER and mitigating self-corrosion. The utilization of corrosion inhibitors in the electrolyte is one of the most practical approaches (Gelman et al., 2015; H. Yang et al., 2020). Inorganic inhibitors, such as potassium fluoride and potassium hydroxide (KOH), embedded in polyacrylic acid hydrogels, have shown efficacy in reducing $Al(OH)_3$ buildup and enhancing discharge performance (M. Wei et al., 2023). Similarly, hydrogen-bond reconfiguration using glycerol-based electrolytes has been demonstrated to suppress hydrogen evolution through dipolar interactions (Lv et al., 2023). Natural and synthetic additives have also shown promise. Cashew phenol polyoxyethylene ether (CPPE) has been reported to significantly reduce hydrogen evolution and protect aluminum surfaces in alkaline media (Pan et al.,

2025). Other additives, including sodium stannate, cerium chloride, benzoic acids, and even botanical extracts like chrysanthemum coronarium, have demonstrated high corrosion inhibition (Saji, 2023). Protective surface coatings such as conductive nanocomposites e.g., polyaniline combined with carbon nanofibers, have emerged as viable techniques to minimize corrosion (Saji, 2023). Hybrid inorganic-organic coatings have also been developed to improve the hydrophobicity of the anode, effectively shielding it from electrolyte-induced degradation (Sinha et al., 2023). Alloying aluminum with elements possessing high hydrogen overpotentials, such as indium (In), magnesium (Mg), tin (Sn), zinc (Zn), and gallium (Ga), can suppress hydrogen evolution (X. Liu et al., 2019). For instance, Al-Mn alloys with 0.1 wt.% Mn have shown optimal discharge performance by minimizing corrosion in sodium hydroxide (NaOH) solutions. However, excessive Mn leads to the formation of Al_6Mn intermetallics, which paradoxically accelerate HER. Advanced Al alloys incorporating ZrO_2 nanoparticles have shown reduced improper dissolution due to grain boundary segregation, and carbonaceous nanomaterial-reinforced Al matrices offer reversible passivation, further inhibiting self-corrosion.

The fabrication method and microstructural features of Al anodes play a critical role in corrosion resistance. Techniques such as extrusion, rolling, 3D printing, laser sintering, and spray forming have been investigated to refine grain size and enhance phase distribution. Studies have demonstrated that spray formed and heat-treated Al alloys exhibited improved uniformity, reduced corrosion rates, and enhanced electrochemical performance (Reddy et al., 2019; Sheng et al., 2020; Siddesh Kumar et al., 2022). Interestingly, the impact of grain size on corrosion is alloy-dependent. Equal channel angular pressing (ECA) reduced the corrosion rate of pure Al but not of Al-Zn and Al-Zn-In alloys. In these cases, ultrafine grains promoted uniform corrosion, whereas coarse grains led to intergranular attack. The corrosion behavior of aluminum is also influenced by its crystallographic orientation. As-cast pure Al anodes with a (001) lattice plane orientation exhibited better corrosion resistance due to a more uniform potential distribution and lower surface energy compared to other orientations (H. Wang et al., 2022).

Aluminum-air batteries hold immense potential due to aluminum's abundance, cost-effectiveness, environmental compatibility, and high theoretical energy density. However, their practical application is still limited by self-corrosion and parasitic hydrogen evolution. A combination of advanced material design (including alloying and coatings), electrolyte engineering, and structural optimization is required to overcome these hurdles. Future work will likely focus on scalable fabrication methods, long-term stability under various environmental conditions, and the development of recyclable, safe, and high-performance Al-air battery systems that can rival lithium-ion technologies in real-world applications.

Aluminum, similar to zinc, possesses a high theoretical specific capacity and forms passivating oxide layers in aqueous media. The corrosion behavior of Al is likewise influenced by electrolyte pH and anode morphology. Development of Al-air batteries focuses on the following points:

- Passivation vs. Corrosion: Just as Zn forms protective ZnO layers in alkaline conditions, Al tends to form $Al(OH)_3$ or Al_2O_3 films, which may hinder electron transfer and reduce active surface area. Strategies to manage this passive layer are critical to improving cycle life.
- Hydrogen Evolution: The HER in Al-based systems remains a challenge, particularly in mildly acidic or neutral media. Tailoring electrolyte composition and incorporating corrosion inhibitors, as demonstrated in Zn systems, can enhance performance.
- Electrode Engineering: Adoption of porous or high-surface-area Al electrodes must balance enhanced energy density with increased susceptibility to corrosion—mirroring findings in Zn-based batteries.
- Alloy and Coating Approaches: Developing Al alloys or composite electrodes with surface modifications (e.g., carbon, polymers and ceramics) can significantly reduce parasitic reactions and improve utilization efficiency.

The development of aluminum-air batteries, as a promising alternative in the metal-air battery family, stands to benefit immensely from the detailed insight of zinc anode corrosion mechanisms and mitigation strategies. While both systems face common challenges such as passivation and hydrogen evolution, the wealth of knowledge for these systems provides a valuable foundation for the rational design of stable and high-performance Al-air systems. Future research should continue to integrate material engineering, electrolyte chemistry, and interface science for surpassing these limitations and realizing the full potential of metal-air technologies for sustainable energy storage.

Figure 4. Corrosion mechanism in Al-air batteries.

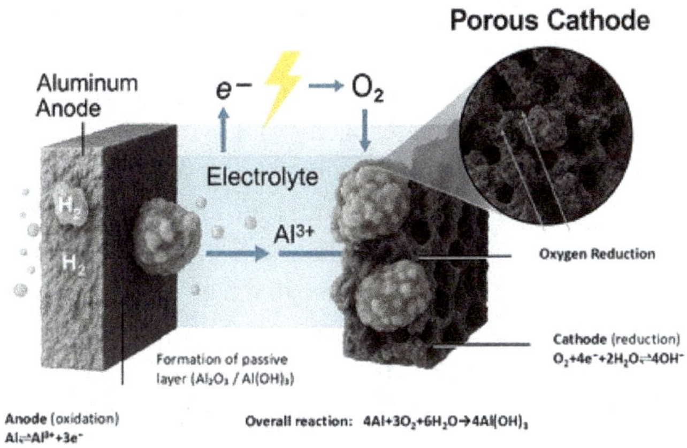

2.2.4 Mg-air batteries

Magnesium-air batteries have attracted substantial interest as promising candidates for high-energy, low-cost, and environmentally benign energy storage systems (X. Liu et al., 2019; Olabi et al., 2021b; Y. Wang et al., 2022). This interest stems from the inherent advantages of magnesium (Mg), including high theoretical voltage (3.1 V), exceptional energy density (6.8 kWh.kg^{-1}), natural abundance, low toxicity, and lightweight characteristics (X. Li et al., 2025). Despite these attractive properties, Mg-air batteries faced several critical challenges, notably high self-corrosion rates, low anodic efficiency, and poor compatibility with aqueous electrolytes. One of the major impediments to the efficient operation of MABs is the negative difference effect (NDE), wherein the self-corrosion and hydrogen evolution reaction (HER) rates escalate with increasing current density and discharge time. Figure 4 illustrates how this degradation manifests through electrochemical dissolution of the Mg substrate and mechanical detachment of metal fragments, commonly known as the chunk effect (CE). Corrosion products formed on the anode surface further compound performance losses by obstructing Mg^{2+} transport and reducing the effective electrochemical area, leading to significant voltage decay over time. (X. Li et al., 2025).

To address these issues, substantial efforts have been directed toward modifying Mg-based anode materials, particularly through alloying strategies. Magnesium alloys such as AZ91 and AZ31 have served as the foundation for many such studies. For instance, Liu et al. demonstrated that AZ91 alloy modified with calcium (Ca), samarium (Sm), and lanthanum (La) exhibited enhanced corrosion resistance and a more positive corrosion potential. The presence of catalytically active Al-(Sm,La,Ca) intermetallic phases promoted anodic reactions under high potentials, while the refined and dispersed second phases improved discharge uniformity and suppressed grain boundary corrosion. This modification led to improved anodic efficiency (61.9%) and a discharge capacity of 1381 mAh.g^{-1} at 40 mA·cm^{-2} (X. Liu et al., 2019). Similarly, Li et al. developed a novel precipitate-free Mg-0.1Sn alloy, rolled at 573 K to achieve a uniform grain structure and basal texture. This material delivered promising results, including a stable cell voltage of 1.626 V at 0.5 mA·cm^{-2} and specific energy of 1730.96 mWh·g^{-1} at 10 mA·cm^{-2}. The reduced chunk effect and suppressed self-corrosion were attributed to the homogeneity of the microstructure (X. Li et al., 2025).

Numerous other alloying elements have also been explored to balance corrosion resistance and discharge activity. Lithium (Li) addition to AZ31 alloy increased anodic efficiency and discharge capacity (62.1% and 1478 mAh·g^{-1}, respectively) due to grain refinement and improved structural uniformity (X. Liu et al., 2018). Zinc (Zn) has also been identified as an effective additive. In Mg-Li-Al-Zn-Ce-Y alloys, Zn reduced micro-galvanic interactions and promoted the formation of Al-Li particles, contributing to anodic efficiencies of around 60.6% at 10 mA·cm^{-2}. Aluminum (Al), another critical component, has been shown to enhance corrosion resistance via the formation of Al-rich β-$Mg_{17}Al_{12}$ phases. For example, as-cast Mg-34%Al exhibited excellent corrosion resistance in saline media, while Mg-Al-Zn and Mg-Al-Sn alloys benefited from grain refinement and reduced hydrogen evolution. The inclusion of tin (Sn) dispersed Mg_2Sn phases along grain boundaries, reducing coarse β-phase precipitation and facilitating uniform dissolution. In, Ga, and Zn additives have been evaluated for their effects on Mg-3Al-based anodes. Mg-3Al-1In alloy achieved high power output (18.5 mW/cm^2) and discharge capacity (1382 mAh.g), demonstrating the value of multi-element alloying in enhancing battery performance. Comparative analysis of die-cast versus hot-rolled Mg-9Al-1Zn alloys revealed superior self-corrosion resistance and discharge performance in the die-cast variant, attributable to its fine, continuous β-phase network and uniform microstructure. Beyond alloying, surface treatments, coatings, and electrolyte modifications represent promising complementary strategies. Conductive nanocomposite coatings, such as those comprising polyaniline (PAni) and functionalized carbon nanotubes (CNTs), have successfully prolonged discharge times and stabilized operating voltages by mitigating anode degradation. The selection of electrolyte or corrosion inhibitor is also crucial; it must inhibit parasitic reactions without impeding Mg^{2+} dissolution necessary for electrochemical operation (Saji, 2023).

Although Mg-air batteries possess substantial theoretical potential as next-generation energy storage systems, their practical deployment demands significant improvements in anode material performance and corrosion management. Alloying, microstructural engineering, and surface/electrolyte optimization are converging as synergistic approaches to unlock the full potential of MABs. Ongoing research focused on understanding the fundamental mechanisms governing corrosion and discharge behavior will be critical to achieving durable and efficient magnesium-based batteries suitable for real-world applications.

Figure 5. Corrosion mechanism in Mg-air batteries.

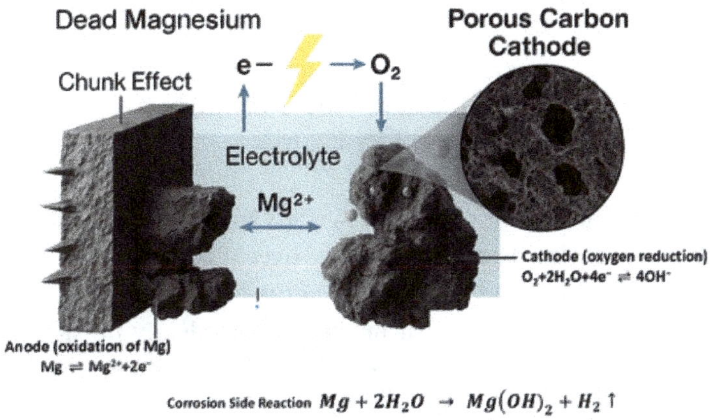

To wrap up, table 1 presents the theoretical performance of the different metal–air batteries, highlighting their potential advantages and limitations (Rai et al., 2025). Each battery type has distinct characteristics that make it suitable for specific applications, but they also face significant challenges that hinder their commercial viability.

Table 1. Comparison of theoretical performance, advantages and limitations of different metal–air batteries.

Battery system	Theoretical energy density (Wh.kg^{-1})	Main advantages	Main limitations
Li–air	11,140	Highest energy density, lightweight	Severe Li corrosion, dendrites, unstable SEI
Zn–air	1350	Low cost, safe, abundant	Hydrogen evolution, ZnO passivation, limited cycle life
Al–air	8100	Abundant, high capacity, low cost	Self-corrosion, hydrogen evolution, poor rechargeability
Mg–air	6800	High voltage, environmentally benign	Rapid self-corrosion, chunk effect, poor anodic efficiency

3. MITIGATION STRATEGIES FOR METAL-AIR BATTERIES CORROSION

When local corrosion begins and spreads in real-world conditions, it results in a significant loss of discharge capacity over time. Moreover, during charging, the corrosion process behaves like a pseudo-platform, causing charging times to be extended. As corrosion current increases, it gradually impairs the delithiation of the cathode. The build-up of corrosion products further elevates the interface impedance, which ultimately leads to decreased current efficiency (S. Yang et al., 2024). Corrosion remains a critical challenge for metal–air batteries, and has a serious impact on the performance, stability, and durability of the metal air batteries. There are several ways to mitigate the effects of corrosion (Dilshad et al., 2025; Hopkins & Rolison, 2021; Saji, 2024; H.-F. Wang & Xu, 2019; S. Yang et al., 2024; You et al., 2023), including the use corrosion resistant anode materials, protective surface coating, including electrolyte additives, corrosion inhibitors, and sealing / battery design strategies (Du et al., 2023; Han et al., 2021; Lei et al., 2020). Proper maintenance and favorable storage conditions are also key to reducing corrosion-induced damage. The most studied mitigation strategy for anode corrosion in Li, Al, Mg and Zn-air batteries is alloying (P. Wu et al., 2021; X. Zhang et al., 2024). Several works studied novel alloy anodes and reported significantly reduced self-corrosion and hydrogen evolution with enhanced anode efficacy. Alloy processing methods, such as extrusion, rolling, 3D printing and laser sintering, were shown to be effective in improving the battery performance via improvement in corrosion resistance attributed to the refinement in microstructure, grain size, and secondary phases distribution. A few works, however, suggested an adverse effect of the grain size reduction strategy. Several studies investigated novel electrolytes, electrolyte additives, corrosion inhibitors and protective coatings for better corrosion resistance and battery performance. Optimizing mixed electrolytes could be helpful to optimize the tradeoff in corrosion inhibition and anodic dissolution. Table 2 below presents a comprehensive overview of the major mitigation strategies currently employed or under investigation. Each method is described in terms of its principles and advantages, while also highlighting the associated challenges and considerations essential for practical application and further development.

Table 2. Mitigation strategies and methods against corrosion in metal–air batteries (MABs).

Strategy	Method	Details	Challenges/Considerations
Material Selection	Corrosion-resistant anode materials	Use of bismuth, indium-zinc alloys, or carbon-based materials to enhance corrosion resistance, chemical stability, and battery life.	May compromise electrical conductivity or energy density; higher material cost.
	Composite anodes	Anodes combining corrosion-resistant and high-performance materials provide improved durability and tolerance to harsh electrolytes.	Fabrication complexity and material compatibility issues.
	High-performance composites	Multifunctional materials combine corrosion resistance with strength and conductivity for advanced battery designs.	High-performance composites
Surface Engineering	Coatings	Passivation layers, ceramic or carbon-based coatings (e.g., graphene) physically block electrolyte-anode interaction, increasing corrosion resistance and efficiency.	Coating must be thin, uniform, conductive and compatible with the anode material; risk of delamination during cycling.
	Nanocoatings	Nanocoatings create protective layers that prevent direct contact between the anode material and the corrosive environment.	Long-term stability and reproducibility can be challenging; costly techniques.
	Polymer-coated anodes	Polymers like polyethylene oxide form protective films that resist corrosion and maintain ionic conductivity for lithium-based systems.	Limited mechanical robustness and thermal stability.

continued on following page

Table 2. Continued

Strategy	Method	Details	Challenges/Considerations
Electrolyte Optimization	Electrolyte formulations	Tailored electrolytes with high ionic and low electronic conductivity reduce side reactions and chemical attack on anodes.	Trade-off between conductivity, stability, and cost; moisture sensitivity.
	Corrosion inhibitors	Introduced into the electrolyte or as part of the anode material and work to suppress corrosion and passivate the anode surface.	Risk of compatibility issues with battery chemistry or electrode materials.
Battery Design & Sealing	Enclosure design	Sealed and robust enclosures reduce exposure to air and moisture, limiting corrosion and enhancing safety.	cost; risk of internal pressure buildup if not properly vented.
	Gaskets and seals	Use of chemical-resistant seals (e.g., silicone) to block environmental ingress and ensure battery longevity.	Material degradation over time; must match electrolyte chemistry.
	Venting mechanisms	Safety vents relieve pressure from gas buildup, preventing rupture or corrosion-related failure.	Must balance gas release with maintaining enclosure integrity.
	Electrolyte containment	Secure containment prevents leakage and evaporation, reducing corrosion of internal components.	Electrolyte compatibility with container materials and seal integrity.
	Physical barriers & diaphragms	Separators and membranes prevent short circuits and cross-contamination, enhancing overall corrosion resistance.	Risk of blockage or degradation over cycles; material cost.
	Electrolyte leak detection	Sensors enable early identification of leaks, allowing fast response to prevent corrosion-related failures.	cost; sensor lifespan and reliability concerns.
	Quality control & certification	Standardized tests ensure corrosion resistance and reliability of materials and systems before use.	Time- and cost-intensive; not always predictive of long-term field performance.
Thermal & Mechanical Control	Temperature management systems	Heat sinks and insulators regulate temperature, minimizing thermal corrosion and preserving battery efficiency.	Cost; Added complexity and weight; requires space and energy.
	Anti-dendrite measures	Use of engineered separators and materials prevents dendritic growth that could lead to corrosion and failure.	Full prevention of dendrite formation is still a challenge; ongoing research area.

4. FUTURE SCOPE

In the coming years, the development of metal–air batteries will increasingly rely on a global and multidisciplinary approach. Physics-based models, particularly those describing electrochemical and transport phenomena at multiple scales, will need to be combined with artificial intelligence tools. This integration will make it possible not only to predict corrosion and aging mechanisms, but also to optimize usage profiles and thermal management in real time, through emerging concepts such as digital twins and hybrid models that merge data-driven methods with physical laws.

At the same time, the wider use of in situ and operando techniques will become indispensable to directly monitor the evolution of the anode–electrolyte interface and to link structural transformations with electrochemical performance. Such experimental insights can feed into machine learning algorithms, providing a deeper understanding of the complex interactions that govern electrode stability.

From a materials perspective, several directions are particularly promising: the development of advanced alloys and functional coatings with self-healing or corrosion-resistant properties, the design of green electrolytes enriched with bio-based corrosion inhibitors, and the application of computational approaches to rapidly explore new families of polymers and protective additives.

Finally, scaling up to industrial production will require a strong focus on material recyclability, energy-efficient manufacturing processes, and overall environmental impact. Artificial intelligence and digital platforms will play a central role in this transition, enabling life-cycle tracking of the devices and accelerating their path from laboratory prototypes to real-world applications, ranging from stationary energy storage to sustainable mobility.

5. CONCLUSION

Metal-air batteries represent a transformative technology bridging the gap between high-capacity energy storage and sustainable materials use. Despite their complexity and current limitations, innovations in materials science, electrochemistry, and cell engineering are rapidly driving metal air-batteries toward practical application. With continued interdisciplinary efforts, metal air-batteries could become central to the future of green, efficient, and clean energy storage systems. In this chapter, we explored the corrosion mechanisms and materials degradation challenges in various metal–air battery systems, including Al–air, Li–air, Mg–air, and Zn–air batteries. The practical deployment of these systems is hindered by severe corrosion phenomena and instability at the metal anodes. Each battery type exhibits distinct degradation pathways depending on the metal's reactivity and the electrolyte environment. Finally, a range of mitigation strategies were reviewed, highlighting the need for integrated approaches to enhance battery durability and performance. Understanding and controlling corrosion remains pivotal to unlocking the full potential of metal–air batteries for sustainable energy storage.

REFERENCES

Abdollahifar, M., & Paolella, A. (2025). "Dead Lithium" Formation and Mitigation Strategies in Anode-Free Li-Metal Batteries. *Batteries & Supercaps*, *8*(3), e202400505. DOI: 10.1002/batt.202400505

Balaish, M., Kraytsberg, A., & Ein-Eli, Y. (2014). A critical review on lithium–air battery electrolytes. *Physical Chemistry Chemical Physics*, *16*(7), 2801. DOI: 10.1039/c3cp54165g

Barman, M., Pal, M., Biswas, R., & Dutta, A. (2025). A comprehensive review of metal-air batteries : Mechanistic aspects, advantages and challenges. *Catalysis Today*, *451*, 115229. DOI: 10.1016/j.cattod.2025.115229

Bhattacharjee, U., Ghosh, S., Bhar, M., & Martha, S. K. (2023). Electrochemical energy storage part I : Development, basic principle and conventional systems. In *Emerging Trends in Energy Storage Systems and Industrial Applications* (pp. 151–188). Elsevier., DOI: 10.1016/B978-0-323-90521-3.00001-6

Brandstätter, H., Hanzu, I., & Wilkening, M. (2016). Myth and Reality about the Origin of Inductive Loops in Impedance Spectra of Lithium-Ion Electrodes—A Critical Experimental Approach. *Electrochimica Acta*, *207*, 218–223. DOI: 10.1016/j.electacta.2016.03.126

Cai, Z., Wang, J., & Sun, Y. (2023a). Anode corrosion in aqueous Zn metal batteries. *eScience, 3*(1), 100093. DOI: 10.1016/j.esci.2023.100093

Cai, Z., Wang, J., & Sun, Y. (2023b). Anode corrosion in aqueous Zn metal batteries. *eScience, 3*(1), 100093. DOI: 10.1016/j.esci.2023.100093

Chen, Y., Xu, J., He, P., Qiao, Y., Guo, S., Yang, H., & Zhou, H. (2022). Metal-air batteries : Progress and perspective. *Science Bulletin*, *67*(23), 2449–2486. DOI: 10.1016/j.scib.2022.11.027

Chung, Y., Lee, S., Yoon, S., Choi, W.-K., & Jeong, S.-K. (2025). Mitigating anodic corrosion in aluminum–air batteries : Effects of alkali metal cation size on electrochemical performance. *Applied Surface Science*, *681*, 161533. DOI: 10.1016/j.apsusc.2024.161533

Detka, K., & Górecki, K. (2023). Selected Technologies of Electrochemical Energy Storage—A Review. *Energies*, *16*(13), 5034. DOI: 10.3390/en16135034

Dilshad, M., Li, T., Lee, S.-L., & Qin, L. (2025). Next-Generation Aluminum-Air Batteries : Integrating New Materials and Technologies for Superior Performance. *ACS Applied Energy Materials*, *8*(6), 3248–3275. DOI: 10.1021/acsaem.4c02926

Dissanayake, K., & Kularatna-Abeywardana, D. (2024). A review of supercapacitors : Materials, technology, challenges, and renewable energy applications. *Journal of Energy Storage*, *96*, 112563. DOI: 10.1016/j.est.2024.112563

Du, P., Liu, D., Chen, X., Xie, H., Qu, X., Wang, D., & Yin, H. (2023). Research progress towards the corrosion and protection of electrodes in energy-storage batteries. *Energy Storage Materials*, *57*, 371–399. DOI: 10.1016/j.ensm.2023.02.028

Fang, Z., Duan, Q., Peng, Q., Wei, Z., Cao, H., Sun, J., & Wang, Q. (2022). Comparative study of chemical discharge strategy to pretreat spent lithium-ion batteries for safe, efficient, and environmentally friendly recycling. *Journal of Cleaner Production*, *359*, 132116. DOI: 10.1016/j.jclepro.2022.132116

Gao, L., Liu, F., Qi, J., Gao, W., & Xu, G. (2025). Recent Advances and Challenges in Hybrid Supercapacitors Based on Metal Oxides and Carbons. *Inorganics*, *13*(2), 49. DOI: 10.3390/inorganics13020049

Gao, M., Li, H., Xu, L., Xue, Q., Wang, X., Bai, Y., & Wu, C. (2021). Lithium metal batteries for high energy density : Fundamental electrochemistry and challenges. *Journal of Energy Chemistry*, *59*, 666–687. DOI: 10.1016/j.jechem.2020.11.034

Gelman, D., Lasman, I., Elfimchev, S., Starosvetsky, D., & Ein-Eli, Y. (2015). Aluminum corrosion mitigation in alkaline electrolytes containing hybrid inorganic/organic inhibitor system for power sources applications. *Journal of Power Sources*, *285*, 100–108. DOI: 10.1016/j.jpowsour.2015.03.048

Guo, X., Tao, Y., Fang, Y., Ren, H., Long, X., Zhang, X., Zhang, W., & Shao, J. (2023a). Surface reconstruction of SrCoO2.52 nanofibers as bifunctional oxygen electrocatalyst for rechargeable Zn-air batteries. *International Journal of Hydrogen Energy*, *48*(83), 32385–32395. DOI: 10.1016/j.ijhydene.2023.04.284

Guo, X., Tao, Y., Fang, Y., Ren, H., Long, X., Zhang, X., Zhang, W., & Shao, J. (2023b). Surface reconstruction of SrCoO2.52 nanofibers as bifunctional oxygen electrocatalyst for rechargeable Zn-air batteries. *International Journal of Hydrogen Energy*, *48*(83), 32385–32395. DOI: 10.1016/j.ijhydene.2023.04.284

Han, Z., Zhang, C., Lin, Q., Zhang, Y., Deng, Y., Han, J., Wu, D., Kang, F., Yang, Q., & Lv, W. (2021). A Protective Layer for Lithium Metal Anode : Why and How. *Small Methods*, *5*(4), 2001035. Advance online publication. DOI: 10.1002/smtd.202001035

Hasan, M. M., Haque, R., Jahirul, M. I., Rasul, M. G., Fattah, I. M. R., Hassan, N. M. S., & Mofijur, M. (2025). Advancing energy storage : The future trajectory of lithium-ion battery technologies. *Journal of Energy Storage*, *120*, 116511. DOI: 10.1016/j.est.2025.116511

Hopkins, B. J., & Rolison, D. R. (2021). Quantifying an acceptable open-circuit corrosion current for aluminum–air batteries. *Materials Advances*, *2*(5), 1595–1599. DOI: 10.1039/D0MA01002B

Kurzweil, P. (2015). Lithium Battery Energy Storage : State of the Art Including Lithium–Air and Lithium–Sulfur Systems. In *Electrochemical Energy Storage for Renewable Sources and Grid Balancing* (p. 269-307). Elsevier. DOI: 10.1016/B978-0-444-62616-5.00016-4

Kwon, H., Kim, H., Hwang, J., Oh, W., Roh, Y., Shin, D., & Kim, H.-T. (2023). Borate–pyran lean electrolyte-based Li-metal batteries with minimal Li corrosion. *Nature Energy*, *9*(1), 57–69. DOI: 10.1038/s41560-023-01405-6

Lee, J., Tai Kim, S., Cao, R., Choi, N., Liu, M., Lee, K. T., & Cho, J. (2011). Metal–Air Batteries with High Energy Density : Li–Air versus Zn–Air. *Advanced Energy Materials*, *1*(1), 34–50. DOI: 10.1002/aenm.201000010

Lee, S. H., Park, J., Lim, H., & Sun, Y. (2017). An Advanced Separator for Li–O_2 Batteries : Maximizing the Effect of Redox Mediators. *Advanced Energy Materials*, *7*(18), 1602417. Advance online publication. DOI: 10.1002/aenm.201602417

Lei, L., Sun, Y., Wang, X., Jiang, Y., & Li, J. (2020). Strategies to Enhance Corrosion Resistance of Zn Electrodes for Next Generation Batteries. *Frontiers in Materials*, *7*, 96. Advance online publication. DOI: 10.3389/fmats.2020.00096

Li, C., Wei, J., Qiu, K., & Wang, Y. (2020). Li–air Battery with a Superhydrophobic Li-Protective Layer. *ACS Applied Materials & Interfaces*, *12*(20), 23010–23016. DOI: 10.1021/acsami.0c05494

Li, X., Cheng, W., Li, J., Shangguan, F., Yu, H., Wang, L., Li, H., Wang, H., Wang, J., & Hou, H. (2025). Electrochemical behavior and discharge performance of as-rolled precipitate-free Mg−Sn alloy as anode for Mg−air batteries. *Transactions of Nonferrous Metals Society of China*, *35*(3), 832–848. DOI: 10.1016/S1003-6326(24)66717-8

Li, Z., Li, J., Li, X., & Zhang, W. (2020). Pseudocapacitance effect in Al-C batteries with expanded graphite positive electrode at different temperatures. *Journal of Power Sources*, *467*, 228323. DOI: 10.1016/j.jpowsour.2020.228323

Liang, Z., Wang, W., & Lu, Y.-C. (2022). The path toward practical Li-air batteries. *Joule*, *6*(11), 2458–2473. DOI: 10.1016/j.joule.2022.10.008

Liu, C., Sun, Y., Nie, J., Dong, D., Xie, J., & Zhao, X. (2020). Stable cycling of a Prussian blue-based Na/Zn hybrid battery in aqueous electrolyte with a wide electrochemical window. *New Journal of Chemistry*, *44*(11), 4639–4646. DOI: 10.1039/C9NJ06024C

Liu, Q., Pan, Z., Wang, E., An, L., & Sun, G. (2020). Aqueous metal-air batteries : Fundamentals and applications. *Energy Storage Materials*, *27*, 478–505. DOI: 10.1016/j.ensm.2019.12.011

Liu, T., Feng, X., Jin, X., Shao, M., Su, Y., Zhang, Y., & Zhang, X. (2019). Protecting the Lithium Metal Anode for a Safe Flexible Lithium-Air Battery in Ambient Air. *Angewandte Chemie*, *131*(50), 18408–18413. DOI: 10.1002/ange.201911229

Liu, X., Liu, S., & Xue, J. (2018). Discharge performance of the magnesium anodes with different phase constitutions for Mg-air batteries. *Journal of Power Sources*, *396*, 667–674. DOI: 10.1016/j.jpowsour.2018.06.085

Liu, X., Xue, J., Zhang, P., & Wang, Z. (2019). Effects of the combinative Ca, Sm and La additions on the electrochemical behaviors and discharge performance of the as-extruded AZ91 anodes for Mg-air batteries. *Journal of Power Sources*, *414*, 174–182. DOI: 10.1016/j.jpowsour.2018.12.092

Lu, Q., Zou, X., Bu, Y., An, L., Wang, Y., & Shao, Z. (2023). What matters in engineering next-generation rechargeable Zn-air batteries? *Next Energy*, *1*(2), 100025. DOI: 10.1016/j.nxener.2023.100025

Lv, C., Zhu, Y., Li, Y., Zhang, Y., Kuang, J., Tang, Y., Li, H., & Wang, H. (2023). Hydrogen-bonds reconstructing electrolyte enabling low-temperature aluminum-air batteries. *Energy Storage Materials*, *59*, 102756. DOI: 10.1016/j.ensm.2023.03.034

Mohammadipour, E., & Gharahcheshmeh, M. H. (2025). Corrosion and Protection in Lithium-Ion and Sodium-Ion Batteries : Principles and Applications of Graphene and Conducting Polymers as Anticorrosion Coatings. *Next Research*, *100789*(4), 100789. Advance online publication. DOI: 10.1016/j.nexres.2025.100789

Olabi, A. G., Sayed, E. T., Wilberforce, T., Jamal, A., Alami, A. H., Elsaid, K., Rahman, S. M. A., Shah, S. K., & Abdelkareem, M. A. (2021a). Metal-Air Batteries—A Review. *Energies*, *14*(21), 7373. DOI: 10.3390/en14217373

Olabi, A. G., Sayed, E. T., Wilberforce, T., Jamal, A., Alami, A. H., Elsaid, K., Rahman, S. M. A., Shah, S. K., & Abdelkareem, M. A. (2021b). Metal-Air Batteries—A Review. *Energies*, *14*(21), 7373. DOI: 10.3390/en14217373

Pan, C., Wang, Z., Cai, S., Zhu, Y., & Zhang, D. (2025). Inhibition effect of cashew phenol polyoxyethylene ether on self-corrosion of anode of alkaline aluminum-air batteries. *Industrial Crops and Products, 228*, 120925. DOI: 10.1016/j.indcrop.2025.120925

Phogat, P., Dey, S., & Wan, M. (2025). Powering the sustainable future : A review of emerging battery technologies and their environmental impact. *RSC Sustainability, 3*(8), 3266–3306. DOI: 10.1039/D5SU00127G

Queiroz, M. B., Duarte, R. C. O., Lima, J. F., & Vilar, E. O. (2024). Comparative evaluation of grid corrosion of lead-acid batteries formed by casting and rolling process. *Journal of Power Sources, 596*, 234113. DOI: 10.1016/j.jpowsour.2024.234113

Rai, S., Prachi, P., Kant, C. R., & Jha, R. (2025). Metal-air batteries : From fundamental mechanisms to practical applications. *Journal of Alloys and Compounds, 1033*, 180805. DOI: 10.1016/j.jallcom.2025.180805

Rani, B., Yadav, J. K., Saini, P., Pandey, A. P., & Dixit, A. (2024). Aluminum–air batteries : Current advances and promises with future directions. *RSC Advances, 14*(25), 17628–17663. DOI: 10.1039/D4RA02219J

Rebrov, E. V., & Gao, P.-Z. (2023). Molecular Catalysts for OER/ORR in Zn–Air Batteries. *Catalysts, 13*(9), 1289. DOI: 10.3390/catal13091289

Reddy, B. V. R., Maity, S. R., & Pandey, K. M. (2019). Characterization of spray formed Al-alloys—A Review. *Reviews on Advanced Materials Science, 58*(1), 147–158. DOI: 10.1515/rams-2019-0013

Saji, V. S. (2023). Corrosion and Materials Degradation in Electrochemical Energy Storage and Conversion Devices. *ChemElectroChem, 202300136*, e202300136. Advance online publication. DOI: 10.1002/celc.202300136

Saji, V. S. (Ed.). (2024). *Corrosion and Degradation in Fuel Cells, Supercapacitors and Batteries*. Springer Nature Switzerland., DOI: 10.1007/978-3-031-57012-4

Salado, M., & Lizundia, E. (2022). Advances, challenges, and environmental impacts in metal–air battery electrolytes. *Materials Today. Energy, 28*, 101064. DOI: 10.1016/j.mtener.2022.101064

Sebbani, I., Ettouhami, M. K., & Boulakhbar, M. (2025). Fuel cells : A technical, environmental, and economic outlook. *Cleaner Energy Systems, 10*, 100168. DOI: 10.1016/j.cles.2024.100168

Serebrennikova, I., Paramasivam, I., Roy, P., Wei, W., Virtanen, S., & Schmuki, P. (2009). Steel corrosion in alkaline batteries. *Electrochimica Acta, 54*(22), 5216–5222. DOI: 10.1016/j.electacta.2009.02.022

Shabeer, Y., Madani, S. S., Panchal, S., Mousavi, M., & Fowler, M. (2025). Different Metal–Air Batteries as Range Extenders for the Electric Vehicle Market : A Comparative Study. *Batteries, 11*(1), 35. DOI: 10.3390/batteries11010035

Sharmoukh, W. (2025). Redox flow batteries as energy storage systems : Materials, viability, and industrial applications. *RSC Advances, 15*(13), 10106–10143. DOI: 10.1039/D5RA00296F

Sheng, X., Yang, Y., Cheng, Y., Li, J., Wu, W., Liu, Y., Li, K., Zhao, Y., & He, G. (2020). Mechanism for enhancing the corrosion resistance in a spray deposition aluminum alloy Al-8.31Zn-2.46Cu-2.07Mg-0.12Zr. *Journal of Alloys and Compounds, 845*, 156198. DOI: 10.1016/j.jallcom.2020.156198

Siddesh Kumar, N. M., Dhruthi, , Pramod, G. K., Samrat, P., & Sadashiva, M. (2022). A Critical Review on Heat Treatment of Aluminium Alloys. *Materials Today: Proceedings, 58*, 71–79. DOI: 10.1016/j.matpr.2021.12.586

Sinha, A. P., Thomas, T. S., & Mandal, D. (2023). An inorganic-organic protective anode interface towards high-performance Al-air battery. *Energy Storage Materials, 63*, 102988. DOI: 10.1016/j.ensm.2023.102988

Sousa, E. G., Campos, G. M., Quaresma, L. S., Mota, T. F. M., De Castilhos Ghisi, N., Gomes, G. C., Santos, R. C. V., De Souza, B. G. R., Guédon, É., De Castro Soares, S., Da Cruz Ferraz Dutra, J., & De Carvalho Azevedo, V. A. (2025). Exploring Bacillus velezensis in a biomedical context : A systematic review. *Academia Molecular Biology and Genomics, 2*(1). Advance online publication. DOI: 10.20935/AcadMolBioGen7598

Wandt, J., Jakes, P., Granwehr, J., Gasteiger, H. A., & Eichel, R. (2016). Singlet Oxygen Formation during the Charging Process of an Aprotic Lithium–Oxygen Battery. *Angewandte Chemie International Edition, 55*(24), 6892–6895. DOI: 10.1002/anie.201602142

Wang, H., Wang, X., Li, M., Zheng, L., Guan, D., Huang, X., Xu, J., & Yu, J. (2020). Porous Materials Applied in Nonaqueous Li–O_2 Batteries : Status and Perspectives. *Advanced Materials, 32*(44), 2002559. Advance online publication. DOI: 10.1002/adma.202002559

Wang, H., Yang, W., Zhang, B., Zhang, Z., Wu, C., & Wu, J. (2022). Effect of Crystal Orientation on the Corrosion Behavior of As-Cast Pure Aluminum Anodes in Air Batteries. *Journal of Materials Engineering and Performance*, *31*(5), 3584–3593. DOI: 10.1007/s11665-021-06531-4

Wang, H.-F., & Xu, Q. (2019). Materials Design for Rechargeable Metal-Air Batteries. *Matter*, *1*(3), 565–595. DOI: 10.1016/j.matt.2019.05.008

Wang, Y., Pan, W., Leong, K. W., Luo, S., Zhao, X., & Leung, D. Y. C. (2023). Solid-state Al-air battery with an ethanol gel electrolyte. *Green Energy & Environment*, *8*(4), 1117–1127. DOI: 10.1016/j.gee.2021.05.011

Wang, Y., Sun, Y., Ren, W., Zhang, D., Yang, Y., Yang, J., Wang, J., Zeng, X., & NuLi, Y. (2022). Challenges and prospects of Mg-air batteries : A review. *Energy Materials*, *2*(4), 200024. DOI: 10.20517/energymater.2022.20

Wei, M., Wang, K., Zhong, L., Pham, T. H. M., Zuo, Y., Wang, H., Zhang, P., Chen, Z., Zhao, S., & Pei, P. (2023). F− competitive attack decomposing parasitic product Al(OH)3 of hydrogel-based Al-air battery. *Energy Storage Materials*, *60*, 102812. DOI: 10.1016/j.ensm.2023.102812

Wei, T.-T., Peng, P., Qi, S.-Y., Zhu, Y.-R., & Yi, T.-F. (2021). Advancement of technology towards high-performance non-aqueous aluminum-ion batteries. *Journal of Energy Chemistry*, *57*, 169–188. DOI: 10.1016/j.jechem.2020.08.035

Wen, H., Liu, Z., Qiao, J., Chen, R., Zhao, R., Wu, J., Qiao, G., & Yang, J. (2020). High energy efficiency and high power density aluminum-air flow battery. *International Journal of Energy Research*, *44*(9), 7568–7579. DOI: 10.1002/er.5485

Wu, L., Zhang, F.-S., Zhang, Z.-Y., & Zhang, C.-C. (2023a). Corrosion behavior and corrosion inhibition performance of spent lithium-ion battery during discharge. *Separation and Purification Technology*, *306*, 122640. DOI: 10.1016/j.seppur.2022.122640

Wu, L., Zhang, F.-S., Zhang, Z.-Y., & Zhang, C.-C. (2023b). Corrosion behavior and corrosion inhibition performance of spent lithium-ion battery during discharge. *Separation and Purification Technology*, *306*, 122640. DOI: 10.1016/j.seppur.2022.122640

Wu, P., Wu, S., Sun, D., Tang, Y., & Wang, H. (2021). A Review of Al Alloy Anodes for Al–Air Batteries in Neutral and Alkaline Aqueous Electrolytes. *Acta Metallurgica Sinica*, *34*(3), 309–320. DOI: 10.1007/s40195-020-01140-x

Xu, Y., Peng, Z., Ding, D., Zhang, W., Gao, Y., Chen, G., Xie, Y., & Liao, Y. (2022). Effect of La Addition on Microstructure and Properties of Al-0.2Fe-0.06Cu Alloy. *Metals*, *12*(2), 211. DOI: 10.3390/met12020211

Yang, H., Li, X., Wang, Y., Gao, L., Li, J., Zhang, D., & Lin, T. (2020). Excellent performance of aluminium anode based on dithiothreitol additives for alkaline aluminium/air batteries. *Journal of Power Sources*, *452*, 227785. DOI: 10.1016/j.jpowsour.2020.227785

Yang, S., Zhong, J., Li, S., & Li, B. (2024). Revisiting aluminum current collector in lithium-ion batteries : Corrosion and countermeasures. *Journal of Energy Chemistry*, *89*, 610–634. DOI: 10.1016/j.jechem.2023.09.036

You, X., Zhang, X., Yu, C., Chen, Y., Li, H., Hou, Y., Tian, L., Yang, N., & Xie, G. (2023). Corrosion protection of magnesium alloys anode by cerium-based anodization coating in magnesium-air battery. *Journal of Rare Earths*, *41*(3), 471–476. DOI: 10.1016/j.jre.2022.04.011

Zhang, S. S., & Jow, T. R. (2002). Aluminum corrosion in electrolyte of Li-ion battery. *Journal of Power Sources*, *109*, 458–464. DOI: 10.1016/S0378-7753(02)00110-6

Zhang, X., Liu, Y., Zhao, X., Cheng, Z., & Mu, X. (2024). Recent Advances and Perspectives of High-Entropy Alloys as Electrocatalysts for Metal-Air Batteries. *Energy & Fuels*, *38*(20), 19236–19252. DOI: 10.1021/acs.energyfuels.4c03386

Zhou, B., Yang, Z., Lu, L., Fang, B., Wilkinson, D. P., Zhang, J., & Rao, Z. (2025). Corrosion mechanisms and mitigation strategies of lithium metal anodes for liquid lithium batteries. *Matter*, *8*(3), 101996. DOI: 10.1016/j.matt.2025.101996

Chapter 4
Effects of Corrosion on the Durability of Battery

Sara Rached
https://orcid.org/0009-0009-0163-1553
Ibn Tofail University, Morocco

Malak Rehioui
https://orcid.org/0000-0001-7833-6426
Faculty of Sciences, Ibn Tofail University, Morocco

Nouhaila Ferraa
https://orcid.org/0000-0003-1817-7798
Faculty of Science, Ibn Tofail University, Morocco

Amar Habsaoui
Faculty of Science, Ibn Tofail University, Morocco

Khaoula Mzioud
https://orcid.org/0009-0007-8781-5024
Faculty of Science, Ibn Tofail University, Morocco

Mohamed Ebn Touhami
Faculty of Science, Ibn Tofail University, Morocco

ABSTRACT

The term "corrosion" refers to the degradation of metallic materials (metals and their alloys) caused by a reaction with environmental factors. This phenomenon causes serious financial and health problems. Corrosion is one of the electrochemical phenomena that manifests itself as the formation of clusters on the surface of metals, where one electrode, known as the "anode", consumes the other, known as the "cathode", which remains dormant. In reality, these clusters form as soon as the metal surface becomes heterogeneous, forming the anode and cathode domains where the two primary electrochemical reactions (oxidation and reduction) take place. For example, dissolution or passivation of active electrode materials and dissolution, oxidation or passivation of current collectors are the main causes of corrosion in batteries. An in-depth analysis of battery corrosion is necessary due to the rapid development of battery research. We begin with an overview of recent developments in corrosion and electrode protection for different battery types.

DOI: 10.4018/979-8-3373-3156-0.ch004

1 INTRODUCTION

Corrosion is a natural process whereby materials, especially metals, lose their structural and functional properties little by little through chemical or electrochemical reactions with their environment (Rached et al., 2023, 2024a). Considering that energy storage systems rely on corrosion for some of their key components, such as electrodes, electrolyte solutions, and protective coatings, all of which are required to enable optimum performance and safety (Kleiner & Ehrenberg, 2019; Rached et al., 2024b; Sanguesa et al., 2021). Corrosion degradation can be caused in a variety of ways: it can be manifested in terms of a cycle life reduction, an increase in internal resistance, a capacity loss, or-even worse-safety risks, such as thermal runaway and short circuiting (Mzioud et al., 2020) (Rached et al., 2023).

Batteries form the base of modern technology and are at the heart of portable electronics, electric vehicles, and renewable energy storage. With the world transitioning towards renewable energy sources such as solar and wind, the demand for effective and long-life energy storage technologies has never been higher (Goodenough & Park, 2013). Such batteries are responsible for storing excess energy generated from intermittent sources and supplying it at the right time to ensure a constant supply of electricity under various conditions. Among other types of batteries, lithium-ion batteries have gained much interest since they stand out due to their superior energy density, long cycle life, and outstanding efficiency (Shirvanimoghaddam et al., 2018).

Despite wide usage, these types of batteries are still prone to corrosion-related degradation, which may substantially reduce their life and effectiveness. Based on an in-depth analysis of the specific corrosion mechanisms that operate in batteries and how these mechanisms contribute to degradation, this review tries to investigate how corrosion impacts battery durability. Such corrosion can accelerate performance loss after successive cycles of charging and discharging, lead to weakening in the internal structures, increase the resistance internally, and result in a total reduction of capacity (W. Li et al., 2022; Song et al., 2022). A deep understanding of such processes allows for the establishment of specific corrosion mitigation strategies that will enhance reliability, efficiency, and safety for the batteries. Applications where long-term performance and safety are particularly crucial, such as high-performance portable electronics, electric cars, and renewable energy storage systems, have a particular relevance to this research. The objective of this study is to provide insights that will guide the creation and maintenance of superior and durable energy storage systems.

2 FUNDAMENTAL PRINCIPLES OF CORROSION

2.1 Description of corrosion mechanisms

The electrochemical breakdown of metals in the presence of their surroundings, known as corrosion, causes material loss and a reduction in performance. In batteries, corrosion mostly damages the electrodes, current collectors, and casings, impairing structural integrity and electrical conductivity. Below is a summary of the primary corrosion types that affect energy storage equipment (Mzioud et al., 2022a),(Mzioud et al., 2020; Rached et al., 2023).

2.1.1 Uniform Corrosion (General Corrosion)

This type of corrosion occurs when metal surfaces oxidize uniformly, such lead plates in lead-acid batteries or steel casings. It gradually decreases wall thickness and mechanical strength, although being predictable (Mzioud et al., 2022b, 2025; Perez, 2024; Zhu et al., 2024).

2.1.2 Galvanic Corrosion (Bimetallic Corrosion)

Galvanic corrosion happens when two different metals are in electrical contact with each other in the presence of an electrolyte, such as when copper and aluminum current collectors are connected. At the interfaces between electrodes and connectors, the less noble metal corrodes more quickly, presenting a significant issue (Ikeuba et al., 2024).

2.1.3 Pitting Corrosion and Crevice Corrosion

These are forms of localized corrosion that target small areas, like sealing joints or the crevices between electrode layers, or that create pits. These types of defects, particularly in zinc and lithium-ion batteries, may cause leaks or short circuits (Frankel, 1998; Koley et al., 2024), (Burstein et al., 2004).

2.1.4 Stress Corrosion Cracking (SCC)

Stress corrosion cracking arises when a corrosive environment and tensile stress combine to cause microcracks. This is important for welded connections and battery casings that are subjected to mechanical or thermal stress (Amiri et al., 2025; Jiang et al., 2023),(Ehrnstén et al., 2024).

2.1.5 Intergranular Corrosion

This internal form of corrosion is specific to batteries and happens at current collectors or electrode-electrolyte contacts. It can result in metal disintegration, gas formation, and capacity fading due to contaminants, moisture intrusion, or unstable SEI layers (Phull, 2003), (Wilson, 1971).

It is a complex interaction of materials, electrochemistry, and environmental factors. Therefore, battery corrosion differs from general corrosion because of a constrained environment and electrochemical cycling effects.

The rare situations where these types of external corrosion could occur are erosion-corrosion, caused by fluid motion or gas release, and microbiologically influenced corrosion, or MIC, mostly in open or degraded systems exposed to humid contaminated environments. But these kinds are nothing compared to the electrochemical and galvanic corrosion processes which give rise to the functioning of batteries (Kuruvila et al., 2018; Rao & Mulky, 2023), (Islam & Farhat, 2017).

Understanding the specific causes of corrosion lays the groundwork for determining the operational, material, and environmental elements that affect its development. These key factors and their impact on battery deterioration are explored in the following section.

2.2 Factors influencing corrosion: materials, environment, temperature, etc.

Most materials, particularly metals, are subject to the natural process of corrosion when subjected to particular environmental factors. The type of the material, the surrounding environment (including salinity and humidity), temperature, and other relevant parameters are the main determinants of corrosion. Every element has a unique impact on the rate and severity of corrosion (**Fig.1**) (Akimov, 1959; Rached et al., 2023).

Figure 1. Factors influencing corrosion.

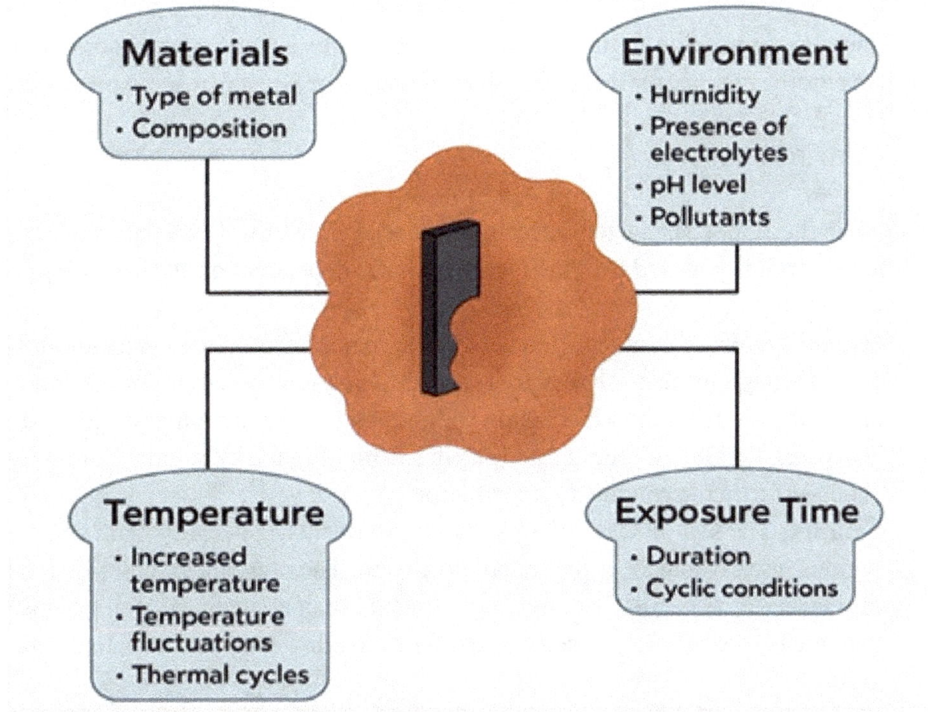

2.2.1 Materials

The materials themselves play an important part in their susceptibility to corrosion. All metals and their alloys are differently reactive; hence, their interaction with a given environment will also differ. For example:

- **Ferrous metals,** mainly iron and steels, are quite susceptible to oxidation, rusting upon coming into contact with oxygen and water. However, in the case of steel, other alloys present within it, such as chromium and nickel in stainless steel, change this by creating a protective oxide layer that enhances corrosion resistance.
- **Nonferrous metals,** including aluminum, copper, and zinc, corrode in a different manner. For instance, while aluminum oxides act to block continued corrosion by forming an impermeable oxide layer on its surface; copper nat-

urally tarnishes (copper carbonate forms a green protective layer that protects it from further corrosion)(Chawla, 1993; Shifler & Aylor, 2005).

Materials' composition and alloys can also affect how resistant they are to corrosion. Brass or chrome plating, for instance, can increase the resistance of some materials.

2.2.2 Environment

One of the important factors affecting material corrosion is the environment that the material is exposed to. This factor is highly dependent on the following:

- **Humidity:** Water content often acts as the major agent in the corrosion process. Through its electrolyte activity, water renews the reactivity of the anode and cathode areas, thus increasing the rate of both oxidation and reduction processes. Cycles of wetting and drying-from climatic or relative humidity changes-further accelerate rust formation.
- **Salinity:** The salt in seawater or salty environments, especially from coastal regions, greatly enhances the rate of corrosion. The presence of chloride ions in seawater is particularly damaging to metals such as steel and iron because they break through the protective oxide layer to allow water to reach the metal surface.
- **pH:** Acidic and alkaline conditions are the factors which change the rate of corrosion. In acidic environments, such as in mines or some industrial processes, metals dissolve easily. Conversely, this process might be slow in an alkaline environment (Rached et al., 2025)(Cai et al., 2018; Soares et al., 2009) (McCafferty, 2010).

2.2.3 Temperature

It is well known that the rate of corrosion is directly affected by temperature. Generally speaking, an increase in temperature normally accelerates chemical reactions, which include corrosion. This is especially evident in industrial or high-temperature settings where such corrosion occurs at a much greater rate. For instance, increased temperatures increase the oxidation rates of alloys used in turbine engines or boilers (Mzioud et al., 2020; Rached et al., 2023). However, temperature not only influences the rate of corrosion but also may change the type of corrosion products. For example, new corrosion phases or specific protective layers can originate due to the oxidation of metals at increased temperatures (Akimov, 1959; Tortorelli & Natesan, 1998; Wasim et al., 2018).

2.2.4 Other Factors

Other factors influencing corrosion include:

- **Impurities and contaminants:** Sulfur, carbon dioxide, among other contaminant particles, either in air or water, accelerate corrosion; this is particularly true when combined with elements that will enhance the corrosion process, like heat or moisture.
- **Mechanical Stress**: Materials that are subjected to mechanical stress include metals in structures. Specific forms of corrosion, such as stress corrosion cracking, can be initiated in such materials. In such corrosion, tension leads to the initiation and propagation of cracks (Harsimran et al., 2021; Wasim & Djukic, 2022).

It is crucial to investigate how these elements lead to different types of corrosion within battery systems and how they impact performance after looking at the main factors that either accelerate or impede corrosion.

2.3 Different types of corrosion affecting batteries

Batteries are susceptible to various forms of corrosion, each of which has a unique impact on safety, dependability, and performance (Fig.2). Alkaline and lead-acid batteries are especially susceptible to electrolyte leakage corrosion. The electrolyte reacts with metals and carbon dioxide in the air to generate visible deposits like white potassium carbonate or green-blue copper compounds on terminals when it leaks sulfuric acid in lead-acid types or potassium hydroxide in alkaline cells. Increased internal resistance and even open circuits may result from this leak. When two dissimilar metals, like copper and zinc, come into electrical contact with one another when an electrolyte is present, a galvanic cell is formed in which the less noble metal corrodes more quickly. This phenomenon is known as galvanic corrosion. This is frequently observed at battery terminals that are attached to various metallic conductors.

Atmospheric corrosion is caused by exposure to the environment- particularly humidity, oxygen, and pollutants. It primarily affects metal casings and terminals, leading to rust (iron oxide) or tarnishing of the metal surface. This type of corrosion is especially problematic in outdoor or marine environments and can be accelerated by salts in the air (X. Li et al., 2025a; Y.-Y. Wang et al., 2023; Zhang et al., 2022; Zhou et al., 2024).

Chemical corrosion occurs when basic or acidic vapors escape batteries and come into contact with nearby materials. For example, sulfuric acid fumes from vented lead-acid batteries can corrode copper, steel, and even concrete. Potassium hydroxide, which aggressively destroys the surfaces of zinc and aluminum and weakens structural integrity, can leak from alkaline batteries. Crevice corrosion is a localized attack that can occur in small, protected areas where electrolyte or moisture may become trapped and stagnant. Because there is less oxygen available in these fissures, a differential aeration cell is formed, which lowers pH and accelerates metal ion dissolution. Deep pitting and structural collapse may result over time, frequently without any obvious symptoms (Kwon et al., 2023; Qu et al., 2022; Saji, 2023; J. Wang et al., 2025).

Figure 2. Different types of corrosion affecting batteries.

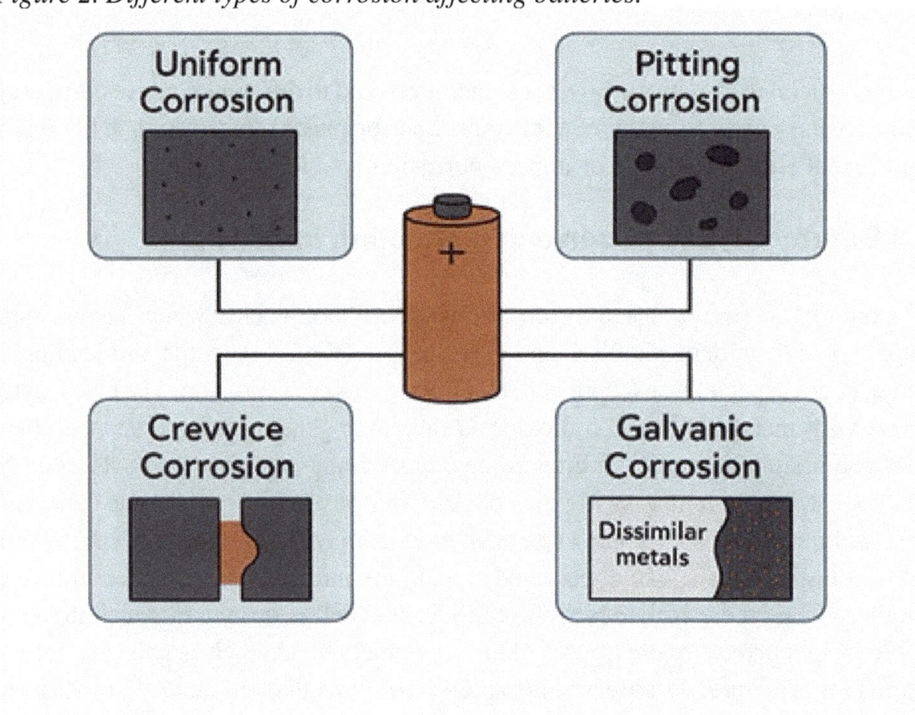

Lastly, internal corrosion is a more hidden but important problem, especially in newer battery chemistries such as lithium-ion. It results from impurities, moisture ingress, or breakdown of internal separators. The corrosion of electrodes or current collectors made from copper and aluminum can lead to gas formation, pressure build-up, and capacity loss. In extreme cases, this leads to thermal runaway where the heat from internal reaction triggers uncontrollable battery failure.

Potential fire and explosion hazards. Understanding this corrosion mechanism is important to improve battery durability and optimize maintenance practice for operational safety, from consumer electronics to electric vehicles and renewable energy systems, by Lee et al., 2024; H. Liu et al., 2023; Wu & Du, 2021a; Zhang et al., 2022).

2.4 Impact of Corrosion on Battery Structure and Components

Corrosion significantly affects the internal and external structures of batteries by threatening their mechanical integrity, safety, and electrochemical performance. Corrosion that occurs on the outside has adverse implications on the mechanical structure of battery casings and terminals, further causing fractures, leakage, and even rupture. This challenge is severe in both lead-acid and alkaline batteries due to the aggressive nature of electrolytes, which can compromise metallic enclosures to the extent that they may bulge or even explode when under pressure. According to Braithwaite et al. 1999 and Qu et al. 2022, such conditions lead to distortions. Corrosion of terminal contacts and connections increases contact resistance, thereby reducing current efficiency and enhancing heat buildup. (Braithwaite et al., 1999a; Qu et al., 2022).

Corrosion attacks electrodes, electrolytes, current collectors, and separators of vital significance internally. At the interface, the electrode/electrolyte corrosion causes an SEI layer to become unstable in lithium-ion batteries, resulting in a rise in internal resistance, the generation of gas, and capacity reduction. The dissolution and subsequent deposition of copper on the cathode due to corrosion of the anode's copper current collector may result in internal short circuits. On nickel-rich cathodes, corrosion also results in surface cracking and a loss of structural cohesiveness that reduces the amount of active material available for electrochemical reactions (J. Feng et al., 2024; Lei et al., 2020; L. Li et al., 2021; Saji, 2023; Shahjalal et al., 2022). The active material shedding and sulfation due to corrosion of lead or zinc electrodes in aqueous or lead-acid systems result in decreased charge retention capacity.

The corrosion-induced processes in the electrolyte can, by way of their byproducts, influence the pH, conductivity, and chemical stability and, therefore, can accelerate further degradation. Moreover, gas production or corrosion particles can weaken the separator, promoting the risk of short circuits. These internal deteriorations, combined with external ones, weaken the general structure of the battery over time and can lead to thermal runaway, gas evolution, swelling, and leakage.

Recent case studies highlight how vital material and operational characteristics are in controlling these effects. For instance, voltage, temperature, electrolyte composition, and the inclusion of compounds all have a significant impact on battery performance and the progression of corrosion. For example, lithium-ion batteries cycled at higher voltages show capacity fade due to rapid transition metal dissolution and SEI layer destabilization. While high temperatures exacerbate corrosion processes, electrolyte breakdown, and internal resistance increase, low temperatures can lead to lithium plating and performance loss (Hossain Ahmed et al., 2015). The selection of the electrolyte serves to enhance its capacity retention during long-term use; LiFSI-based formulations, for instance, reduce copper and aluminum corrosion (Wu & Du, 2021b). Phosphate-based additives also enhance cycle life by preventing dendritic growth and stabilizing the SEI, highlighting the importance of carefully engineered chemical treatments (Phung et al., 2025).

These results show that to arrive at an optimal design of battery, material selection, and protection, detailed knowledge of structural deterioration, electrochemical effects, and operational factors is required. Case study insights combined with traditional corrosion analysis can be developed into batteries that are safer, more energy-efficient, and have longer cycle lives, which could be used in consumer electronics, electric vehicles, and renewable energy storage systems (J. Feng et al., 2024; X. Li et al., 2025a; Phillip et al., 2020; Saji, 2023; Z. Wang et al., 2024).

2.4.1 Degradation of conductive materials and loss of performance

The degradation of the conducting materials in a battery has a direct and considerable impact on its electrochemical performance. Conductive materials, which include copper and aluminum current collectors along with carbon-based additives in the electrodes, are essential to ensure effective electron transport.

These materials may degrade over time due to oxidation, corrosion, and other unwanted electrochemical reactions that may interfere with continuity. For example, pits and cracks that develop through aluminum corrosion of the cathode current collector can increase internal resistance and reduce surface conductivity. Internal short circuits and irreversible damage may also be caused by dendritic deposits, which can be produced at the cathode by the dissolution of copper from the anode under conditions of over-discharge. In lithium-ion batteries, conductive carbon particles can also become electrically isolated due to the degradation of binder and electrolyte, resulting in reduced electron mobility and active surface area (Holm, 2013; W. Li et al., 2022; Malucci, 2003; Williams et al., 2024).

It decreases the capacity of the battery to carry current effectively and leads to capacity fade, voltage drop, and poor cycling stability. Under extreme conditions, the battery is unable to maintain safe operating conditions with high loads or temperature fluctuations. In high-power applications, such as energy storage devices or electric vehicles, such loss of conductivity can lead to thermal instability, hot spots, and inhomogeneous distribution of current. Energy output also suffers accordingly. The degradation of conducting material should be prevented by appropriate material selection, protective coatings, and use of stable electrolyte compositions to retain better battery performance and extend its operational life (Byun et al., 2024; C. Fan et al., 2024; Y. Li et al., 2024; Mansir & Okonkwo, 2025).

2.4.2 Impact on the storage capacity and lifespan of the battery

Corrosion reduces the life span and storage capacity of batteries significantly. The ability of a battery to store and deliver charge efficiently deteriorates with time as internal parts, such as electrodes, current collectors, and separators, degrade due to corrosive processes.

Capacity fade is caused by the loss of active material arising from electrode corrosion, which directly lowers the number of available sites to be used by electrochemical reactions. For example, the dissolution of transition metals from the cathode and degradation of the solid electrolyte interphase layer significantly affects long-term capacity retention in lithium-ion batteries. In lead-acid batteries, sulfation and active material shedding due to lead plate corrosion decrease the effective surface area, further limiting the capacity of the battery to retain charge (L. Guo et al., 2021; Pender et al., 2020; Tran & Fowler, 2020).

Besides, corrosion increases internal resistance and, therefore, further degradation through rising temperatures during operation and lowered efficiency in charge/discharge. These effects will finally reduce the cycle life of the battery, which impairs its ability to function consistently through many cycles of charging and discharging. In high-demand applications such as backup power or electric vehicles, the loss of capacity or lifetime can have serious safety and economic consequences. In this regard, corrosion control becomes a crucial step to retain high energy density, operational stability, and long service life for all types of battery systems (Geslin et al., 2025; Rahman & Alharbi, 2024; Velasco-Arellano et al., 2024).

Corrosion can affect the life span and storage capacity of a battery in many ways:

2.4.2.1 Reduced Storage Capacity:

- Internal Short Circuits: Corrosion of internal components like electrodes or connectors can lead to partial or complete short circuits, reducing the overall capacity of the battery to hold charge.
- Increased Internal Resistance: As corrosion progresses, the internal resistance of the battery increases. This results in a decrease in effective storage capacity, as more energy is lost as heat.
- Degraded Electrodes: The battery's capacity to store and release energy effectively can be hampered by corrosion, especially on the anode or cathode (Agubra & Fergus, 2013; Islam & Farhat, 2017; Y. Li et al., 2024; Lim et al., 2025).

2.4.2.2 Shortened Lifespan:

- Electrode Breakdown: Corrosion, over time, could spell deterioration of the electrodes inside the battery, preventing them from functioning optimally. The consequence may be that the battery can support fewer charge and discharge cycles.
- Electrolyte Leakages: Corrosion sometimes causes the release of certain gases or electrolytes that might cause damage to other internal parts. A leaking electrolyte shortens the life span because this is directly related to the charging and discharging capabilities of the battery.
- Swelling or Rupture: Physical changes in the battery due to corrosion may result in swelling over time, which might lead to failure or rupture. This will drastically reduce the life span of the battery and render it useless (Allerth & Niklasson, 2022; Chang et al., 2024).

2.4.2.3 Loss of Efficiency:

Corrosion buildup in batteries leads to a decrease in energy efficiency, which reduces the overall system efficiency. As the battery becomes less effective in storing power, it also takes longer to charge (Y. Chen et al., 2021).

2.4.2.4 Thermal Instability:

Heat can be generated through corrosion-induced rises in internal resistance. If thermal instability occurs, then there is a vicious circle of more corrosion and extra capacity loss, further accelerating the degradation rate (W. Shen et al., 2022).

2.4.2.5 Preventive Measures:

The negative consequences of this can be softened by the application of corrosion inhibitors or protective coatings for very high-performance and critical batteries. The effective service life of a battery can be extended, and the risk of corrosion reduced, by regular maintenance and careful handling (Etaiw et al., 2023).

It also creates serious safety risks. Internal corrosion can initiate a chain of degradation mechanisms and lead to short circuits, thermal runaway, even fire or explosion. For example, when the corrosion of current collectors leads to dissolved metallic ions diffusing through the electrolyte, depositing on the cathode, such as copper, conductive dendritic bridges may arise from this and may lead to internal short circuits and localized overheating (Spotnitz & Franklin, 2003).

Similarly, the corrosion-induced breakdown of SEI exposes the active materials to the electrolyte reaction; exothermic side reactions resulting in gas and heat release have been observed (X. Feng et al., 2018). These exothermic reactions increase the rate of gas evolution, pressure buildup, and venting, which with the rising temperature further accelerate and may lead to thermal runaway.

External corrosion of the battery shell, terminals, or seals can be accompanied by the leakage of electrolyte and ingress of moisture, which further destabilizes internal components. The possibilities of electrical shorting, failure of insulation, and even rupture or explosion thus become higher. In addition, the possibility of toxic gas emissions is also higher (James Abraham et al., 2021; Larsson et al., 2017). Corrosion and electrochemical reactions at high temperatures tend to build upon themselves, as there is positive feedback between heat production and material deterioration.

Most corrosion-driven safety failures are due to interrelated processes, such as metal dissolution, SEI breakdown, gas generation, and structural deformation. To design corrosion-resistant materials, advanced protective coatings, and real-time monitoring systems that identify early signs of degradation, these interactions need to be understood. Reinforcement of these preventive measures must be taken to avoid catastrophic failure and ensure the safe operation of lithium-ion batteries for electric vehicles, aerospace systems, and renewable energy storage applications.

Corrosion directly impacts the electrochemical performance of batteries by destroying important internal components such as electrodes, current collectors, and separators. In lithium-ion systems, the corrosion of copper and aluminum current collectors and destabilization of the SEI layer increase internal resistance, diminish active surface area, and impede lithium-ion transport, leading to capacity fading and reduced energy efficiency. Likewise, in lead-acid and alkaline batteries, sulfation, shedding of active material, and electrolyte contamination due to lead or zinc electrode corrosion reduce cycle life and storage capacity.

Protective coatings, electrolyte additives, and corrosion-resistant materials are important means for effective prevention of corrosion to maintain long-term efficiency and operational stability in a wide variety of chemistries. In general, performance degradation is faster through a combined effect of localized structural damage, gas evolution, and material degradation (Braithwaite et al., 1999b; Delgado-Sanchez & Lillo-Bravo, 2020a, 2020b; Yeung et al., 2015).

2.5 Specific Corrosion Mechanisms in Batteries

2.5.1 Corrosion of metal electrodes (lithium, zinc, nickel).

The corrosion of metal electrodes such as lithium, zinc, and nickel have a significant effect on the stability and performance of electrochemical systems, particularly batteries. When exposed to air and moisture, lithium readily corrodes due to its high reactivity, forming compounds like lithium hydroxide (LiOH) and lithium carbonate (Li_2CO_3). Dendritic development and capacity loss may result from the solid electrolyte interphase (SEI), a passivation layer that develops in lithium-based batteries due to corrosion. (Fig.3) and (Fig.4) (Pei et al., 2017).

Commonly found in alkaline and zinc-air batteries, zinc oxidizes to produce Zn^{2+} ions, which then create zinc hydroxide ($Zn(OH)_2$) and zinc oxide (ZnO). It also forms soluble complexes such as $Zn(OH)_4^{2-}$ in alkaline conditions, and one of the main parasitic reactions is hydrogen evolution. Systems containing NiMH and NiCd use nickel metal, which resists corrosion. However, the nickel could further corrode with the appearance of either nickel hydroxide, $Ni(OH)_2$, or nickel oxide, NiO, in oxidizing or acidic conditions. Corrosion properties of these metals are greatly affected by such parameters as temperature, voltage, impurity content, and electrolyte composition, often resulting in efficiency losses, structural breakdown, and shortened life expectancy for the battery (C. Gu et al., 2021; Q. Wang et al., 2024).

Electrochemical Corrosion Reactions:

(a) The Lithium Electrode:

Lithium metal easily oxidizes when exposed to air or electrolyte:
Anodic reaction: $Li_{(s)} \rightarrow Li^{+}+e^{-}$
In presence of moisture or CO_2:
$2Li+2H_2O \rightarrow 2LiOH+ H_2$
$2LiOH+ CO_2 \rightarrow Li_2CO_3+ H_2O$
Overall corrosion causes Li_2CO_3, LiF, and organic components to form SEI.

(b) Zinc Electrode (in Alkaline Medium):

Zinc oxidation and hydroxide/oxide formation:
Anodic reaction: $Zn_{(s)} \rightarrow Zn^{2+}+2e$
$Zn^{2+}+2OH- \rightarrow Zn(OH)_2$
$Zn(OH)_2 \rightarrow ZnO + H_2O$
In strong alkaline media, soluble zincate forms:
$ZnO+H_2O+2OH^{-} \rightarrow [Zn(OH)_4]^{2-}$
Hydrogen evolution (undesired side reaction):
$2H_2O+2e^{-} \rightarrow H_2 + 2OH^{-}$

(c) Nickel Electrode (Ni, NiMH, NiCd Batteries)

Nickel corrosion typically involves oxidation to nickel hydroxide or oxide:
Anodic reaction: $Ni(s) \rightarrow Ni^{2+}+2e^{-}$
$Ni^{2+}+2OH^{-} \rightarrow Ni(OH)_2$
Under more oxydative conditions:
$2Ni(OH)_2 \rightarrow 2NiO+2H_2O$

Cathodic reactions (oxygen or hydrogen reduction) can accompany the corrosion:
$O_2+2H_2O+4e^- \rightarrow 4OH^-$

Figure 3. Zinc Electrochemical Behavior and Corrosion Mechanisms in Aqueous Environments (X. Li et al., 2025b).

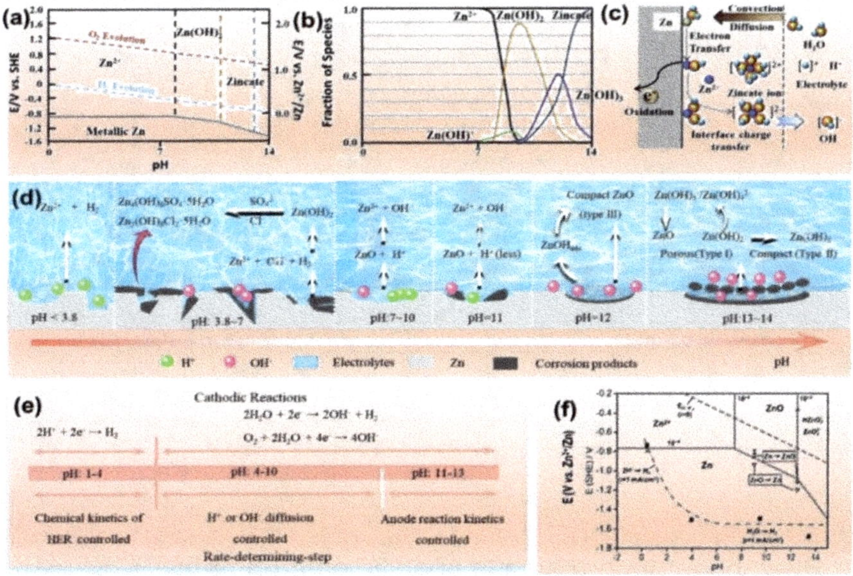

Figure 4. Schematic Illustration of Corrosion and Interfacial Reactions in Lithium- and Zinc-Based Batteries (Du et al., 2023).

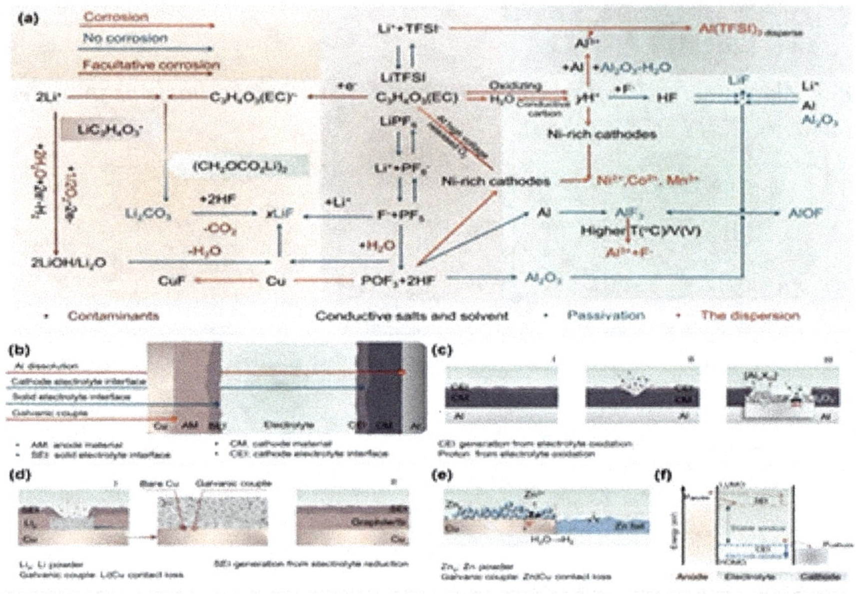

2.5.2 Prevention of corrosion in specific environments (automotive, portable devices, etc.).

Several protective techniques that are adapted to the particular exposure conditions of each application must be integrated in order to avoid corrosion in particular environments, such as the automotive industry and portable electronic devices. Vehicles used in the automotive sector frequently operate in challenging outdoor conditions where they are exposed to moisture, road salts, acidic pollutants, and extreme temperature fluctuations. These elements hasten localized and uniform corrosion, particularly in exhaust lines, brake systems, and underbody parts.

Manufacturers employ multilayer protective coatings, such as zinc galvanization followed by epoxy primers and polyurethane topcoats, and corrosion-resistant materials, such as stainless steel or coated carbon steel, to counteract this. Weld joints and inside body panels are treated with cavity protection sprays and wax-based sealants. Resistance is further increased by sophisticated treatments like cathodic protection (especially in fuel tanks and pipes) and the addition of corrosion inhibitors to engine coolants and lubricants (Benavides-Guerrero et al., 2022; Bhat et al., 2024; Gharieh et al., 2025; Iravani et al., 2025).

Internal circuit boards, battery terminals, and connectors in portable electronic devices such as laptops, tablets, and smartphones may get degraded due to corrosion from sweat, salt residues, humidity, and accidental contact with water. For that reason, manufacturers apply conformal coatings to circuit boards since they provide a barrier against moisture without significantly affecting the electrical function. Typically, connectors are nickel- or gold-plated to avoid oxidation, while device housings can be sealed using rubber gaskets or adhesive-based waterproofing techniques that meet the IP rating of IP67 or IP68. Additionally, desiccant and moisture-absorbing packs may be included inside the enclosure or packaging to protect fragile parts during storage and transport (He et al., 2025; Rautio et al., 2022; Su et al., 2024; Suppa, 2007).

Careful material selection, environmental exposure management, and the use of corrosion inhibitors or other passivating treatments are important in both fields. Research on innovative solutions also focuses on self-healing coatings and barriers based on nanomaterials for better long-term corrosion resistance. In the end, these techniques guarantee the product's safety, dependability, and extended service life in its natural operating environment in addition to preventing surface deterioration (J. Wang et al., 2022; Yang et al., 2023).

2.6 Solutions and Protection Methods Against Corrosion

Preventive measures, material selection, and protective technologies suited to particular applications and conditions are all part of corrosion protection **(Fig.5** and **Table 1**).

Advanced protective techniques have been developed recently to further improve the corrosion resistance and longevity of batteries. Stabilizing the solid electrolyte interphase (SEI) layer is a crucial strategy since it reduces parasitic reactions and stops ongoing electrolyte breakdown, therefore increasing cycle life. When applied to electrodes and current collectors, nanostructured coatings like graphene, TiO_2, or Al_2O_3 offer robust physical defenses against electrolyte attack while preserving electrical conductivity.

Furthermore, by avoiding direct metal-electrolyte contact, solid electrolytes (sulfide or oxide-based systems) stop liquid leakage and lessen interior corrosion. Long-term defense against mechanical and electrochemical deterioration is guaranteed by self-healing coatings, which can self-heal microcracks or flaws. In order to identify early-stage deterioration and enable prompt maintenance, sophisticated corrosion monitoring systems that combine real-time sensors and predictive analytics are also being used more and more. When combined, these cutting-edge techniques not only slow down corrosion but also greatly increase battery safety, robustness, and operational dependability especially in extreme or hot environments (Bi et al., 2019; Jin et al., 2023; B. Liu et al., 2024a; Park et al., 2022; Vinodhini & Xavier, 2025).

For example, solid electrolytes and nanostructured coatings are used in the automotive industry to reduce corrosion in EV battery packs under extreme temperature swings and high voltage cycling. Conformal coatings and moisture barriers are applied to lithium-ion cells by manufacturers of portable electronics to protect against corrosive contact with humidity or perspiration in service; this enables reliable operation over years of daily use. Similarly, advanced corrosion monitoring, protective coatings, and SEI stabilization are employed in grid-scale energy storage systems to ensure long-term operating safety, minimizing maintenance costs. These examples demonstrate how advanced corrosion prevention methods are applied across multiple industries to extend battery life, improve safety, and optimize performance (Haghbin et al., 2025; Han et al., 2023; Palakkool et al., 2025) .

Figure 5. Solution and protection methods against corrosion.

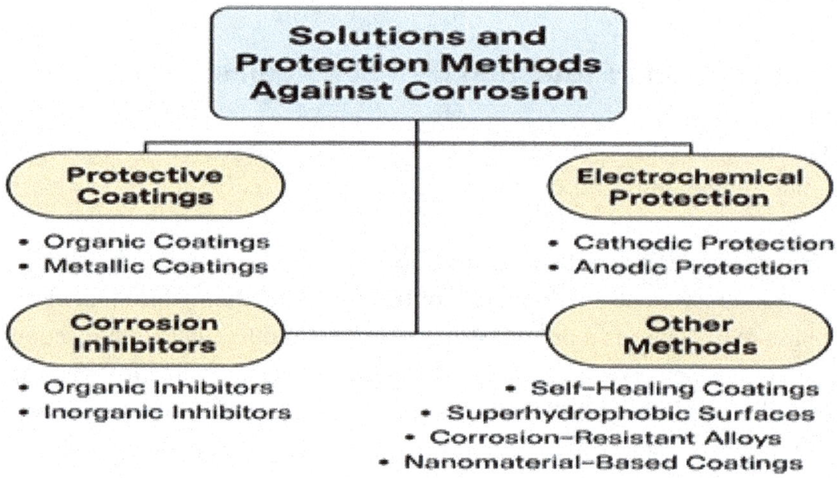

Table 1. Corrosion Protection Methods and Their Effects.

Protection Method	Description	Measurable Impact on Durability	Reference
Protective Coatings	Organic paints, metallic coatings (zinc), and polymers that act as barriers.	Reduces corrosion rate; extends cycle life by preventing material degradation.	(J. Wang et al., 2022)
Corrosion Inhibitors	Chemicals (amines, imidazoles) added to slow electrochemical reactions.	Reduces internal resistance increase; improves long-term capacity retention.	(S. Wang et al., 2024)
Cathodic/Anodic Protection	Electrochemical methods that shift potential to reduce corrosion tendency.	Prevents metal dissolution; improves structural stability and cycle life.	(J. Wang et al., 2022)
Self-Healing Coatings	Coatings that autonomously repair microcracks or damage.	Maintains surface integrity; prevents localized corrosion; stabilizes performance over time.	(Alves et al., 2021)
Nanomaterial-Based Barriers	Use of nanocomposites to improve mechanical and barrier properties.	Reduces electrolyte attack; enhances conductivity stability; prolongs cycle life.	(J.-S. Li et al., 2022)
Superhydrophobic Surfaces	Water-repellent coatings that prevent moisture accumulation.	Minimizes moisture-induced corrosion; reduces capacity fade and internal resistance.	(J. Fan et al., 2025)

Current collectors, electrodes, terminals, and casings are just a few of the interfaces in battery systems that can corrode, particularly when exposed to moisture, electrolytes, or high temperatures (Table 2) (Chyrkin et al., 2023; Ren et al., 2022; Q. Shen et al., 2023).

Table 2. Battery-Specific Corrosion Protection Methods and Their Effects.

Method	Description	Measurable Impact on Durability	Reference
Corrosion-Resistant Materials	Use of stainless steel, aluminum, and carbon-coated metals for casings and contacts.	Reduces material degradation; maintains structural integrity and electrical performance.	(B. Liu et al., 2024b)
Protective Coatings	Ceramic or polymer coatings on electrodes/separators to prevent electrolyte attack.	Minimizes electrolyte attack; stabilizes internal resistance and capacity retention.	(T. Adebanjo et al., 2025)
SEI Layer Stabilization	Formation of a stable SEI layer on lithium anodes to prevent further degradation.	Prevents dendrite formation; reduces internal resistance; extends cycle life.	(Nogales et al., 2024)

continued on following page

Table 2. Continued

Method	Description	Measurable Impact on Durability	Reference
Electrolyte Additives	Phosphates, ionic liquids, or PEG added to minimize anode corrosion and side reactions.	Improves capacity retention; reduces internal resistance growth; enhances overall performance.	(W. Gu et al., 2025)
Encapsulation & Moisture Barriers	Use of sealed enclosures and desiccants to prevent external moisture ingress.	Prevents external corrosion; maintains voltage stability and long-term reliability.	(C. Li et al., 2020)
Nanomaterials & Surface Treatment	Nano-coatings and surface functionalization to resist chemical and electrochemical attack.	Reduces localized corrosion; stabilizes electrode and current collector integrity.	(Abdeen et al., 2019)

2.6.1 Selection of corrosion-resistant materials.

A key tactic in preventing corrosion is the choice of materials that are resistant to it, particularly in settings where there are chemicals, moisture, or electrochemical reactions. Operating variables, including pH, temperature, humidity, and the presence of corrosive species (chlorides, sulfates), determine the choice of material. Because to its chromium-rich passive oxide layer, stainless steel finds extensive usage in structural, medicinal, and maritime applications. Alloys based on titanium and nickel provide exceptional resistance in extremely salty or acidic conditions. Because of their conductivity and resistance to electrolyte corrosion, materials such as nickel or stainless steel for terminals and carbon-coated aluminum for cathodes are chosen in battery systems.

Non-load-bearing components also employ polymers, ceramics, and composite materials, particularly when chemical stability and electrical insulation are needed. In order to balance performance with real-world limitations, the selection procedure also takes into account mechanical strength, weight, cost, and galvanic compatibility. High-entropy alloys, nanostructured surfaces, and self-passivating materials are examples of recent developments that provide customized corrosion resistance for cutting-edge industrial and energy applications (Chyrkin et al., 2023; Craig et al., 2023; Larché & Dézerville, 2011; Shokri & Sanavi Fard, 2022).

2.6.2 Improvement of coatings and internal component protections

Figure 6. Modern Approaches to Enhancing Corrosion Resistance in Materials and Devices.

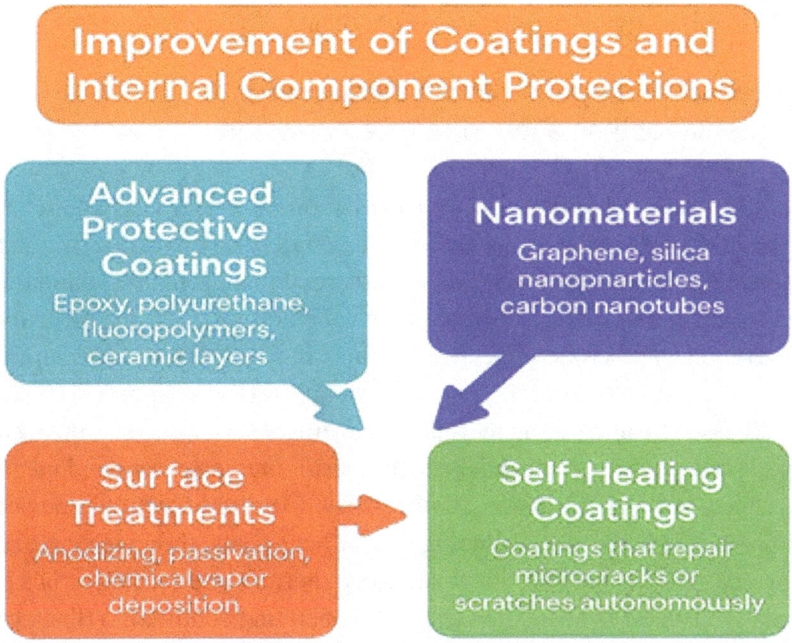

In many industrial applications, such as electronics, automotive, marine, and energy storage systems, improving coatings and internal component safeguards is an essential tactic in the fight against corrosion. Epoxy resins, polyurethane, fluoropolymers, and ceramic-based layers are examples of advanced protective coatings that offer strong defenses against chemical, oxygen, and moisture damage. Encapsulation resins and conformal coatings protect delicate parts of electronic and battery systems from moisture and contaminants while maintaining electrical insulation.

Nanomaterials that increase mechanical strength, thermal stability, and corrosion resistance such as graphene, silica nanoparticles, or carbon nanotubes are frequently used into new-generation coatings. Chemical vapor deposition, passivation, anodizing, and other surface treatments that alter the surface chemistry to reduce reactivity are becoming increasingly common ways of protecting internal components. The development of self-healing coatings, which repair scratches or microcracks without assistance, further extends service life. All these developments taken together are providing multifunctional mechanical, chemical, thermal, and electrical protection by increasing robustness and reliability in extreme or changeable conditions.

Encapsulation resins and conformal coatings prevent moisture and contaminants from reaching sensitive areas in electronic and battery systems while maintaining electrical insulation (X. Chen et al., 2024; Farooq et al., 2022; Idumah et al., 2020; T. Liu et al., 2021; Takada et al., 2023; Zafar et al., 2024).

2.6.3 Thermal managements techniques and environmental optimization to reduce corrosion

Since corrosion is highly sensitive to environmental and thermal factors, it is now essential to stop material degradation in a variety of industries by combining environmental optimization strategies with thermal management techniques.

High temperatures speed up electrochemical reactions, raise the diffusion rates of corrosive ions, and erode protective coatings in high-performance applications like automobile engines, aircraft parts, and energy storage systems (like batteries). This is addressed by integrating thermal management systems to stabilize temperatures and lessen thermal gradients that could result in microcracks and mechanical strains, which serve as corrosion starting sites. Compact electronic devices and battery packs frequently use technologies like heat pipes, thermoelectric coolers, phase change materials (PCMs), and advanced thermal interface materials (TIMs) to efficiently disperse heat and preserve ideal operating conditions (Bhattacharyya et al., 2025; Luo et al., 2022; Mahek et al., 2025).

Environmental optimization is altering the surrounding atmosphere to reduce the amount of oxygen, water vapor, chlorides, and other corrosive chemicals in addition to direct temperature control. For example, in closed systems, reactive gases are displaced by inert gas purging (e.g., with nitrogen or argon). In battery manufacturing and storage, dry rooms and settings with regulated humidity are essential for preventing terminal corrosion and electrolyte breakdown. To counteract forceful salt-laden air, structures in maritime and coastal applications are generally housed in dehumidified enclosures and have cathodic protection systems and salt-resistant coatings (Dobó et al., 2023; Y. Liu et al., 2021).

Furthermore, smart corrosion monitoring systems, which integrate IoT sensors, machine learning, and predictive analytics, have become powerful tools for the real-time detection of hot spots, humidity fluctuations, and early signs of corrosion activity. By facilitating prompt intervention and preventative maintenance, these technologies lower long-term expenses while enhancing safety. In order to promote self-healing, thermoresponsive, and adaptive barrier capabilities that regenerate or activate under particular thermal or chemical triggers, materials and coatings are being developed to react dynamically to environmental stimuli. Overall, efficient temperature regulation and environmental control work together to lower corrosion rates while simultaneously increasing energy efficiency, safety, and the lifespan of vital systems, infrastructure, and components (Fu et al., 2023; Narayana et al., 2024; Taffese et al., 2019) .

After having a thorough grasp of the various protection techniques, the next stage is to assess the practical advantages of corrosion inhibition, especially how these techniques improve battery safety, longevity, and dependability in a range of applications with a human focus.

2.7 Impact of Corrosion Inhibition on Human Applications

In a wide range of human-focused technologies, corrosion inhibition is essential for improving the longevity, performance, and safety of batteries. Its incorporation into a range of applications, from large-scale energy storage to personal electronics, provides benefits linked to both practicality and safety. (**Figure 7**).

Improved Safety in Personal Electronics

Wearable technology, cell phones, and computers that are corroded can have hazardous consequences like overheating, leaks, and even fires. By creating impenetrable barriers against heat, moisture, and humidity, corrosion inhibitors lower these risks. As the internal conditions inside the battery cells stabilize, the user's safety becomes crucial because there is a decreased chance of thermal runaways and short circuits. (Zanoni & Pavan, 1993).

Extended Battery Life in Electric Vehicles (EVs)

High-performance batteries are necessary for electric vehicles because internal component corrosion can lower the batteries' energy storage capacity, effectiveness, and safety. By reducing battery waste and the consumption of necessary raw materials, corrosion inhibition techniques increase vehicle dependability, decrease maintenance and replacement frequency, and prolong battery life. This makes vehicles more sustainable and improves user experience (Adany et al., 2013).

Figure 7. The Human Impact of Corrosion Inhibition in Battery Technology.

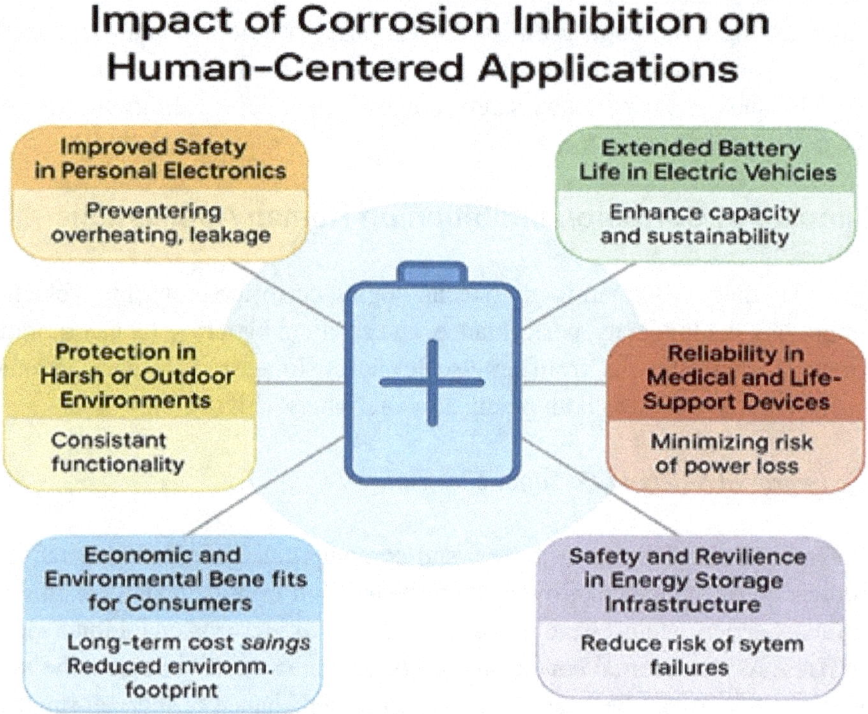

Reliability in Medical and Life-Support Devices

Reliable battery performance is essential for life-sustaining medical devices like insulin pumps, pacemakers, and hearing aids. The safety and effectiveness of patient care may be impacted by corrosion-related malfunctions in these medical devices. Because consistent and prolonged battery performance lowers the likelihood of power loss in life-supporting equipment, corrosion inhibitors increase overall system reliability (Taghipour, 2011).

Protection in Harsh or Outdoor Environments

Drones, fitness trackers, and solar-powered sensors are examples of devices that are susceptible to dust, moisture, and temperature fluctuations; these elements can hasten corrosion and impair functionality. In order to guarantee dependable performance and extend operational life in harsh environments, corrosion inhibitors can shield internal battery systems from external stresses (Lakhouil et al., 2021).

Economic and Environmental Benefits for Consumers

Over time, corrosion-resistant battery technologies tend to reduce the frequency of device failures and replacements, therefore saving money for consumers. Devices, electric vehicles, and home energy storage systems with longer-lasting batteries experience fewer interruptions and have lower ownership costs. Additionally, fewer batteries being disposed of means a smaller environmental impact, which advances sustainability objectives (Mungwari et al., 2024).

Safety and Resilience in Energy Storage Infrastructure

Massive energy storage systems used for renewable energy integration or grid support are especially susceptible to corrosion-related performance losses. By maintaining the integrity and functionality of battery banks, inhibitors reduce the likelihood of system-wide issues or outages. This ensures energy security and dependability for communities and industrial areas that depend on a consistent supply of electricity (Azzuni & Breyer, 2018).

2.8 Comparison of Corrosion Effects Across Battery Types

The manifestations of corrosion in various battery chemistries result from electrode material, electrolyte, and operating conditions. The primary attack point of corrosion in lead-acid batteries is the lead plates, bringing about sulfation, shedding of active material, and capacity loss. In zinc and alkaline batteries, pitting and electrolyte leakage increase internal resistance and localized damage. In lithium-ion batteries, degradation of the SEI layer and corrosion of the copper and aluminum current collectors lead to capacity fade, gas production, and the potential for thermal runaway. Even nickel-based systems, although quite resistant to corrosion, may undergo surface oxidation under hot or acidic conditions (NiMH and NiCd). Such differences, indeed, allow the formulation of appropriate protective methods-coatings, electrolyte additives, or proper material selection-that can diminish such corrosion mechanisms and improve performance in various applications (L. Guo et al., 2021; Ma et al., 2017; Palakkool et al., 2025; Pei et al., 2017; Yu et al., 2019).

2.9 Comparative Effectiveness of Corrosion Protection Strategies

The efficiency of corrosion mitigation strategies is influenced by battery chemistry and operating conditions. Stabilizing the SEI layer and adding electrolyte additives like phosphates and ionic liquids to lithium-ion systems can effectively decrease dendritic formation, stop transition metal dissolution, and lengthen cycle life. Because they lessen sulfation and active material loss, protective coatings on electrodes and casings are especially beneficial for lead-acid batteries.

Alkaline and zinc-based batteries are best protected by corrosion inhibitors and nanostructured coatings, which stop pitting and electrolyte leakage. Nickel-based systems, like NiMH and NiCd, can be made more durable by surface passivation and temperature control. In the end, corrosion prevention techniques related to specific battery chemistry, electrolyte composition, and operating conditions will determine how long a battery lasts, how well it performs, and how safe it is (El-Sayed et al., 2012; W. Guo et al., 2020; Sullivan et al., 2022; Yan et al., 2020; Yuan et al., 2024).

3 CONCLUSION

Corrosion continues to significantly reduce battery life by degrading critical components, such as internal connections, terminals, and current collectors in a way that results in capacity loss, increased internal resistance, and possible safety hazards. Although protective techniques like environmental sealing, corrosion-resistant materials, improved coatings, and thermal management have evolved, a number of research gaps still exist. Among the top ones are how batteries respond to extreme operating conditions, such as high temperatures, overcharges, or fast cycling; developing standardized test protocols that can reliably predict long-term durability; and relating accelerated aging research to real performance. Future studies should focus on combining multidisciplinary approaches in materials science, electrochemistry, nanotechnology, and real-time monitoring to realize batteries with high corrosion resistance.

Additionally, promising new avenues for safety and dependability have been made possible by the development of solid-state electrolytes, self-healing coatings, and intelligent corrosion detection systems. To enable batteries that satisfy the demands of electric vehicles, portable electronics, and renewable energy storage while maintaining long-term performance, safety, and sustainability, these gaps must be filled.

REFERENCES

Abdeen, D. H., El Hachach, M., Koc, M., & Atieh, M. A. (2019). A Review on the Corrosion Behaviour of Nanocoatings on Metallic Substrates. *Materials (Basel), 12*(2), 210. DOI: 10.3390/ma12020210

Adany, R., Aurbach, D., & Kraus, S. (2013). Switching algorithms for extending battery life in Electric Vehicles. *Journal of Power Sources, 231*, 50–59. https://www.sciencedirect.com/science/article/pii/S0378775312019295?casa_token=1RqnZHgoNgAAAAAA:3umKrt7N8BNw7BXlpMLiKUrGii2hyfu5mzkFKRSZyp6_HS_9kQEcMFaSywZZbL00ctnlrCl-o2Lj. DOI: 10.1016/j.jpowsour.2012.12.075

Adebanjo, T. I., Eko, J., G. Agbeyegbe, A., F. Yuk, S., V. Cowart, S., A. Nagelli, E., John Burpo, F., L. Allen, J., T. Tran, D., Bhattarai, N., Shah, K., Hwang, J.-Y., & Hohyun Sun, H. (2025). A comprehensive review of lithium-ion battery components degradation and operational considerations : A safety perspective. *Energy Advances, 4*(7), 820-877. DOI: 10.1039/D5YA00065C

Agubra, V., & Fergus, J. (2013). Lithium ion battery anode aging mechanisms. *Materials (Basel), 6*(4), 1310–1325. https://www.mdpi.com/1996-1944/6/4/1310. DOI: 10.3390/ma6041310

Akimov, G. V. (1959). Factors influencing corrosion. *Corrosion, 15*(9), 23–36. https://meridian.allenpress.com/corrosion/article-abstract/15/9/23/157446. DOI: 10.5006/0010-9312-15.9.23

Allerth, D., & Niklasson, A. (2022). *How is Electrification Affecting the Onboard Fire Safety?-An investigation on ships using lithium-ion batteries for ship propulsion.* https://odr.chalmers.se/handle/20.500.12380/304856

Alves, P. M., Barrias, C. C., Gomes, P., & Martins, M. C. L. (2021). Smart biomaterial-based systems for intrinsic stimuli-responsive chronic wound management. *Materials Today. Chemistry, 22*, 100623. DOI: 10.1016/j.mtchem.2021.100623

Amiri, A. A., Toor, I. U., & Shams, A. (2025). Understanding Stress Corrosion Cracking (SCC), Affecting Variables and Prevention Strategies in Nuclear Power Plants—A Review. *Arabian Journal for Science and Engineering, 50*(5), 2957–2990. DOI: 10.1007/s13369-024-09241-w

Azzuni, A., & Breyer, C. (2018). Energy security and energy storage technologies. *Energy Procedia, 155*, 237–258. https://www.sciencedirect.com/science/article/pii/S1876610218310075. DOI: 10.1016/j.egypro.2018.11.053

Benavides-Guerrero, J., Banerjee, D., Gedamu, D., Gerlein, L. F., & Cloutier, S. G. (2022). Conductive, Anti-Corrosion, Self-Healing Smart Coating Technology Incorporating Graphene-Based Nanocomposite Matrix. *Frontiers in Materials*, *9*, 835855. Advance online publication. DOI: 10.3389/fmats.2022.835855

Bhat, S. I., Mobin, M., Islam, S., Zehra, S., & Shahid-ul-Islam, . (2024). Recent advances in anticorrosive coatings based on sustainable polymers : Challenges and perspectives. *Surface and Coatings Technology*, *480*, 130596. DOI: 10.1016/j.surfcoat.2024.130596

Bhattacharyya, S., Dinh, Q. T., & McGordon, A. (2025). Optimising thermoelectric coolers for battery thermal management in light electric vehicles. *Applied Energy*, *386*, 125516. https://www.sciencedirect.com/science/article/pii/S0306261925002466. DOI: 10.1016/j.apenergy.2025.125516

Bi, P., Li, H., Zhao, G., Ran, M., Cao, L., Guo, H., & Xue, Y. (2019). Robust Super-Hydrophobic Coating Prepared by Electrochemical Surface Engineering for Corrosion Protection. *Coatings*, *9*(7), 452. DOI: 10.3390/coatings9070452

Braithwaite, J. W., Gonzales, A., Nagasubramanian, G., Lucero, S. J., Peebles, D. E., Ohlhausen, J. A., & Cieslak, W. R. (1999a). Corrosion of lithium-ion battery current collectors. *Journal of the Electrochemical Society*, *146*(2), 448–456. https://iopscience.iop.org/article/10.1149/1.1391627/meta. DOI: 10.1149/1.1391627

Braithwaite, J. W., Gonzales, A., Nagasubramanian, G., Lucero, S. J., Peebles, D. E., Ohlhausen, J. A., & Cieslak, W. R. (1999b). Corrosion of lithium-ion battery current collectors. *Journal of the Electrochemical Society*, *146*(2), 448–456. Advance online publication. DOI: 10.1149/1.1391627

Burstein, G. T., Liu, C., Souto, R. M., & Vines, S. P. (2004). Origins of pitting corrosion. *Corrosion Engineering, Science and Technology*, *39*(1), 25–30. DOI: 10.1179/147842204225016859

Byun, J., Lee, C. R., Kim, W., Lee, M. A., Jang, H. Y., Hwang, C., Song, J. H., Yu, J., Back, S., Kim, K. J., & Kim, H. (2024). Surface Work Function Modifier to Modulate Electrolyte Decomposition on Negative Electrode in Lithium-Ion Batteries. *Advanced Functional Materials*, *34*(32), 2401620. DOI: 10.1002/adfm.202401620

Cai, Y., Zhao, Y., Ma, X., Zhou, K., & Chen, Y. (2018). Influence of environmental factors on atmospheric corrosion in dynamic environment. *Corrosion Science*, *137*, 163–175. https://www.sciencedirect.com/science/article/pii/S0010938X17313513?casa_token=m0peGjjbbRwAAAAA:3jYkTGFbu6NRybSXQbIGLOskxHX7SqWT89SqcDTfMmyMxBxmOhYs8gDzMPxffwVIvoIKJ-K8TJg. DOI: 10.1016/j.corsci.2018.03.042

Chang, W., Qian, H., Li, Z., Mol, A., & Zhang, D. (2024). Application and prospect of localized electrochemical techniques for microbiologically influenced corrosion : A review. *Corrosion Science, 112246*, 112246. https://www.sciencedirect.com/science/article/pii/S0010938X24004414?casa_token=CQZiuSzhXZ4AAAAA:ni2G9u34nvSEiaYVCxf-u0e-HCSlpCg152RIgpz-WZEKYMGwHGBEIUxu6RvYYyXZ3JkuE-2QE501. DOI: 10.1016/j.corsci.2024.112246

Chawla, S. L. (1993). *Materials selection for corrosion control*. ASM international. https://books.google.com/books?hl=fr&lr=&id=_NXYRgHnnqkC&oi=fnd&pg=PA1&dq=Factors+influencing+corrosion:+materials&ots=p_KnhOX7Hs&sig=8B6QhCezXvREr3PuG8GObC1RDHw

Chen, X., Zheng, B., Zhou, S., Shi, C., Liang, Y., & Hu, L. (2024). Development and Application of Intelligent Coating Technology : A Review. *Coatings, 14*(5), 5. Advance online publication. DOI: 10.3390/coatings14050597

Chen, Y., Kang, Y., Zhao, Y., Wang, L., Liu, J., Li, Y., Liang, Z., He, X., Li, X., Tavajohi, N., & Li, B. (2021). A review of lithium-ion battery safety concerns : The issues, strategies, and testing standards. *Journal of Energy Chemistry, 59*, 83–99. https://www.sciencedirect.com/science/article/pii/S2095495620307075. DOI: 10.1016/j.jechem.2020.10.017

Chyrkin, A., Nowak, W. J., Gunduz, K. O., Fedorova, I., Sattari, M., Froitzheim, J., Halvarsson, M., & Stiller, K. M. (2023). Intergranular oxidation of additively manufactured Ni-base alloy 625 : The role of Si. *Corrosion Science, 219*, 111234. https://www.sciencedirect.com/science/article/pii/S0010938X23002767. DOI: 10.1016/j.corsci.2023.111234

Craig, B., Rowe, A., Warmack, M., Doll, T. E., Stevens, C., & Connors, K. C. (2023). Guidelines for the selection of corrosion resistant alloys for CCS and CCUS injection wells. *International Journal of Greenhouse Gas Control, 129*, 103988. https://www.sciencedirect.com/science/article/pii/S1750583623001585?casa_token=Qh_Pu3LA9s4AAAAA:XFvFYLkNtu79L2aJB7GFFGHULCnTFEAK4hfiyeJoPsRITSNvVs7CuKlABjILfQGygteExM1fOUk. DOI: 10.1016/j.ijggc.2023.103988

Delgado-Sanchez, J.-M., & Lillo-Bravo, I. (2020a). Influence of Degradation Processes in Lead–Acid Batteries on the Technoeconomic Analysis of Photovoltaic Systems. *Energies, 13*(16), 4075. DOI: 10.3390/en13164075

Delgado-Sanchez, J.-M., & Lillo-Bravo, I. (2020b). Influence of Degradation Processes in Lead–Acid Batteries on the Technoeconomic Analysis of Photovoltaic Systems. *Energies, 13*(16), 4075. DOI: 10.3390/en13164075

Dobó, Z., Dinh, T., & Kulcsár, T. (2023). A review on recycling of spent lithium-ion batteries. *Energy Reports*, *9*, 6362–6395. https://www.sciencedirect.com/science/article/pii/S2352484723010089. DOI: 10.1016/j.egyr.2023.05.264

Du, P., Liu, D., Chen, X., Xie, H., Qu, X., Wang, D., & Yin, H. (2023). Research progress towards the corrosion and protection of electrodes in energy-storage batteries. *Energy Storage Materials*, *57*, 371–399. DOI: 10.1016/j.ensm.2023.02.028

Ehrnstén, U., Andresen, P. L., & Que, Z. (2024). A review of stress corrosion cracking of austenitic stainless steels in PWR primary water. *Journal of Nuclear Materials*, *588*, 154815. https://www.sciencedirect.com/science/article/pii/S0022311523005822. DOI: 10.1016/j.jnucmat.2023.154815

El-Sayed, A.-R., Mohran, H. S., & Abd El-Lateef, H. M. (2012). Corrosion Study of Zinc, Nickel, and Zinc-Nickel Alloys in Alkaline Solutions by Tafel Plot and Impedance Techniques. *Metallurgical and Materials Transactions. A, Physical Metallurgy and Materials Science*, *43*(2), 619–632. DOI: 10.1007/s11661-011-0908-4

Etaiw, S. E. H., Hassan, G. S., El-Hossiany, A. A., & Fouda, A. S. (2023). Nano-metal–organic frameworks as corrosion inhibitors for strengthening anti-corrosion behavior of carbon steel in a sulfuric acid environment : From synthesis to applications. *RSC Advances*, *13*(22), 15222–15235. https://pubs.rsc.org/en/content/articlehtml/2023/ra/d3ra01644g. DOI: 10.1039/D3RA01644G

Fan, C., Liu, K., Zhu, T., & Peng, Q. (2024). Understanding of Lithium-ion battery degradation using multisine-based nonlinear characterization method. *Energy*, *290*, 130230. https://www.sciencedirect.com/science/article/pii/S036054422400001X ?casa_token=JbG92Cn9BR4AAAAA:V5sugqraQChQTZXHnivtYEe36K jeG6pMjBweEV8hEwbbvg-WOA1yuQ2CwBGQ74W3HW7mHVKMjw. DOI: 10.1016/j.energy.2024.130230

Fan, J., Zhang, M., & Li, H. (2025). Fabrication of Durable Superhydrophobic Surfaces with a Mesh Structure and Drag Reduction by Chemical Etching Technology. *Coatings*, *15*(4), 4. Advance online publication. DOI: 10.3390/coatings15040402

Farooq, S. A., Raina, A., Mohan, S., Arvind Singh, R., Jayalakshmi, S., & Irfan Ul Haq, M. (2022). Nanostructured Coatings : Review on Processing Techniques, Corrosion Behaviour and Tribological Performance. *Nanomaterials (Basel, Switzerland)*, *12*(8), 8. Advance online publication. DOI: 10.3390/nano12081323

Feng, J., Fang, D., Li, J., & Li, J. (2024). A Review of Degradation Mechanisms and Recent Achievements for Na-Ion Layered Transition Metal Oxides. *Batteries & Supercaps*, *7*(3), e202300476. DOI: 10.1002/batt.202300476

Feng, X., Ouyang, M., Liu, X., Lu, L., Xia, Y., & He, X. (2018). Thermal runaway mechanism of lithium ion battery for electric vehicles : A review. *Energy Storage Materials*, *10*, 246–267. https://www.sciencedirect.com/science/article/pii/S2405829716303464?casa_token=ZzIBj4upEYgAAAAA:2ttR2OUE6Cz13QIh1RNnfX6cFsoI6TIxfl9IFGU0tTkqBail44R0flfNuPlRri4KNcawonpCK-St. DOI: 10.1016/j.ensm.2017.05.013

Frankel, G. S. (1998). Pitting corrosion of metals : A review of the critical factors. *Journal of the Electrochemical Society*, *145*(6), 2186–2198. https://iopscience.iop.org/article/10.1149/1.1838615/meta. DOI: 10.1149/1.1838615

Fu, X., Li, W., Li, L., Wang, J., Lu, B., & Wu, J. (2023). Cloud-based pipe corrosion monitoring using electromechanical impedance instrumented piezoelectric ring sensor. *Automation in Construction*, *156*, 105124. https://www.sciencedirect.com/science/article/pii/S0926580523003849?casa_token=Q4rPAJip4psAAAAA:LN41wW7AUBdGV-nEAWGGAOZYWm5R9DO2Ocju6WvQL54rAjb7wvKVEKE3f9vHDmMyv3g6nGuNuPz6. DOI: 10.1016/j.autcon.2023.105124

Geslin, A., Xu, L., Ganapathi, D., Moy, K., Chueh, W. C., & Onori, S. (2025). Dynamic cycling enhances battery lifetime. *Nature Energy*, *10*(2), 172–180. https://www.nature.com/articles/s41560-024-01675-8

Gharieh, A., Sharifian, A., & Dadkhah, S. (2025). Enhanced long-term corrosion resistance and self-healing of epoxy coating with HQ-Zn-PA nanocomposite. *Scientific Reports*, *15*(1), 8154. DOI: 10.1038/s41598-025-92158-7

Goodenough, J. B., & Park, K.-S. (2013). The Li-Ion Rechargeable Battery : A Perspective. *Journal of the American Chemical Society*, *135*(4), 1167–1176. DOI: 10.1021/ja3091438

Gu, C., Yin, L., Li, S., Zhang, B., Liu, X., & Yan, T. (2021). Differential capacitance of ionic liquid and mixture with organic solvent. *Electrochimica Acta*, *367*, 137517. DOI: 10.1016/j.electacta.2020.137517

Gu, W., He, D., Qin, Y., Fu, C., Lu, J., Wang, T., Wang, G., & Sun, B. (2025). From efficiency to sustainability : Organic additives for interfacial regulation in lithium metal batteries. *Chemical Science (Cambridge)*, *16*(39), 18050–18091. DOI: 10.1039/D5SC03975D

Guo, L., Thornton, D. B., Koronfel, M. A., Stephens, I. E., & Ryan, M. P. (2021). Degradation in lithium ion battery current collectors. *Journal of Physics: Energy*, *3*(3), 032015. https://iopscience.iop.org/article/10.1088/2515-7655/ac0c04/meta. DOI: 10.1088/2515-7655/ac0c04

Guo, W., Meng, Y., Hu, Y., Wu, X., Ju, Z., & Zhuang, Q. (2020). Surface and Interface Modification of Electrode Materials for Lithium-Ion Batteries With Organic Liquid Electrolyte. *Frontiers in Energy Research*, *8*, 170. Advance online publication. DOI: 10.3389/fenrg.2020.00170

Haghbin, S., Rezaei Larijani, M., Zolghadri, M., & Hedayati Kia, S. (2025). Automotive battery pack standards and design characteristics : A review. *Discover Applied Sciences*, *7*(7), 731. DOI: 10.1007/s42452-025-07339-w

Han, C., Chen, G., Ma, Y., Ma, J., Shui, X., Dong, S., Xu, G., Zhou, X., Cui, Z., Qiao, L., & Cui, G. (2023). Strategies towards inhibition of aluminum current collector corrosion in lithium batteries. *Energy Materials*, *3*(6), N/A, N/A. DOI: 10.20517/energymater.2023.53

Harsimran, S., Santosh, K., & Rakesh, K. (2021). Overview of corrosion and its control : A critical review. *Proc. Eng. Sci*, *3*(1), 13–24. https://www.researchgate.net/profile/Santosh-Kumar-101/publication/350056220_OVERVIEW_OF_CORROSION_AND_ITS_CONTROL_A_CRITICAL_REVIEW/links/6113dc071ca20f6f8616d092/OVERVIEW-OF-CORROSION-AND-ITS-CONTROL-A-CRITICAL-REVIEW.pdf?_sg%5B0%5D=started_experiment_milestone&origin=journalDetail. DOI: 10.24874/PES03.01.002

He, Z., Kwon, D., & Pecht, M. (2025). Evaluation of IEC 60529 as a standard for liquid protection assessment of portable electronics. *e-Prime - Advances in Electrical Engineering. Electronics and Energy*, *12*, 100952. DOI: 10.1016/j.prime.2025.100952

Holm, R. (2013). *Electric contacts : Theory and application*. Springer Science & Business Media. https://books.google.com/books?hl=fr&lr=&id=yfzuCAAAQBAJ&oi=fnd&pg=PR13&dq=Degradation+Mechanisms+in+Electrical+Contacts%22+Authors:+Holm,+R.&ots=Lgxjjm0i8s&sig=73wOyctWU-XF_6NGmBbkqkmWBeA

Hossain Ahmed, S., Kang, X., & Bade Shrestha, S. O. (2015). Effects of Temperature on Internal Resistances of Lithium-Ion Batteries. *Journal of Energy Resources Technology*, *137*(031901), 031901. Advance online publication. DOI: 10.1115/1.4028698

Idumah, C. I., Obele, C. M., Emmanuel, E. O., & Hassan, A. (2020). Recently Emerging Nanotechnological Advancements in Polymer Nanocomposite Coatings for Anti-corrosion, Anti-fouling and Self-healing. *Surfaces and Interfaces*, *21*, 100734. DOI: 10.1016/j.surfin.2020.100734

Ikeuba, A. I., Njoku, C. N., Ekerenam, O. O., Njoku, D. I., Udoh, I. I., Daniel, E. F., Uzoma, P. C., Etim, I.-I. N., & Okonkwo, B. O. (2024). A review of the electrochemical and galvanic corrosion behavior of important intermetallic compounds in the context of aluminum alloys. *RSC Advances*, *14*(43), 31921–31953. https://pubs.rsc.org/en/content/articlehtml/2024/ra/d4ra06070a. DOI: 10.1039/D4RA06070A

Iravani, D., Farhadian, A., Javidparvar, A. A., & Abrari, H. (2025). MOFs Corrosion Mitigation Potential in Automotive Industries. In *Advancing Corrosion Control with Metal–Organic Frameworks : Beyond Rust* (Vol. 1503, p. 207-222). American Chemical Society. DOI: 10.1021/bk-2025-1503.ch010

Islam, M. A., & Farhat, Z. (2017). Erosion-corrosion mechanism and comparison of erosion-corrosion performance of API steels. *Wear*, *376*, 533–541. https://www.sciencedirect.com/science/article/pii/S004316481730042X?casa_token=RxJIFKK_1iUAAAAA:HYAgo1d1wjAP9COALOj4Ze-yeh-O7_O6Zwvc1IMdph72rK34DJXK4DxsZoHg-9sAHqOSdxo11lWj. DOI: 10.1016/j.wear.2016.12.058

James Abraham, J., Arro, C. R. A., Tariq, H. A., Kahraman, R., Al-Qaradawi, S., Al tahtamouni, T. M., & Shakoor, R. A. (2021). Sodium and lithium incorporated cathode materials for energy storage applications—A focused review. *Journal of Power Sources*, *506*, 230098. DOI: 10.1016/j.jpowsour.2021.230098

Jiang, J., Geng, X., & Zhang, X. (2023). Stress corrosion cracking of magnesium alloys : A review. *Journal of Magnesium and Alloys*, *11*(6), 1906–1930. https://www.sciencedirect.com/science/article/pii/S2213956723001160. DOI: 10.1016/j.jma.2023.05.011

Jin, C., Huang, Y., Li, L., Wei, G., Li, H., Shang, Q., Ju, Z., Lu, G., Zheng, J., Sheng, O., & Tao, X. (2023). A corrosion inhibiting layer to tackle the irreversible lithium loss in lithium metal batteries. *Nature Communications*, *14*(1), 8269. https://www.nature.com/articles/s41467-023-44161-7. DOI: 10.1038/s41467-023-44161-7

Kleiner, K., & Ehrenberg, H. (2019). Challenges Considering the Degradation of Cell Components in Commercial Lithium-Ion Cells : A Review and Evaluation of Present Systems. In Eichel, R.-A. (Ed.), *Electrochemical Energy Storage : Next Generation Battery Concepts* (pp. 169–213). Springer International Publishing., DOI: 10.1007/978-3-030-26130-6_5

Koley, I., Dhar, A., & Kumar, A. (2024). Corrosion in friction stir-welded joints. In *Friction Stir Spot Welding* (pp. 205–220). CRC Press., https://www.taylorfrancis.com/chapters/edit/10.1201/9781003432289-12/corrosion-friction-stir-welded-joints-ishita-koley-arindam-dhar-avinash-kumar DOI: 10.1201/9781003432289-12

Kuruvila, R., Thirumalai Kumaran, S., Khan, M. A., & Uthayakumar, M. (2018). A brief review on the erosion-corrosion behavior of engineering materials. *Corrosion Reviews*, *36*(5), 435–447. DOI: 10.1515/corrrev-2018-0022

Kwon, H., Baek, J., & Kim, H.-T. (2023). Building lithium metal batteries under lean electrolyte conditions : Challenges and progress. *Energy Storage Materials*, *55*, 708–726. DOI: 10.1016/j.ensm.2022.12.016

Lakhouil, S., Boukheir, S., Chbihi, A., & Naamane, S. (2021). Outdoor exposure approach in a harsh environment toward the study of corrosion aspects of primary solar reflectors. *Solar Energy*, *230*, 1070–1084. https://www.sciencedirect.com/science/article/pii/S0038092X2100966X?casa_token=fCu4OkGyQkIAAAAA:sdFjgtWkQ8MHczaM2Vy6G6mhTG2K0oVoMlbBRotTUaTkldj7kkwRGxr5OW0wfYI5AmibCB_C7CW8. DOI: 10.1016/j.solener.2021.11.013

Larché, N., & Dézerville, P. (2011). Review of material selection and corrosion in seawater reverse osmosis desalination plants. *Desalination and Water Treatment*, *31*(1), 121–133. DOI: 10.5004/dwt.2011.2362

Larsson, F., Andersson, P., Blomqvist, P., & Mellander, B.-E. (2017). Toxic fluoride gas emissions from lithium-ion battery fires. *Scientific Reports*, *7*(1), 10018. https://www.nature.com/articles/s41598-017-09784-z. DOI: 10.1038/s41598-017-09784-z

Lee, S., Song, G., Yun, B., Kim, T., Choi, S. H., Kim, H., Doo, S. W., & Lee, K. T. (2024). Revealing the Nanoscopic Corrosive Degradation Mechanism of Nickel-Rich Layered Oxide Cathodes at Low State-of-Charge Levels : Corrosion Cracking and Pitting. *ACS Nano*, *18*(15), 10566–10581. DOI: 10.1021/acsnano.4c00202

Lei, L., Sun, Y., Wang, X., Jiang, Y., & Li, J. (2020). Strategies to enhance corrosion resistance of Zn electrodes for next generation batteries. *Frontiers in Materials*, *7*, 96. https://www.frontiersin.org/articles/10.3389/fmats.2020.00096/full. DOI: 10.3389/fmats.2020.00096

Li, C., Cauwe, M., Mader, L., Schaubroeck, D., & Op de Beeck, M. (2020). Accelerated Hermeticity Testing of Biocompatible Moisture Barriers Used for the Encapsulation of Implantable Medical Devices. *Coatings*, *10*(1), 19. DOI: 10.3390/coatings10010019

Li, J.-S., Li, M.-X., Hua, Z.-M., Hu, Z.-T., & Wang, H.-Y. (2022). Effect of micro-alloyed Ce on the microstructure evolution and mechanical properties of rolled Mg–0.6Al–0.5Mn–0.2Ca alloy sheets. *Journal of Materials Research and Technology*, *19*, 3088–3099. DOI: 10.1016/j.jmrt.2022.06.064

Li, L., Chakik, M., & Prakash, R. (2021). A review of corrosion in aircraft structures and graphene-based sensors for advanced corrosion monitoring. *Sensors (Basel), 21*(9), 2908. https://www.mdpi.com/1424-8220/21/9/2908. DOI: 10.3390/s21092908

Li, W., Li, H., He, Z., Ji, W., Zeng, J., Li, X., Zhang, Y., Zhang, P., & Zhao, J. (2022). Electrochemical failure results inevitable capacity degradation in Li-ion batteries—A review. *Energies, 15*(23), 9165. https://www.mdpi.com/1996-1073/15/23/9165. DOI: 10.3390/en15239165

Li, X., Liu, P., Han, C., Cai, T., Cui, Y., Xing, W., & Zhi, C. (2025a). Corrosion of metallic anodes in aqueous batteries. *Energy & Environmental Science, 18*(5), 2050–2094. DOI: 10.1039/D5EE00075K

Li, X., Liu, P., Han, C., Cai, T., Cui, Y., Xing, W., & Zhi, C. (2025b). Corrosion of metallic anodes in aqueous batteries. *Energy & Environmental Science, 18*(5), 2050–2094. DOI: 10.1039/D5EE00075K

Li, Y., Guo, W., Stroe, D.-I., Zhao, H., Kristensen, P. K., Jensen, L. R., Pedersen, K., & Gurevich, L. (2024). Evolution of aging mechanisms and performance degradation of lithium-ion battery from moderate to severe capacity loss scenarios. *Chemical Engineering Journal, 498*, 155588. https://www.sciencedirect.com/science/article/pii/S1385894724070797. DOI: 10.1016/j.cej.2024.155588

Lim, J. J. N., Cai, Y., & Srinivasan, M. (2025). A Universal Highly Concentrated Electrolyte for Improved Cycling Stability in Li (Ni1-x-yMnxCoy) O2-NMC-Based Batteries. *Energies, 18*(4), 974. https://www.mdpi.com/1996-1073/18/4/974. DOI: 10.3390/en18040974

Liu, B., Lv, T., Zhou, A., Zhu, X., Lin, Z., Lin, T., & Suo, L. (2024a). Aluminum corrosion–passivation regulation prolongs aqueous batteries life. *Nature Communications, 15*(1), 2922. https://www.nature.com/articles/s41467-024-47145-3. DOI: 10.1038/s41467-024-47145-3

Liu, B., Lv, T., Zhou, A., Zhu, X., Lin, Z., Lin, T., & Suo, L. (2024b). Aluminum corrosion–passivation regulation prolongs aqueous batteries life. *Nature Communications, 15*(1), 2922. DOI: 10.1038/s41467-024-47145-3

Liu, H., Hwang, J., Matsumoto, K., & Hagiwara, R. (2023). Systematic Study of Aluminum Corrosion in Ionic Liquid Electrolytes for Sodium-Ion Batteries : Impact of Temperature and Concentration. *ACS Applied Materials & Interfaces, 15*(29), 35062–35071. DOI: 10.1021/acsami.3c06812

Liu, T., Ma, L., Wang, X., Wang, J., Qian, H., Zhang, D., & Li, X. (2021). Self-healing corrosion protective coatings based on micro/nanocarriers : A review. *Corrosion Communications, 1*, 18–25. DOI: 10.1016/j.corcom.2021.05.004

Liu, Y., Zhang, R., Wang, J., & Wang, Y. (2021). Current and future lithium-ion battery manufacturing. *iScience, 24*(4), 102332. https://www.cell.com/iscience/fulltext/S2589-0042(21)00300-X. DOI: 10.1016/j.isci.2021.102332

Luo, J., Zou, D., Wang, Y., Wang, S., & Huang, L. (2022). Battery thermal management systems (BTMs) based on phase change material (PCM) : A comprehensive review. *Chemical Engineering Journal, 430*, 132741. https://www.sciencedirect.com/science/article/pii/S1385894721043199?casa_token=EXHZGhN-J_8AAAAA:_RalLgCkjJjtwSlf4xig3QNejQ8Ia_bIRM9ehjUaOOWTz7cpwcPFz3r4eYzmx8RtWfD-N1AOlGME. DOI: 10.1016/j.cej.2021.132741

Ma, T., Xu, G.-L., Li, Y., Wang, L., He, X., Zheng, J., Liu, J., Engelhard, M. H., Zapol, P., Curtiss, L. A., Jorne, J., Amine, K., & Chen, Z. (2017). Revisiting the Corrosion of the Aluminum Current Collector in Lithium-Ion Batteries. *Journal of Physical Chemistry Letters, 8*(5). *The Journal of Physical Chemistry Letters, 125564*(5), 1072–1077. Advance online publication. DOI: 10.1021/acs.jpclett.6b02933

Mahek, M. K., Ramadan, M., Ghazal, M., Riaz, F., Choi, D. S., & Alkhedher, M. (2025). Advanced thermal management with heat pipes in lithium-ion battery systems : Innovations and AI-driven optimization. *Next Energy, 7*, 100223. https://www.sciencedirect.com/science/article/pii/S2949821X24001285

Malucci, R. D. (2003). Fretting corrosion degradation, threshold behavior and contact instability [electrical contacts]. *Proceedings of the Forty-Ninth IEEE Holm Conference on Electrical Contacts, 2003.*, 2-15. https://ieeexplore.ieee.org/abstract/document/1246472/

Mansir, I. B., & Okonkwo, P. C. (2025). Component Degradation in Lithium-Ion Batteries and Their Sustainability : A Concise Overview. *Sustainability (2071-1050), 17*(3). https://search.ebscohost.com/login.aspx?direct=true&profile=ehost&scope=site&authtype=crawler&jrnl=20711050&AN=182983794&h=wUuJEjiGFgSmhLcfMbtuCt6pIxPlI5b08UTxY2hWBWhEo9saeIS7oj87ZbvU9eNjGbdgpVS5au3yH52KhYGaxA%3D%3D&crl=c

McCafferty, E. (2010). *Introduction to corrosion science*. Springer Science & Business Media. https://books.google.com/books?hl=fr&lr=&id=g9oXcE-gQKUC&oi=fnd&pg=PR5&dq=%E2%80%A2%09McCafferty,+E.+(2010).+Introduction+to+Corrosion+Science.+Springer,+pp.+58-63.&ots=6UGkFXZqAc&sig=LZbT0jwmjrEV1lOUHuTOEJNanVo

Mungwari, C. P., Obadele, B. A., & King'ondu, C. K. (2024). Phytochemicals as green and sustainable corrosion inhibitors for mild steel and aluminium. *Results in Surfaces and Interfaces*, 100374. https://www.sciencedirect.com/science/article/pii/S2666845924001946

Mzioud, K., Ferraa, N., Aribou, Z., Rached, S., Ouakki, M., Habsaoui, A., & Touhami, M. E. (2025). Green Corrosion Protection : Plant Extracts as Sustainable Solutions. In *Emerging Technologies and Industrial Applications of Corrosion Science* (p. 215-254). IGI Global Scientific Publishing. https://www.igi-global.com/chapter/green-corrosion-protection/376959

Mzioud, K., Habsaoui, A., Ouakki, M., Galai, M., El Fartah, S., & Ebn Touhami, M. (2020). Inhibition of copper corrosion by the essential oil of Allium sativum in 0.5M H2SO4 solutions. *SN Applied Sciences*, *2*(9), 1611. DOI: 10.1007/s42452-020-03393-8

Mzioud, K., Habsaoui, A., Rached, S., Lachhab, R., Dkhireche, N., Ouakki, M., Galai, M., El Fartah, S., & Touhami, M. E. (2022a). Synergistic Effect from Allium Sativum Essential Oil and Diethylthiourea for Corrosion Inhibition of Carbon Steel in 0.5 M H2SO4 Medium. In Vaseashta, A., Achour, M. E., Mabrouki, M., Fasquelle, D., & Tachafine, A. (Éds.), *Proceedings of the Sixth International Symposium on Dielectric Materials and Applications (ISyDMA'6)* (p. 251-266). Springer International Publishing. DOI: 10.1007/978-3-031-11397-0_23

Mzioud, K., Habsaoui, A., Rached, S., Lachhab, R., Dkhireche, N., Ouakki, M., Galai, M., El Fartah, S., & Touhami, M. E. (2022b). Synergistic Effect from Allium Sativum Essential Oil and Diethylthiourea for Corrosion Inhibition of Carbon Steel in 0.5 M H2SO4 Medium. In Vaseashta, A., Achour, M. E., Mabrouki, M., Fasquelle, D., & Tachafine, A. (Éds.), *Proceedings of the Sixth International Symposium on Dielectric Materials and Applications (ISyDMA'6)* (p. 251-266). Springer International Publishing. DOI: 10.1007/978-3-031-11397-0_23

Narayana, T. L., Venkatesh, C., Kiran, A., Kumar, A., Khan, S. B., Almusharraf, A., & Quasim, M. T. (2024). Advances in real time smart monitoring of environmental parameters using IoT and sensors. *Heliyon*, *10*(7), e28195. https://www.cell.com/heliyon/fulltext/S2405-8440(24)04226-9. DOI: 10.1016/j.heliyon.2024.e28195

Nogales, P. M., Lee, S., Yang, S., Yang, I., Choi, S. H., Park, S.-M., Lee, J. H., Kim, C. J., An, J.-C., & Jeong, S.-K. (2024). Stabilizing the Solid Electrolyte Interphase of SiOx Negative Electrodes : The Role of Fluoroethylene Carbonate in Enhancing Electrochemical Performance. *Batteries*, *10*(11), 385. DOI: 10.3390/batteries10110385

Palakkool, N. M., Taverne, M. P. C., Bell, O. G., Jones, C. P., Mar, J. D., Ho, D. T., Barrioz, V., Qu, Y., Ren, Z., Huang, C.-C., & Ho, Y.-L. D. (2025). *Conformally-coated Nickel-Carbon Nitride on Structured Pyrolytic Carbon Electrodes for Enhanced Hydrogen Evolution* (No. arXiv:2505.06627). arXiv. https://doi.org//arXiv.2505.06627DOI: 10.48550

Park, J. W., Park, D. H., Go, S., Nam, D.-H., Oh, J., Han, Y.-K., & Lee, H. (2022). Malonatophosphate as an SEI- and CEI-forming additive that outperforms malonatoborate for thermally robust lithium-ion batteries. *Energy Storage Materials, 50,* 75–85. DOI: 10.1016/j.ensm.2022.05.009

Pei, A., Zheng, G., Shi, F., Li, Y., & Cui, Y. (2017). Nanoscale Nucleation and Growth of Electrodeposited Lithium Metal. *Nano Letters, 17*(2), 1132–1139. DOI: 10.1021/acs.nanolett.6b04755

Pender, J. P., Jha, G., Youn, D. H., Ziegler, J. M., Andoni, I., Choi, E. J., Heller, A., Dunn, B. S., Weiss, P. S., Penner, R. M., & Mullins, C. B. (2020). Electrode Degradation in Lithium-Ion Batteries. *ACS Nano, 14*(2), 1243–1295. DOI: 10.1021/acsnano.9b04365

Perez, N. (2024). Electrochemical Corrosion. In N. Perez, *Materials Science : Theory and Engineering* (p. 835-898). Springer Nature Switzerland. DOI: 10.1007/978-3-031-57152-7_16

Phillip, N. D., Westover, A. S., Daniel, C., & Veith, G. M. (2020). Structural Degradation of High Voltage Lithium Nickel Manganese Cobalt Oxide (NMC) Cathodes in Solid-State Batteries and Implications for Next Generation Energy Storage. *ACS Applied Energy Materials, 3*(2), 1768–1774. DOI: 10.1021/acsaem.9b02230

Phull, B. (2003). Evaluating intergranular corrosion. *Corrosion: Fundamentals, Testing, and Protection*, 568-571. https://dl.asminternational.org/handbooks/edited-volume/46/chapter-abstract/539380/Evaluating-Intergranular-Corrosion

Phung, M. T., Vu, T. T. P., Lee, S. B., Kong, I., Kim, M., Akhtar, M. S., & Yang, O.-B. (2025). Coupling of Ammonium Dihydrogen Phosphate Additives with LiPF6 Electrolytes for Improving Thermal Stability and Performance of Lithium-Ion Batteries. *Micromachines, 16*(9), 966. DOI: 10.3390/mi16090966

Qu, J. G., Jiang, Z. Y., & Zhang, J. F. (2022). Investigation on lithium-ion battery degradation induced by combined effect of current rate and operating temperature during fast charging. *Journal of Energy Storage, 52,* 104811. DOI: 10.1016/j.est.2022.104811

Rached, S., Habsaoui, A., Mzioud, K., Lachhab, R., Haida, S., Errahmany, N., Galai, M., & Touhami, M. E. (2023). Valorization of the green corrosion inhibitor Marrubium vulgare L.: Electrochemical, thermodynamic, theoretical & surface studies. *Chemical Data Collections, 48*, 101099. https://www.sciencedirect.com/science/article/pii/S2405830023001106?casa_token=mRSqRD11L84AAAAA:BkWI4_NN5bUHncl569L4SzScDUKOXqhaxYxIWOReFDl-QLGS5aUyd7K8LcTaL1ItwnBdzQMjVL8

Rached, S., Mzioud, K., Er-rajy, M., Habsaoui, A., Lachhab, R., Hammouti, B., & Ebn Touhami, M. (2025). Combined experimental and computational evaluation of *Mentha pulegium* L., for sustainable corrosion protection. *Hybrid Advances, 11*, 100551. DOI: 10.1016/j.hybadv.2025.100551

Rached, S., Mzioud, K., Habsaoui, A., Galai, M., Dahmani, K., Ouakki, M., Fartah, S. E. L., Dkhireche, N., & Touhami, M. E. (2024a). Inhibition of Copper Corrosion in Sulfuric Acid by Mentha Pulegium L. *Portugaliae Electrochimica Acta, 42*(2), 137–153. https://www.peacta.org/articles_upload/v4202a05_137_153.pdf. DOI: 10.4152/pea.2024420205

Rached, S., Mzioud, K., Habsaoui, A., Galai, M., Dahmani, K., Ouakki, M., Fartah, S. E. L., Dkhireche, N., & Touhami, M. E. (2024b). Inhibition of Copper Corrosion in Sulfuric Acid by Mentha Pulegium L. *Portugaliae Electrochimica Acta, 42*(2), 137–153. https://www.peacta.org/articles_upload/v4202a05_137_153.pdf. DOI: 10.4152/pea.2024420205

Rahman, T., & Alharbi, T. (2024). Exploring lithium-Ion battery degradation : A concise review of critical factors, impacts, data-driven degradation estimation techniques, and sustainable directions for energy storage systems. *Batteries, 10*(7), 220. https://www.mdpi.com/2313-0105/10/7/220

Rao, P., & Mulky, L. (2023). Erosion-Corrosion of Materials in Industrial Equipment : A Review. *ChemElectroChem, 10*(16), e202300152. DOI: 10.1002/celc.202300152

Rautio, T., Torbati-Sarraf, H., Allam, T., Järvenpää, A., & Hamada, A. (2022). Enhancement of electrical conductivity and corrosion resistance by gold-nickel coating of additively manufactured AlSi10Mg alloy. *journal of materials research and technology, 17*, 521-536. https://www.sciencedirect.com/science/article/pii/S2238785422000229

Ren, H., Meng, X., Lin, Y., & Shao, Z. (2022). Microstructure formation mechanism of catalyst layer and its effect on fuel cell performance : Effect of dispersion medium composition. *Journal of Energy Chemistry*, *73*, 588–598. https://www.sciencedirect.com/science/article/pii/S2095495622003412?casa_token=OAvLjzp-pyQAAAAA:TWRXklAl3xgID0HT5eBpr1uATEKfmVuqRtDcr_sw2pYbA9KG1x2z_FHQQNgoVxIU7J07WuBAyrE. DOI: 10.1016/j.jechem.2022.06.034

Saji, V. S. (2023). Corrosion and Materials Degradation in Electrochemical Energy Storage and Conversion Devices. *ChemElectroChem*, *10*(11), e202300136. DOI: 10.1002/celc.202300136

Sanguesa, J. A., Torres-Sanz, V., Garrido, P., Martinez, F. J., & Marquez-Barja, J. M. (2021). A Review on Electric Vehicles : Technologies and Challenges. *Smart Cities*, *4*(1), 1. Advance online publication. DOI: 10.3390/smartcities4010022

Shahjalal, M., Roy, P. K., Shams, T., Fly, A., Chowdhury, J. I., Ahmed, M. R., & Liu, K. (2022). A review on second-life of Li-ion batteries : Prospects, challenges, and issues. *Energy*, *241*, 122881. https://www.sciencedirect.com/science/article/pii/S0360544221031303?casa_token=MC3QtHzgehUAAAAA:Eoy7g8lhLiSNAqrfiY6zXJXMs4bP1Xa7bs0hoQU6yx_c3adT6vDd0UUybmSSHH5uOL3n3XSrko0. DOI: 10.1016/j.energy.2021.122881

Shen, Q., Wang, Y., Han, G., Li, X., Yuan, T., Sun, H., Gong, Y., & Chen, T. (2023). Recent Progress in Electrolyte Additives for Highly Reversible Zinc Anodes in Aqueous Zinc Batteries. *Batteries*, *9*(5), 5. Advance online publication. DOI: 10.3390/batteries9050284

Shen, W., Wang, N., Zhang, J., Wang, F., & Zhang, G. (2022). Heat Generation and Degradation Mechanism of Lithium-Ion Batteries during High-Temperature Aging. *ACS Omega*, *7*(49), 44733–44742. DOI: 10.1021/acsomega.2c04093

Shifler, D. A., & Aylor, D. M. (2005). Factors Affecting Corrosion Performance and Testing of Materials and Components in Sea Water. *NACE CORROSION*, NACE-05224. https://onepetro.org/NACECORR/proceedings-abstract/CORR05/All-CORR05/115133

Shirvanimoghaddam, M., Shirvanimoghaddam, K., Abolhasani, M. M., Farhangi, M., Barsari, V. Z., Liu, H., Dohler, M., & Naebe, M. (2018). Paving the path to a green and self-powered internet of things. *Electrical Engineering and Systems Science*. https://www.researchgate.net/profile/Mahyar-Shirvanimoghaddam/publication/321604366_Paving_the_Path_to_a_Green_and_Self-Powered_Internet_of_Things/links/5a2c7ede45851552ae7c9b36/Paving-the-Path-to-a-Green-and-Self-Powered-Internet-of-Things.pdf

Shokri, A., & Sanavi Fard, M. (2022). Corrosion in seawater desalination industry : A critical analysis of impacts and mitigation strategies. *Chemosphere*, *307*, 135640. DOI: 10.1016/j.chemosphere.2022.135640

Soares, C. G., Garbatov, Y., Zayed, A., & Wang, G. (2009). Influence of environmental factors on corrosion of ship structures in marine atmosphere. *Corrosion Science*, *51*(9), 2014–2026. https://www.sciencedirect.com/science/article/pii/S0010938X09002315?casa_token=z_J86PdPa1QAAAAA:Qq5JFujA_ZkgTbZNx55V9DloWL9fcmr7XMY4YHDoGA548zthUN_ladjw2VNioO3xLYZydyHwwlk. DOI: 10.1016/j.corsci.2009.05.028

Song, L., Zheng, Y., Xiao, Z., Wang, C., & Long, T. (2022). Review on Thermal Runaway of Lithium-Ion Batteries for Electric Vehicles. *Journal of Electronic Materials*, *51*(1), 30–46. DOI: 10.1007/s11664-021-09281-0

Spotnitz, R., & Franklin, J. (2003). Abuse behavior of high-power, lithium-ion cells. *Journal of Power Sources*, *113*(1), 81–100. https://www.sciencedirect.com/science/article/pii/S0378775302004883?casa_token=SMVVrGYC9lYAAAAA:rNUbAIfdab-JXETsYg756z5fo710OCM_k4MzwolPEA4gcX8c4uqZW2a2CD_kVXuuH27oIb-nK0BP. DOI: 10.1016/S0378-7753(02)00488-3

Su, X., Geng, Y., Huang, L., Li, S., Wang, Q., Xu, Z., & Tian, S. (2024). Review on dehumidification technology in low and extremely low humidity industrial environments. *Energy*, *302*, 131793. DOI: 10.1016/j.energy.2024.131793

Sullivan, M., Tang, P., & Meng, X. (2022). Atomic and Molecular Layer Deposition as Surface Engineering Techniques for Emerging Alkali Metal Rechargeable Batteries. *Molecules (Basel, Switzerland)*, *27*(19), 6170. DOI: 10.3390/molecules27196170

Suppa, M. (2007). Conformal coatings and their increasing importance for a safe operation of electronic assemblies. *Circuit World*, *33*(4), 60–67. https://www.emerald.com/insight/content/doi/10.1108/03056120710836954/full/html. DOI: 10.1108/03056120710836954

Taffese, W. Z., Nigussie, E., & Isoaho, J. (2019). Internet of things based durability monitoring and assessment of reinforced concrete structures. *Procedia Computer Science*, *155*, 672–679. https://www.sciencedirect.com/science/article/pii/S1877050919310129. DOI: 10.1016/j.procs.2019.08.096

Taghipour, S. (2011). *Reliability and maintenance of medical devices* [PhD Thesis]. https://utoronto.scholaris.ca/items/dbb4c9ad-ddf0-4016-a563-aedc0055a233

Takada, R., Hirasawa, K., Takahashi, H., & Chiba, M. (2023). Self-Healing Coatings Consisting of an Outer Electrodeposited Epoxy Resin Layer and an Inner Porous Anodic Oxide Layer with Healing Agents for the Corrosion Protection of Al Alloys. *Corrosion and Materials Degradation*, *4*(3), 3. Advance online publication. DOI: 10.3390/cmd4030027

Tortorelli, P. F., & Natesan, K. (1998). Critical factors affecting the high-temperature corrosion performance of iron aluminides. *Materials Science and Engineering A*, *258*(1-2), 115–125. https://www.sciencedirect.com/science/article/pii/S0921509398009241. DOI: 10.1016/S0921-5093(98)00924-1

Tran, M.-K., & Fowler, M. (2020). A review of lithium-ion battery fault diagnostic algorithms : Current progress and future challenges. *Algorithms*, *13*(3), 62. https://www.mdpi.com/1999-4893/13/3/62. DOI: 10.3390/a13030062

Velasco-Arellano, H., Castillo-Magallanes, N., Visairo-Cruz, N., Núñez-Gutiérrez, C. A., & Lázaro, I. (2024). Parametric Correlation Analysis between Equivalent Electric Circuit Model and Mechanistic Model Interpretation for Battery Internal Aging. *World Electric Vehicle Journal*, *15*(7), 291. https://www.mdpi.com/2032-6653/15/7/291

Vinodhini, S. P., & Xavier, J. R. (2025). Recent progress in graphene-based nanocomposites for enhanced energy storage and corrosion protection. *Journal of Materials Science*, *60*(34), 14837–14879. DOI: 10.1007/s10853-025-11307-5

Wang, J., Chen, Y., Zhang, Y., Zhang, Y., Li, J., Liu, J., Liu, Y., & Li, W. (2022). Microstructure evolution and acid corrosion behavior of CoCrFeNiCu1−xMox high-entropy alloy coatings fabricated by coaxial direct laser deposition. *Corrosion Science*, *198*, 110108. DOI: 10.1016/j.corsci.2022.110108

Wang, J., Zhou, Y., Zhuo, Y., Fang, K., Wang, S., Zhao, B., Zhou, J., & Wang, H. (2025). The challenges and strategies towards high-performance anode-free post-lithium metal batteries. *Chemical Science*. https://pubs.rsc.org/en/content/articlehtml/2025/sc/d4sc06630h

Wang, Q., Zhao, J., Yang, X., Li, J., Wu, C., Shen, D., Cheng, C., & Xu, L.-H. (2024). Tuning the electronic metal-carbon interactions in Lignin-based carbon-supported ruthenium-based electrocatalysts for enhanced hydrogen evolution reactions. *Journal of Colloid and Interface Science*, *664*, 251–262. DOI: 10.1016/j.jcis.2024.03.019

Wang, S., Zheng, Z., Long, J., Wang, J., Zheng, K., Ke, Z., Luo, Z., Pokrovsky, A. I., & Khina, B. B. (2024). Recent advances in wear-resistant steel matrix composites : A review of reinforcement particle selection and preparation processes. *Journal of Materials Research and Technology*, *29*, 1779–1797. DOI: 10.1016/j.jmrt.2024.01.195

Wang, Y.-Y., Zhang, X.-Q., Zhou, M.-Y., & Huang, J.-Q. (2023). Mechanism, quantitative characterization, and inhibition of corrosion in lithium batteries. *Nano Research Energy*, *2*(1), e9120046–e9120046. DOI: 10.26599/NRE.2023.9120046

Wang, Z., Zhao, Q., Wang, S., Song, Y., Shi, B., & He, J. (2024). Aging and post-aging thermal safety of lithium-ion batteries under complex operating conditions : A comprehensive review. *Journal of Power Sources*, *623*, 235453. https://www.sciencedirect.com/science/article/pii/S0378775324014058?casa_token=owJ4UR09y9IAAAAA:rh6w7hzS40RDaJTuc_HFYYXd4j3k-1AdXgUb4BSWj9V0yFPbNb9AWtx6Xualo4t6nD0B8Dzb3pCp. DOI: 10.1016/j.jpowsour.2024.235453

Wasim, M., & Djukic, M. B. (2022). External corrosion of oil and gas pipelines : A review of failure mechanisms and predictive preventions. *Journal of Natural Gas Science and Engineering*, *100*, 104467. https://www.sciencedirect.com/science/article/pii/S1875510022000580?casa_token=JybaTL5yuUwAAAAA:YMOgKkToaB7jndzTlzm315u5w9_kS8tZwd0rduG6v9evNezoj-2mgv4lP3imQifE8bHzxUUHfwk. DOI: 10.1016/j.jngse.2022.104467

Wasim, M., Shoaib, S., Mubarak, N. M., Inamuddin, , & Asiri, A. M. (2018). Factors influencing corrosion of metal pipes in soils. *Environmental Chemistry Letters*, *16*(3), 861–879. DOI: 10.1007/s10311-018-0731-x

Williams, D. M., Jimkoski, M., & Williams, M. K. (2024). Electrochemical Models for Connector Degradation. *2024 IEEE 69th Holm Conference on Electrical Contacts (HOLM)*, 1-7. DOI: 10.1109/HOLM56222.2024.10768585

Wilson, F. G. (1971). Mechanism of Intergranular Corrosion of Austenitic Stainless Steels—Literature Review. *British Corrosion Journal*, *6*(3), 100–108. DOI: 10.1179/000705971798323874

Wu, X., & Du, Z. (2021a). Study of the corrosion behavior of LiFSI based electrolyte for Li-ion cells. *Electrochemistry Communications*, *129*, 107088. https://www.sciencedirect.com/science/article/pii/S1388248121001727. DOI: 10.1016/j.elecom.2021.107088

Wu, X., & Du, Z. (2021b). Study of the corrosion behavior of LiFSI based electrolyte for Li-ion cells. *Electrochemistry Communications*, *129*, 107088. DOI: 10.1016/j.elecom.2021.107088

Yan, C., Xu, R., Xiao, Y., Ding, J.-F., Xu, L., Li, B.-Q., & Huang, J.-Q. (2020). Toward Critical Electrode/Electrolyte Interfaces in Rechargeable Batteries. *Advanced Functional Materials*, *30*(23), 1909887. DOI: 10.1002/adfm.201909887

Yang, X., Zhang, C., Yang, X., & Xu, Z. (2023). Water-Mediated attraction between Like-charged species involved in calcium phosphate nucleation. *Journal of Molecular Liquids*, *378*, 121585. DOI: 10.1016/j.molliq.2023.121585

Yeung, K. K., Zhang, X., Kwok, S. C. T., Ciucci, F., & Yuen, M. M. F. (2015). Enhanced cycle life of lead-acid battery using graphene as a sulfation suppression additive in negative active material. *RSC Advances*, *5*(87), 71314–71321. DOI: 10.1039/C5RA11114E

Yu, Y., Zuo, Y., Zhang, Z., Wu, L., Ning, C., & Zuo, C. (2019). Al2O3 Coatings on Zinc for Anti-Corrosion in Alkaline Solution by Electrospinning. *Coatings*, *9*(11), 692. DOI: 10.3390/coatings9110692

Yuan, Y., Li, Z., & Pu, S. D. (2024). Corrosion and Degradation in Aqueous Zn-based Batteries. In Saji, V. S. (Ed.), *Corrosion and Degradation in Fuel Cells, Supercapacitors and Batteries* (pp. 325–352). Springer Nature Switzerland., DOI: 10.1007/978-3-031-57012-4_14

Zafar, S., Kahraman, R., & Shakoor, R. A. (2024). Recent developments and future prospective of polyurethane coatings for corrosion protection – a focused review. *European Polymer Journal*, *220*, 113421. DOI: 10.1016/j.eurpolymj.2024.113421

Zanoni, E., & Pavan, P. (1993). Improving the reliability and safety of automotive electronics. *IEEE Micro*, *13*(1), 30–48. https://ieeexplore.ieee.org/abstract/document/210523/. DOI: 10.1109/40.210523

Zhang, G., Wei, X., Chen, S., Han, G., Zhu, J., & Dai, H. (2022). Investigation the Degradation Mechanisms of Lithium-Ion Batteries under Low-Temperature High-Rate Cycling. *ACS Applied Energy Materials*, *5*(5), 6462–6471. DOI: 10.1021/acsaem.2c00957

Zhou, B., Stoševski, I., Bonakdarpour, A., & Wilkinson, D. P. (2024). Suppressing Chemical and Galvanic Corrosion in Anode-Free Lithium Metal Batteries Through Electrolyte Design. *Advanced Functional Materials*, *34*(7), 2311212. DOI: 10.1002/adfm.202311212

Zhu, W., Ren, Y., Yu, Z., Tang, F., Xu, Y., & Xu, Y. (2024). Characteristics of the non-uniform corrosion of the steel bars extracted from the marine transportation infrastructures. *Journal of Building Engineering*, *87*, 109108. https://www.sciencedirect.com/science/article/pii/S2352710224006764?casa_token=bRu8t6E2mM4AAAAA:hJziApLJG3vZrO3f5k9HD8t8vXbijfhCmL5hSjxpF_Cl6CEMJRtmo8kvhLGfhVF2rHwWwrEZ_7Ul. DOI: 10.1016/j.jobe.2024.109108

Chapter 5

Recent Advances in Supercapacitor Materials:
A Review on Performance Enhancement, Flexibility, and Scalability for Next-Generation Energy Storage

Zakia Aribou
https://orcid.org/0000-0002-5422-7062
Ibn Tofail University, Morocco

Nouhaila Ferraa
https://orcid.org/0000-0003-1817-7798
Ibn Tofail University, Morocco

Khaoula Mzioud
https://orcid.org/0009-0007-8781-5024
Ibn Tofail University, Morocco

Said Bakkali
https://orcid.org/0000-0001-6484-1878
University of Hassan II Casablanca, Morocco

Moussa Ouakki
Ibn Tofail University, Morocco

Mohamed Ebn Touhami
Ibn Tofail University, Morocco

ABSTRACT

Supercapacitors are attracting attention as high-performance energy storage devices, especially for flexible electronics. They offer high power density, long cycle life, and fast charge/discharge, making them strong alternatives to batteries in applications requiring quick energy delivery. This study reviews recent advances in supercapacitor materials, focusing on enhancing electrode performance through nanostructures and conductive composites. These innovations have significantly improved specific capacitance and electrochemical stability. The development of heat-resistant separators and highly conductive electrolytes has further improved

DOI: 10.4018/979-8-3373-3156-0.ch005

performance under harsh conditions. Research is also advancing flexible supercapacitors to meet demand for wearable and portable electronics. Their improved mechanical flexibility and solid-state performance enable new tech integrations. However, large-scale production faces challenges such as high manufacturing costs and complex material synthesis. Continued efforts are needed to reduce costs and enable commercial use across energy storage sectors

1. INTRODUCTION

The rapid growth of electronic technologies and digitalization has led to a continuous increase in worldwide energy consumption. This growth was driven by urbanization, the use of more and more connected devices, developing transportation infrastructures, and rapid industrialization(Mahmood et al. 2024; Reenu et al. 2024a; Wu et al. 2016a). Even with advances within renewable energy development, oil, natural gas, and coal continue to be heavily utilized by society to meet demand for energy. (Komal Zafar et al. 2024; Strielkowski et al. 2021; Wang and Azam 2024). Even worse, the consumption of such resources makes a greater contribution to greenhouse gas emission and global warming(Wang and Azam 2024). Scientifically speaking, moving into a future with no dependence on fossil fuels, the scientific community is in pursuit of greener and cleaner energy solutions. Of these, renewable energy sources, namely, solar, wind, and hydropower, offer a desirable trade-off(Adedoja, Sadiku, and Hamam 2023; Dissanayake and Kularatna-Abeywardana 2024; Olabi et al. 2022). That said, renewables are intermittent by nature and the development of efficient, stable energy storage solutions is crucial to smoothing the intermittency of a renewable system. Energy storage technologies are thus being conceived as the most important factors in smart, sustainable energy grids of the future(Ayorinde et al. 2024; Muralee Gopi et al. 2025a; Olabi et al. 2022).

Among different energy storage technologies, electrochemical devices occupy a special position in this space. These devices refer to batteries, conventional capacitors, and supercapacitors, which exhibit quite distinct characteristics (Muralee Gopi et al. 2025a; Zhang, Gu, and Chen 2023a). Batteries, for instance, offer high energy density (600 Wh/kg) and are suitable for storage with longer durations but come with a shorter life span and a slower charging period than both methods of energy storage, as in the literature (Meena et al. 2023). In contrast, conventional capacitors offer ultra-fast charging and discharging but store very little energy (0.01–0.05 Wh/kg) (Meena et al. 2023). Supercapacitors combine both methods during storage and power density period, showing energy densities ranging from 10-100 Wh/kg, with supercapacitors exhibiting significantly higher power densities than even batteries (Meena et al. 2023; Pandolfo and Hollenkamp 2006). As shown in the Ragone diagram (Fig. 1) and data in Table 1, supercapacitors have performance characteristics between batteries and capacitors, which gives them an advantage in applications requiring high power and fast cycling.

Figure 1. **Ragone plot for various storage devices**

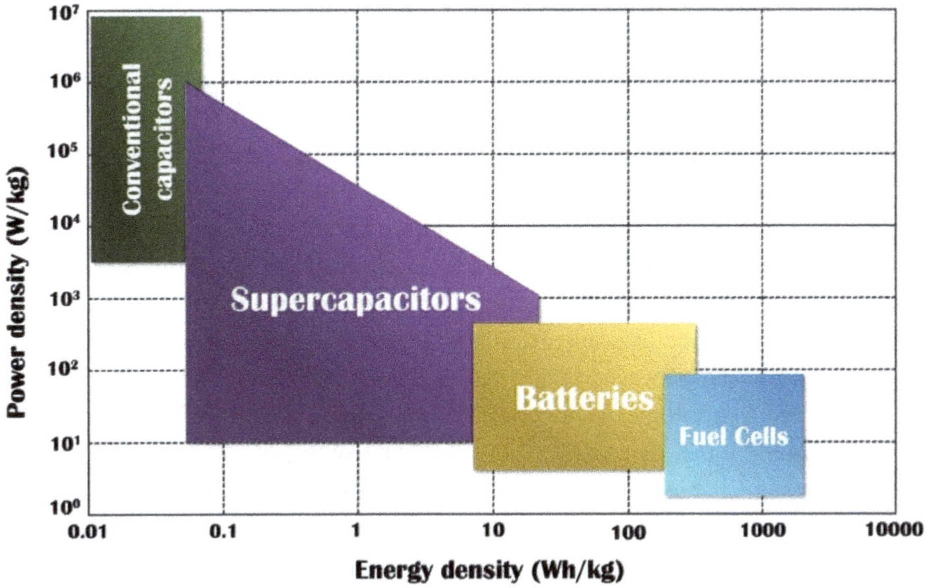

Table 1. ***Comparison of essential characteristics of the battery, supercapacitor, and conventional capacitor(Meena et al. 2023)***

Characteristic	Conventional capacitor	Supercapacitor	Battery
Charging/discharging time	Picoseconds to milliseconds	Milliseconds to seconds	1 to 10 h
Energy density	0.01 to 0.05 Wh/kg	1 to 10 Wh/kg	8 to 600 Wh/kg
Power density	0.25 to 10,000 Wh/kg	1 to 120 Wh/kg	0.005 to 0.4 Wh/kg
Weight	1 to 10 kg	0.001 to 0.230 kg	1 to >10 kg
Operating voltage	2.3 to 2.75 V	6 to 800 V	1.25 to 4.2 V
Life	>100,000 cycles	50,000 + hrs unlimited cycles	150 to 1500 cycles

The operation of supercapacitors depends on two main mechanisms: electrostatic charge storage via the formation of the electrical double layer (EDLC), a concept introduced by Helmholtz in the 19th century, and pseudocapacitance, which involves quick and reversible redox reactions at the electrode surface (Czagany et al. 2024; Rashid Khan and Latif Ahmad 2025a). The modern development of supercapacitors started in the 20th century with early electrochemical prototypes created by General Electric (Dutta et al. 2023). Later, Standard Oil commercialized these devices, and NEC popularized the term "supercapacitors" (Lakshmi and Vedhanarayanan 2023a; Shuja et al. 2024). Since then, significant progress has been made, especially with new electrode materials like carbon nanotubes, graphene, and conductive polymers, which have greatly improved energy storage capacity, cyclability, and device flexibility.

Nowadays, supercapacitors have a wide range of applications, including portable electronics, electric vehicles, energy recovery systems, and smart grids. Their remarkable cycle life (>100,000 cycles) and temperature range of −50 to 200 °C make them highly versatile (Ding et al. 2024a; Li et al. 2023). With rapid charge/discharge times from seconds to minutes, a higher power density of >10 kW kg^{-1}, and higher environmental compatibility than conventional lithium-ion batteries, supercapacitors have emerged as appealing contenders for the next generation in energy storage technologies (Ding et al. 2024a; Heligman and Manthiram 2021). Supercapacitors in the form of flexible solid-state supercapacitors represent some of the most promising innovations, marrying mechanical flexibility with stable electrochemical performance that can be matched for wearable, portable, and embedded systems (Ding et al. 2024b; Muralee Gopi et al. 2025a; Ramadoss et al. 2022; Shuja et al. 2024; Zhai et al. 2020). However, there is still much work to be done regarding optimization of the electrode/electrolyte interface, further development of safe/stable solid-state electrolytes, maintaining performance under extreme conditions, and suitable methods of manufacturing that will be scalable and low cost.

This review presents a comprehensive analysis of the fundamental principles of operation for supercapacitors in light of promising electrode materials such as carbon-based compounds, transition metal oxides, conducting polymers, and emerging variants like perovskites. In this review, performance metrics from different device geometries, especially that of flexible solid-state supercapacitors, are analyzed considering their limitations, recent developments, and possibilities of future applications in sustainable and flexible energy storage. Based on previous works (Ding et al. 2024b; Muralee Gopi et al. 2025a; Ramadoss et al. 2022; Shuja et al. 2024; Zhai et al. 2020), the current work connects past achievements with the present trend and future directions, noting the milestones achieved thus far while addressing certain technical challenges. Additionally, it covers variant types of supercapacitors, provides an in-depth electrode material selection analysis, and emphasizes the importance of electrolytes in supercapacitors. Finally, advantages and disadvantages of the materials and configurations are discussed along with recommendations for further study.

2. TYPES OF SUPERCAPACITORS

The performance of a supercapacitor is greatly influenced by the nature and characteristics of each component. There are three primary categories of supercapacitors (Fig. 2) based on energy storage mechanism or type of active materials used: (a) Electrical double-layer capacitors (EDLCs) that accumulate energy via electrostatic charging in the region between the electrode and the electrolyte; (b) pseudocapacitors that take advantage of fast, reversible faradaic redox reactions at or near the surface of the electrode; and (c) hybrid supercapacitors that utilize both the electrostatic and faradaic energy storage mechanisms to achieve enhanced energy and power performance.

Figure 2. ***Categorization of supercapacitors according to their charge storage mechanism.***

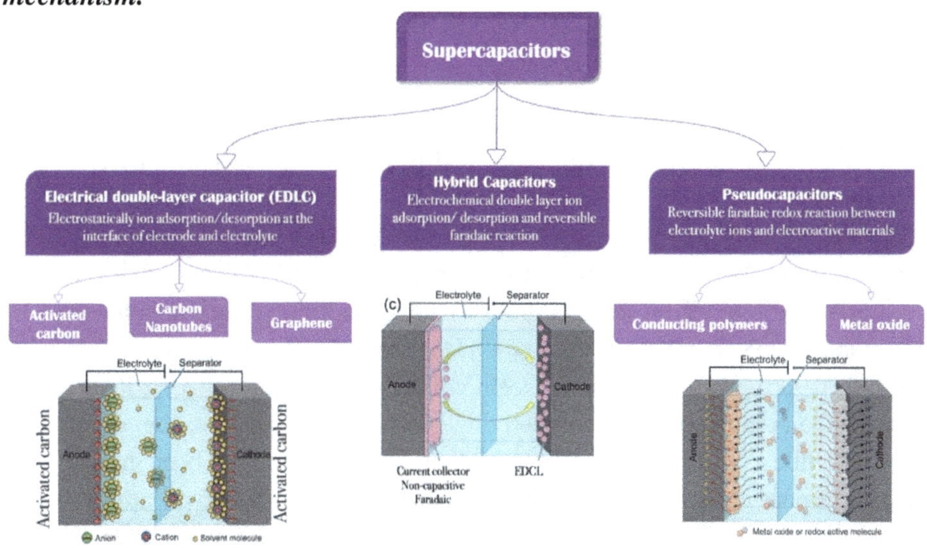

2.1. Electric Double-Layer Capacitor (EDLC)

Electrical double-layer capacitors (EDLC), as part of the family of supercapacitors, typically utilize carbon-based electrodes, such as activated carbon, carbon nanotubes, or graphene. Their working principle is based on an electrostatic mechanism (non - Faradic), that is, there is no redox reaction or electron transfer across the electrode/ electrolyte interface (Daraghmeh et al. 2020; Olabi et al. 2023; Zhang, Gu, and Chen 2023b). Energy is stored only by physical adsorption of ions at the electrode surface through the establishment of an electrical double layer. A classical EDLC typically consists of two porous conducting electrodes, separated by an ionically conductive but electrically insulating separator, all within an electrolyte solution (Shuaibu et al. 2025). Once a voltage has been put across the electrodes, opposite charges will develop at the respective electrodes, attracting oppositely charged ions from the electrolyte solution into the double layer geometry; during charging, cations move towards the cathode and anions towards the anode, and this behavior is reversed during discharge.

The complete reversibility of this process relies only on the electrostatic interactions that gives the EDLCs high charge/discharge rates, power density, and cycling stability. The electrochemical double layer formed at the electrode–electrolyte interface consists of two oppositely charged layers which are separated by a nano meter scale distance. The specific capacitance (C) of the EDLC is strongly related to the accessible surface area (A) of the electrodes and the distance (d) separating charge layers, shown by the following expression (Zhai et al. 2022):

$$C = \varepsilon_r \varepsilon_0 \, A/d \qquad (1)$$

where ε_0 is the vacuum permittivity and εr is the relative permittivity of the electrolyte.

Initially, it was generally accepted that C should increase in a linear relationship with A. However, experimental studies since then have shown that this isn't necessarily the case. Even some carbons that exhibit high surface area have lower capacitance than some that exhibit low surface area (Huang, Sumpter, and Meunier 2008; Shi 1996). Though this may seem contrary to accepted theory, it serves to illustrate the complexity of the underlying mechanisms. For instance, ε_0 remains constant, while ε_r may be affected by pore characteristics, and has been measured to decrease as pore width decreases (Prehal et al. 2017). Thus, effective capacitance is a function of factors beyond simply surface area, including, but not limited to, pore size distribution, ion accessibility, and dielectric properties within confined spaces.

You can also calculate the electrical capacitance of an electrode with the following relationship:

$$C = \frac{Q}{V} \qquad (2)$$

where Q is the accumulated charge, and V is the applied potential, which is the potential difference between the electrodes. The relationship between response current and changes in applied potential over time is described by (Kar 2021):

$$I = \frac{dQ}{dt} = C\left(\frac{dV}{dt}\right) \tag{3}$$

where I is the response current, time, and C is the capacitance.

In order to gain insight into double-layer formation in EDLCs, several proposed theoretical models have been discussed including the Helmholtz, Gouy–Chapman, and Stern models. The first, Helmholtz model, depicts the double layer as a compact structure similar to a capacitor, based on electrostatic attraction between oppose charge, it arises when placing electrical conductor in the electrolyte solution, generating a rigid layer of counter-ions termed as Helmholtz layer in the vicinity of the electrode–electrolyte interface. The second, Gouy–Chapman model, adds that layer of ions (or diffuse layer) forms as the ions move thermally in the electrolyte, we don't see it as a rigid adsorbed layer of ions but instead ions distributed in a gradient away from the electrode surface and into the bulk of the electrolyte. The ions in the Gouy–Chapman layer will assumed to be distributed in conformity with the so-called Boltzmann equation:

$$n(r) = Nexp(-\frac{V(r)}{k_b T}) \tag{4}$$

where n(r) is the ion density at distance r, V(r) is the electric potential, k is Boltzmann's constant, and T is temperature. However, the Gouy–Chapman model fails at high surface charge densities or ion concentration.

The Stern model incorporates both the Helmholtz and Gouy–Chapman approaches into a more realistic model of the double layer by splitting the interface into 2 regions, the compact Stern layer, which is found nearest the surface of the electrode, and the outer diffuse layer. Within the Stern layer, 2 planes are specified, the inner Helmholtz plane (IHP), which goes through the center of specifically adsorbed (not solvated) ions, and the outer Helmholtz plane (OHP), which includes solvated ions. This composite model describes the capacitance of electrochemical double layer capacitors (EDLCs) more accurately, because it takes into account both electrstatic and thermodynamic effects (Fig. 3).

Figure 3. ***Schematic of EDLC models: Helmholtz, Gouy & Chapman, and Gouy & Chapman–Stern***

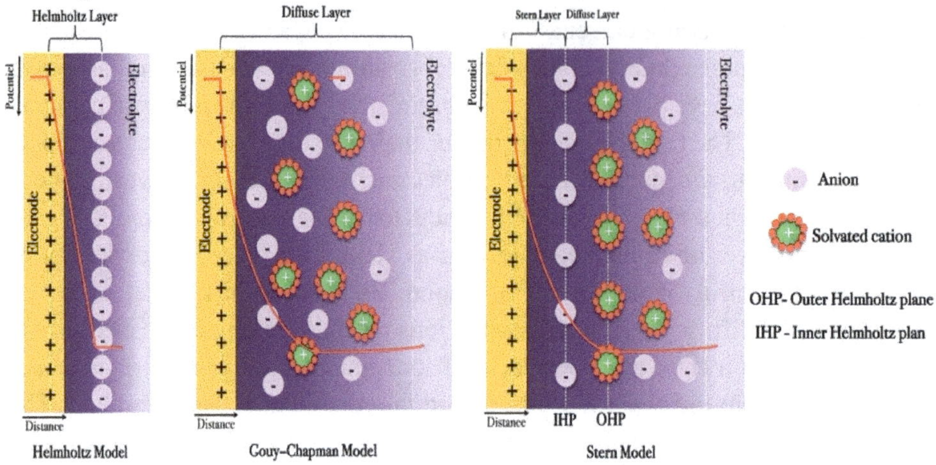

Building on this theoretical framework, the practical charge storage mechanism in EDLCs is governed by the physical adsorption of electrolyte ions onto the surface of electrically conductive electrodes with high surface area and porosity. This simple yet efficient mechanism gives EDLCs distinct advantages over batteries, including higher power density, excellent cycling performance, low internal resistance, and extremely fast charging typically within seconds(Simon and Burke 2008; Zhao and Zheng 2015). However, despite these benefits, EDLC development is currently constrained by relatively low energy density.

Current research seeks to address this challenge by investigating electrode materials with different architectures, morphologies, and compositions. An EDLC, as a type of supercapacitor, is usually based on two porous and conductive electrodes that are separated by an ion-conductive but electron-insulating separator and immersed in an electrolyte, while subject to a voltage. When a voltage is applied, opposite charges accumulate on the internal surfaces of the electrodes while some counter-ions are attracted from the electrolyte to compensate for the charge, forming electric double layers (Meena et al. 2023). During charging, cations and anions (counter-ions) are migrating towards the corresponding opposite charged electrodes, while the directions of the ionic migration would reverse if the process was reversed to a discharge. However, while the energy density of EDLCs is around 30 Wh/kg (Simon, Gogotsi, and Dunn 2014) which remains less than that of batteries but is still significantly higher than conventional capacitors. Their power density, on the other hand, can reach as high as 10 kW/kg, a great advantage over batteries for applications requiring rapid energy delivery. Furthermore, EDLCs have excellent cycling stability, often attaining coulombic efficiencies at or exceeding 99.9% (Simon and Burke 2008).

Research into new electrode materials, optimization of pore size distribution, and designing new suitable electrode architectures is ongoing to further improve the energy density. Energy density can be estimated for commercial EDLCs using the relation given below(Meena et al. 2023):

$$E^{EDLC} = \left(\frac{C_{-} \times C_{+}}{C_{-} + C_{+}}\right) \times \left(V_{max}^{EDLC}\right)^{2} \tag{5}$$

where C- and C+ are the capacitances of the negative and positive electrodes, respectively, and V is the maximum operating voltage. For this energy to be at a maximum, it would be optimal if the two electrodes had identical capacitances, i.e., C- = C+ = C. In general, the equivalent capacitance of such a system is given by the relation (Meena et al. 2023):

$$E^{EDLC} = \frac{1}{2} C \times \left(V_{max}^{EDLC}\right)^{2} \tag{6}$$

While much experimental effort is being made to enhance the energy density of EDLCs, the theoretical models that describe the nanoscale mechanisms of charge storage remain underdeveloped. In turn, this is a barrier to the detailed understanding of electrochemical phenomena occurring in complex pore systems. In this respect, a further decisive role is played by the electrode material choice.

2.2. Pseudocapacitors

Charge is stored in pseudocapacitors through faradaic mechanisms, whereas this is not the case in electrical double-layer capacitors (EDLCs), which use no-faradaic charge storage. These faradaic mechanisms utilize reversible redox (reduction–oxidation) reactions at the electrode–electrolyte interface, like in batteries((Kar 2021; Reenu et al. 2024b). As a result of this process, electrons are transferred from one side of the interface to the other, and the materials in the electrode experience chemical changes. To have the best cycling stability, the ideal faradaic mechanism should be reversible(Kar 2021).To take advantage of these faradaic mechanisms in pseudocapacitors, the choice of electrode material is very important. Typically, both transition metal oxides and conducting polymers are used. Pseudocapacitors have the benefit of showing higher specific capacitances and energy densities than the EDLCs, but lower power densities and less long-term stability (González et al. 2016).

These types of materials can also be compared theoretically given pseudocapacitance equations. Theoretical pseudocapacitance (Cpseudo) for a pseudocapacitor might be described through the equation(Zhi, Fu, and Ding 2012):

$$C = \frac{n \times F \times M}{V} \tag{7}$$

where n is the average number of electrons transferred, F is Faraday's constant, M is the molar mass of the active material, and V is the operating voltage window.

In practical cases, several mechanisms can simultaneously contribute to the overall pseudocapacitive behavior of an electrode. They include subpotential deposition, surface or bulk redox reactions of transition metal oxides, intercalation pseudocapacitance (Conway and Pell 2003), and reversible electrochemical doping/de-doping of conducting polymers. Carbonaceous supports, metal oxides, and conducting polymers such as PANI and PPy are among the most used active materials (González et al. 2016; Muralee Gopi et al. 2025a). Though they significantly augment capacitance, they also present some trade-offs. Even though such mechanisms increase capacitance, they come with some compromise. Pseudocapacitance can increase the specific capacitance of an EDLC-type charge storage device by one to two orders of magnitude when integrating faradaic processes will enhance the specific capacitance of EDLC-type charge storage type device; however, the charge storage time of redox reactions is much slower than the adsorption/desorption times of ions, which severely limits the power density(Chuang et al. 2010). The structure does make it that much more important to use care when choosing the materials and designing the electrode. Pseudocapacitive-type electrodes often mechanically change dimensions (swell and shrink) during cycling, which can reduce the mechanical stability, and ultimately lifecycle, of the intended use (Z. Zhu et al. 2011). Transition metal oxides are valued for their multiple oxidation states and high theoretical capacitance. Conductive polymers have excellent electrical conductivity and low ESR, while poor mechanical brittleness hampers their long-term durability. By comparison, such polymers like polyacetylene are able to demonstrate enhanced cyclic stability; their lifetime may reach up to 10,000 charge/discharge cycles(Masood et al. 2024).

Pseudocapacitors offer higher energy density compared to EDLCs by virtue of fast and reversible redox reactions; however, at the cost of lower power density and cycling stability. Compared to lithium-ion batteries based on the diffusion-limited redox processes in bulk electrodes, pseudocapacitors rely on surface or near-surface reactions, offering much faster charge-discharge rates. This puts them at an advantageous position for hybrid energy storage systems in case material selection and electrode architecture are optimized correctly.

2.3. Hybrid Supercapacitors

Hybrid supercapacitors combine two fundamentally different energy storage mechanisms: faradaic intercalation in which ions are inserted into the crystalline structures of the host and non-faradaic surface adsorption, typical of carbon-based materials(Zhang et al. 2023b). This dual concept is realized by the coupling of two different electrodes: a battery-type electrode (faradaic) and a capacitive electrode (non-faradaic), thus obtaining a compromise between high energy density and high power density(Afif et al. 2019). Performance of the device is also elevated by the use of three-dimensional mesoporous materials which enable fast ion transport, enhance electrolyte accessibility, and co-contribute to both cyclic stability and low price(Chen, Paul, and Dai 2017; Ma et al. 2015). Such a hybridization of the materials makes it possible to use the properties of one supplement which are lacking in another, for example, a conductive polymer can compensate for the low electrical conductivity of a metal oxide, while activated carbon can both strengthen the structural stability and increase the surface area of the composite(Namisnyk and Zhu n.d.; Reenu et al. 2024a; Zhang et al. 2023b).

Figure 4 . Hybrid Supercapacitor Concept

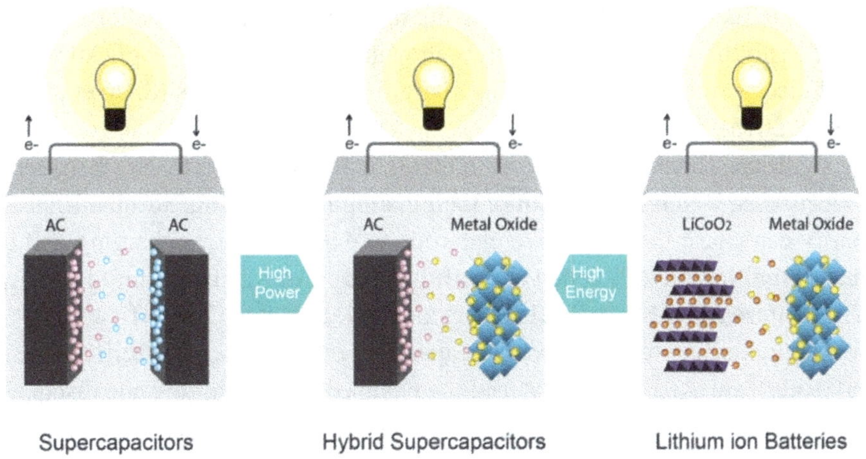

Manyexperiments have confirmed that one of the best ways to achieve supercapacitors with high specific capacitance and durable long-term performance is to couple electroactive polymers with carbons(Kulkarni 2024; Shah, Aziz, and Yamani 2022; Shown et al. 2015; Singh et al. 2025; Vinayagam et al. 2025). This material-combination strategy extends to the electrode architecture itself, where homogeneous composite structures can be formed by blending multiple active materials at the nanoscale. This combination principle also extends to the structure of the electrodes themselves, where homogeneous composites can be formed from separate active materials(Adedoja et al. 2023; Loh et al. 2024; Ma et al. 2015). Depending on the nature of the materials used and the architecture of the electrodes, there are generally three main families of hybrid supercapacitors. Asymmetrical hybrids that join a capacitive electrode with a pseudocapacitive one to get faradaic and non-faradaic mechanisms operating in the same device simultaneously. Guaranteeing very high energy densities at the same time maintaining a long-life span is the result of this kind of hybridization. Battery-type hybrids, or directly a supercapacitor electrode with a lithium-ion or sodium-ion battery electrode (lithium-ion or sodium-ion), only mix one battery and supercapacitor to move the performance of the former closer while still maintaining high power response (Cai et al. 2014; Lim, Jo, and Lee n.d.; Liu, Shi, and Zhitomirsky 2014; Yun et al. 2019). are the second devices to call the name of the game "energy storage". At last, composite electrodes consist of such materials as conducting polymers, carbon, and metal oxides, chemically bound into one structure which gives a good compromise between physical and chemical .

The hybrid configuration lithium-ion or sodium-ion capacitors (LIC or NIC) which utilize lithium or sodium intercalating materials on load the negative electrode has been extensively studied in literature (Adedoja et al. 2023). The goal of such systems is to merge a single system having the energy density of batteries with the power density of supercapacitors, thus showing a potential for application in the field of electric mobility or as rapidly charging storage systems (Shao et al. 2018; Zhu et al. 2015).

A significant challenge however remains is the kinetic disparity between the two electrodes, which is the result of their distinguished storage mechanisms. This disparity can cause the capacitive electrode to experience overcharging and contribute to limiting the efficiency of the full device (Amiri, Bruno, and Polycarpou 2023; Chatterjee and Nandi 2021; Fan et al. 2024; Loh et al. 2024).To address this, increasingly fast intercalating pseudocapacitive materials, e.g. specific transition metal oxides, are used. These materials can help equalize charging and discharging rates safely between the electrodes, while allowing for good power capability. Even though it may not qualify as pseudocapacitive, lithium or sodium can be operated at rates for hybrid supercapacitors. A number of recent reports have provided insight into the performance of these systems demonstrating promise for a variety of applications shown in **Table 2.**

Table 2. Summary of various electrolytes

Electrode	Specific Capacitance (F/g)	Retention with cycle	Reference
CoNi LDH	740.8	96.8%@5000	(Chen et al. 2024)
$FeNi_2S_4$@Co_9S_8	1308C/g	93.75%@8500	(Mohammadi Zardkhoshoui, Ameri, and Saeed Hosseiny Davarani 2023)
DyCoO3@rGO	1418 F/g	93.22%@5000	(El-Bahy et al. 2025)
MnSe@Y_2O_3	123.91	70.16 %@5000	(Bharathi et al. 2024)
TiO_2 with activated carbon	61.1	93% @10,000	(Choi et al. 2016)
Sn0.95$Al_{0.05}H_{0.05}P_2O_7$-PTFE	210	83% @7000	(Hibino et al. 2015)
Fe_3O_4/graphene nanocomposite//3D graphene	1000 mAhg−1 704 mAhg-1	70% @1000	(Zhang et al. 2013)
Graphite/CA	N/A	85% @10 000	(Khomenko, Raymundo-Piñero, and Béguin 2008)
CA //CA	186	70% @1000	(Yun et al. 2015)
$H_2Ti1_2O_{25}$//CA	72.1	95,3% @20 000	(Lee and Kim 2018)
Co_9S_8@NiCo-LDH	1335	88,21% @10 000	(Liu et al. 2022)
rGO/PANI/eCFC	1145	94% @5000	(P. Yu et al. 2014)
RuO_2 with GO	570	97.9% @1000	(Wu et al. 2010)
$NiFe_2O_4$//$CoFe_2O_4$ nanosheet	330	88.6% @10,000	(Dasha Kumar et al. 2024)
rGO - V_2O_5 NS//RGO	195	94% @3000	(K. Zhang et al. 2021)
$ZnCo_2O_4$@MnO_2// Fe_2O_3 in 1 M KOH	161 at 1 A/g	91% @5000	(Tholkappiyan and Vishista 2015)
Mn_3O_4//CNT/graphene	72.6 at 0.5 A/g	86% @10,000	(Zou et al. 2023)
N-CNFs/Ni$(OH)_2$//N-CNFs	N/A	84% @5000	(Fu et al. 2022)
Ni$(OH)_2$-CNS//AC	198	93% @10,000	(Wang et al. 2023)

Hybrid supercapacitors have a couple of issues besides their advantages. First of all, their cyclability is lower by about a factor of ten as compared to pure EDLCs that are to a large extent attributed to the gradual degradation of pseudocapacitive materials of the electrode. Their total performance is also very dependent on how well the electrodes and electrolyte are integrated and on the internal losses resulting from the system resistance. The present goal of the research on hybrid supercapacitors is to achieve an energy density in the range 20-30 Wh/kg while keeping high power density, long cycle life, and good thermal and chemical stability(Muzaffar et al. 2019).To achieve these objectives, researchers are turning to the study of non-aqueous redox materials, novel electrochemical pairs, and more efficient cell .

3. COMPONENTS OF A SUPERCAPACITOR

Electrochemical supercapacitors resemble conventional batteries in structure; they consist of two electrodes (anode and cathode) separated by a porous membrane, immersed in an electrolyte, and connected to current collectors, as illustrated in Fig. 5. In general, two porous electrodes face each other on either side of a thin separator, with the electrolyte effecting ion transport between the two. Each of these components is crucial in determining the overall performance of the device, mainly in terms of energy storage capacity, power and energy density, charge-discharge kinetics, and long-term stability. The electrodes are usually made of conducting materials with high specific surface area and optimized porosity, their design aimed at maximizing the accessibility to the electrolyte to achieve maximum electrical double-layer formation. Common electrode materials used include activated carbon, graphene, CNTs, and conducting polymers. The separator prevents electrical short circuits due to physical separation of the electrodes, yet allows for rapid ionic conduction. The nature of the electrolyte, whether aqueous, organic, or based on ionic liquids, directly influences the electrochemical window, internal resistance, and therefore the performance of the device in general. Organic electrolytes usually offer larger operating voltage, while some of these electrolytes can also contribute to improving the pseudocapacitance due to fast, reversible redox reactions at or near the electrode surface. In the following sections, a critical review of recent developments in materials and engineering strategies concerning each constituent component of electrochemical supercapacitors is presented.

Figure 5. ***Electrochemical supercapacitors***

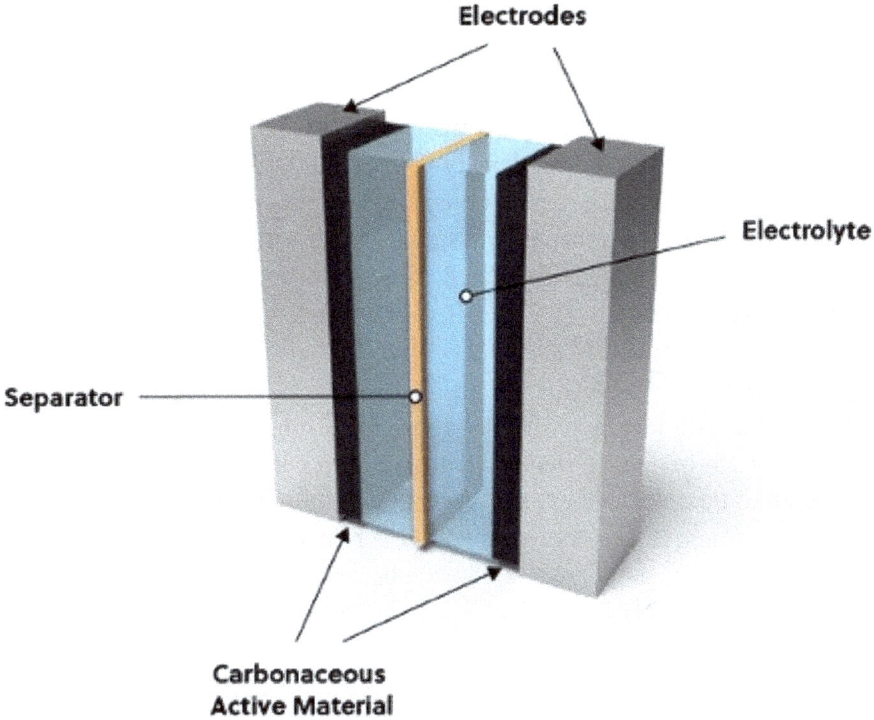

3.1. Electrodes

Electrodes in a supercapacitor are usually made of thin films of conductive materials that are deposited on metallic current collectors. The choice of electrode material is an important factor that has a direct influence on the electrochemical properties, such as specific capacitance, energy and power densities, and cycling stability over time. To enable optimal performance, candidate materials must meet a series of requirements, such as high electrical conductivity, large specific surface area, robust thermal and chemical stability, electrochemical inertness, corrosion resistance, environmental compatibility, and cost-effectiveness (Molahalli et al. 2024). In recent years, a growing amount of research has focused on the development of advanced electrode materials, including graphene and its derivatives, transition metal oxides and hydroxides, carbonaceous materials, 2D materials, conducting polymers, and a variety of hybrid and composite architectures. These materials can be broadly divided into two categories based on their charge storage mechanism: electrostatic (electric double-layer) materials and faradaic (redox-active) materials.

Currently, three major families dominate the field of carbon-based materials, metal oxides, and conductive polymers; each of these offers specific advantages tailored to particular application needs(Kumar et al. 2024). In addition, the pore architecture in electrodes also plays a very important role in determining their device performance. A smaller pore can provide enhanced specific capacitance and energy density due to increased surface accessibility. However, this usually comes with a higher ESR, which then decreases the power density. Consequently, pore size distribution needs to be selected based on the applications; larger pores are good for high-power applications using fast ion transport and smaller pores are better for devices that need to maximize energy storage (Tudu et al. 2024).

3.1.1. Carbon-Based Materials

Oneof the main factors in the development of supercapacitor (SC) technology has been carbon-basedmaterials, which have been particularly instrumental in the fabrication of flexible devices and the generation of scalable systems for industrial production (Wang, Zhang, and Zhang 2012) are materials whose nature can differ substantially, for example, activated (Heidarinejad et al. 2020; Simpson 2008), carbon nanotubes (CNTs) (De Volder et al. 2013), graphene derivatives (Gao et al. 2012; Zhai et al. 2013), and any other types of carbon. The mentioned materials have a unique combination of high specific surface area, excellent electrical conductivity, exceptional chemical stability, low density, and wide thermal operating range among others that give them the high ability to be used as electrodes in high-performance supercapacitors. One of their major strengths is put in their connected porous architecture which makes ion transport within the electrode can be done quickly. This is what enables one to achieve charge storage efficiency on the basis of electric double-layer formation via very fast ion adsorption and desorption at the electrode–electrolyte interface.

Thus, carbon-based supercapacitors can be powered at a high rate and can quickly release the accumulated energy(Borenstein et al. 2017; Zhong, Zhang, and Li 2022). Any attempt to judge how well carbon materials might be used in a supercapacitor (SC) must include a consideration of their specific surface area, pore size distribution, structural integrity, and electrical conductivity. Besides, these substances give the freedom to introduce surface functionalities and to have controlled pore structures, thus enabling the setting of electrochemical performance parameters as per the need of the application (Wang et al. 2021; Zhong et al. 2022). Besides these advantages, that are based on the intrinsic qualities of carbon materials, the plentiful supply, reasonable price, and environment-friendly characteristics of carbon-based materials justify their vital role in the sustainable development of energy storage technologies. Whether it is for small portable electronic devices or large-scale energy storage systems, the use of carbon-based materials is what is leading to the next-generation supercapacitors' flexible and industrial deployment .

a. Carbone active

Among the carbon-based materials that have been used for electrical double layer supercapacitors (EDLC), activated carbon (AC) is still the one that is most commonly used, mainly because of its large specific surface area, moderate electrical conductivity, good chemical stability and relatively low cost (Dong et al. 2024; Kierzek and Gryglewicz 2020).It is a material that is most commonly used for both the positive and negative electrodes in commercial EDLCs(Dong et al. 2024). Coal, coke, and other fossil sources can be used to produce AC, or it can be produced from renewable natural materials like wood, coconut husks, or other agricultural residues((Naji and Tye 2022; Ukanwa et al. 2019). The variety of the precursors thus can be utilized to an almost limitless extent for modification of the material's final properties depending on the application requirements.

The production of activated carbon requires two main operations: first the carbonization, then the activation (Naji and Tye 2022). Carbonization refers to the thermal decomposition of organic materials of the source in an inert atmosphere leading to the formation of amorphous carbon (Abdullah, Tan, and Lim 2011). Activation is a process to develop the porosity of the carbon material, either by physical means or by chemical means (Kar 2021). Physical activation, the first method and still by far the most popular in industrial practice, generally involves the use of heat-reactive gases like steam CO_2, or air at a high temperature (700–1200 °C) (Kar 2021; Molina-Sabio et al. 1996, 1996; Rodríguez-Reinoso, Molina-Sabio, and González 1995; Zhang et al. 2004). The process is inexpensive, since the price of old reagents is very cheap and there is almost no need for post-treatment (Yi et al. 2021). Unfortunately, it only produces a limited development of pores and lower activation efficiency. On the other hand, chemical activation is the process of carbonization followed by the treatment of the precursor with $ZnCl_2$, H_3PO_4, or alkali metal compounds preferably at a low temperature (400–700 °C) (Kar 2021; Pandolfo and Hollenkamp 2006; Yi et al. 2021).In general, this method is associated with higher yields and well-developed pore structures (Yi et al. 2021), albeit with washing steps in acid for the removal of residual activating agents, which makes the overall process costlier than it seems. Typically, it shows a high specific surface area ranging from 1000 to 2000 m^2/g, although some studies have reported extreme values of up to 3000 m^2/g (Kuptajit et al. 2021). The pores of the final product can be classified as micropores (<2 nm), mesopores (2-50 nm), and macropores (>50 nm), having different roles in the electrochemical behavior of the material (Lozano-Castelló et al. 2002).

Still,a high specific surface area alone does not necessarily lead to high electrochemical performance (Pandolfo and Hollenkamp 2006). Many studies have been reported that there is no direct linear relationship between the specific capacity and specific surface area of activated carbons. Overly developed microporosity can hinder the ion transport in electrolytes, mainly large ions, thus seriously reducing the capacity of the device (Dujearic-Stephane et al. 2021; Lobato et al. 2017; Sevilla and Mokaya 2014; Vaquero et al. 2012). The energy density and specific power can both be increased as a result of the optimized pore distribution that has more mesopores(Kar 2021). In addition, a large part of the SSA can still be there without being utilized due to the limited accessibility, in particular, if the pore texture is not highly controllable. This limitation emphasizes the need for porosity engineering which is specifically designed for the ions in the .

Developing high-performance, inexpensive supercapacitor materials is an absolute necessity if we want to meet the demands of next-generation technologies such as flexible electronics and hybrid vehicles. On the other hand, the use of activated carbon (AC) is favored in the commercial market because of its chemical stability, easily adjustable porosity, and conductivity. However, the application of advanced materials such as CNTs and graphene is constrained by complicated and expensive production .As a result, researchers are increasingly turning to bio-waste as a sustainable carbon source. Bio-derived activated carbons not only reduce reliance on fossil fuels but also offer excellent electrochemical properties depending on the precursor and activation method. Various agricultural and plant wastes such as coconut shells, peanut shells, watermelon rind, and others have been successfully used to produce porous carbons with high surface area and capacitance. The **table 3** compares the electrochemical performance of different activated carbons reported in the literature.

Table 3. Comparison of Activated Carbons as Electrode Materials

Source	Activation	Surface Area (m^2/g)	Specific Capacitance	Cycle Stability	Ref
cashew nut shells	KOH	2034	106 at 0.5 A/g	87% after10000 cycles	(Chaiammart et al. 2024)
Coconut shell	Steam	1532	228 at 5 mV/s	93% after 3000 cycles	(Mi et al. 2012)
Peanut shell	$ZnCl_2$	1549	340 at 1 A/g	95.3% after 10000 cycles	(Xiao et al. 2018)
Watermelon rind	KOH	2277	333.4 at 1 A/g	96.8% after 10000 cycles	(Mo et al. 2016)
Tea waste buds	KOH	1610	332 at 1 A/g	97.8% after 100000 cycles	(Abrar Khan et al. 2020)
Withered rose flower	KOH/ KNO_3	1980	350 at 1 A/g	96.5% after 15000 cycles	(A. Khan et al. 2020)
Albizia flowers	KOH	2757.6	406 at 0.5 A/g	97% after 5000 cycles	(Wu et al. 2019)
Jujube fruit	NaOH	1135	460 at 1 A/g	92.2% after 130000 cycles	(Yang et al. 2020)
American Poplar fruit	KOH	942	423 at 1 A/g	97% after 200000 cycles	(Kumar et al. 2020)

continued on following page

Table 3. Continued

Source	Activation	Surface Area (m^2/g)	Specific Capacitance	Cycle Stability	Ref
Rice husk	KOH	3145	367 at 5 mV/s	~100% after 30000 cycles	(Gao et al. 2015)
Soybean pods	NaOH	2612	352.6 at 0.5 A/g	94.2% after 50000 cycles	(Kong et al. 2020)
Orange peel	KOH	2160	460 at 1 A/g	98% after 10000 cycles	(Subramani et al. 2017)
Mangosteen peels	NaOH	2623	357 at 1 A/g	94.5% after 130000 cycles	(Yang et al. 2019)

b. Nanotubes de carbone

Carbon nanotubes (CNTs) are cylindrical carbonaceous materials, classified into two types: single-walled (SWCNT) and multi-walled (MWCNT), depending on the number of graphene layers of which they are composed(Rahman et al. 2019). Their cylindrical structure, unlike spherical fullerenes, gives them a unique combination of properties: high electrical conductivity, high mechanical strength, large specific surface area and excellent chemical stability(Ahmad 1999; Muralee Gopi et al. 2025b; Saleh et al. 2025; Z. Yang et al. 2015).

These characteristics make CNTs interesting candidates for supercapacitors, in particular thanks to their porous structure and conductivity. The mesopores present in CNTs ensure good charge distribution and efficient use of the active surface area. Chunming et al (Niu et al. 1997) demonstrated that electrodes based on catalytically grown CNTs, with a diameter of around 80 Å, displayed a specific capacitance of 102 F/g at 1 Hz, and 49 F/g at 100 Hz. Besides, CNT thin films were able to reach power densities of more than 20 kW/kg. Different treatments can be used to optimize the performance of CNTs(Reenu et al. 2024a). For instance, thermal annealing enhances purity, increases hydrophobicity, and gets rid of unwanted functional groups (Barisci et al. 2003). An et al. (Liu et al. 2008) have demonstrated that electrochemical oxidation creates nanometric mesopores and oxygenated groups, which results in a specific capacitance of 113 F/g and a kink frequency that is twice as high as that of activated carbon. However, carbon nanotubes have a relatively small specific surface area compared to activated carbons (<500 m2/g), and, therefore, their energy density is limited(Childress, Ferri, and Rao 2018). To break this barrier, there are different ways to address the problem, such as activation, chemical functionalization, thermal treatment, and surface modification ((Liu et al. 2006; Pan, Feng, and Lin 2006; Zhou et al. 2005).

Much of the recent research has focused on the chemical enhancement of carbon nanotubes (CNTs). Yan et al (Yan et al. 2023) achieved this by oxidizing the material and introducing hydroxyl (-OH) groups, which increased the specific capacity of TCNTs from 55 to 118 F/g at 1 A/g, nearly doubling it, while maintaining excellent stability over 10,000 cycles. In addition, Li et al. (Li et al. 2022) reported that combining polyaniline (PANI) with hydroxylated TCNTs not only increased the specific capacity but also the mechanical strength, and that the material could deliver 97 mAh/g. In summary, Huo et al (Huo et al. 2023) fabricated polypyrrole-derived CNTs and were able to achieve a very high specific surface area of 3537 m^2/g.

c. Graphene and graphene derivatives

Graphene and its derivatives are considered best to be used as electrode materials in high-performance supercapacitors. This is mainly because of their exceptional physico-chemical properties. Graphene which consists of a single layer of sp2-hybridized carbon atoms arranged in a hexagonal lattice has a very large specific surface area (up to 2620 m^2/g), excellent thermal (~5000 W-m-1. -K-1) and electrical conductivity, high chemical and thermal stability, and a wide electrochemical window (Brownson, Kampouris, and Banks 2012; Tortora and Bussetti 2025). These features make it a perfect material for energy storage, as it facilitates rapid ionic diffusion and provides an efficient response to charge/discharge . GO and rGO are among the most broadly utilized derivatives that are acknowledged for their capability to produce stable, conductive structures (Hung et al. 2025; Tarcan et al. 2020). The method involves the oxidation of graphite to graphite oxide (GO), followed by chemical reduction with a reagent such as hydrazine to get reduced graphene (rGO) (De Silva et al. 2017). Nevertheless, the biggest challenge is the eventuality of graphene that will restack due to van der Waals forces, thus drastically limiting the active surface area available (Li and Östling 2013). To prevent this occurrence, there are different means by which bends in graphene sheets, mesopores creation, molecules intercalation or composites preparation can be involved so that the interlamellar spacing can be widened and ionic diffusion can be .

Graphene-based supercapacitors show specific capacities ranging from 75 F/g in ionic electrolytes to 205 F/g in aqueous media(Vivekchand et al. 2008). In the H_2SO_4, capacities of 117 F/g have been reported, and up to 233 F/g in a flexible graphene/polyaniline composite(D.-W. Wang et al. 2009). Some three-dimensional rGO-based structures have achieved 352 F/g at 5 mV/s(González et al. 2016). Moreover, devices prepared from weakly agglomerated graphene showed excellent stability, with 90% capacity retention after 1200 cycles and energy density reaching 28.5 Wh/kg(Y. Wang et al. 2009).

Another potential method of enhancing the performance of graphene is by functionalizing it through heteroatomic doping. The insertion of atoms like N, S, B, P, or F may change the electron density and create more active sites, thus increasing the pseudocapacitive capacity(Ma et al. 2021). For example, nitrogen-doped reduced graphene (rGO-N) can achieve a specific capacitance of 261 F/g((Youn et al. 2015). Besides, derivatives like GCOOH, cyanographene, or graphitic carbon nitride (g-

C_3N_4) also show that they have good potential, especially in aqueous media where electrostatic interactions are favorable(Kumar et al. 2024). Moreover, non-covalent changes, for instance, thermomechanical exfoliation or triamide (TMA) treatment, can lead to better dispersion and functionalization without a decrease in the material's conductivity.

In order to fully harness the power of graphene, the development of hybrid composites with pseudocapacitive or conductive materials has been notably successful. The combination with metal oxides (MnO_2, $Ni(OH)_2$, $Co(OH)_2$, RuO_2) not only avoids restacking but also capacity is raised due to the pseudocapacitive contribution of the oxides. As a matter of fact, Gr-CNT@Co composite from GO and ZIF-67 recorded a specific capacitance of 1108 F/g along with 78% stability after 6000 cycles (W. Zhang et al. 2021). The same way, coupling with conductive polymers like polyaniline or polypyrrole creates fascinating synergies. Hence, a composite based on PPy/graphene aerogel/MoO_3 experienced its capacity to go up from 379 to 1788 F/g, thereby demonstrating the combined effect of the hierarchical structure and electrochemical properties of the components (Xin et al. 2020).

In the end, quite a few publications have exemplified the exceptional stability as well of some graphene-based units. Bulk hydrodispersible graphene retains up to 95.3% of its initial capacity after 60,000 cycles at 3 A/g ((Šedajová et al. 2020). Another set of nanoplatelets, which were deliberately prepared to avert agglomeration, demonstrated excellent capacitance retention over 10,000 cycles and volumetric density reaching 22.2 F/cm^3 (Haridas et al. 2024). These evidences support the fact that, due to its outstanding characteristics, high structural modularity and countless ways to be combined with other materials, graphene is the material of the future supercapacitors' design which are efficient, durable and tailored to the needs of the modern energy storage .

3.1.2. Transition metal oxides (TMOs)

Transition metal oxides (TMOs) are mainly considered to be the most suitable electrode materials for supercapacitors, as they have intrinsic electrochemical advantages. One of the main reasons is that they have multiple oxidation states, a large specific surface area, a lot of electroactive sites, and a high theoretical specific capacitance (González et al. 2016). The different nanostructured morphologies of these such as nanofilms, nanowires, nanotubes, nanosheets, nanoflowers, and hollow nanospheres are able to efficiently transport electrons by shortening the ions diffusion paths during quick charge-discharge cycles(Chandra Sekhar, Ramulu, and Yu 2022; Liang et al. 2021). Besides, the material's low fabrication cost and the use of low-corrosion electrolytes make the product more attractive. However, the situations of uncontrolled volume expansion during ion transport in the bulk phase may limit sufficiently the performance of TMOs(Chandra Sekhar et al. 2022). As a solution for such problems, hierarchically nanostructured TMOs have shown increased performance to power density, specific capacitance, and cyclic stability(Iro, Subramani, and Dash 2016; Mondal, Velpula, and Sangaranarayanan 2024).

TMOs are particularly useful in energy storage because they support fast and reversible redox processes. Of these, RuO_2 is amongst the best performing, with a specific capacitance of up to 1000 $F{\cdot}g^{-1}$ under standard conditions among the pseudocapacitive materials (González et al. 2016; Miller and Simon 2008; Mondal et al. 2024). For its amorphous or hydrated $RuO_2{\cdot}xH_2O$ form, the best reported value of the specific capacitance was around 900 $F{\cdot}g_{-1}$ (Long et al. 1999), although its theoretical maximum specific capacitance may be as high as 1358 F g-1 (Naoi and Simon 2008)., at an electrical conductivity of 3×102 Ω^{-1} cm^{-1}(Naoi and Simon 2008). These properties are largely attributed to the presence of inter-particle and inter-layer hydrous regions, which facilitate protonic conduction and enable both high power and high energy performance in supercapacitors(González et al. 2016; Sugimoto et al. 2005). It is characterized by excellent properties including a wide potential window, highly reversible redox behavior, high proton conductivity, very good thermal stability, metallic-level electrical conductivity (~10^4 $S{\cdot}cm^{-1}$), cycle life, and high-rate (González et al. 2016; Lee et al. 2010). Its pseudocapacitive behavior arises mainly from the rapid redox transitions associated with the intercalation of protons in acid media or changes in oxidation state at the surface in alkaline media. However, its high cost and toxicity limit practical applications to specialized domains, such as in space applications (Lakshmi and Vedhanarayanan 2023b). To improve the practical value of this material, nanostructured hydrated RuO_2 synthesized by sol–gel or electrochemical deposition methods displayed higher ionic conductivity and larger electroactive surface areas (Jang et al. 2003). Moreover, composites of RuO_2-based with more inexpensive oxides like SnO_2, VO_2, and TiO_2 or carbon-based materials such as graphene also maintain high performance with minimal Ru content.

Anenvironmentally-friendly and cost-effective alternative would be manganese dioxide (MnO_2), recognized for both its abundance and nontoxicity (Pappu et al. 2022; Patil and Jadhav 2024). MnO_2 has a theoretical specific capacitance of about 1370 $F{\cdot}g^{-1}$ (Shen et al. 2012a; Zhu et al. 2016), however, the practical values are only in the range of 200–300 $F{\cdot}g^{-1}$ because of the material's low intrinsic conductivity (~10^{-5} to 10^{-6} $S{\cdot}cm^{-1}$) and limited active site utilization (Lu et al. 2016; Shen et al. 2012b). The performance of the material depends on the crystallinity and morphology to a great extent, as an excess of crystallinity hampers ion diffusion, whereas an overly amorphous structure decreases conductivity. To enhance the electrochemical performance of manganese dioxide-based electrodes, a lot of methods such as nanostructuring (e.g., hollow nanospheres, nanoflowers, nanosheets) and making composites with conductive materials like carbon, graphene, conductive polymers, metals, and conductive metal oxide have been employed (Alkallas et al. 2024; Du and Zhang 2023; Mao et al. 2016; Phattharasupakun et al. 2016; Xiao et al. 2014). Despite these methods improving the performance, the stability during cycling is still varying greatly even as a result of similar compositions of electrodes which syntheses methods and microstructures thereby influencing the effect cannot be pinpointed conclusively. As an example, different synthesis methods of MnO2/rGO lead to distinct capacitance retention where one method results in 88.3% retention after 1,000 cycles and another in 95.4% after 15,000 cycles (Yan et al. 2010). Comparably, the same extent of stability variation has been found in other systems such as MnO_2/CNT-CNF composites (Huang et al. 2021; Wang et al. 2015). These results point out that microstructural factors not only have a substantial effect on specific capacitance and rate capability but also they have a significant say in the issue of electrode materials' long-term stability. Nevertheless, the mechanisms behind the structural aspects of granule and oxide-based materials that account for the decrease of capacity are in a state of a dilemma, and moreover, those aspects have not been very thoroughly investigated.

Other TMOs such as cobalt oxides (CoO, Co_3O_4) also show potential. In particular, Co_3O_4, with its mixed Co^{2+}/Co^{3+} oxidation states, has a high specific capacitance of 3560 Fg^{-1}, and is an eco-friendly material(Lee, Chin, and Sow 2014; Yuan et al. 2011). Although Co_3O_4 exhibits excellent theoretical specific capacitance, practical values in real devices are significantly lower due to limitations in electron transport and ion diffusion. To address this, extensive research has focused on engineering Co_3O_4 nanostructures to enhance electrode–electrolyte interactions and improve electrochemical performance(Xiao et al. 2012; Zhong et al. 2012).

Various synthesis methods have yielded promising results: mesoporous nanocrystalline Co_3O_4 (~3 nm) achieved 401 $F{\cdot}g^{-1}$ using a polyacrylamide template(Cao, Lu, and Li 2005). crater-like microspheres templated by MCM-41 delivered 102 $F{\cdot}g^{-1}$ with 74% retention over 500 cycles; and sol–gel/freeze-drying techniques produced ultrafine Co_3O_4 with 742.3 $F{\cdot}g^{-1}$(Wang et al. 2012). Additionally, nanocubes and nanotubes formed with mesoporous carbon supports or calcination yielded capacitances around 350 $F{\cdot}g^{-1}$(Liu et al. 2013). Biomorphic synthesis has also enabled unique cylindrical mesoporous Co_3O_4 structures, further underscoring the importance of morphology and porosity in enhancing electrochemical performance. Examples include mesoporous and nanosheet Co_3O_4 structures, as well as hybrids grown on Ni foam, achieving capacitances as high as 2735 $F{\cdot}g^{-1}$ (Yuan et al. 2012). while binary oxides like $NiCo_2O_4$ balance performance and durability (over 10,000 cycles) (Chen et al. 2013; Gaikwad et al. 2024; Wu et al. 2016b; Xiong et al. 2025). MnO_2 is particularly appreciated for its low cost and efficiency in aqueous electrolytes, making it an ideal candidate for large-scale applications (J.-C. Wang et al. 2016).

Iron oxides (FeO, Fe_3O_4, Fe_2O_3, FeOOH) are also under extensive investigation, particularly as negative electrodes(Kumar et al. 2024; Xia et al. 2016; Xu et al. 2019; Yedluri and Kim 2019). Their natural abundance, environmental friendliness, and diverse oxidation states offer clear advantages. Nevertheless, their low electrical conductivity and poor dimensional stability necessitate structural engineering or hybridization with conductive materials to achieve practical performance. Vanadium pentoxide (V_2O_5) features a layered crystal structure and multiple accessible oxidation states (V^{2+} to V^{5+}), allowing high charge storage capacity and good ion diffusion(Kar 2021; Mondal et al. 2024). Its low toxicity, stability, and abundance are appealing, but high resistivity and structural fragility limit long-term performance. Nanoscale modifications and composite architectures are effective in addressing these issues.

Transition metal oxides present a broad platform in developing advanced supercapacitor electrodes due to their tunable redox chemistry, diverse morphologies, and favorable environmental and economic profiles. However, intrinsic drawbacks such as low conductivity, structural instability, and poor coulombic efficiency self-suggest hybrid strategies. Embedding TMOs into conductive matrices such as graphene, carbon nanotubes, or conductive polymers is a very vital strategy for the enhancement of electronic conductivity, mechanical robustness, and cycling stability. Such composite systems are opening routes toward next-generation energy storage devices, meeting the growing demands from emerging applications including portable electronics, electric vehicles, and renewable energy systems.

Whereas transition metal oxides offer eminent pseudocapacitive behavior based on multi-redox states and tunable oxidation states, their actual application has often been prevented by poor electrical conductivity and inadequate mechanical flexibility. In line with the efforts to overcome these limitations, organic materials are being explored that can offer high electrical conductivity and structural resilience. Conductive polymers represent one such class of materials, which complement or act as alternatives to TMOs by marrying redox activity with mechanical adaptability and ease of processing. Their intrinsic conductivity and flexibility make them ideal candidates for hybridization with metal oxides or for use as standalone pseudocapacitive electrodes, thus bridging a gap between inorganic rigidity and organic versatility.

3.1.3. Conducting polymers

Conductive polymers (CPs) are a key class of materials central to the development of pseudocapacitors owing to their high specific capacitance, high conductivity, lower cost, and relative ease of synthesis. Discovered in the late 1970s, mainly with polyacetylene $[(CH)_2]$, these materials are distinguished by the presence of conjugated double bonds (C=C) along their molecular chain (Jang 2006; Kiran Raj et al. 2023; Paramshetti et al. 2023). This architecture promotes the delocalized electrons, facilitates charge transport, especially when these polymers are doped. CPs can be primarily understood as conjugated organic polymers which can transport electrons through the alternating π-bonds (Mdluli et al. 2022; Sumdani et al. 2021). Unlike saturated polymers such as polyethylene, in which electrons are trapped in σ-bonds, conjugated polymers offer enhanced conduction through electron delocalization. However, the development of PC-based devices is limited by the lack of stable and efficient n-type conjugated materials, as well as their unstable behavior during repeated electrochemical cycling(Mudila et al. 2019).

Therefore, conductive polymers are regarded as very potent materials for pseudo-capacitors. They promise high charge density, simple fabrication, and low cost as compared to electrical double-layer capacitors (Mondal et al. 2024). The main conductive polymers used as electrodes in energy storage devices are polyaniline (PANI), polypyrrole (PPy) and polythiophene (PTh). These materials are highly attractive, particularly, due to their high conductivity, flexibility, low cost, and simple fabrication. The explosion of research in the field by various researchers has resulted in numerous significant efficiency-enhancing modifications having been made.

Among the polymers, polyaniline (PANI) stands out as one of the most widely researched, most of all, due to its flexibility, very high specific capacitance, easy synthesis, cheapness, high doping level (0.5), and controllable conductivity (Reenu et al. 2024b; H. Wang, Lin, and Shen 2016). Nevertheless, the cause of its rapid degradation, during repeated charging and discharging cycles, has been its weakness. There have been suggestions of the need for protic solvents or ionic liquids to achieve the best performance (Adedoja et al. 2023; Wu et al. 2005). The foremost tactic now is the PANI hybridization with carbonaceous materials and metal oxides as an approach, which has demonstrated promising outcomes to mitigate these . For example, Sivakumar et al.(Sivakkumar et al. 2007) synthesized PANI nanofibers by interfacial polymerization, achieving a specific capacitance of 554 F/g at 1.0 A/g, although cyclic stability remained low. PANI-MWCNT composites then improved the specific capacity to 606 F/g.(Sivakkumar et al. 2007). Furthermore, Li et al.(Li et al. 2009) estimated the theoretical maximum specific capacity of PANI at 2000 F/g, although experimental values remained significantly lower.

Polypyrrole (PPy) is also widely studied for its excellent properties: high flexibility, easy manufacture, good cyclic stability, and relatively high specific capacity. Its high density also gives it excellent volumetric capacity(Reenu et al. 2024b). However, it has a number of limitations, such as difficulty of doping and a relatively lower specific capacitance. Moreover, doping with multicharged anions can induce physical crosslinking. Yang et al. (Q. Yang, Hou, and Huang 2015) proposed the synthesis of self-supporting PPy films, with or without surfactant, and achieved a maximum specific capacitance of 261 F/g, with 75% capacitive retention. Rajesh et al. (Rajesh et al. 2016)reported phytic acid-doped PPy thin films with a maximum specific capacitance of 343 F/g and excellent stability after 4000 cycles (91% retention at 10 A/g).

Polythiophene (PTh) is also used as an electrode material. It is flexible, easy to synthesize, cyclically stable and environmentally friendly(Simon and Gogotsi 2008; Snook, Kao, and Best 2011). However, it suffers from weaknesses, notably lower conductivity and lower specific capacitance compared with other conductive polymers. Laforgue et al.(Laforgue et al. 1999) have chemically synthesized PTh and PFPT, reporting specific capacitances of 7 mAh/g and 40 mAh/g respectively. Patil et al.(Patil, Jagadale, and Lokhande 2012) also developed PTh thin films by a successive ionic layer adsorption method, using $FeCl_3$ as the oxidizing agent, and reported a capacity of 252 $F.g^{-1}$. Further work using the chemical bath deposition method achieved a maximum specific capacity of 300 $F.g^{-1}$. Its performance can be significantly enhanced by the addition of surfactants such as Triton X-100, enabling an increase in specific capacity from 78 $F.g^{-1}$ to 117 $F.g^{-1}$(Senthilkumar, Thenamirtham, and Kalai Selvan 2011).

Conductive polymers have been favored for their numerous benefits, but they are still plagued with significant drawbacks especially in their low cyclic stability. The cycling of a device multiple times will inevitably cause it to undergo mechanical stress, so it will swell and shrink thus the material's structure will become less and less intact. Besides, their mechanical properties are most of the time weaker than those of carbonaceous materials, which in turn, their durability is negatively impacted. To solve such problems, a lot of measures have been devised, among them the production of hybrid composites that consist of conductive polymers and inorganic or carbon nanomaterials. These mixtures have the effect of the stability, specific capacitance and the total conductivity of the material being better. As an example, the fabrication of mechanically robust all-solid-state supercapacitors has been achieved through the in situ polymerization of PANI on cellulose nanofibers (CNF) and graphite nanoplatelets (GNP).

The choice and design of electrode materials ranging from carbon-based structures to transition metal oxides, conducting polymers, and MXenes, can significantly influence the electrochemical process, specific capacitance, and supercapacitors' cycling stability. A comparative summary to the main materials with the highlighted advantages, limitations, and typical usages is given in Table 4. Their individual advantages notwithstanding, the overall electrochemical performance of these devices still depends largely on the electrolyte which accompanies them and determines ion transport, potential window, and interfacial stability. Hence, it is very important to optimize not only the electrolyte composition but also the interaction of the electrolyte with the electrode surface in order to be able to use the full capacity of these advanced materials. The next section provides an overview of the main types of electrolytes used in supercapacitors, highlighting their properties, limitations, and impact on device performance.

Table 4. Comparative overview of main electrode materials for supercapacitors

Material Type	Advantages	Limitations	Typical Applications
Carbon-based (graphene, CNTs, activated carbon)	High surface area, good electrical conductivity, chemical stability, lightweight	Restacking of graphene sheets, moderate capacitance for EDLC	EDLCs, flexible supercapacitors
Transition Metal Oxides (RuO_2, MnO_2, NiO, Co_3O_4)	High pseudocapacitance, multiple redox states	Low conductivity, limited cycle life, expensive (RuO_2)	Pseudocapacitors, hybrid SCs
Conducting Polymers (PPy, PANI, PEDOT)	High pseudocapacitance, facile synthesis, flexibility	Poor cycling stability, volume change during redox	Flexible SCs, hybrid electrodes
MXenes (Ti_3C_2Tx, Mo_2CTx)	High conductivity, hydrophilic surface, tunable interlayer spacing	Oxidation in air/water, synthesis challenges	High-rate SCs, hybrid SCs

3.2. Electrolytes

Electrolytes are a vital part of electrochemical supercapacitors (ES), they allow ion movement between the electrodes during the charging and discharging stages. This ion movement is what mainly determines the device's electrochemical performance such as specific capacity, energy efficiency, cyclic stability, voltage range, and safety. Besides that, the chemical reaction of the electrolyte with the active electrode material which is important to lead both the supercapacitor's longevity and its performance. Accordingly, the problem in question is to find a way to optimize capacitance, voltage and specific surface area to increase energy density, at the same time, keep system stability intact. Today, there are many kinds of electrolytes that have been invented to fulfill various application demands. Those types can be divided into four main groups: aqueous electrolytes, organic electrolytes, ionic liquids, and solid or gel electrolytes. Each group has different properties in terms of ionic conductivity, electrochemical stability, cost, and safety, which are the factors that directly determine their choice for different .

Aqueous electrolytes are widely used due to their excellent ionic conductivity, often greater than 1 S/cm, and it also has a minimum pore size requirement in comparison to the organic electrolyte(Pell and Conway 2001). Commonly employed aqueous electrolytes include acidic (such as H_2SO_4), alkaline (KOH, NaOH, LiOH) and neutral (Na_2SO_4, K_2SO_4, Li_2SO_4, $NaNO_3$, KCl) solutions(Poonam et al. 2019). Acidic and alkaline electrolytes generally offer high specific capacitance thanks to the high mobility of H^+ and OH^- ions. KOH, in particular, is effective in reducing the equivalent series resistance of the supercapacitor thanks to the small size of the K^+ cation, which promotes high ion mobility(Barzegar et al. 2015; Kar 2021).

Neutral electrolytes, although less efficient in terms of capacity, are more stable and can participate in useful faradaic processes depending on the electrode materials used(Mandal et al. 2025). Despite their advantages, aqueous electrolytes have one major limitation: their narrow electrochemical stability window, generally restricted to 1.0-1.23 V due to water decomposition(Hou et al. 2017; Lu et al. 2024). This limits the energy density of supercapacitors. In addition, their operating temperature range is narrow, and leakage can occur due to the liquid nature of the system. To overcome these limitations, research is focusing on the development of highly concentrated aqueous electrolytes, such as "water-in-salt electrolytes", which enable the stability window to be widened while maintaining

In comparison, organic electrolytes enable operation at much higher voltages, up to 2.5-3.0 V, considerably increasing energy density(Lu et al. 2024). They are prepared by the dissolution of conductive salts in organic solvents. Unfortunately, organic electrolytes possess certain disadvantages such as high cost, with a lower specific capacitance, higher internal resistance, reduced ionic conductivity, and increased flammability. In addition, they are toxic, difficult to fabricate, prone to capacitance degradation with time, and have higher self-discharge rates. Commonly used organic electrolytes consist of salts such as tetraethylammonium tetrafluoroborate ($TEABF_4$) dissolved in organic solvents such as acetonitrile or propylene carbonate(Lakshmi and Vedhanarayanan 2023b; Lu et al. 2024; Zhong et al. 2015). These large voltage windows have been attributed to organic solvents' much greater electrochemical stability. The organic electrolyte, however, has certain disadvantages, including lower ionic conductivity, normally in a range of 10-2 to 10-3 S/cm compared to aqueous solution, and become more flammable, and thus strict safety standards must be applied (Sharma and Chand 2023). Commercial activated carbon-based supercapacitors often use $TEABF_4$/ACN to enable nominal voltages of 2.7 V, which practically justifies these electrolytes despite challenges(Lai 2019).

Concurrently, solid electrolytes and polymer gels are becoming increasingly popular, mainly for flexible and portable devices. Solid electrolytes can be in the form of conductive polymers such as polyethylene oxide (PEO) or gels with a liquid electrolyte entrapped in a polymer matrix like the PVA/H_3PO_4 or PVA/KOH systems (Alipoori et al. 2020, 2021; Zhao et al. 2023). Such electrolytes boast several advantages, which include excellent mechanical flexibility, improved safety by the absence of leakage, and compatibility with flexible device designs(Alipoori et al. 2020; Samantaray et al. 2023). Nevertheless, their ionic conductivities remain lower than those of liquid electrolytes, affecting power delivery. Such limitations are currently being overcome by incorporating nanomaterials or optimizing polymer networks. These can be done by strategies such as incorporating nanomaterials or optimizing polymer networks to tackle these challenges, and these have been summarized in Table 5.

Table 5. Summary of various electrolytes

Electrode	Specific Capacitance	Retention with cycle	Refs.
PVA-H_3PO_4 / [EMIM]BF_4 (Gel polymer electrolyte)	271 F/g	Over 3000 cycles	(Jang et al. 2016)
PVA-H_3PO_4 (PANI/graphene)	261 F/g	92% after 10,000 cycles	(Xie et al. 2018)
PVDF-HFP / $Mg(Tf)_2$ / EMITf (peanut shell electrode)	189 F/g	98% after 10,000 cycles	(Yadav et al. 2018)
1 M Et_4NBF_4/ propylene carbonate	160 F/g	96.4 after 1000 cycles	(Izadi-Najafabadi et al. 2010)
KOH	205 F/g	97.1% after 1000 cycles	(LI et al. 2013)
1 M $LiClO_4$	N/A	85% after 1000 cycles	(Bai et al. 2013)
1M H_2SO_4	215 F/g	99% after 1500 cycles	(Gueon and Moon 2015)
1-butyl-3- methylimidazolium hexafluorophosphate ($BMIPF_6$)	348F/g	120% after 3000 cycles	(Chen et al. 2011)
1-ethyl-3- methylimidazolium tetrafluoroborate/acetonitrile ($EMIMBF_4$/ AN)	166 F/g	97% after 10 000 cycles	(Y. Zhu et al. 2011)
1-ethyl-3- methylimidazolium bis(trifluoromethylsulfonyl) imide, [EMIM][TFSI] and Acetronitrile in 1:1 ratio	174 F/g	94% after 1000 cycles	(Kim et al. 2013)
PVA-H_3PO_4, 1M H_2SO_4	300 F/g	93% after 10000	(D. Yu, Goh, Wang, et al. 2014)

The choice of electrolyte affects several essential parameters of supercapacitors. Specific capacitance depends on the size of ions, the ion mobility, and the extent of penetration into electrode pores, whereas internal resistance changes the maximum power that potentially can be delivered, and is directly related to ionic conductivity and the electrolyte's viscosity. Similarly, cyclic stability is influenced by the chemical compatibility of the electrolyte with the active material, as well as any secondary reactions that may take place during charge-discharge cycles.

The electrolyte is a key component that controls ion transport, electrochemical stability, and the overall energy-power performance of a supercapacitor. Nevertheless, an optimal device performance also involves the separator that physically separates the two electrodes allowing efficient ionic flow. The separator's characteristics, such as porosity, ionic permeability, mechanical strength, and chemical compatibility with the electrolyte, play an important role in ensuring safe operation, preventing short circuits, and stabilizing the device over time. The upcoming section will focus on the various separators used in supercapacitors and their influence on electrochemical performance.

3.3. Separators

In supercapacitors, separators serve a critical dual function as electrical insulators and ion-conducting membranes (Kar 2021). The main purpose is to prevent direct electrode-to-electrode contact to avoid dangerous sending currents while the device allows ions to flow through the separator. Separators additionally allow ions to depart from one electrode and travel through the membrane. Separators are chemically inert, have a high mechanical strength, optimal thickness, porosity, and good wettability of the electrolyte (Verma et al. 2020). These two roles are important for a device to operate safely and effectively during the lifetime of the device. Damage to a separator will cause significant performance drops or even irrecoverable short circuits. A properly purchased separator needs to have several properties: chemical inertness, high mechanical strength, good wettability of the electrolyte, sufficient thickness and porosity, as well as good thermal and electrochemical stability (Kar 2021; Verma et al. 2020). These properties aid compatibility with the electrolyte and electrodes while providing a stable and long-lived performance under various conditions.

Supercapacitors commonly employ several types of separators, such as polymer membrane-based, paper-based (cellulose), and inorganic or ceramic-based membranes. Frequently used materials include polypropylene, polyvinylidene fluoride (PVDF), polytetrafluoroethylene (PTFE), cellulose derivatives (e.g., nitrate or acetate), glass fiber, Nafion, oxidized graphene, and Celgard membranes. Based on their composition, separators can be broadly classified into three categories: polymeric, cellulose-based, and ceramic or inorganic materials.

Synthetic polymer separators are popular materials due to their low costs and chemical stability. Nonetheless, synthetic polymer separators show lower dielectric constant, which limits the power and energy density achievable. The dielectric constant disadvantage can be overcome by incorporating ceramic fillers to polymer matrices, for instance, incorporating titanium dioxide (TiO_2) into poly(vinyl alcohol) will enhance ionic conductivity, porosity, and mechanical strength (Shuja et al. 2024). For example, poly(vinyl butyral) (PVB) separators reinforced with one-dimensional aluminum oxide nanorods exhibit enhanced properties, including high porosity, mechanical strength exceeding 30 MPa, thermal stability up to 200 °C, electrolyte uptake greater than 200 wt% without swelling, and ionic conductivity as high as 14 $mS \cdot cm^{-1}$(Liu et al. 2020). Parallel to this, there is an increasing interest in biopolymer-based separators owing to their sustainability. Cellulose and its derivatives, together with biodegradable materials, such as membranes prepared from eggshell, provide environmentally friendly alternatives to this technology (Bon et al. 2018; Teng et al. 2020; Verma et al. 2023). Cellulose-based paper separators are particularly attractive for large-scale supercapacitor production owing to their high porosity and mechanical robustness, especially in systems using organic electrolytes.

The performance of a separator is typically assessed using several physico-chemical parameters. The MacMullin number quantifies the ratio of the resistance of the electrolyte-saturated separator to the resistance of the electrolyte alone. The Gurley number measures air permeability, while Porosity is calculated using the following equation(Kar 2021):

$$\text{Porosity} = \frac{(W - W_0)}{\rho_e \times V} \times 100\% \tag{8}$$

whereW is the weight of the separator immersed in the electrolyte, W_0 is the weight of the separator before immersing in the electrolyte, ρ_e the density of the electrolyte, and V the volume of the separator. Tortuosity refers to how complicated the ion transport pathways are in the separator. In the end, the selection and design of the separator have a major influence on the total performance of supercapacitors. It is necessary to combine advanced materials and improve the fabrication processes as two main approaches for the creation of the following generation of energy storage devices.

4. FLEXIBLE SOLID-STATE SUPERCAPACITORS

Flexible Solid-State Capacitors (FSSCs) offer a significant advancement in energy storage technology based on the needs of mobile, flexible, and connected devices. Their structure allows FSSCs to sustain their electrochemical performances with mechanical changes (bending, stretching, twisting) applied to them, unlike traditional solid-state supercapacitors, in which their mechanical flexibility is the result of incorporating flexible elements within every material field of the device (current collector, substrate, electrodes, and/or electrolytes). Their mechanical flexibility makes FSSCs a suitable option for applications in wearable electronics, flexible sensors, smart textiles, and connected devices. Following this concept, the emergence of flexible electronic systems increases the advantages of FSSCs: lightweight structure and conformability to reduce weight, flexibility, and promote portable and wearable electronics. As previously discussed, the composites can be designed based on the device specifications and development of performance criteria. Specifically, the use of a polymer-based composite material is of particular interest because of their material structure, which can enhance the following properties: specific capacitance, flexibility/ease of processing/manufacturing. Choosing materials with these innovations leads to a larger available energy storage capacity to maintain mechanical flexibility towards the next generation of energy storage technology for applications today.

Researchers have shifted away from using rigid metal current collectors and binders which are not conductive in order to transform these material advantages into fully flexible devices, and are using light weight and flexible alternatives instead. Functional materials based on carbon nanostructures, metal oxides and conductive polymers are showing high electrical conductivity, electrochemical compatibility and mechanical flexibility. The choice of substrate is also a crucial factor in device performance. Substrates such as carbon occurred fabrics, paper and textiles have the advantages of low weight, inherent porosity and large surface area, allowing for the direct growth of active materials and subsequently enhance overall conductivity (as noted in Table 6). Textile-based flexible energy storage devices intended for wearable applications can be categorized broadly as 1D fiber/yarn-shaped devices or 2D fabric-shaped devices. Furthermore, a few 3D-shaped energy storage devices have been reported in the literature (Levitt et al. 2020).

Table 6. Summary of Textile-Based Supercapacitors

Substrate	capacitance	Capacitance retention	Flexibility	Refs.
Cotton knit fabric	Electrode 481 and 1433mFcm^{-2} at 5 mV s^{-1} and 1 mAcm^{-2} respectively, device 101 and 450 mFcm^{-2} at 5mVs^{-1} and1mAcm^{-2} respectively	30% after 500 cycles (gel electrolyte), above 53% at 5000 cycle (aqueous electrolyte)	Electrode ≈78%–91% after stretched 1000 times, device capacitance enhancedto 160% at 5 mA cm−2	(Wang et al. 2019)
mesh fabric	266.6 Fg^{-1}	95.4% after 5000 cycles	N/A	(Khan, Grabher, and Hossain 2023)
Poly-cotton (65/35) 2/1 twill fabric	2.7 mF cm^{-2}	≈98% after 15 000 cycles	98% after bending 150 cycles at 180°	(Afroj et al. 2020)
rGO fiber yarns	409 Fg^{-1}	No decay after 5000 cycles	N/A	(Aboutalebi et al. 2014)
Carbon nanotube and rGO composite yarn	116.3 mFcm^{-2}, 45 Fcm^{-3} at 26.7 mA cm^{-3}	93% after 10 000 cycles	>97% after 1000 bending cycles at 90°	(D. Yu, Goh, Zhang, et al. 2014)
Cotton fabric	464Fg^{-1} at 0.25Ag^{-1}	97% after 10 000 cycles	95.6% after folding 180° for 2000 cycles	(Zhao et al. 2015)

continued on following page

Table 6. Continued

Substrate	capacitance	Capacitance retention	Flexibility	Refs.
Polyester fabric	26 pF cm−2	N/A	Sustains 100 cycles of repeated bending	(Qiang et al. 2019)
Polypropylene fabric	13.3 mF cm^{-2} (79.9 F g^{-1}) at 0.1 mA cm^{-2}	Almost 100% after 5000 cycles	N/A	(Stempien et al. 2019)
Cotton yarn	74.0 mF cm^{-2}	N/A	97% after 200 cycles	(Zhang et al. 2019)
Silk woven fabric	32 mF cm^{-2} at 25mV s^{-1}, 41 mF cm^{-2} at 0.75 mA cm^{-2}	45% retention after 1000 cycles	N/A	(Pullanchiyodan, Manjakkal, and Dahiya 2021)

One of the most representative examples of the innovations is the creation of wearable textile pseudocapacitor electrodes made from polyaniline (PANI) deposited on activated carbon fiber felts (ACFF) (Li and Chen 2017). Although the ACFF/PANI composite provides higher specific capacitance than textile-only structures, its conductivity remains limited. In order to overcome this, very conductive carbon nanotubes (CNTs) were added, resulting in the ACFF/PANI/CNT composite (Li and Chen 2017). This hierarchically porous architecture with both mesopores and micropores has excellent electrolyte access, a surface capacitance of around 5119 mF·cm^{-2} at 1 mV·s^{-1}, and greatly improved energy density and cycling stability.

Following a comparable design approach, CNT aerogel fibrous electrodes were produced for use in flexible pseudocapacitors (Li et al. 2018). Through hydrating CNT fibers to fabricate hydrogels, and subsequently processing the hydrogels to form aerogels, the authors produced lightweight, high-performing electrodes. When paired with a gel electrolyte of P(VDF-HFP) and $EMIMBF_4$, the symmetric supercapacitor achieved a specific capacitance of 160.8 F·g^{-1} at 0.5 mA, and maintained 99.7% capacitance retention after 3000 cycles. The device demonstrated an energy density of 29.6 Wh·kg^{-1} at a power density of 27.3 kW·kg^{-1}. While only minor performance losses were observed during bending, the device retained 92.9% of the original capacity after 2000 cycles, and the capabilities were preserved to temperatures as high as 80 degrees Celsius, which may have been particularly a function of the non-volatile electrolyte used (Khan et al. 2024).

Alongside these material and structural changes, research has also been oriented towards the development of self-supporting electrode architectures that do not require metallic current collectors, thus, impeding the devices' compactness. The most attractive one may be the immobilization of carbon fiber fabric with Co3O4 nanorods covered with polypyrrole (PPy) .(Khan et al. 2024; Ma et al. 2018). The asymmetric configuration is responsible for the extension of the operating voltage window, surface conductivity (up to 6670 mF·cm^{-2} at 2 mA·cm^{-2}) and surface capacitance (2470 mF·cm^{-2} at 4 mA· cm^{-2}) are still excellent. The electrode exhibited 97% capacitance retention after 2000 cycles and a capacitance of 70% at the severed half of the electrode, which is proof of its mechanical durability. Besides that, the linear voltage increase in the series configuration is the evidence of its potential as a power source for the forthcoming portable electronics.

In addition to composing materials, the procedures of mechanical tests to assess FSSCs are critical for determining the functional flexibility and electrochemical stability of the FSSCs. Mechanical tests usually involve cycles of flexing, twisting or stretching under controlled bending radii of between 2 and 10 mm. The bending radius determines the amount of mechanical strain applied to the different layers of the electrode and the electrolyte, with smaller bending radii resulting in more strain and greater influence on ion transport and charge storage. The number of cycles is an additional measure of mechanical stability, and like most high performance FSSCs, devices operating at a bending radius of 5 mm report to operate well with greater than 90% of capacity remaining after a million cycles of bending, for example CNT/rGO based yarn supercapacitors retain 97% of capacitance after being bent (90°) for a thousand cycles (P. Yu et al. 2014), while cotton-fabric based devices retain 95.6% of original capacity after folding (180°) for two thousand cycles (Zhao et al. 2015). During these test sequences, electrochemical analysis is conducted using cyclic voltammetry (CV) and/or galvanostatic charge-discharge (GCD) and electrochemical impedance spectroscopy (EIS) will be conducted before and after the mechanical tests, as assessment of performance under different mechanical stress and subsequent recovery of performance. CV can be plotted continuously to provides the gradation of values between stress applied, while the shift in equivalent series resistance (ESR) from the EIS analysis provides further information on the mechanical stresses influence on performance. Although the above testing demonstrates good performance and electrochemical stability, relating mechanical flexibility with electrochemical performance provides a more appropriate assessment of FSSC reliability for mobile and flexible electronics.

Though structural and mechanical optimization is essential, gel-based electrolytes play a significant role in the aspect of flexibility and safety as well. These solid-state electrolytes, usually polymer-based (Table 7), are generally made of materials like PVA and thus are able to prevent leakage and still have a high ionic conductivity. Moreover, it is possible to design them for more features such as being able to resist a freeze, thermal endurance, self-discharge, and self-healing, which in turn increases the safety and the lifetime of flexible supercapacitors by an order of magnitude.

Table 7. Summary of polymer-based as solid-state electrolytes

Electrolyte Composition	Additive Type	Performance	Refs.
rGO/PVA + DMSO/ H_2O/H_2SO_4	Organic solvent (DMSO)	252.0 F g^{-1}; 93% retention at −60 °C	(Zheng et al. 2022)
PVA + EG/LiCl (binary organo-hydrogel)	Organic solvent + inorganic salt	70.6% retention at −40 °C; 11.7% decay after 5000 cycles at −20 °C	(Bai et al. 2024; Feng et al. 2021; Rong et al. 2018)
PVA + Glycerol/LiCl	Organic solvent + inorganic salt	87.5% retention after 5000 cycles at −20 °C	(Hou et al. 2021)
PEG diglycidyl ether-based organo-hydrogel	Organic crosslinker	150 F g^{-1} at −30 °C	(Duan et al. 2024; Luo et al. 2024)
PVA/PANI/glycerin/ phytic acid	N/A	97.5% (1000 cycles) and 95% (after 10 healing cycles)	(Lai et al. 2020)
PVdF-HFP/EMITf/ $Al(Tf_{)3}$	Ionic liquid	75.5 F g^{-1} at −20 °C; 306.5 F g^{-1} at 60 °C	(Liu et al. 2021)
EMIM-BF_4/ zwitterionic gel (PVA/ PAS)	Ionic liquid	254.0 mF cm^{-2} at 50 °C; 221.6 mF cm^{-2} at −20 °C	(Wang et al. 2022)
Cellulose/PVA + H_2SO_4	Inorganic salt	454.9 F cm^{-3} at −40 °C (~83% of RT)	(Wu, Yu, and Zhong 2021)

In spite of these significant developments, some issues remain; among them, the self-discharge issue is still prevailing and limits long-term energy retention. Overcoming this challenge needs meticulous engineering of the electrode-electrolyte interface besides optimization of material properties. However, continuous research in the fields of polymer composites, self-supporting electrodes, and multifunctional electrolytes does show a bright future for flexible, high-performance energy storage systems. In summary, the above discussion identifies the intrinsic relationship between mechanical flexibility and electrochemical stability, again emphasizing the imperative of correlating mechanical performance metrics with practical device reliability, an essential aspect pointed out by the reviewer.

5. ELECTROCHEMICAL CHARACTERIZATION TECHNIQUES

The performance of supercapacitors depends on many parameters, like power density, lifetime, energy density, ESR, and cell architecture. The properties mentioned above have been widely evaluated by a set of electrochemical techniques. Some of the important techniques include cyclic voltammetry, galvanostatic charge/discharge, and electrochemical impedance spectroscopy. Each one provides complementary information on the behavior of energy storage systems.

5.1. Cyclic voltammetry

Cyclic voltammetry (CV) is one of the most widely used electrochemical methods for appraising electrode material performances in energy storage systems, particularly supercapacitors. It records the evolution of current in response to a potential sweep, identifying the oxidation and reduction processes of electroactive species. This method can offer qualitative information on such aspects as reversibility and the nature of storage mechanisms, along with quantitative data like specific capacity and electrochemical stability.

CV is commonly performed experimentally in a standard three-electrode electrochemical cell, typically with a working electrode coated by the active material being studied, along with a reference electrode, such as Ag/AgCl or SCE, and an auxiliary electrode, usually platinum or carbon. The potential is linearly swept between two limits, for example between 0 and 0.8 V, and then reversed (Alsufyani et al. 2024). This polarization cycle generates a current response measured as a function of the applied potential. This potential sweep leads to electrochemical reactions on the surface of the working electrode. The electroactive species present in the electrolyte can undergo oxidation transformations, corresponding to the loss of electrons, or reduction, corresponding to the gain of electrons. These reactions can be generally represented by the following equation(Alsufyani et al. 2024):

Oxidized species (Ox) + $ne^{-} \leftrightharpoons$ Reduced species

Graphically, the electrochemical response takes the form of a triangular waveform voltammogram. Typically, this shows an oxidation wave (anodic peak) and a reduction wave (cathodic peak). When the peaks are symmetrical and close in potential, this indicates a rapid, reversible process.

*Figure 6. **Voltammetric response of various capacitors and the estimation of specific capacitance from cyclic voltammetry. Typical representation of cyclic voltammograms of (A) electrochemical double layer capacitor (EDLC), (B) pseudocapacitors, (C) hybrid capa- citors, (D) estimation of specific capacitance using cyclic voltammetry.***

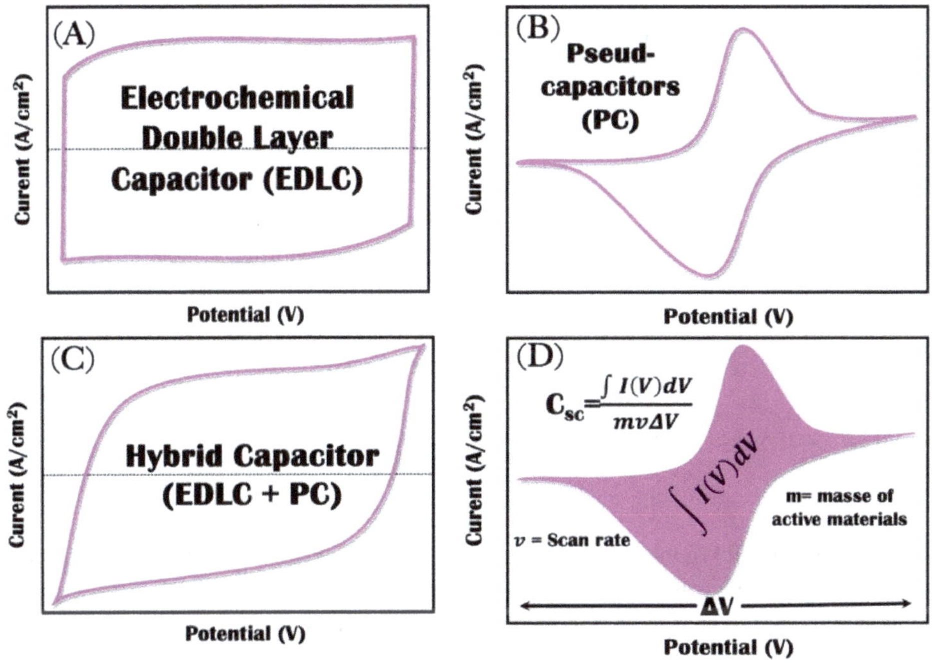

The shape of the voltammogram reflects the nature of the energy storage mechanism. Thus, a quasi-rectangular curve is characteristic of double-layer capacitive behavior (EDLC), where capacitance is due to simple charge separation at the electrode/electrolyte interface (Fig. 6.A). On the other hand, the presence of well-defined redox peaks reflects pseudocapacitive behavior, associated with reversible faradic reactions (Fig. 6.B). In some cases, a voltammogram showing both a quasi-rectangular shape and a large current amplitude may indicate a mixed contribution from EDLC and pseudocapacitive mechanisms (Fig. 6.C).

In pseudocapacitive supercapacitors, these redox reactions may be confined to the surface of the active material or its proximate layers, sometimes resulting in a CV curve shape similar to that of EDLCs (Fig. 6.C). Other pseudocapacitive materials show more pronounced redox peaks, linked to oxidation/reduction processes such as the deposition or dissolution of metal oxides (Fig. 6.B), or to changes in oxidation states during reversible intercalation(Yan et al. 2020). VC can also be used to quantify the specific capacitance (C_{SC}) of materials, calculated according to the following formula(Srinivasan and Weidner 2000):

$$C_{SC} = \frac{\int I(V)dV}{m\,v\,V} \tag{9}$$

where ν is the sweep velocity in V/s, m the mass of active material in grams, ΔV the potential window in volts, and the integral corresponds to the area under the voltammogram (calculation of Csc from cyclic voltammetry is illustrated in Fig. 6.D). The integral corresponds to the area under the voltammogram curve, representative of the amount of charge exchanged. A large area under the voltammogram and a high current amplitude are therefore desirable, as they indicate a higher storage capacity (Fig. 6.D). Cyclic voltammetry is a powerful method for elucidating the electrochemical mechanisms at work in supercapacitor electrodes, characterizing both the reversibility of processes, the rate of charge transfer, and the capacitive or pseudocapacitive contribution of the materials tested.

5.2. Galvanostatic charge-discharge (GCD)

Galvanostatic charge/discharge (GCD) is an electrochemical technique for analyzing the performance of batteries and supercapacitors. It involves applying a constant current to the device, while recording the voltage response during the charge and discharge phases. Fig.7 shows a potential profile on the applied current.

Figure 7. Galvanostatic charge-discharge curve of different capacitors and the esti- mation of specific capacitance from charge-discharge experiments. Typical galvanostatic charge discharge curve for (A) EDLC, (B) pseudocapacitor, (C) estimation of specific capacitance from charge-discharge techniques, and (D) variation of specific capacitance with applied current density for two different polymers.

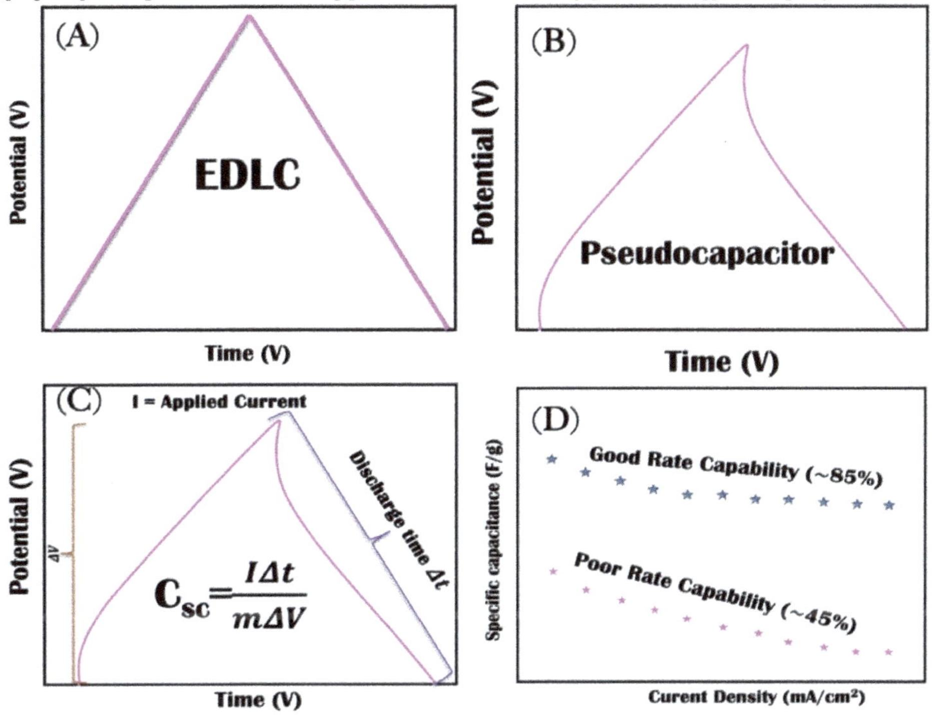

The resulting curve ideally has a symmetrical triangular shape, characteristic of pure capacitive behavior, typical of electrical double-layer supercapacitors (EDLC). This symmetry is an important indicator of the reversibility of the electrochemical process. On the other hand, the presence of curvatures or plateaus indicates the intervention of faradaic processes, as in the case of pseudocapacitors. These redox reactions at the electrode/electrolyte interfaces induce a non-linear response, particularly visible at low current densities where the discharge time (Δt) is sufficient to allow full development of pseudocapacitive processes.

The GCD technique provides access to several essential parameters: specific capacitance, cyclic stability, reversibility, and the energy and power densities of the material under test. The mass specific capacitance CSC, expressed in farads per gram (F/g), can be calculated from the following relationship(Wu et al. 2007):

$$C_{SC} = \frac{I.dt}{m.\ V} \tag{10}$$

where I is the applied current (in A) and Δt the discharge time (in s), m the mass of active material (g), and ΔV the potential window (V).

From this capacity, energy density E (Wh/kg) and power density P (W/kg) can be estimated using the following formulas(Alsufyani et al. 2024):

$$E\left(\frac{Wh}{kg}\right) = \frac{1}{2x3.6} C_{SC}(\ V)^2 \text{ and } P(W/kg) = E \times \frac{3600}{t} \tag{11}$$

Both quantities are highly dependent on the operating voltage, which in turn is influenced by the nature of the electrodes and electrolyte. In addition, capacitance can be expressed per unit area (F/cm^2) or volume (F/cm^3), to assess performance in miniaturized devices. Areal capacitance refers to the total geometric surface area of the electrodes, while volumetric capacitance depends on the density of the active material.

GCD tests can also be used to analyze throughput capacity, i.e. the supercapacitor's ability to maintain its performance under high currents. For flexible devices, tests can be carried out at different bending angles (from 0° to 180°), to assess electrochemical flexibility. In addition, cyclic stability is assessed over several hundred or thousand cycles, using either GCD or cyclic voltametry. Finally, the energy stored in a supercapacitor can be expressed more generally by the following integral(Ding et al. 2024c):

$$E = \int V dq = \frac{1}{2} C_{SC} V^2 \tag{12}$$

where q is the accumulated charge and V is the voltage. This expression expresses the work required to charge the capacitor.

Compared with batteries, supercapacitors feature very high-power density, exceptionally rapid charging and discharging, and remarkable longevity, making them particularly suitable for applications requiring frequent and rapid cycling(Rashid Khan and Latif Ahmad 2025b). To facilitate comparison of energy performance, energy and power densities obtained experimentally are often plotted on a Ragone diagram. The following diagram highlights the position that supercapacitors assume, somewhere between traditional capacitors with very high-power density but low energy density, and batteries that can offer high energy density but more limited power.

5.3. Electrochemical impedance spectroscopy

Electrochemical Impedance Spectroscopy (EIS) is one of the most powerful tools for the investigation of the electrical response of electrochemical systems. It consists of the application of a low-amplitude AC over a wide frequency range, generally from 0.01 Hz to 100 kHz, usually between ±5 mV and ±10 mV (Alsufyani et al. 2024). By using this technique, it is possible to study electrochemical processes as a function of time, including charge transfer, mass transport, and diffusion phenomena.

The ability of EIS to distinguish between fast and slow mechanisms at the electrode/electrolyte interface, by sweeping frequencies from megahertz (MHz) to millihertz (mHz), means a single frequency sweep is sufficient to acquire useful data regarding the capacitive/resistive nature. Electrical capacitance can be estimated through impedance data, in particular from the linear part of the Bode diagram (log |Z| as function of log f) using (Alsufyani et al. 2024):

$$C = \frac{1}{2ðf|Z|} \tag{13}$$

where f is the frequency in Hz, and |Z| is the absolute value of the complex impedance in Ω. The Bode diagram shows the magnitude of impedance and phase angle as functions of frequency. It points out the gradual reduction of capacitance with the increment in frequency. This trend is explained by the inability of electrolytic ions to penetrate electrode micropores at high frequencies, giving the supercapacitor a behavior close to that of a pure resistor in this frequency range.

Furthermore, Nyquist diagrams are very commonly used for the graphical interpretation of the electrochemical response of the system. It plots the imaginary part of the impedance Z″(f) against its real part Z'(f) (Magar, Hassan, and Mulchandani 2021). (Magar, Hassan, and Mulchandani 2021). It helps largely in the estimation of charge transfer resistance, which, in turn, is related to the diameter of the plot's semicircle. It can also depict the two most important phenomena: ion diffusion and ion penetration through electrodes. In low-frequency regions, a pure capacitor should behave like a vertical axis in the Nyquist plot, but tilt angles ranging from 45° to 90° are observed because of ionic diffusion and the Warburg phenomenon (Raza et al. 2018). These diagrams, along with analysis of phase and amplitude, provide a comprehensive description of electrochemical processes in supercapacitors.

Although EIS is very useful for the characterization, it has the disadvantage of using a low amplitude of alternative potential (5 to 10 mV), which may be insufficient in the analysis of systems that are under the action of high currents(Magar et al. 2021; Wang et al. n.d.). In any case, EIS is a basic tool for the determination of the resistive and capacitive properties of different materials and for the optimization of electrochemical devices. In order to interpret the results of complex EIS, simplified analog circuit models are used. The Randles cell model is the most used to describe the electrode/electrolyte interface of an electrochemical cell. The model comprises three basic elements: the electrolyte resistance Rs, the double-layer capacitance Cdl, and the polarization resistance Rp. Each of those elements represents one important aspect of the electrochemical behavior. The value of Rs depends on temperature and the geometry of the charge carrier, as well as on the nature of the ions in the electrolyte solution. Double-layer capacitance, or Cdl, describes the accumulation of charges at the electrode/electrolyte interface, increasing its value with the active surface of the electrode. Polarization resistance, or Rp, reflects the obstacles to charge transfer at the electrode/electrolyte interface, depending on the type of electrochemical reactions that take place. These parameters are extracted through Bode and Nyquist diagrams, which provide an overview of system performance and stability.

6. TECHNOLOGICAL CHALLENGES AND FUTURE PROSPECTS

Among the most important advantages of supercapacitors are the high-power density, long cycle life, excellent energy efficiency, and a wide operating temperature range-all making them perfect candidates for applications needing fast charge/discharge cycles. Yet, despite such strengths, their typically low energy density-5-20 Wh/kg compared with 150-250 Wh/kg for lithium-ion batteries-stands out as their major limitation in systems demanding extended energy supply. While they can generate high-power outputs up to 10 kW/kg and support millions of cycles, the mentioned modest energy capacity puts a cap on their wider commercial acceptance.

Besides energy density, there are many other technical challenges. The economic barrier is created by high production costs due to the use of advanced or rare materials. Supercapacitors also face some critical problems associated with high leakage currents and voltage constraints, necessitating series configurations and complex balancing circuits. Innovation in supercapacitor technology is also much slower compared to battery technologies mainly because of the fewer studies that have been conducted on these systems. From a performance point of view, another key issue involves equivalent series resistance (ESR), which generates energy losses and heat during operation. Although pseudocapacitive materials exhibit high specific capacitance, many suffer from poor electrochemical stability upon contact with both acidic aqueous and organic electrolytes. Moreover, the mismatch between the pore structure of electrodes and the size of electrolyte ions may impede ion accessibility and lower the efficiency of charge storage and overall performance.

These limitations are compounded by manufacturing challenges. Most current fabrication processes are complex, expensive, and ill-suited for large-scale production. Consistent control of porosity, pore size distribution, and electrode morphology is indispensable in ensuring the maximum active surface area but remains a challenging task. Besides, despite the increasing concerns for the ecological impact brought about by energy storage systems, environmental sustainability is not yet fully integrated into material design.

To overcome these difficulties, continuous efforts are being made by using novel materials, optimized electrolytes, and scalable fabrication methods. The main strategy is the use of nanostructured electrode materials including graphene, MXenes, MOFs, and COFs, which possess large specific surface areas, excellent electrical conductivity, and suitable electrochemical stability. Three-dimensional hierarchical structures of such materials allow for more efficient transport of ions and electrons, which greatly enhances capacitance. It is difficult to fabricate these materials at the industrial level because their synthesis methodologies are mostly costly and complicated. Concurrently, advanced electrolyte development has also been playing an important role in the improved performance of supercapacitors. Accordingly, electrolytes with a wider electrochemical window, such as ionic liquids and polymer-based systems, enable devices to work at higher voltages. Other functional electrolytes having self-healing, freeze-resistant, and low self-discharge properties are also under investigation for reliable operations under harsh conditions and for portable applications.

The environment is also driving a rethink on research priorities. The use of renewable and biodegradable materials for the development of electrodes has offered a greener alternative to conventional materials in developing electrodes. Biomass-derived activated carbons, notably those originating from coconut shells, sawdust, corn stalks, and rice husks, hold great promises owing to their low cost and ecological values. Other promising bio-based electrode materials include lignin, algae, cellulose, silk, and biodegradable polymers like PLA with no sacrifice in electrochemical performance. With the hybrid supercapacitor that combines EDLC and pseudocapacitance mechanisms, a very bright future for supercapacitors can be foreseen.

The hybridization synergistically integrates the high-power capability of EDLC with the greatly improved energy storage capacity of pseudocapacitors. An approach using ternary composite materials and hierarchical heterostructures enhances conductivity, ion diffusion, and structural stability. The integration of state-of-the-art fabrication techniques, such as 3D printing and AI-assisted synthesis, is enabling the design and creation of high-performance supercapacitors with power efficiency, intended for industrial production. These enabling technologies, coupled with greater focuses on environmental sustainability and economic viability, will be the path to next-generation supercapacitors that could secure a prime position in the future landscape of energy storage.

7. CONCLUSION

In the context of a fast-tracked energy transition, supercapacitors are gaining attention as a potential alternative or complement to traditional batteries because of their rapid charging abilities, safety features, and prolonged lifespan. In general, two concepts can be distinguished: electric double-layer capacitors, which are known to be highly stable, and pseudocapacitors, whose higher capacity is usually associated with low durability. Currently, a hybrid approach seems to represent the most promising path toward finding a balance between power, energy density, and stability. The review has highlighted the recent advances and remaining technological challenges in terms of the fundamental parts constituting flexible supercapacitors: current collectors, self-supporting electrodes, and functional electrolytes. Various novel material systems have been explored in enhancing the mechanical, chemical, and electrochemical performance of the device by carbon, metal, and polymer-based composite materials. Indeed, the perfect current collector is still quite a challenge to create. It needs to possess high electrical conductivity, mechanical strength, durability in various environments, compatibility at the interface, and should be easy to integrate in thin-film formats. Films made of metals show great promise, yet further optimization of their structure is necessary. Self-supporting electrodes offer a viable alternative by minimizing interfacial resistance and reducing the distance for charge transport. They have to meet these demands by combining excellent electrical conductivity with mechanical flexibility, while also being highly electrochemically active. Composite materials made from the combination of polymers with carbon, metals, or inorganic components satisfy these stringent demands. In terms of electrolytes, polymer gels, particularly organohydrogels, are leading in flexible designs.

Current efforts are focused on improving mechanical stability, freeze resistance, self-healing, and ionic conductivity over a wide temperature range. The ideal electrolyte should also possess low electronic conductivity and good environmental durability while being processable into thin flexible films suitable for wearables. A significant challenge still remains: self-discharge, which limits practical uses of supercapacitors due to spontaneous voltage drop.

Tackling this problem often entails sacrifices in charge transfer speed, adversely affecting capacity and cycle stability. Striking an optimal balance between reducing self-discharge and preserving high electrochemical efficiency is a crucial goal. In the future, further development will be realized with the incorporation of materials with optimized architectures, tight control of electrochemical interfaces, and using advanced modeling and characterization tools. New fabrication technologies, such as 3D printing, roll-to-roll coating, and AI-assisted material design, are in position to enable a new class of more efficient, customized, and scalable devices.

Large-scale deployment already occurs in commercial systems, from Maxwell Technologies ultracapacitors in hybrid buses and industrial energy recovery to Skeleton Technologies graphene-based ultracapacitors for high-power applications, and Nesscap devices for grid stabilization. Specific cost-performance metrics of different devices illustrate common compromises. Carbon-based EDLCs achieve 5–10 Wh/kg with >10,000 cycles for a moderate cost, while RuO2-based pseudocapacitors exhibit very high capacitance (700–1200 F/g) but remain prohibitively costly for mass production. Possible strategies toward sustainable manufacturing include using biomass-derived carbons, synthesizing low-cost carbon matrices with small amounts of high-performance pseudocapacitive materials, electrolyte optimization for safety and low cost, and using scalable fabrication techniques that result in minimal waste generation.

REFERENCES

Abdullah, M. O., Ivy, A. W. T., & Lim, L. S. (2011). 'Automobile Adsorption Air-Conditioning System Using Oil Palm Biomass-Based Activated Carbon: A Review'. *Renewable & Sustainable Energy Reviews, 15*(4), 2061–2072. DOI: 10.1016/j.rser.2011.01.012

Aboutalebi, S. H., Jalili, R., Esrafilzadeh, D., Salari, M., Gholamvand, Z., Yamini, S. A., Konstantinov, K., Shepherd, R. L., Chen, J., Moulton, S. E., Innis, P. C., Minett, A. I., Razal, J. M., & Wallace, G. G. (2014). 'High-Performance Multifunctional Graphene Yarns: Toward Wearable All-Carbon Energy Storage Textiles'. *ACS Nano, 8*(3), 2456–2466. DOI: 10.1021/nn406026z

Adedoja, O. S., Sadiku, E. R., & Hamam, Y. (2023). 'An Overview of the Emerging Technologies and Composite Materials for Supercapacitors in Energy Storage Applications'. *Polymers, 15*(10), 2272. DOI: 10.3390/polym15102272

Afif, A., Rahman, S. M. H., Tasfiah Azad, A., Zaini, J., Islan, M. A., & Azad, A. K. (2019). 'Advanced Materials and Technologies for Hybrid Supercapacitors for Energy Storage – A Review'. *Journal of Energy Storage, 25*, 100852. DOI: 10.1016/j.est.2019.100852

Afroj, S., Tan, S., Abdelkader, A. M., Novoselov, K. S., & Karim, N. (2020). 'Highly Conductive, Scalable, and Machine Washable Graphene-Based E-Textiles for Multifunctional Wearable Electronic Applications'. *Advanced Functional Materials, 30*(23), 2000293. DOI: 10.1002/adfm.202000293

Ahmad, S. (1999). 'Carbon Nanostructures Fullerenes and Carbon Nanotubes'. *IETE Technical Review, 16*(3–4), 297–310. DOI: 10.1080/02564602.1999.11416845

Alipoori, S., Mazinani, S., Aboutalebi, S. H., & Sharif, F. (2020). 'Review of PVA-Based Gel Polymer Electrolytes in Flexible Solid-State Supercapacitors: Opportunities and Challenges'. *Journal of Energy Storage, 27*, 101072. DOI: 10.1016/j.est.2019.101072

Alipoori, S., Torkzadeh, M. M., Mazinani, S., Aboutalebi, S. H., & Sharif, F. (2021). 'Performance-Tuning of PVA-Based Gel Electrolytes by Acid/PVA Ratio and PVA Molecular Weight'. *SN Applied Sciences, 3*(3), 310. DOI: 10.1007/s42452-021-04182-7

Alkallas, F. H., Esayed, A. M., Amira, B. G. T., & Rabia, M. (2024). 'Basic and Acidic Electrolyte Mediums Impact on MnO2-Mn2O3/Poly-2-Methylaniline Hexagonal Nanocomposite Pseudo-Supercapacitor'. *Physica Scripta, 99*(6), 065972. DOI: 10.1088/1402-4896/ad3f85

Alsufyani, S. M., Jafer, R., Iqbal, J., Alwafi, R., Bashir, S., Ramesh, S., & Ramesh, K. (2024). 'Carbon and Polymer-Based Conducting Platforms Incorporated with Electroactive Metal-Oxides/Sulphides for Energy Storage Devices'. *Journal of Energy Storage*, *84*, 110713. DOI: 10.1016/j.est.2024.110713

Amiri, A., Bruno, A., & Polycarpou, A. A. (2023). Configuration-dependent stretchable all-solid-state supercapacitors and hybrid supercapacitors. *Carbon Energy*, *5*(5), e320.

Ayorinde, O., Etukudoh, E., Nwokediegwu, Z., Ibekwe, K., Umoh, A., & Hamdan, A. (2024). 'Renewable Energy Projects in Africa: A Review of Climate Finance Strategies'. *International Journal of Science and Research Archive*, *11*(1), 923–932. DOI: 10.30574/ijsra.2024.11.1.0170

Bai, X., Hu, X., Zhou, S., Yan, J., Sun, C., Chen, P., & Li, L. (2013). 'In Situ Polymerization and Characterization of Grafted Poly (3,4-Ethylenedioxythiophene)/ Multiwalled Carbon Nanotubes Composite with High Electrochemical Performances'. *Electrochimica Acta*, *87*, 394–400. DOI: 10.1016/j.electacta.2012.09.079

Bai, Y., Li, K., Wang, L., Gao, Y., Li, X., Yang, X., & Lü, W. (2024). 'Flexible Low-Temperature Zinc Ion Supercapacitor Based on Gel Electrolyte with α-MnO2@rGO Electrode'. *Journal of Power Sources*, *591*, 233878. Advance online publication. DOI: 10.1016/j.jpowsour.2023.233878

Barisci, J. N., Wallace, G. G., Chattopadhyay, D., Papadimitrakopoulos, F., & Baughman, R. H. (2003). 'Electrochemical Properties of Single-Wall Carbon Nanotube Electrodes'. *Journal of the Electrochemical Society*, *150*(9), E409–E415. DOI: 10.1149/1.1593045

Barzegar, F., Momodu, D. Y., Fashedemi, O. O., Bello, A., Dangbegnon, J. K., & Manyala, N. (2015). 'Investigation of Different Aqueous Electrolytes on the Electrochemical Performance of Activated Carbon-Based Supercapacitors'. *RSC Advances*, *5*(130), 107482–107487. DOI: 10.1039/C5RA21962K

Bharathi, A., Devi, V. K., Sambasivam, S., Albaqami, M. D., Kumar, M. K., Velsankar, K., & Sudhahar, S. (2024). 'Investigating MnSe@Y2O3 Nanocomposite as an Electrode for Asymmetric Hybrid Supercapacitor'. *Journal of Alloys and Compounds*, *1009*, 176867. DOI: 10.1016/j.jallcom.2024.176867

Bon, C. Y., Mohammed, L., Kim, S., Manasi, M., Isheunesu, P., Lee, K. S., & Ko, J. M. (2018). 'Flexible Poly(Vinyl Alcohol)-Ceramic Composite Separators for Supercapacitor Applications'. *Journal of Industrial and Engineering Chemistry*, *68*, 173–179. DOI: 10.1016/j.jiec.2018.07.043

Borenstein, A., Hanna, O., Attias, R., Luski, S., Brousse, T., & Aurbach, D. (2017). 'Carbon-Based Composite Materials for Supercapacitor Electrodes: A Review'. *Journal of Materials Chemistry. A, Materials for Energy and Sustainability*, *5*(25), 12653–12672. DOI: 10.1039/C7TA00863E

Brownson, D. A. C., Kampouris, D. K., & Banks, C. E. (2012). 'Graphene Electrochemistry: Fundamental Concepts through to Prominent Applications'. *Chemical Society Reviews*, *41*(21), 6944. DOI: 10.1039/c2cs35105f

Cai, Y., Zhao, B., Wang, J., & Shao, Z. (2014). 'Non-Aqueous Hybrid Supercapacitors Fabricated with Mesoporous TiO2 Microspheres and Activated Carbon Electrodes with Superior Performance'. *Journal of Power Sources*, *253*, 80–89. DOI: 10.1016/j.jpowsour.2013.11.097

Cao, L., Lu, M., & Li, H.-L. (2005). 'Preparation of Mesoporous Nanocrystalline Co[Sub 3]O[Sub 4] and Its Applicability of Porosity to the Formation of Electrochemical Capacitance'. *Journal of the Electrochemical Society*, *152*(5), A871. DOI: 10.1149/1.1883354

Chaiammart, N., Vignesh, V., Thu, M. M., Eiad-ua, A., Maiyalagan, T., & Panomsuwan, G. (2024). 'Chemically Activated Carbons Derived from Cashew Nut Shells as Potential Electrode Materials for Electrochemical Supercapacitors'. *Carbon Resources Conversion*, *100267*. Advance online publication. DOI: 10.1016/j.crcon.2024.100267

Chandra Sekhar, S., Ramulu, B., & Yu, J. S. (2022). Transition metal oxides for supercapacitors. In *Nanostructured Materials for Supercapacitors* (pp. 267–292). Springer International Publishing.

Chatterjee, D. P., & Nandi, A. K. (2021). 'A Review on the Recent Advances in Hybrid Supercapacitors'. *Journal of Materials Chemistry. A, Materials for Energy and Sustainability*, *9*(29), 15880–15918. DOI: 10.1039/D1TA02505H

Chen, H., Bao, E., Sun, H., Ren, X., Han, X., Wang, Y., Zhang, Z., Luo, C., & Xu, C. (2024). 'Sonochemical Synthesis of CoNi Layered Double Hydroxide as a Cathode Material for Assembling High Performance Hybrid Supercapacitor'. *Journal of Colloid and Interface Science*, *664*, 117–127. DOI: 10.1016/j.jcis.2024.03.003

Chen, X., Paul, R., & Dai, L. (2017). 'Carbon-Based Supercapacitors for Efficient Energy Storage'. *National Science Review*, *4*(3), 453–489. DOI: 10.1093/nsr/nwx009

Chen, Y., Qu, B., Hu, L., Xu, Z., Li, Q., & Wang, T. (2013). 'High-Performance Supercapacitor and Lithium-Ion Battery Based on 3D Hierarchical NH4F-Induced Nickel Cobaltate Nanosheet–Nanowire Cluster Arrays as Self-Supported Electrodes'. *Nanoscale*, *5*(20), 9812–9820. DOI: 10.1039/c3nr02972g

Chen, Y., Zhang, X., Zhang, D., Yu, P., & Ma, Y. (2011). 'High Performance Supercapacitors Based on Reduced Graphene Oxide in Aqueous and Ionic Liquid Electrolytes'. *Carbon*, *49*(2), 573–580. DOI: 10.1016/j.carbon.2010.09.060

Childress, A., Ferri, K., & Rao, A. M. (2018). 'Enhanced Supercapacitor Performance with Binder-Free Helically Coiled Carbon Nanotube Electrodes'. *Carbon*, *140*, 377–384. DOI: 10.1016/j.carbon.2018.08.064

Choi, H.-J., Kim, J. H., Kim, H.-K., Lee, S.-H., & Lee, Y.-H. (2016). 'Improving the Electrochemical Performance of Hybrid Supercapacitor Using Well-Organized Urchin-like TiO2 and Activated Carbon'. *Electrochimica Acta*, *208*, 202–210. DOI: 10.1016/j.electacta.2016.05.039

Chuang, C.-M., Huang, C.-W., Teng, H., & Ting, J.-M. (2010). 'Effects of Carbon Nanotube Grafting on the Performance of Electric Double Layer Capacitors'. *Energy & Fuels*, *24*(12), 6476–6482. DOI: 10.1021/ef101208x

Conway, B. E., & Pell, W. G. (2003). 'Double-Layer and Pseudocapacitance Types of Electrochemical Capacitors and Their Applications to the Development of Hybrid Devices'. *Journal of Solid State Electrochemistry*, *7*(9), 637–644. DOI: 10.1007/s10008-003-0395-7

Czagany, M., Hompoth, S., Keshri, A. K., Pandit, N., Galambos, I., Gacsi, Z., & Baumli, P. (2024). 'Supercapacitors: An Efficient Way for Energy Storage Application'. *Materials (Basel)*, *17*(3), 702. DOI: 10.3390/ma17030702

Daraghmeh, A., Hussain, S., Haq, A. U., Saadeddin, I., Servera, L., & Ruiz, J. M. (2020). 'Carbon Nanocomposite Electrodes for Electrical Double Layer Capacitor'. *Journal of Energy Storage*, *32*, 101798. DOI: 10.1016/j.est.2020.101798

De Silva, K. K. H., Huang, H. H., Joshi, R. K., & Yoshimura, M. (2017). 'Chemical Reduction of Graphene Oxide Using Green Reductants'. *Carbon*, *119*, 190–199. DOI: 10.1016/j.carbon.2017.04.025

De Volder, M. F. L., Tawfick, S. H., Baughman, R. H., & John Hart, A. (2013). 'Carbon Nanotubes: Present and Future Commercial Applications'. *Science*, *339*(6119), 535–539. DOI: 10.1126/science.1222453

Ding, J., Yang, Y., Poisson, J., He, Y., Zhang, H., Zhang, Y., Bao, Y., Chen, S., Yong, M. C., & Zhang, K. (2024). 'Recent Advances in Biopolymer-Based Hydrogel Electrolytes for Flexible Supercapacitors'. *ACS Energy Letters*, *9*(4), 1803–1825. DOI: 10.1021/acsenergylett.3c02567

Dissanayake, K., & Kularatna-Abeywardana, D. (2024). 'A Review of Supercapacitors: Materials, Technology, Challenges, and Renewable Energy Applications'. *Journal of Energy Storage*, *96*, 112563. DOI: 10.1016/j.est.2024.112563

Dong, S., Li, Z., Yi, Z., Li, H., Tian, Y., & Liu, S. (2024). 'Recycling of Activated Carbons from Spent Supercapacitors to Refabricate Improved Supercapacitors'. *Journal of Energy Storage*, *102*, 114182. DOI: 10.1016/j.est.2024.114182

Du, M., & Zhang, K. (2023). 'Nanoporous Conducting Polymer Nanowire Network-Encapsulated MnO2-Based Flexible Supercapacitor with Enhanced Rate Capability and Cycling Stability'. *ACS Applied Materials & Interfaces*, *15*(18), 22563–22573. DOI: 10.1021/acsami.3c03028

Duan, Y., Long, J., Li, Y., Tian, X., Li, J., Fang, Z., Wang, J., & Huo, P. (2024). 'Lignin/Soy Protein Isolate-Based Hydrogel Polymer Electrolytes for Flexible Solid-State Supercapacitors with Low Temperature Resistance'. *Journal of Solid State Electrochemistry*, *28*(7), 2021–2033. DOI: 10.1007/s10008-023-05726-0

Dujearic-Stephane, K., Gupta, M., Kumar, A., Sharma, V., Pandit, S., Bocchetta, P., & Kumar, Y. (2021). 'The Effect of Modifications of Activated Carbon Materials on the Capacitive Performance: Surface, Microstructure, and Wettability'. *Journal of Composites Science*, *5*(3), 66. DOI: 10.3390/jcs5030066

Dutta, A., Mitra, S., Basak, M., & Banerjee, T. (2023). 'A Comprehensive Review on Batteries and Supercapacitors: Development and Challenges since Their Inception'. *Energy Storage*, *5*(1), e339. DOI: 10.1002/est2.339

El-Bahy, Z. M., AlMasoud, N., Ali, M., Alomar, T. S., Al-wallan, A. A., & Tahir, H. M. (2025). 'A Novel DyCoO3@rGO Nanocomposite Electrode Material for Hybrid Supercapacitor Devices'. *Journal of Alloys and Compounds*, *1010*, 178091. DOI: 10.1016/j.jallcom.2024.178091

Fan, W., Wang, F., Xiong, X., Song, B., Wang, T., Cheng, X., Zhu, Z., He, J., Liu, Y., & Wu, Y. (2024). 'Recent Advances in Functional Materials and Devices for Zn-Ion Hybrid Supercapacitors'. *NPG Asia Materials*, *16*(1), 18. DOI: 10.1038/s41427-024-00537-9

Feng, E., Li, J., Zheng, G., Yan, Z., Li, X., Gao, W., Ma, X., & Yang, Z. (2021). 'Long-Term Anti-Freezing Active Organohydrogel Based Superior Flexible Supercapacitor and Strain Sensor'. *ACS Sustainable Chemistry & Engineering*, *9*(21), 7267–7276. DOI: 10.1021/acssuschemeng.1c01209

Fu, C., Feng, L., Yin, H., Li, Y., Xie, Y., Feng, Y., Zhao, Y., Cao, L., Huang, J., & Liu, Y. (2022). 'Heterogeneous Ni3P/Ni Nanoparticles with Optimized Ni Active Sites Anchored in N-Doped Mesoporous Nanofibers for Boosting pH-Universal Hydrogen Evolution'. *Nanoscale*, *14*(39), 14779–14788. DOI: 10.1039/D2NR04053K

Gaikwad, P., Tiwari, N., Kamat, R., Mane, S. M., & Kulkarni, S. B. (2024). 'A Comprehensive Review on the Progress of Transition Metal Oxides Materials as a Supercapacitor Electrode'. *Materials Science and Engineering B*, *307*, 117544. DOI: 10.1016/j.mseb.2024.117544

Gao, H., Xiao, F., Ching, C. B., & Duan, H. (2012). 'High-Performance Asymmetric Supercapacitor Based on Graphene Hydrogel and Nanostructured MnO 2'. *ACS Applied Materials & Interfaces*, *4*(5), 2801–2810. DOI: 10.1021/am300455d

Gao, Yu., Li, L., Jin, Y., Wang, Y., Yuan, C., Wei, Y., Chen, G., Ge, J., & Lu, H. (2015). 'Porous Carbon Made from Rice Husk as Electrode Material for Electrochemical Double Layer Capacitor'. *Applied Energy*, *153*, 41–47. DOI: 10.1016/j.apenergy.2014.12.070

González, A., Goikolea, E., Barrena, J. A., & Mysyk, R. (2016). 'Review on Supercapacitors: Technologies and Materials'. *Renewable & Sustainable Energy Reviews*, *58*, 1189–1206. DOI: 10.1016/j.rser.2015.12.249

Gopi, M. (2025a). 'Supercapacitors: A Promising Solution for Sustainable Energy Storage and Diverse Applications'. *Journal of Energy Storage*, *114*, 115729. DOI: 10.1016/j.est.2025.115729

Gueon, D., & Moon, J. H. (2015). 'Nitrogen-Doped Carbon Nanotube Spherical Particles for Supercapacitor Applications: Emulsion-Assisted Compact Packing and Capacitance Enhancement'. *ACS Applied Materials & Interfaces*, *7*(36), 20083–20089. DOI: 10.1021/acsami.5b05231

Haridas, H., Arsath, K. A. K., Sellathurai, A., Barz, D. P. J., & Kontopoulou, M. (2024). 'Noncovalent Functionalization of Graphene Nanoplatelets and Their Applications in Supercapacitors'. *ACS Applied Materials & Interfaces*, *16*(13), 16630–16640. DOI: 10.1021/acsami.3c18174

Heidarinejad, Z., Dehghani, M. H., Heidari, M., Javedan, G., Ali, I., & Sillanpää, M. (2020). 'Methods for Preparation and Activation of Activated Carbon: A Review'. *Environmental Chemistry Letters, 18*(2), 393–415. DOI: 10.1007/s10311-019-00955-0

Heligman, B. T., & Manthiram, A. (2021). 'Elemental Foil Anodes for Lithium-Ion Batteries'. *ACS Energy Letters, 6*(8), 2666–2672. DOI: 10.1021/acsenergylett.1c01145

Hibino, T., Kobayashi, K., Nagao, M., & Kawasaki, S. (2015). 'High-Temperature Supercapacitor with a Proton-Conducting Metal Pyrophosphate Electrolyte'. *Scientific Reports, 5*(1), 7903. DOI: 10.1038/srep07903

Hou, X., Zhang, Q., Wang, L., Gao, G., & Lü, W. (2021). 'Low-Temperature-Resistant Flexible Solid Supercapacitors Based on Organohydrogel Electrolytes and Microvoid-Incorporated Reduced Graphene Oxide Electrodes'. *ACS Applied Materials & Interfaces, 13*(10), 12432–12441. DOI: 10.1021/acsami.0c18741

Hou, Z., Zhang, X., Li, X., Zhu, Y., Liang, J., & Qian, Y. (2017). 'Surfactant Widens the Electrochemical Window of an Aqueous Electrolyte for Better Rechargeable Aqueous Sodium/Zinc Battery'. *Journal of Materials Chemistry. A, Materials for Energy and Sustainability, 5*(2), 730–738. DOI: 10.1039/C6TA08736A

Huang, C.-L., Chiang, L.-M., Su, C.-A., & Li, Y.-Y. (2021). 'MnO2/Carbon Nanotube-Embedded Carbon Nanofibers as Core–Shell Cables for High Performing Asymmetric Flexible Supercapacitors'. *Journal of Industrial and Engineering Chemistry, 103*, 142–153. DOI: 10.1016/j.jiec.2021.07.026

Huang, J., Sumpter, B. G., & Meunier, V. (2008). 'Theoretical Model for Nanoporous Carbon Supercapacitors'. *Angewandte Chemie International Edition, 47*(3), 520–524. DOI: 10.1002/anie.200703864

Hung, C.-K., Thiruppathi, A. R., McGuire, C., Jiang, D.-T., & Chen, A. (2025). 'Expanded Graphite/Reduced Graphene Oxide Hybrid Architecture Functionalized with RuO2 Nanoclusters for High Performance Energy Storage'. *Journal of Energy Storage, 106*, 114776. DOI: 10.1016/j.est.2024.114776

Huo, J., Wang, X., Zhang, X., Zhang, L., Yue, G., & Guo, S. (2023). 'Porous Polypyrrole-Derived Carbon Nanotubes as a Cathode Material for Zinc-Ion Hybrid Supercapacitors'. *Journal of Energy Storage, 73*, 108925. DOI: 10.1016/j.est.2023.108925

Iro, Z. S., Subramani, C., & Dash, S. S. (2016). 'A Brief Review on Electrode Materials for Supercapacitor'. *International Journal of Electrochemical Science, 11*(12), 10628–10643. DOI: 10.20964/2016.12.50

Izadi-Najafabadi, A., Yasuda, S., Kobashi, K., Yamada, T., Futaba, D. N., Hatori, H., Yumura, M., Iijima, S., & Hata, K. (2010). 'Extracting the Full Potential of Single-Walled Carbon Nanotubes as Durable Supercapacitor Electrodes Operable at 4 V with High Power and Energy Density'. *Advanced Materials*, *22*(35). Advance online publication. DOI: 10.1002/adma.200904349

Jang, H. S., Raj, C. J., Lee, W.-G., Kim, B. C., & Yu, K. H. (2016). 'Enhanced Supercapacitive Performances of Functionalized Activated Carbon in Novel Gel Polymer Electrolytes with Ionic Liquid Redox-Mediated Poly(Vinyl Alcohol)/Phosphoric Acid'. *RSC Advances*, *6*(79), 75376–75383. DOI: 10.1039/C6RA15070E

Jang, J. 2006. 'Conducting Polymer Nanomaterials and Their Applications'. Pp. 189–260 in *Emissive Materials Nanomaterials*. Vol. 199, *Advances in Polymer Science*. Berlin, Heidelberg: Springer Berlin Heidelberg. DOI: 10.1007/12_075

Jang, J. H., Han, S., Hyeon, T., & Oh, S. M. (2003). 'Electrochemical Capacitor Performance of Hydrous Ruthenium Oxide/Mesoporous Carbon Composite Electrodes'. *Journal of Power Sources*, *123*(1), 79–85. DOI: 10.1016/S0378-7753(03)00459-2

Kar, K. K. (Ed.). (2021). *Handbook of Nanocomposite Supercapacitor Materials III: Selection* (Vol. 313). Springer International Publishing. DOI: 10.1007/978-3-030-68364-1

Khan, A., Arumugam Senthil, R., Pan, J., Sun, Y., & Liu, X. (2020). 'Hierarchically Porous Biomass Carbon Derived from Natural Withered Rose Flowers as High-Performance Material for Advanced Supercapacitors'. *Batteries & Supercaps*, *3*(8), 731–737. DOI: 10.1002/batt.202000046

Khan, A., Grabher, G., & Hossain, G. (2023). 'Smart-Textile Supercapacitor for Wearable Energy Storage System'. *Journal of Energy Storage*, *73*, 108963. DOI: 10.1016/j.est.2023.108963

Khan, A., Senthil, R. A., Pan, J., Osman, S., Sun, Y., & Shu, X. (2020). 'A New Biomass Derived Rod-like Porous Carbon from Tea-Waste as Inexpensive and Sustainable Energy Material for Advanced Supercapacitor Application'. *Electrochimica Acta*, *335*, 135588. DOI: 10.1016/j.electacta.2019.135588

Khan, H. A., Tawalbeh, M., Aljawrneh, B., Abuwatfa, W., Al-Othman, A., Sadeghifar, H., & Olabi, A. G. (2024). 'A Comprehensive Review on Supercapacitors: Their Promise to Flexibility, High Temperature, Materials, Design, and Challenges'. *Energy*, *295*, 131043. DOI: 10.1016/j.energy.2024.131043

Khomenko, V., Raymundo-Piñero, E., & Béguin, F. (2008). 'High-Energy Density Graphite/AC Capacitor in Organic Electrolyte'. *Journal of Power Sources, 177*(2), 643–651. DOI: 10.1016/j.jpowsour.2007.11.101

Kierzek, K., & Gryglewicz, G. (2020). 'Activated Carbons and Their Evaluation in Electric Double Layer Capacitors'. *Molecules (Basel, Switzerland), 25*(18), 4255. DOI: 10.3390/molecules25184255

Kim, T., Jung, G., Yoo, S., Suh, K. S., & Ruoff, R. S. (2013). Activated graphene-based carbons as supercapacitor electrodes with macro-and mesopores. *ACS Nano, 7*(8), 6899–6905.

Kiran Raj, G., Singh, E., Hani, U., Ramesh, K. V. R. N. S., Talath, S., Garg, A., Savadatti, K., Bhatt, T., Madhuchandra, K., & Osmani, R. A. M. (2023). 'Conductive Polymers and Composites-Based Systems: An Incipient Stride in Drug Delivery and Therapeutics Realm'. *Journal of Controlled Release, 355*, 709–729. DOI: 10.1016/j.jconrel.2023.02.017

Kong, X., Zhang, Y., Zhang, P., Song, X., & Xu, H. (2020). 'Synthesis of Natural Nitrogen-Rich Soybean Pod Carbon with Ion Channels for Low Cost and Large Areal Capacitance Supercapacitor'. *Applied Surface Science, 516*, 146162. DOI: 10.1016/j.apsusc.2020.146162

Kulkarni, V. V. (2024). Synthesis Processes of Carbonaceous Material/Conducting Polymer Nanocomposites in Relation to Grafting and Electrochemical Properties for Supercapacitor Application. *RE:view.*

Kumar, D., Kulurumotlakatla, N. R., Ramachandran, T., Durga, I. K., Khan, M. S., Kumar, Y. A., Rao, S. S., & Joo, S. W. (2024). 'Dual NiFe2O4//CoFe2O4 Nanosheets Compound Dynamism for Supercharged Supercapacitors'. *Materials Science and Engineering B, 306*, 117444. DOI: 10.1016/j.mseb.2024.117444

Kumar, T. R., Senthil, R. A., Pan, Z., Pan, J., & Sun, Y. (2020). 'A Tubular-like Porous Carbon Derived from Waste American Poplar Fruit as Advanced Electrode Material for High-Performance Supercapacitor'. *Journal of Energy Storage, 32*, 101903. DOI: 10.1016/j.est.2020.101903

Kumar, Y. A., Roy, N., Ramachandran, T., Hussien, M., Moniruzzaman, M., & Joo, S. W. (2024). 'Shaping the Future of Energy: The Rise of Supercapacitors Progress in the Last Five Years'. *Journal of Energy Storage, 98*, 113040. DOI: 10.1016/j.est.2024.113040

Kuptajit, P., Sano, N., Nakagawa, K., & Suzuki, T. (2021). 'A Study on Pore Formation of High Surface Area Activated Carbon Prepared by Microwave-Induced Plasma with KOH (MiWP-KOH) Activation: Effect of Temperature-Elevation Rate'. *Chemical Engineering and Processing*, *167*, 108511. DOI: 10.1016/j.cep.2021.108511

Laforgue, A., Simon, P., Sarrazin, C., & Fauvarque, J.-F. (1999). 'Polythiophene-Based Supercapacitors'. *Journal of Power Sources*, *80*(1), 142–148. DOI: 10.1016/S0378-7753(98)00258-4

Lai, C. W. (2019). 'Organic Electrolytes for Supercapacitors'. *Materials Research Foundations* 61.

Lai, F., Fang, Z., Cao, L., Li, W., Lin, Z., & Zhang, P. (2020). 'Self-Healing Flexible and Strong Hydrogel Nanocomposites Based on Polyaniline for Supercapacitors'. *Ionics*, *26*(6), 3015–3025. DOI: 10.1007/s11581-020-03438-3

Lakshmi, K. C., & Vedhanarayanan, B. (2023). 'High-Performance Supercapacitors: A Comprehensive Review on Paradigm Shift of Conventional Energy Storage Devices'. *Batteries*, *9*(4), 202. DOI: 10.3390/batteries9040202

Lee, H., Cho, M. S., Kim, I. H., Nam, J. D., & Lee, Y. (2010). 'RuOx/Polypyrrole Nanocomposite Electrode for Electrochemical Capacitors'. *Synthetic Metals*, *160*(9), 1055–1059. DOI: 10.1016/j.synthmet.2010.02.026

Lee, K. K., Chin, W. S., & Sow, C. H. (2014). 'Cobalt-Based Compounds and Composites as Electrode Materials for High-Performance Electrochemical Capacitors'. *Journal of Materials Chemistry. A, Materials for Energy and Sustainability*, *2*(41), 17212–17248. DOI: 10.1039/C4TA02074J

Lee, S.-H., & Kim, J.-M. (2018). 'Punched H2Ti12O25 Anode and Activated Carbon Cathode for High Energy/High Power Hybrid Supercapacitors'. *Energy*, *150*, 816–821. DOI: 10.1016/j.energy.2018.03.038

Levitt, A., Hegh, D., Phillips, P., Uzun, S., Anayee, M., Razal, J. M., Gogotsi, Y., & Dion, G. (2020). '3D Knitted Energy Storage Textiles Using MXene-Coated Yarns'. *Materials Today*, *34*, 17–29. DOI: 10.1016/j.mattod.2020.02.005

Li, H., Wang, J., Chu, Q., Wang, Z., Zhang, F., & Wang, S. (2009). 'Theoretical and Experimental Specific Capacitance of Polyaniline in Sulfuric Acid'. *Journal of Power Sources*, *190*(2), 578–586. DOI: 10.1016/j.jpowsour.2009.01.052

Li, J., & Östling, M. (2013). 'Prevention of Graphene Restacking for Performance Boost of Supercapacitors—A Review'. *Crystals*, *3*(1), 163–190. DOI: 10.3390/cryst3010163

Li, M., Zhou, S., Cheng, L., Mo, F., Chen, L., Yu, S., & Wei, J. (2023). '3D Printed Supercapacitor: Techniques, Materials, Designs, and Applications'. *Advanced Functional Materials*, *33*(1), 2208034. DOI: 10.1002/adfm.202208034

Li, X., Li, Y., Xie, S., Zhou, Y., Rong, J., & Dong, L. (2022). 'Zinc-Based Energy Storage with Functionalized Carbon Nanotube/Polyaniline Nanocomposite Cathodes'. *Chemical Engineering Journal*, *427*, 131799. Advance online publication. DOI: 10.1016/j.cej.2021.131799

Li, Y., & Chen, C. (2017). 'Polyaniline/Carbon Nanotubes-Decorated Activated Carbon Fiber Felt as High-Performance, Free-Standing and Flexible Supercapacitor Electrodes'. *Journal of Materials Science*, *52*(20), 12348–12357. DOI: 10.1007/s10853-017-1291-3

Li, Y., Kang, Z., Yan, X., Cao, S., Li, M., Guo, Y., Huan, Y., Wen, X., & Zhang, Y. (2018). 'A Three-Dimensional Reticulate CNT-Aerogel for a High Mechanical Flexibility Fiber Supercapacitor'. *Nanoscale*, *10*(19), 9360–9368. DOI: 10.1039/C8NR01991F

Liang, R., Du, Y., Xiao, P., Cheng, J., Yuan, S., Chen, Y., Yuan, J., & Chen, J. (2021). 'Transition Metal Oxide Electrode Materials for Supercapacitors: A Review of Recent Developments'. *Nanomaterials (Basel, Switzerland)*, *11*(5), 1248. DOI: 10.3390/nano11051248

Lim, E., Jo, C., & Lee, J. (2016). A mini review of designed mesoporous materials for energy-storage applications: From electric double-layer capacitors to hybrid supercapacitors. *Nanoscale*, *8*(15), 7827–7833.

Liu, C. G., Fang, H. T., Li, F., Liu, M., & Cheng, H. M. (2006). 'Single-Walled Carbon Nanotubes Modified by Electrochemical Treatment for Application in Electrochemical Capacitors'. *Journal of Power Sources*, *160*(1), 758–761. DOI: 10.1016/j.jpowsour.2006.01.072

Liu, C. G., Liu, M., Li, F., & Cheng, H. M. (2008). 'Frequency Response Characteristic of Single-Walled Carbon Nanotubes as Supercapacitor Electrode Material'. *Applied Physics Letters*, *92*(14), 143108. Advance online publication. DOI: 10.1063/1.2907501

Liu, J., Khanam, Z., Ahmed, S., Wang, H., Wang, T., & Song, S. (2021). 'A Study of Low-Temperature Solid-State Supercapacitors Based on Al-Ion Conducting Polymer Electrolyte and Graphene Electrodes'. *Journal of Power Sources*, *488*, 229461. Advance online publication. DOI: 10.1016/j.jpowsour.2021.229461

Liu, M., Turcheniuk, K., Fu, W., Yang, Y., Liu, M., & Yushin, G. (2020). 'Scalable, Safe, High-Rate Supercapacitor Separators Based on the Al2O3 Nanowire Polyvinyl Butyral Nonwoven Membranes'. *Nano Energy*, *71*, 104627. DOI: 10.1016/j.nanoen.2020.104627

Liu, M., Wang, L., Yu, X., Zhang, H., Zhang, H., Li, S., & Huang, F. (2022). 'Introducing Oxygen Vacancies for Improving the Electrochemical Performance of Co9S8@NiCo-LDH Nanotube Arrays in Flexible All-Solid Battery-Capacitor Hybrid Supercapacitors'. *Energy*, *238*, 121767. DOI: 10.1016/j.energy.2021.121767

Liu, X., Long, Q., Jiang, C., Zhan, B., Li, C., Liu, S., Zhao, Q., Huang, W., & Dong, X. (2013). 'Facile and Green Synthesis of Mesoporous Co3O4 Nanocubes and Their Applications for Supercapacitors'. *Nanoscale*, *5*(14), 6525. DOI: 10.1039/c3nr00495c

Liu, Y., Shi, K., & Zhitomirsky, I. (2014). 'New Colloidal Route for Electrostatic Assembly of Oxide Nanoparticle – Carbon Nanotube Composites'. *Colloids and Surfaces. A, Physicochemical and Engineering Aspects*, *446*, 15–22. DOI: 10.1016/j.colsurfa.2014.01.032

Lobato, B., Suárez, L., Guardia, L., & Centeno, T. A. (2017). 'Capacitance and Surface of Carbons in Supercapacitors'. *Carbon*, *122*, 434–445. DOI: 10.1016/j.carbon.2017.06.083

Loh, K. H., Liew, J., Liu, L., Goh, Z. L., Pershaanaa, M., Kamarulazam, F., Bashir, S., Ramesh, K., & Ramesh, S. (2024). 'A Comprehensive Review on Fundamentals and Components of Zinc-Ion Hybrid Supercapacitors'. *Journal of Energy Storage*, *81*, 110370. DOI: 10.1016/j.est.2023.110370

Long, J. W., Swider, K. E., Merzbacher, C. I., & Rolison, D. R. (1999). 'Voltammetric Characterization of Ruthenium Oxide-Based Aerogels and Other RuO_2 Solids: The Nature of Capacitance in Nanostructured Materials'. *Langmuir*, *15*(3), 780–785. DOI: 10.1021/la980785a

Lozano-Castelló, D., Alcañiz-Monge, J., de la Casa-Lillo, M. A., Cazorla-Amorós, D., & Linares-Solano, A. (2002). 'Advances in the Study of Methane Storage in Porous Carbonaceous Materials'. *Fuel*, *81*(14), 1777–1803. DOI: 10.1016/S0016-2361(02)00124-2

Lu, J., Zhang, J., Wang, X., Zhang, J., Tian, Z., Zhu, E., Yang, L., Guan, X., Ren, H., Wu, J., Li, X., & Wang, G. (2024). 'A Review of Advanced Electrolytes for Supercapacitors'. *Journal of Energy Storage*, *103*, 114338. DOI: 10.1016/j.est.2024.114338

Lu, S., Yan, D., Chen, L., Zhu, G., Xu, H., & Yu, A. (2016). 'One-Pot Fabrication of Hierarchical Ag/MnO2 Nanoflowers for Electrochemical Capacitor Electrodes'. *Materials Letters*, *168*, 40–43. DOI: 10.1016/j.matlet.2016.01.021

Luo, N., Wang, J., Zhang, D., Zhao, Y., Wei, Y., Liu, Y., Zhang, Y., Han, S., Kong, X., & Huo, P. (2024). 'Inorganic Nanoparticle-Enhanced Double-Network Hydrogel Electrolytes for Supercapacitor with Superior Low-Temperature Adaptability'. *Chemical Engineering Journal*, *479*, 147741. Advance online publication. DOI: 10.1016/j.cej.2023.147741

Ma, L., Fan, H., Wei, X., Chen, S., Hu, Q., Liu, Y., Zhi, C., Lu, W., Zapien, J. A., & Huang, H. (2018). 'Towards High Areal Capacitance, Rate Capability, and Tailorable Supercapacitors: Co3O4@polypyrrole Core-Shell Nanorod Bundle Array Electrodes'. *Journal of Materials Chemistry. A, Materials for Energy and Sustainability*, *6*(39), 19058–19065. DOI: 10.1039/C8TA07477A

Ma, T., Liao, L., Zhang, X., Lou, S., Gao, S., Gao, Y., & Fang, T. (2021). 'Hierarchical Pores from Microscale to Macroscale Boost Ultrahigh Lithium Intercalation Pseudocapacitance of Biomass Carbon'. *Journal of Energy Storage*, *33*, 102068. DOI: 10.1016/j.est.2020.102068

Ma, Y., Chang, H., Zhang, M., & Chen, Y. (2015). 'Graphene-Based Materials for Lithium-Ion Hybrid Supercapacitors'. *Advanced Materials*, *27*(36), 5296–5308. DOI: 10.1002/adma.201501622

Magar, H. S., Hassan, R. Y. A., & Mulchandani, A. (2021). 'Electrochemical Impedance Spectroscopy (Eis): Principles, Construction, and Biosensing Applications'. *Sensors (Basel)*, *21*(19), 6578. Advance online publication. DOI: 10.3390/s21196578

Mahmood, Mou, Prangon Chowdhury, Rahbaar Yeassin, Mahmudul Hasan, Tanvir Ahmad, and Nahid-Ur-Rahman Chowdhury. 2024. 'Impacts of Digitalization on Smart Grids, Renewable Energy, and Demand Response: An Updated Review of Current Applications'. *Energy Conversion and Management: X* 24:100790. .DOI: 10.1016/j.ecmx.2024.100790

Mandal, S., Mendhe, A. B., Rakhade, H. M., Barse, N. S., Roy, M., Rosaiah, P., Park, T., Lee, H.-S., Mendhe, A. C., & Kim, D. (2025). 'Recent Advancement and Design in Supercapacitor Hybrid Electrode Materials: Bridging the Gap between Energy and Power Density'. *Chemical Engineering Journal Advances*, *21*, 100690. DOI: 10.1016/j.ceja.2024.100690

Mao, C., Liu, S., Pang, L., Sun, Q., Liu, Y., Xu, M., & Lu, Z. (2016). 'Ultrathin MnO2 Nanosheets Grown on Fungal Conidium-Derived Hollow Carbon Spheres as Supercapacitor Electrodes'. *RSC Advances*, *6*(7), 5184–5191. DOI: 10.1039/C5RA22193E

Masood, M., Hussain, S., Sohail, M., Rehman, A., Uzzaman, M. A., Alnaser, I. A., Karim, M. R., & Wahab, M. A. (2024). 'Recent Progress, Challenges, and Opportunities of Conducting Polymers for Energy Storage Applications'. *ChemistrySelect*, *9*(23), e202302876. DOI: 10.1002/slct.202302876

Mdluli, S. B., Ramoroka, M. E., Yussuf, S. T., Modibane, K. D., John-Denk, V. S., & Iwuoha, E. I. (2022). 'π-Conjugated Polymers and Their Application in Organic and Hybrid Organic-Silicon Solar Cells'. *Polymers*, *14*(4), 716. DOI: 10.3390/polym14040716

Meena, D., Kumar, R., Gupta, S., Khan, O., Gupta, D., & Singh, M. (2023). 'Energy Storage in the 21st Century: A Comprehensive Review on Factors Enhancing the next-Generation Supercapacitor Mechanisms'. *Journal of Energy Storage*, *72*, 109323. DOI: 10.1016/j.est.2023.109323

Mi, J., Wang, X.-R., Fan, R.-J., Qu, W.-H., & Li, W.-C. (2012). 'Coconut-Shell-Based Porous Carbons with a Tunable Micro/Mesopore Ratio for High-Performance Supercapacitors'. *Energy & Fuels*, *26*(8), 5321–5329. DOI: 10.1021/ef3009234

Miller, J. R., & Simon, P. (2008). 'Electrochemical Capacitors for Energy Management'. *Science*, *321*(5889), 651–652. DOI: 10.1126/science.1158736

Mo, R. J., Zhao, Y., Wu, M., Xiao, H. M., Kuga, S., Huang, Y., Li, J. P., & Fu, S. Y. (2016). 'Activated Carbon from Nitrogen Rich Watermelon Rind for High-Performance Supercapacitors'. *RSC Advances*, *6*(64), 59333–59342. DOI: 10.1039/C6RA10719B

Molahalli, V., Sharma, A., Bijapur, K., Soman, G., Chattham, N., & Hegde, G. (2024). 'Low-Cost Bio-Waste Carbon Nanocomposites for Sustainable Electrochemical Devices: A Systematic Review'. *Materials Today. Communications*, *38*, 108034. DOI: 10.1016/j.mtcomm.2024.108034

Molina-Sabio, M., Gonzalez, M. T., Rodriguez-Reinoso, F., & Sepúlveda-Escribano, A. (1996). 'Effect of Steam and Carbon Dioxide Activation in the Micropore Size Distribution of Activated Carbon'. *Carbon*, *34*(4), 505–509. DOI: 10.1016/0008-6223(96)00006-1

Mondal, S., Velpula, D., & Sangaranarayanan, M. V. 2024. 'Electrochemical Supercapacitors: An Overview on Analysis and Modeling'. Pp. 255–82 in *Polymer Electrolyte-Based Electrochemical Devices*. Elsevier.

Mudila, H., Prasher, P., Kumar, M., Kumar, A., Zaidi, M. G. H., & Kumar, A. (2019). 'Critical Analysis of Polyindole and Its Composites in Supercapacitor Application'. *Materials for Renewable and Sustainable Energy*, *8*(2), 9. DOI: 10.1007/s40243-019-0149-9

Muzaffar, A., Ahamed, M. B., Deshmukh, K., & Thirumalai, J. (2019). 'A Review on Recent Advances in Hybrid Supercapacitors: Design, Fabrication and Applications'. *Renewable & Sustainable Energy Reviews*, *101*, 123–145. DOI: 10.1016/j.rser.2018.10.026

Naji, Samah Zaki, and Ching Thian Tye. 2022. 'A Review of the Synthesis of Activated Carbon for Biodiesel Production: Precursor, Preparation, and Modification'. *Energy Conversion and Management: X* 13:100152. .DOI: 10.1016/j.ecmx.2021.100152

Namisnyk, A. M., and J. G. Zhu. n.d. 'Faculty of Engineering University of Technology, Sydney'.

Naoi, K., & Simon, P. (2008). 'New Materials and New Confgurations for Advanced Electrochemical Capacitors'. *The Electrochemical Society Interface*, *17*(1), 34–37. DOI: 10.1149/2.F04081IF

Niu, C., Sichel, E. K., Hoch, R., Moy, D., & Tennent, H. (1997). 'High Power Electrochemical Capacitors Based on Carbon Nanotube Electrodes'. *Applied Physics Letters*, *70*(11), 1480–1482. DOI: 10.1063/1.118568

Olabi, A. G., Abbas, Q., Abdelkareem, M. A., Alami, A. H., Mirzaeian, M., & Sayed, E. T. (2023). 'Carbon-Based Materials for Supercapacitors: Recent Progress, Challenges and Barriers'. *Batteries*, *9*(1), 19. DOI: 10.3390/batteries9010019

Olabi, A. G., Abbas, Q., Al Makky, A., & Abdelkareem, M. A. (2022). 'Supercapacitors as next Generation Energy Storage Devices: Properties and Applications'. *Energy*, *248*, 123617. DOI: 10.1016/j.energy.2022.123617

Pan, H., Feng, Y. P., & Lin, J. Y. (2006). 'Ab Initio Study of F-and Cl-Functionalized Single Wall Carbon Nanotubes'. *Journal of Physics Condensed Matter*, *18*(22), 5175–5184. DOI: 10.1088/0953-8984/18/22/017

Pandolfo, A. G., & Hollenkamp, A. F. (2006). 'Carbon Properties and Their Role in Supercapacitors'. *Journal of Power Sources*, *157*(1), 11–27. DOI: 10.1016/j.jpowsour.2006.02.065

Pappu, S., Rao, T. N., Martha, S. K., & Bulusu, S. V. (2022). 'Electrodeposited Manganese Oxide Based Redox Mediator Driven 2.2 V High Energy Density Aqueous Supercapacitor'. *Energy*, *243*, 122751. DOI: 10.1016/j.energy.2021.122751

Paramshetti, S., Angolkar, M., Al Fatease, A., Alshahrani, S. M., Hani, U., Garg, A., Ravi, G., & Riyaz Ali, M. O. (2023). 'Revolutionizing Drug Delivery and Therapeutics: The Biomedical Applications of Conductive Polymers and Composites-Based Systems'. *Pharmaceutics*, *15*(4), 1204. DOI: 10.3390/pharmaceutics15041204

Patil, B. H., Jagadale, A. D., & Lokhande, C. D. (2012). 'Synthesis of Polythiophene Thin Films by Simple Successive Ionic Layer Adsorption and Reaction (SILAR) Method for Supercapacitor Application'. *Synthetic Metals*, *162*(15), 1400–1405. DOI: 10.1016/j.synthmet.2012.05.023

Patil, P. H., & Jadhav, S. A. (2024). 'Manganese Dioxide (MnO2) and Biomass-Derived Carbon-Based Electroactive Composite Materials for Supercapacitor Applications'. *RSC Applied Interfaces*, *1*(4), 624–647. DOI: 10.1039/D4LF00085D

Pell, W. G., & Conway, B. E. (2001). 'Voltammetry at a de Levie Brush Electrode as a Model for Electrochemical Supercapacitor Behaviour'. *Journal of Electroanalytical Chemistry (Lausanne, Switzerland)*, *500*(1), 121–133. DOI: 10.1016/S0022-0728(00)00423-X

Phattharasupakun, N., Wutthiprom, J., Chiochan, P., Suktha, P., Suksomboon, M., Kalasina, S., & Sawangphruk, M. (2016). 'Turning Conductive Carbon Nanospheres into Nanosheets for High-Performance Supercapacitors of MnO2 Nanorods'. *Chemical Communications (Cambridge)*, *52*(12), 2585–2588. DOI: 10.1039/C5CC09648K

Poonam, K. S., Arora, A., & Tripathi, S. K. (2019). 'Review of Supercapacitors: Materials and Devices'. *Journal of Energy Storage*, *21*, 801–825. DOI: 10.1016/j.est.2019.01.010

Prehal, C., Koczwara, C., Jäckel, N., Schreiber, A., Burian, M., Amenitsch, H., Hartmann, M. A., Presser, V., & Paris, O. (2017). 'Quantification of Ion Confinement and Desolvation in Nanoporous Carbon Supercapacitors with Modelling and in Situ X-Ray Scattering'. *Nature Energy*, *2*(3), 16215. DOI: 10.1038/nenergy.2016.215

Pullanchiyodan, A., Manjakkal, L., & Dahiya, R. (2021). 'Metal Coated Fabric Based Asymmetric Supercapacitor for Wearable Applications'. *IEEE Sensors Journal*, *21*(23), 26208–26214. DOI: 10.1109/JSEN.2021.3058894

Qiang, S., Carey, T., Arbab, A., Song, W., Wang, C., & Torrisi, F. (2019). 'Wearable Solid-State Capacitors Based on Two-Dimensional Material All-Textile Heterostructures'. *Nanoscale*, *11*(20), 9912–9919. DOI: 10.1039/C9NR00463G

Rahman, Gul, Najaf, Zainab, Mehmood, Asad, & Bilal, Salma, Anwar ul Haq Ali Shah, Shabeer Ahmad Mian, and Ghulam Ali. (2019). 'An Overview of the Recent Progress in the Synthesis and Applications of Carbon Nanotubes'. *C*, *5*(1), 3.

Rajesh, M., Raj, C. J., Kim, B. C., Cho, B.-B., Ko, J. M., & Yu, K. H. (2016). 'Supercapacitive Studies on Electropolymerized Natural Organic Phosphate Doped Polypyrrole Thin Films'. *Electrochimica Acta*, *220*, 373–383. DOI: 10.1016/j.electacta.2016.10.118

Ramadoss, A., Wong, K. K., Swain, N., Mohanty, A., K, K., K, S., & Schmidt Mende, L. (2022). 'Flexible, Lightweight, and Ultrabendable RuO_2–MnO_2/Graphite Sheets for Supercapacitors'. *Energy & Fuels*, *36*(18), 11194–11204. DOI: 10.1021/acs.energyfuels.2c01336

Rashid Khan, H., & Latif Ahmad, A. (2025a). 'Supercapacitors: Overcoming Current Limitations and Charting the Course for next-Generation Energy Storage'. *Journal of Industrial and Engineering Chemistry*, *141*, 46–66. DOI: 10.1016/j.jiec.2024.07.014

Rashid Khan, H., & Latif Ahmad, A. (2025b). 'Supercapacitors: Overcoming Current Limitations and Charting the Course for next-Generation Energy Storage'. *Journal of Industrial and Engineering Chemistry*, *141*, 46–66. DOI: 10.1016/j.jiec.2024.07.014

Raza, W., Ali, F., Raza, N., Luo, Y., Kim, K. H., Yang, J., Kumar, S., Mehmood, A., & Kwon, E. E. (2018). 'Recent Advancements in Supercapacitor Technology'. *Nano Energy*, *52*, 441–473. DOI: 10.1016/j.nanoen.2018.08.013

Reenu, S., Phor, L., Kumar, A., & Chahal, S. (2024a). 'Electrode Materials for Supercapacitors: A Comprehensive Review of Advancements and Performance'. *Journal of Energy Storage*, *84*, 110698. DOI: 10.1016/j.est.2024.110698

Reenu, S., Phor, L., Kumar, A., & Chahal, S. (2024b). 'Electrode Materials for Supercapacitors: A Comprehensive Review of Advancements and Performance'. *Journal of Energy Storage*, *84*, 110698. DOI: 10.1016/j.est.2024.110698

Rodríguez-Reinoso, F., Molina-Sabio, M., & González, M. T. (1995). 'The Use of Steam and CO2 as Activating Agents in the Preparation of Activated Carbons'. *Carbon*, *33*(1), 15–23. DOI: 10.1016/0008-6223(94)00100-E

Rong, Q., Lei, W., Huang, J., & Liu, M. (2018). 'Low Temperature Tolerant Organohydrogel Electrolytes for Flexible Solid-State Supercapacitors'. *Advanced Energy Materials*, *8*(31), 1801967. Advance online publication. DOI: 10.1002/aenm.201801967

Saleh, M., Gul, A., Nasir, A., Moses, T. O., Nural, Y., & Yabalak, E. (2025). 'Comprehensive Review of Carbon-Based Nanostructures: Properties, Synthesis, Characterization, and Cross-Disciplinary Applications'. *Journal of Industrial and Engineering Chemistry*, *146*, 176–212. DOI: 10.1016/j.jiec.2024.11.052

Samantaray, S., Mohanty, D., Hung, I.-M., Moniruzzaman, M., & Satpathy, S. K. (2023). 'Unleashing Recent Electrolyte Materials for Next-Generation Supercapacitor Applications: A Comprehensive Review'. *Journal of Energy Storage*, *72*, 108352. DOI: 10.1016/j.est.2023.108352

Šedajová, V., Jakubec, P., Bakandritsos, A., Ranc, V., & Otyepka, M. (2020). 'New Limits for Stability of Supercapacitor Electrode Material Based on Graphene Derivative'. *Nanomaterials (Basel, Switzerland)*, *10*(9), 1731. DOI: 10.3390/nano10091731

Senthilkumar, B., Thenamirtham, P., & Kalai Selvan, R. (2011). 'Structural and Electrochemical Properties of Polythiophene'. *Applied Surface Science*, *257*(21), 9063–9067. DOI: 10.1016/j.apsusc.2011.05.100

Sevilla, M., & Mokaya, R. (2014). 'Energy Storage Applications of Activated Carbons: Supercapacitors and Hydrogen Storage'. *Energy & Environmental Science*, *7*(4), 1250–1280. DOI: 10.1039/C3EE43525C

Shah, S. S., Aziz, M. A., & Yamani, Z. H. (2022). 'Recent Progress in Carbonaceous and Redox-Active Nanoarchitectures for Hybrid Supercapacitors: Performance Evaluation, Challenges, and Future Prospects'. *Chemical Record (New York, N.Y.)*, *22*(7), e202200018. DOI: 10.1002/tcr.202200018

Shao, Y., El-Kady, M. F., Sun, J., Li, Y., Zhang, Q., Zhu, M., Wang, H., Dunn, B., & Kaner, R. B. (2018). 'Design and Mechanisms of Asymmetric Supercapacitors'. *Chemical Reviews*, *118*(18), 9233–9280. DOI: 10.1021/acs.chemrev.8b00252

Sharma, S., & Chand, P. (2023). 'Supercapacitor and Electrochemical Techniques: A Brief Review'. *Results in Chemistry*, *5*, 100885. DOI: 10.1016/j.rechem.2023.100885

Shen, J., Liu, A., Tu, Y., Wang, H., Jiang, R., Ouyang, J., & Chen, Y. (2012a). 'Asymmetric Deposition of Manganese Oxide in Single Walled Carbon Nanotube Films as Electrodes for Flexible High Frequency Response Electrochemical Capacitors'. *Electrochimica Acta*, *78*, 122–132. DOI: 10.1016/j.electacta.2012.05.138

Shen, J., Liu, A., Tu, Y., Wang, H., Jiang, R., Ouyang, J., & Chen, Y. (2012b). 'Asymmetric Deposition of Manganese Oxide in Single Walled Carbon Nanotube Films as Electrodes for Flexible High Frequency Response Electrochemical Capacitors'. *Electrochimica Acta*, *78*, 122–132. DOI: 10.1016/j.electacta.2012.05.138

Shi, H. (1996). 'Activated Carbons and Double Layer Capacitance'. *Electrochimica Acta, 41*(10), 1633–1639. DOI: 10.1016/0013-4686(95)00416-5

Shown, I., Ganguly, A., Chen, L.-C., & Chen, K.-H. (2015). 'Conducting Polymer-Based Flexible Supercapacitor'. *Energy Science & Engineering*, *3*(1), 2–26. DOI: 10.1002/ese3.50

Shuaibu, A. D., Shah, S. S., Alzahrani, A. S., & Aziz, M. A. (2025). 'Advancing Gel Polymer Electrolytes for Next-Generation High-Performance Solid-State Supercapacitors: A Comprehensive Review'. *Journal of Energy Storage*, *107*, 114851. DOI: 10.1016/j.est.2024.114851

Shuja, A., Khan, H. R., Murtaza, I., Ashraf, S., Abid, Y., Farid, F., & Sajid, F. (2024). 'Supercapacitors for Energy Storage Applications: Materials, Devices and Future Directions: A Comprehensive Review'. *Journal of Alloys and Compounds*, *1009*, 176924. DOI: 10.1016/j.jallcom.2024.176924

Simon, P., & Burke, A. (2008). 'Nanostructured Carbons: Double-Layer Capacitance and More'. *The Electrochemical Society Interface*, *17*(1), 38–43. DOI: 10.1149/2.F05081IF

Simon, P., & Gogotsi, Y. (2008). 'Materials for Electrochemical Capacitors'. *Nature Materials*, *7*(11), 845–854. DOI: 10.1038/nmat2297

Simon, P., Gogotsi, Y., & Dunn, B. (2014). Where Do Batteries End and Supercapacitors Begin? *Science*, *343*(6176), 1210–1211. DOI: 10.1126/science.1249625

Simpson, D. R. (2008). 'Biofilm Processes in Biologically Active Carbon Water Purification'. *Water Research*, *42*(12), 2839–2848. DOI: 10.1016/j.watres.2008.02.025

Singh, P. N., Mohamed, M. G., Kotp, M. G., Mondal, T., Chaganti, S. V., Ibrahim, M., Sharma, S. U., Ye, Y., & Kuo, S.-W. (2025). 'Nitrogen- and Sulfur-Rich Microporous Carbons Derived from Conjugated Microporous Polymers for CO_2 Uptake, Supercapacitor Energy Storage, and Electrochemical Hydrogen Production'. *ACS Applied Polymer Materials*, *7*(5), 3324–3336. DOI: 10.1021/acsapm.5c00012

Sivakkumar, S. R., Kim, W. J., Choi, J. A., MacFarlane, D. R., Forsyth, M., & Kim, D. W. (2007). 'Electrochemical Performance of Polyaniline Nanofibres and Polyaniline/Multi-Walled Carbon Nanotube Composite as an Electrode Material for Aqueous Redox Supercapacitors'. *Journal of Power Sources*, *171*(2), 1062–1068. DOI: 10.1016/j.jpowsour.2007.05.103

Snook, G. A., Kao, P., & Best, A. S. (2011). 'Conducting-Polymer-Based Supercapacitor Devices and Electrodes'. *Journal of Power Sources*, *196*(1), 1–12. DOI: 10.1016/j.jpowsour.2010.06.084

Srinivasan, V., & Weidner, J. W. (2000). 'Studies on the Capacitance of Nickel Oxide Films: Effect of Heating Temperature and Electrolyte Concentration'. *Journal of the Electrochemical Society*, *147*(3), 880–885. DOI: 10.1149/1.1393286

Stempien, Z., Khalid, M., Kozicki, M., Kozanecki, M., Varela, H., Filipczak, P., Pawlak, R., Korzeniewska, E., & Sąsiadek, E. (2019). 'In-Situ Deposition of Reduced Graphene Oxide Layers on Textile Surfaces by the Reactive Inkjet Printing Technique and Their Use in Supercapacitor Applications'. *Synthetic Metals*, *256*, 116144. DOI: 10.1016/j.synthmet.2019.116144

Strielkowski, W., Civín, L., Tarkhanova, E., Tvaronavičienė, M., & Petrenko, Y. (2021). 'Renewable Energy in the Sustainable Development of Electrical Power Sector: A Review'. *Energies*, *14*(24), 8240. DOI: 10.3390/en14248240

Subramani, K., Sudhan, N., Karnan, M., & Sathish, M. (2017). 'Orange Peel Derived Activated Carbon for Fabrication of High-Energy and High-Rate Supercapacitors'. *ChemistrySelect*, *2*(35), 11384–11392. DOI: 10.1002/slct.201701857

Sugimoto, W., Iwata, H., Yokoshima, K., Murakami, Y., & Takasu, Y. (2005). 'Proton and Electron Conductivity in Hydrous Ruthenium Oxides Evaluated by Electrochemical Impedance Spectroscopy: The Origin of Large Capacitance'. *The Journal of Physical Chemistry B*, *109*(15), 7330–7338. DOI: 10.1021/jp044252o

Sumdani, G., Islam, M. R., Yahaya, A., & Safie, S. (2021). 'Recent Advancements in Synthesis, Properties, and Applications of Conductive Polymers for Electrochemical Energy Storage Devices: A Review'. *Polymer Engineering and Science*, *62*(2), 269–303. Advance online publication. DOI: 10.1002/pen.25859

Tarcan, R., Todor-Boer, O., Petrovai, I., Leordean, C., Astilean, S., & Botiz, I. (2020). 'Reduced Graphene Oxide Today'. *Journal of Materials Chemistry. C, Materials for Optical and Electronic Devices*, *8*(4), 1198–1224. DOI: 10.1039/C9TC04916A

Teng, G., Lin, S., Xu, D., Heng, Y., & Hu, D. (2020). 'Renewable Cellulose Separator with Good Thermal Stability Prepared via Phase Inversion for High-Performance Supercapacitors'. *Journal of Materials Science Materials in Electronics*, *31*(10), 7916–7926. DOI: 10.1007/s10854-020-03330-w

Tholkappiyan, R., & Vishista, K. (2015). 'Tuning the Composition and Magnetostructure of Dysprosium Iron Garnets by Co-Substitution: An XRD, FT-IR, XPS and VSM Study'. *Applied Surface Science*, *351*, 1016–1024. DOI: 10.1016/j.apsusc.2015.05.193

Tortora, L., & Bussetti, G. (Eds.). (2025). *Surface and Interface Modification of Graphite and Graphene-Based Materials for Energy and Sensor Applications.* MDPI. DOI: 10.3390/books978-3-7258-3310-8

Tudu, S., Layek, R., Ali, M. S., Dutta, K., Mallik, A., Karmakar, S., Kumbhakar, P., & Chattopadhyay, D. (2024). 'Fabrication of CH3NH3SnCl3 Perovskite Nanocrystal-Based Electrode for Supercapacitor Devices Application'. *Journal of Alloys and Compounds, 971*, 172554. DOI: 10.1016/j.jallcom.2023.172554

Ukanwa, K. S., Patchigolla, K., Sakrabani, R., Anthony, E., & Mandavgane, S. (2019). 'A Review of Chemicals to Produce Activated Carbon from Agricultural Waste Biomass'. *Sustainability (Basel), 11*(22), 6204. DOI: 10.3390/su11226204

Vaquero, S., Díaz, R., Anderson, M., Palma, J., & Marcilla, R. (2012). 'Insights into the Influence of Pore Size Distribution and Surface Functionalities in the Behaviour of Carbon Supercapacitors'. *Electrochimica Acta, 86*, 241–247. DOI: 10.1016/j.electacta.2012.08.006

Verma, C., Goni, L. K. M. O., Yaagoob, I. Y., Vashisht, H., Mazumder, M. A. J., & Alfantazi, A. (2023). 'Polymeric Surfactants as Ideal Substitutes for Sustainable Corrosion Protection: A Perspective on Colloidal and Interface Properties'. *Advances in Colloid and Interface Science, 318*, 102966. DOI: 10.1016/j.cis.2023.102966

Verma, K. D., Sinha, P., Banerjee, S., Kar, K. K., & Ghorai, M. K. (2020). Characteristics of Separator Materials for Supercapacitors. In Kar, K. K. (Ed.), *Handbook of Nanocomposite Supercapacitor Materials I: Characteristics* (pp. 315–326). Springer International Publishing. DOI: 10.1007/978-3-030-43009-2_11

Vinayagam, M., Babu, R. S., Sivasamy, A., & de Barros, A. L. F. (2025). 'Improved Capacitive Performance of Conducting Polymer with Incorporation of Biomass Derived Activated Carbon for Supercapacitors'. *Diamond and Related Materials, 154*, 112165. DOI: 10.1016/j.diamond.2025.112165

Vivekchand, S. R. C., Rout, C. S., Subrahmanyam, K. S., Govindaraj, A., & Rao, C. N. R. (2008). 'Graphene-Based Electrochemical Supercapacitors'. *Journal of Chemical Sciences, 120*(1), 9–13. DOI: 10.1007/s12039-008-0002-7

Wang, B., Song, W., Gu, P., Fan, L., Yin, Y., & Wang, C. (2019). 'A Stretchable and Hydrophobic Polypyrrole/Knitted Cotton Fabric Electrode for All-Solid-State Supercapacitor with Excellent Strain Capacitance'. *Electrochimica Acta, 297*, 794–804. DOI: 10.1016/j.electacta.2018.12.042

Wang, D.-W., Li, F., Zhao, J., Ren, W., Chen, Z.-G., Tan, J., Wu, Z.-S., Gentle, I., Gao, Q. L., & Cheng, H.-M. (2009). 'Fabrication of Graphene/Polyaniline Composite Paper *via In Situ* Anodic Electropolymerization for High-Performance Flexible Electrode'. *ACS Nano*, *3*(7), 1745–1752. DOI: 10.1021/nn900297m

Wang, G., Xu, Z., Li, Z., Ding, Y., Ge, R., Xiang, M., Wang, G., & Yan, Z. (2023). 'Ni(OH)2/CoS Heterostructure Grown on Carbon Cloth for Robust Supercapacitor and Methanol Electrocatalytic Oxidation'. *Electrochimica Acta*, *443*, 141980. DOI: 10.1016/j.electacta.2023.141980

Wang, H., Lin, J., & Ze, X. S. (2016). 'Polyaniline (PANi) Based Electrode Materials for Energy Storage and Conversion'. *Journal of Science. Advanced Materials and Devices*, *1*(3), 225–255. DOI: 10.1016/j.jsamd.2016.08.001

Wang, J., & Azam, W. (2024). 'Natural Resource Scarcity, Fossil Fuel Energy Consumption, and Total Greenhouse Gas Emissions in Top Emitting Countries'. *Geoscience Frontiers*, *15*(2), 101757. DOI: 10.1016/j.gsf.2023.101757

Wang, J.-C., Yao, H.-C., Fan, Z.-Y., Zhang, L., Wang, J.-S., Zang, S.-Q., & Li, Z.-J. (2016). 'Indirect Z-Scheme BiOI/g-C3N4 Photocatalysts with Enhanced Photoreduction CO2 Activity under Visible Light Irradiation'. *ACS Applied Materials & Interfaces*, *8*(6), 3765–3775. DOI: 10.1021/acsami.5b09901

Wang, S., Zhang, D., He, X., Yuan, J., Que, W., Yang, Y., Protsak, I., Huang, X., Zhang, C., Lu, T., Pal, P., Liu, S., Zheng, S. Y., & Yang, J. (2022). 'Polyzwitterionic Double-Network Ionogel Electrolytes for Supercapacitors with Cryogenic-Effective Stability'. *Chemical Engineering Journal*, *438*, 135607. Advance online publication. DOI: 10.1016/j.cej.2022.135607

Wang, Shangshang, Jianbo Zhang, Oumaïma Gharbi, Vincent Vivier, Ming Gao, and Mark E. Orazem. n.d. 'Electrochemical Impedance Spectroscopy'.

Wang, T., Song, D., Zhao, H., Chen, J., Zhao, C., Chen, L., Chen, W., Zhou, J., & Xie, E. (2015). 'Facilitated Transport Channels in Carbon Nanotube/Carbon Nanofiber Hierarchical Composites Decorated with Manganese Dioxide for Flexible Supercapacitors'. *Journal of Power Sources*, *274*, 709–717. DOI: 10.1016/j.jpowsour.2014.10.102

Wang, X., Sumboja, A., Khoo, E., Yan, C., & Lee, P. S. (2012). 'Cryogel Synthesis of Hierarchical Interconnected Macro-/Mesoporous Co3O4 with Superb Electrochemical Energy Storage'. *The Journal of Physical Chemistry. C, Nanomaterials and Interfaces*, *116*(7), 4930–4935. DOI: 10.1021/jp211339t

Wang, Y., Shi, Z., Huang, Y., Ma, Y., Wang, C., Chen, M., & Chen, Y. (2009). 'Supercapacitor Devices Based on Graphene Materials'. *The Journal of Physical Chemistry. C, Nanomaterials and Interfaces*, *113*(30), 13103–13107. DOI: 10.1021/jp902214f

Wang, Y., Zhang, L., Hou, H., Xu, W., Duan, G., He, S., Liu, K., & Jiang, S. (2021). 'Recent Progress in Carbon-Based Materials for Supercapacitor Electrodes: A Review'. *Journal of Materials Science*, *56*(1), 173–200. DOI: 10.1007/s10853-020-05157-6

Wu, D., Yu, C., & Zhong, W. (2021). 'Bioinspired Strengthening and Toughening of Carbon Nanotube@polyaniline/Graphene Film Using Electroactive Biomass as Glue for Flexible Supercapacitors with High Rate Performance and Volumetric Capacitance, and Low-Temperature Tolerance'. *Journal of Materials Chemistry. A, Materials for Energy and Sustainability*, *9*(34), 18356–18368. DOI: 10.1039/D1TA05729D

Wu, F., Gao, J., Zhai, X., Xie, M., Sun, Y., Kang, H., Tian, Q., & Qiu, H. (2019). 'Hierarchical Porous Carbon Microrods Derived from Albizia Flowers for High Performance Supercapacitors'. *Carbon*, *147*, 242–251. DOI: 10.1016/j.carbon.2019.02.072

Wu, M., Snook, G. A., Gupta, V., Shaffer, M., Fray, D. J., & Chen, G. Z. (2005). 'Electrochemical Fabrication and Capacitance of Composite Films of Carbon Nanotubes and Polyaniline'. *Journal of Materials Chemistry*, *15*(23), 2297–2303. DOI: 10.1039/b418835g

Wu, Q.-F., He, K.-X., Mi, H.-Y., & Zhang, X.-G. (2007). 'Electrochemical Capacitance of Polypyrrole Nanowire Prepared by Using Cetyltrimethylammonium Bromide (CTAB) as Soft Template'. *Materials Chemistry and Physics*, *101*(2–3), 367–371. DOI: 10.1016/j.matchemphys.2006.06.013

Wu, Z., Zhu, Y., Ji, X., & Banks, C. E. (2016a). Transition Metal Oxides as Supercapacitor Materials. In Ozoemena, K. I., & Chen, S. (Eds.), *Nanomaterials in Advanced Batteries and Supercapacitors, Nanostructure Science and Technology* (pp. 317–344). Springer International Publishing. DOI: 10.1007/978-3-319-26082-2_9

Wu, Z., Zhu, Y., Ji, X., & Banks, C. E. (2016b). Transition Metal Oxides as Supercapacitor Materials. In Ozoemena, K. I., & Chen, S. (Eds.), *Nanomaterials in Advanced Batteries and Supercapacitors, Nanostructure Science and Technology* (pp. 317–344). Springer International Publishing. DOI: 10.1007/978-3-319-26082-2_9

Wu, Z.-S., Wang, D.-W., Ren, W., Zhao, J., Zhou, G., Li, F., & Cheng, H.-M. (2010). 'Anchoring Hydrous RuO_2 on Graphene Sheets for High-Performance Electrochemical Capacitors'. *Advanced Functional Materials*, *20*(20), 3595–3602. DOI: 10.1002/adfm.201001054

Xia, Q., Xu, M., Xia, H., & Xie, J. (2016). 'Nanostructured Iron Oxide/Hydroxide-Based Electrode Materials for Supercapacitors'. *ChemNanoMat : Chemistry of Nanomaterials for Energy, Biology and More*, *2*(7), 588–600. DOI: 10.1002/cnma.201600110

Xiao, K., Li, J. W., Chen, G. F., Liu, Z. Q., Li, N., & Su, Y. Z. (2014). 'Amorphous MnO2 Supported on 3D-Ni Nanodendrites for Large Areal Capacitance Supercapacitors'. *Electrochimica Acta*, *149*, 341–348. DOI: 10.1016/j.electacta.2014.10.117

Xiao, Y., Zhang, A., Liu, S., Zhao, J., Fang, S., Jia, D., & Li, F. (2012). 'Free-Standing and Porous Hierarchical Nanoarchitectures Constructed with Cobalt Cobaltite Nanowalls for Supercapacitors with High Specific Capacitances'. *Journal of Power Sources*, *219*, 140–146. DOI: 10.1016/j.jpowsour.2012.07.030

Xiao, Z., Chen, W., Liu, K., Cui, P., & Zhan, D. (2018). 'Porous Biomass Carbon Derived from Peanut Shells as Electrode Materials with Enhanced Electrochemical Performance for Supercapacitors'. *International Journal of Electrochemical Science*, *13*(6), 5370–5381. DOI: 10.20964/2018.06.54

Xie, Q., Huang, X., Zhang, Y., Wu, S., & Zhao, P. (2018). 'High Performance Aqueous Symmetric Supercapacitors Based on Advanced Carbon Electrodes and Hydrophilic Poly(Vinylidene Fluoride) Porous Separator'. *Applied Surface Science*, *443*, 412–420. DOI: 10.1016/j.apsusc.2018.02.274

Xin, Q., Guo, Z., Zang, Y., & Lin, J. (2020). 'Graphene Aerogel-Coated MoO3 Nanoparticle/Polypyrrole Ternary Composites for High-Performance Supercapacitor'. *Journal of Materials Science Materials in Electronics*, *31*(19), 17110–17119. DOI: 10.1007/s10854-020-04271-0

Xiong, W., Zhao, L., Jie, O., Tian, Y., Wang, L., Li, M., Wang, Y., Cheng, M., Sheng, Q., Li, Z., Luo, J., & Luo, Y. (2025). 'Surface-Modified Composites of Metal–Organic Framework and Wood-Derived Carbon for High-Performance Supercapacitors'. *Journal of Colloid and Interface Science*, *679*, 243–252. DOI: 10.1016/j.jcis.2024.09.247

Xu, B., Zheng, M., Tang, H., Chen, Z., Chi, Y., Wang, L., Zhang, L., Chen, Y., & Pang, H. (2019). 'Iron Oxide-Based Nanomaterials for Supercapacitors'. *Nanotechnology*, *30*(20), 204002. DOI: 10.1088/1361-6528/ab009f

Yadav, N., Singh, M. K., Yadav, N., & Hashmi, S. A. (2018). 'High Performance Quasi-Solid-State Supercapacitors with Peanut-Shell-Derived Porous Carbon'. *Journal of Power Sources*, *402*, 133–146. DOI: 10.1016/j.jpowsour.2018.09.032

Yan, J., Fan, Z., Wei, T., Qian, W., Zhang, M., & Wei, F. (2010). 'Fast and Reversible Surface Redox Reaction of Graphene–MnO2 Composites as Supercapacitor Electrodes'. *Carbon*, *48*(13), 3825–3833. DOI: 10.1016/j.carbon.2010.06.047

Yan, L., Huang, J., Dong, X., Guo, Z., Wang, Z., & Wang, Y. (2020). 'Energizing Hybrid Supercapacitors by Using Mn2+-Based Active Electrolyte'. *Journal of Materials Chemistry. A, Materials for Energy and Sustainability*, *8*(30), 15051–15057. DOI: 10.1039/D0TA04864J

Yan, S., Tang, Y., Liu, L., Gao, Y., Zhang, Y., & Yang, C. (2023). 'Amorphous Thin-Walled Carbon Nanotubes Modified by Simple Oxidation for Zinc-Ion Hybrid Supercapacitors'. *ACS Applied Energy Materials*, *6*(8), 4144–4149. DOI: 10.1021/acsaem.2c03688

Yang, Qinghao, Zhenzhong Hou, and Tianzhu Huang. 2015. 'Self-assembled Polypyrrole Film by Interfacial Polymerization for Supercapacitor Applications'. *Journal of Applied Polymer Science* 132(11):app.41615. .DOI: 10.1002/app.41615

Yang, V., Senthil, R. A., Pan, J., Khan, A., Osman, S., Wang, L., Jiang, W., & Sun, Y. (2019). 'Highly Ordered Hierarchical Porous Carbon Derived from Biomass Waste Mangosteen Peel as Superior Cathode Material for High Performance Supercapacitor'. *Journal of Electroanalytical Chemistry (Lausanne, Switzerland)*, *855*, 113616. DOI: 10.1016/j.jelechem.2019.113616

Yang, V., Senthil, R. A., Pan, J., Kumar, T. R., Sun, Y., & Liu, X. (2020). 'Hierarchical Porous Carbon Derived from Jujube Fruits as Sustainable and Ultrahigh Capacitance Material for Advanced Supercapacitors'. *Journal of Colloid and Interface Science*, *579*, 347–356. DOI: 10.1016/j.jcis.2020.06.080

Yang, Z., Ren, J., Zhang, Z., Chen, X., Guan, G., Qiu, L., Zhang, Y., & Peng, H. (2015). 'Recent Advancement of Nanostructured Carbon for Energy Applications'. *Chemical Reviews*, *115*(11), 5159–5223. DOI: 10.1021/cr5006217

Yedluri, A. K., & Kim, H.-J. (2019). 'Enhanced Electrochemical Performance of Nanoplate Nickel Cobaltite (NiCo2O4) Supercapacitor Applications'. *RSC Advances*, *9*(2), 1115–1122. DOI: 10.1039/C8RA09081E

Yi, H., Nakabayashi, K., Yoon, S.-H., & Miyawaki, J. (2021). 'Pressurized Physical Activation: A Simple Production Method for Activated Carbon with a Highly Developed Pore Structure'. *Carbon*, *183*, 735–742. DOI: 10.1016/j.carbon.2021.07.061

Youn, H.-C., Bak, S.-M., Kim, M.-S., Jaye, C., Fischer, D. A., Lee, C.-W., Yang, X.-Q., Roh, K. C., & Kim, K.-B. (2015). 'High-Surface-Area Nitrogen-Doped Reduced Graphene Oxide for Electric Double-Layer Capacitors'. *ChemSusChem*, *8*(11), 1875–1884. DOI: 10.1002/cssc.201500122

Yu, D., Goh, K., Wang, H., Wei, L., Jiang, W., Zhang, Q., Dai, L., & Chen, Y. (2014). 'Scalable Synthesis of Hierarchically Structured Carbon Nanotube–Graphene Fibres for Capacitive Energy Storage'. *Nature Nanotechnology*, *9*(7), 555–562. DOI: 10.1038/nnano.2014.93

Yu, D., Goh, K., Zhang, Q., Wei, L., Wang, H., Jiang, W., & Chen, Y. (2014). 'Controlled Functionalization of Carbonaceous Fibers for Asymmetric Solid-State Micro-Supercapacitors with High Volumetric Energy Density'. *Advanced Materials*, *26*(39), 6790–6797. DOI: 10.1002/adma.201403061

Yu, P., Li, Y., Zhao, X., Wu, L., & Zhang, Q. (2014). 'Graphene-Wrapped Polyaniline Nanowire Arrays on Nitrogen-Doped Carbon Fabric as Novel Flexible Hybrid Electrode Materials for High-Performance Supercapacitor'. *Langmuir*, *30*(18), 5306–5313. DOI: 10.1021/la404765z

Yuan, C., Yang, L., Hou, L., Shen, L., Zhang, F., Li, D., & Zhang, X. (2011). 'Large-Scale Co3O4 Nanoparticles Growing on Nickel Sheets via a One-Step Strategy and Their Ultra-Highly Reversible Redox Reaction toward Supercapacitors'. *Journal of Materials Chemistry*, *21*(45), 18183–18185. DOI: 10.1039/c1jm14173b

Yuan, C., Yang, L., Hou, L., Shen, L., Zhang, X., & Lou, X. W. D. (2012). Growth of ultrathin mesoporous Co 3 O 4 nanosheet arrays on Ni foam for high-performance electrochemical capacitors. *Energy & Environmental Science*, *5*(7), 7883–7887.

Yun, X., Li, J., Luo, Z., Tang, J., & Zhu, Y. (2019). 'Advanced Aqueous Energy Storage Devices Based on Flower-like Nanosheets-Assembled Ni0.85Se Microspheres and Porous Fe2O3 Nanospheres'. *Electrochimica Acta*, *302*, 449–458. DOI: 10.1016/j.electacta.2019.02.038

Yun, Y. S., Cho, S. Y., Kim, H., Jin, H.-J., & Kang, K. (2015). 'Ultra-Thin Hollow Carbon Nanospheres for Pseudocapacitive Sodium-Ion Storage'. *ChemElectroChem*, *2*(3), 359–365. DOI: 10.1002/celc.201402359

Zafar, K., Hafiza, S. Z., Masood, M., Sohail, M., Syed, S. A. S., Karim, M. R., & O'Mullane, A. (2024). 'Recent Advances on Nitrogen-Doped Porous Carbons Towards Electrochemical Supercapacitor Applications'. *Chemical Record (New York, N.Y.)*, *24*(1), e202300161. DOI: 10.1002/tcr.202300161

Zardkhoshoui, M., Akbar, B. A., & Saied, S. H. D. (2023). 'A Hybrid Supercapacitor Assembled by Reduced Graphene Oxide Encapsulated Lollipop-like FeNi2S4@ Co9S8 Nanoarrays'. *Chemical Engineering Journal*, *470*, 144132. DOI: 10.1016/j.cej.2023.144132

Zhai, S., Fan, Z., Jin, K., Man Zhou, H. Z., Zhao, Y., Ge, F., Li, X., & Cai, Z. (2020). 'Synthesis of Zinc Sulfide/Copper Sulfide/Porous Carbonized Cotton Nanocomposites for Flexible Supercapacitor and Recyclable Photocatalysis with High Performance'. *Journal of Colloid and Interface Science*, *575*, 306–316. DOI: 10.1016/j.jcis.2020.04.073

Zhai, T., Wang, F., Yu, M., Xie, S., Liang, C., Li, C., Xiao, F., Tang, R., Wu, Q., Lu, X., & Tong, Y. (2013). '3D MnO2–Graphene Composites with Large Areal Capacitance for High-Performance Asymmetric Supercapacitors'. *Nanoscale*, *5*(15), 6790–6796. DOI: 10.1039/c3nr01589k

Zhai, Z., Zhang, L., Du, T., Ren, B., Xu, Y., Wang, S., Miao, J., & Liu, Z. (2022). 'A Review of Carbon Materials for Supercapacitors'. *Materials & Design*, *221*, 111017. DOI: 10.1016/j.matdes.2022.111017

Zhang, C., Chen, Z., Rao, W., Fan, L., Xia, Z., Xu, W., & Xu, J. (2019). 'A High-Performance All-Solid-State Yarn Supercapacitor Based on Polypyrrole-Coated Stainless Steel/Cotton Blended Yarns'. *Cellulose (London, England)*, *26*(2), 1169–1181. DOI: 10.1007/s10570-018-2126-3

Zhang, F., Zhang, T., Yang, X., Zhang, L., Leng, K., Huang, Y., & Chen, Y. (2013). 'A High-Performance Supercapacitor-Battery Hybrid Energy Storage Device Based on Graphene-Enhanced Electrode Materials with Ultrahigh Energy Density'. *Energy & Environmental Science*, *6*(5), 1623. DOI: 10.1039/c3ee40509e

Zhang, J., Gu, M., & Chen, X. (2023). 'Supercapacitors for Renewable Energy Applications: A Review'. *Micro and Nano Engineering*, *21*, 100229. DOI: 10.1016/j.mne.2023.100229

Zhang, K., Lin, Y., Chen, L., Huang, J., Wang, L., Peng, M., Tang, X., Hu, T., Yuan, K., & Chen, Y. (2021). 'Enabling 2.4-V Aqueous Supercapacitors through the Rational Design of an Integrated Electrode of Hollow Vanadium Trioxide/ Carbon Nanospheres'. *Science China Materials*, *64*(9), 2163–2172. DOI: 10.1007/s40843-020-1608-4

Zhang, T., Walawender, W. P., Fan, L. T., Fan, M., Daugaard, D., & Brown, R. C. (2004). 'Preparation of Activated Carbon from Forest and Agricultural Residues through CO2 Activation'. *Chemical Engineering Journal*, *105*(1), 53–59. DOI: 10.1016/j.cej.2004.06.011

Zhang, W., Wang, Y., Guo, X., Liu, Y., Zheng, Y., Zhang, M., Li, R., Peng, Z., Xie, H., & Zhao, Y. (2021). 'Graphene-Carbon Nanotube @ Cobalt Derivatives from ZIF-67 for All-Solid-State Asymmetric Supercapacitor'. *Applied Surface Science, 568*, 150929. DOI: 10.1016/j.apsusc.2021.150929

Zhao, C., & Zheng, W. (2015). 'A Review for Aqueous Electrochemical Supercapacitors'. *Frontiers in Energy Research, 3*. Advance online publication. DOI: 10.3389/fenrg.2015.00023

Zhao, J., Lu, Y., Liu, Y., Liu, L., Yin, J., Sun, B., Wang, G., & Zhang, Y. (2023). 'A Self-Healing PVA-Linked Phytic Acid Hydrogel-Based Electrolyte for High-Performance Flexible Supercapacitors'. *Nanomaterials (Basel, Switzerland), 13*(3), 380. DOI: 10.3390/nano13030380

Zhao, X., Zheng, B., Huang, T., & Gao, C. (2015). 'Graphene-Based Single Fiber Supercapacitor with a Coaxial Structure'. *Nanoscale, 7*(21), 9399–9404. DOI: 10.1039/C5NR01737H

Zheng, H., Guan, R., Liu, Q., Ou, K., Li, D. S., Fang, J., Fu, Q., & Sun, Y. (2022). 'A Flexible Supercapacitor with High Capacitance Retention at an Ultra-Low Temperature of -65.0°C'. *Electrochimica Acta, 424*, 140644. Advance online publication. DOI: 10.1016/j.electacta.2022.140644

Zhi, H. M., L. Fu, and Z. J. Ding. (2012). 'Continuous Measurement for Chromium and Manganese in the Metal Bonded Tools and Its Reliability Verification'. *Jingangshi Yu Moliao Moju Gongcheng/Diamond and Abrasives Engineering* 32(5):72–75.

Zhong, C., Deng, Y., Hu, W., Qiao, J., Zhang, L., & Zhang, J. (2015). 'A Review of Electrolyte Materials and Compositions for Electrochemical Supercapacitors'. *Chemical Society Reviews, 44*(21), 7484–7539. DOI: 10.1039/C5CS00303B

Zhong, J.-H., Wang, A.-L., Li, G.-R., Wang, J.-W., Ou, Y.-N., & Tong, Y.-X. (2012). 'Co3O4/Ni(OH)2 Composite Mesoporous Nanosheet Networks as a Promising Electrode for Supercapacitor Applications'. *Journal of Materials Chemistry, 22*(12), 5656–5665. DOI: 10.1039/c2jm15863a

Zhong, M., Zhang, M., & Li, X. (2022). 'Carbon Nanomaterials and Their Composites for Supercapacitors'. *Carbon Energy, 4*(5), 950–985. DOI: 10.1002/cey2.219

Zhou, C., Kumar, S., Doyle, C. D., & Tour, J. M. (2005). 'Functionalized Single Wall Carbon Nanotubes Treated with Pyrrole for Electrochemical Supercapacitor Membranes'. *Chemistry of Materials, 17*(8), 1997–2002. DOI: 10.1021/cm047882b

Zhu, K., Wang, Y., Tang, J. A., Qiu, H., Meng, X., Gao, Z., Chen, G., Wei, Y., & Gao, Y. (2016). 'In Situ Growth of MnO2 Nanosheets on Activated Carbon Fibers: A Low-Cost Electrode for High Performance Supercapacitors'. *RSC Advances*, *6*(18), 14819–14825. DOI: 10.1039/C5RA24692J

Zhu, Y., Murali, S., Stoller, M. D., Ganesh, K. J., Cai, W., Ferreira, P. J., Pirkle, A., Wallace, R. M., Cychosz, K. A., Thommes, M., Su, D., Stach, E. A., & Ruoff, R. S. (2011). 'Carbon-Based Supercapacitors Produced by Activation of Graphene'. *Science*, *332*(6037), 1537–1541. DOI: 10.1126/science.1200770

Zhu, Y., Wu, Z., Jing, M., Yang, X., Song, W., & Ji, X. (2015). 'Mesoporous NiCo2S4 Nanoparticles as High-Performance Electrode Materials for Supercapacitors'. *Journal of Power Sources*, *273*, 584–590. DOI: 10.1016/j.jpowsour.2014.09.144

Zhu, Z. Z., Wang, G. C., Sun, M. Q., Li, X. W., & Li, C. Z. (2011). Fabrication and electrochemical characterization of polyaniline nanorods modified with sulfonated carbon nanotubes for supercapacitor applications. *Electrochimica Acta*, *56*(3), 1366–1372.

Zou, W., Fang, H., Ma, T., Zhao, Y., Wang, L., Jia, X., & Zhang, L. (2023). 'Three-Dimensional Nanoporous CNT@Mn3O4 Hybrid Anode: High Pseudocapacitive Contribution and Superior Lithium Storage'. *Batteries*, *9*(7), 389. DOI: 10.3390/batteries9070389

Chapter 6

Design and Optimisation of Metal-Air Batteries: A Critical Review of Mechanisms and Performance

Moussa Ouakki
https://orcid.org/0000-0002-6265-4734
Laboratory of Engineering, Sustainability, and Innovation, École Nationale Supérieure de Chimie (ENSC), Morocco

ABSTRACT

Metal-air batteries (Li-air, Zn-air, Na-air, and Al-air) represent a promising alternative for energy storage due to their high energy density and potential for sustainable applications. However, their performance is severely limited by challenges associated with metallic anodes, including corrosion, passivation, and uncontrolled hydrogen evolution, which reduce cycle life and faradaic efficiency. This critical review examines the electrochemical mechanisms associated with metallic anodes in metal-air batteries, comparing the performance of Li-air systems (3500 Wh/kg), Zn-air (300 cycle), Na-air (1000 Wh/kg), and Al-air (2790 Wh/kg). Recent solutions, such as the use of alloyed anodes, surface modifications, and the incorporation of inhibitor additives in alkaline electrolytes, have significantly improved the performance and durability of metal-air batteries. The findings from this analysis highlight recent advancements and persistent challenges in the optimisation of metallic anodes, paving the way for more efficient and durable metal-air batteries for large-scale applications.

DOI: 10.4018/979-8-3373-3156-0.ch006

INTRODUCTION

The global energy transition, driven by the urgent need to reduce greenhouse gas emissions and limit dependence on fossil fuels, is largely based on the development and massive integration of renewable energies such as wind and solar power. However, the intermittent nature of these energy sources poses a major challenge achieving efficient and sustainable energy storage (Gaele et al., 2022; Ostergaard et al., 2020; Shittu et al., 2022). In this context, energy storage is technologies have become essential pillars for ensuring grid stability, mitigating production fluctuations, and flexibly meeting growing energy demand (Branco et al., 2018; Areola et al., 2025; Di Somma et al., 2022).

Among these challenges, electrochemical energy storage systems particularly batteries occupy a central role. Their high efficiency, modularity, and scalability make them suitable for diverse applications, ranging from portable electronics to large-scale stationary storage (Shabeer et al., 2025). Since Alessandro Volta's pioneering work in the 19th century, battery technologies have undergone significant evolution, giving rise to several families such as lithium-ion, lead-acid, and redox flow batteries (Olabi et al., 2021). Despite their diversity, all these systems rely on the same fundamental mechanism: the reversible transfer of ions between an anode and a cathode through an electrolyte, enabling energy storage and release.

Within this technological landscape, metal–air batteries (MABs) have recently attracted increasing attention due to their exceptionally high theoretical energy densities and potential contribution to carbon-neutral energy systems. For example, lithium-air batteries can reach up to 3500 Wh/kg (Ye et al., 2024; Bruce et al., 2012; Imanishi et al., 2019), around ten times more than conventional lithium-ion batteries, whose density varies between 100 and 300 Wh/kg (Shabeer et al., 2025). This remarkable energy capacity makes MABs promising candidates for both large-scale stationary storage and long-range electric vehicles (EVs). Furthermore, the recyclability and natural abundance of metals such as zinc and aluminum reinforce their strategic importance in sustainable energy transitions. Unlike traditional systems, MABs utilize atmospheric oxygen as the cathodic reactant, reducing overall weight and enhancing energy density (Khan et al., 2025).

Recent advances in materials science and electrochemistry have significantly improved the performance and cycling stability of MABs. Innovations in metallic anode engineering (Li, Zn, Na, Al) and electrolyte optimization have proven particularly effective (Shinde et al., 2023; Ji et al., 2022; Chen et al., 2025; Che et al., 2025). Consequently, understanding the electrochemical mechanisms and degradation pathways of these anodes remains a key step toward developing efficient and durable systems.

The principle behind MABs is both innovative and efficient. These batteries derive their energy from the electrochemical reaction between a reactive metal (e.g., zinc, aluminum, lithium, iron, magnesium) used as the anode and oxygen from the ambient air, captured at the cathode (Olabi et al., 2021; Zhang et al., 2019). Although the concept of metal–air batteries dates back further, the first commercial zinc–air (Zn–air) batteries appeared in 1932 (Heise et al., 1933). These primary Zn–air batteries typically offer high energy densities, ranging from 200 to 500 Wh/kg, and achieve over 300 cycles with coulombic efficiencies exceeding 90% (Zhang et al., 2019). However, they are characterized by relatively low power output. Despite this limitation, Zn–air batteries are widely used in low-power applications such as hearing aids and telecommunication devices, including pagers (Zhang et al., 2019).

Building on these early innovations, recent research has increasingly focused on advanced MAB systems, particularly lithium-air batteries, which boast the highest theoretical energy density (has an energy density of 13000 W h kg^{-1}) among known electrochemical technologies (Olabi et al., 2021). Despite their promise, these systems face considerable technological challenges. One of the major obstacles is the formation of solid by-products, such as lithium peroxide (Li_2O_2), which can clog cathode pores and limit battery life to only a few tens or hundreds of cycles (Olabi et al., 2021). Additionally, the electrolytes used in these systems must resist aggressive reactive species like superoxide radicals, complicating their design. Consequently, lithium-air batteries currently suffer from low overall efficiency (3500 Wh/kg < 50 cycles) due to significant overvoltages during charge and discharge cycles (Yang et al., 2024; Imanishi et al., 2019; Zahoor et al., 2019).

The scope of metal-air technologies, however, extends beyond lithium. Other variants, such as aluminum-air and iron-air batteries, are drawing attention, particularly for stationary energy storage applications or as range extenders for EVs. is the Al–air system demonstrated by the company Phinergy in 2013 (Shabeer et al., 2025; Rani et al., 2024). Phinergy demonstrated the potential of Al–air batteries by powering an electric vehicle with a system offering up to 1,000 miles of range using 50 aluminum plates, with water refills every 200 miles. They also developed a hybrid EV combining Al–air and lithium-ion technologies to significantly extend driving range, proposed applications in aviation, and signed an MoU with Hindalco for battery development (Rani et al., 2024). Their vision includes battery-swapping stations and use in remote sensors and surveillance systems.

This hybrid configuration optimally leverages the strengths of both technologies: the lithium-ion battery provides instant power and urban mobility, while the heavier but more energy-dense metal-air battery is reserved for long-distance driving. Although promising, this hybrid architecture introduces new challenges. Key issues include cost, recyclability, safety, and compatibility with existing infrastructures. Nevertheless, the relative abundance of the materials involved (e.g., zinc, aluminum, iron), their moderate cost, and their recyclability potential reinforce the appeal of MABs from both environmental and economic sustainability perspectives (Olabi et al., 2021). Despite these advantages, no metal-air technology has yet reached the level of technological maturity required for widespread industrial adoption. In particular, their role as range extenders for electric vehicles remains underexplored in the literature, especially from a comparative perspective across different metal-air chemistries. However, advancing this approach could strategically address the current limitations of EV range without relying on larger lithium-ion batteries, which are often more expensive, heavier, and dependent on critical raw materials.

This context sets the stage for the present study, which provides an in-depth comparative analysis of the main metal-air battery technologies. The objective is to evaluate their suitability as range extenders for electric vehicles. This summary is based on a detailed analysis of performance parameters (energy density, efficiency, service life), operating principles, technological barriers, and associated costs. Furthermore, it examines the potential impact of hydrogen evolution in certain configurations and explores strategies to optimize charge/discharge cycles. The study is structured as follows: Section 1: Comparative analysis of the electrochemical performance of metal anodes. Section 2: Hydrogen evolution and its impact on the durability of metal-air batteries. Section 3: Performance optimization and comparison of charge-discharge cycles.

This chapter is based exclusively on a critical review and comparative analysis of previously published studies. All numerical data, figures, and performance metrics presented in this work were extracted from peer-reviewed scientific sources published between 2012 and 2025. Literature was collected using reliable academic databases such as ScienceDirect, Wiley, and Springer. Only studies providing reproducible experimental or simulation data were considered. Each data point and performance metric reported in the tables and figures is directly cited from the original publication to ensure accuracy and transparency.

I. COMPARATIVE ANALYSIS OF METAL ANODES

Rechargeable batteries have become indispensable to modern energy storage technologies, powering a broad spectrum of applications from electric vehicles (EVs) (Khan et al., 2023; Nagmani et al., 2022) aimed at reducing reliance on fossil fuels and lowering greenhouse gas emissions, to stationary energy storage systems that support the integration and stability of renewable energy sources into power grids (Barman et al., 2025; C. Zhao et al., 2018). Within these systems, every component plays a critical role; among them, the anode is particularly pivotal, as it directly impacts the battery's capacity, energy density, cycle life, and overall safety (W.-J. Zhang, 2011).

Traditionally, graphite has been the predominant anode material in commercial lithium-ion batteries due to its moderate specific capacity (~372 mAh·g^{-1}), structural and chemical stability, and low production cost (Bai et al., 2015; Feyzi et al., 2024). However, as the demand grows for batteries with higher energy density, extended longevity, and faster charging capabilities, the limitations of graphite are becoming increasingly evident (Xu et al., 2022; Zhong et al., 2023). Its relatively low specific capacity constrains the overall energy density achievable by lithium-ion systems. In response to these challenges, research efforts have increasingly focused on metallic anodes, which offer substantially higher specific capacities and more favorable redox potentials (Barman et al., 2025; Nzereogu et al., 2022; Shi et al., 2024; Sun et al., 2025; Uzun et al., 2025; Wang et al., 2017). Table1 summarizes the electrochemical properties of selected metal anodes for lithium-ion batteries, highlighting their performance relative to graphite (Wang et al., 2017; W.-J. Zhang, 2011).

Table 1. Comparative analysis of the electrochemical properties of metal-based anodes for lithium-ion batteries (Wang et al., 2017; W.-J. Zhang, 2011). Reported values correspond to average performance metrics under comparable experimental conditions.

Metal anode	Density (g.cm^{-3})	Specific capacity (mAh.g^{-1})	Theoretical charge density (mAh.cm^{-3})	Volume expansion (%)	Potential vs Li$^+$/Li (V)	Lithiated phase	Energy gain vs graphite (%)
Li	0.53	3862	2047	100	0	Li	23
C (graphite)	2.25	372	837	12	0.05	LiC_6	0
Al	2.7	993	2681	97	0.3	LiAl	25
Sn	7.29	994	7246	260	0.6	$Li_{4.4}Sn$	30
Zn	7.14	410	5855	71	0.157	LiZn	29
Mg	1.3	3350	4355	100	0.1	Li_3Mg	42
Sb	6.7	660	4422	200	0.9	Li_3Sb	10
Bi	9.78	385	3765	215	0.8	Li_3Bi	13

In recent years, extensive research efforts have been devoted to improving the performance and stability of metal–air batteries through material innovations and design optimization. Shabeer et al. (Shabeer et al., 2025) compared Li–air, Zn–air, Na–air, and Al–air batteries as potential range extenders for electric vehicles, reporting that Zn–air systems currently exhibit the best compromise between cost, safety, and cycling stability, achieving more than 300 stable cycles with coulombic efficiencies exceeding 90%. Similarly, Barman et al. (Barman et al., 2025) emphasized that controlling anode corrosion, dendritic growth, and hydrogen evolution remains a critical challenge in improving rechargeability and long-term durability. Recent studies have also focused on nano-engineered anode architectures including 3D porous Zn foams, Mg-based alloys, and Al composite frameworks that improve ion transport and reduce passivation phenomena (Nazir et al., 2024). On the cathode side, the development of bifunctional oxygen electrocatalysts based on transition-metal oxides (e.g., Co_3O_4, $MnCo_2O_4$) and heteroatom-doped carbons has considerably enhanced oxygen reduction and evolution kinetics, improving the overall round-trip efficiency of rechargeable configurations (Lee et al., 2016). Collectively, these studies reveal that combining optimized metallic anodes, advanced electrolytes, and durable catalytic interfaces is essential to achieving the next generation of high-performance, sustainable metal–air batteries.

Pure metals and their alloys including lithium (Li), sodium (Na), magnesium (Mg), zinc (Zn), and aluminum (Al) are promising candidates for next-generation anodes, owing to their potential to enhance both energy performance and charging rates (Shinde et al., 2023). The selection of a suitable metal for anode applications is governed by a combination of interrelated electrochemical and physical parameters, such as specific capacity, volumetric charge density, insertion potential vs. Li^+/Li, material density, volumetric expansion upon lithiation, and theoretical energy gain relative to graphite. These factors collectively determine the feasibility and effectiveness of a given material for integration into advanced lithium-ion battery systems.

From a gravimetric standpoint, alkali and alkaline earth metals like lithium and magnesium exhibit exceptionally high theoretical specific capacities, far exceeding that of graphite (Romio et al., 2022; Yanagisawa, 1999; S. Yang et al., 2022). This superior capacity stems from their intrinsic ability to form lithium-rich phases through alloying or electrochemical deposition, thereby enabling the storage of a large number of Li^+ ions per unit mass (Behara et al., 2024; Y. Liu & Holze, 2022). However, practical battery design is often constrained by volumetric limitations; thus, the material's density must also be considered. A comparison of theoretical volumetric charge densities reveals that heavier metals such as tin, zinc, and antimony surpass lithium in terms of energy storage per unit volume (Hossain et al., 2023; Nazir et al., 2024). This renders them particularly advantageous for applications requiring device miniaturization, such as portable electronics and microscale integrated devices (H. Liu et al., 2020; Muhammad Saqib et al., 2024).

It is important to note that the charge density values presented in Table 1 are calculated based on the pristine densities of the respective metals (W.-J. Zhang, 2011). Even when considering for the full volume expansion of lithiated products, these metal-based anodes retain charge densities that are two to five times higher than those of graphite (Lyu et al., 2024; H. Zhang et al., 2021; W.-J. Zhang, 2011). Another significant advantage of metallic anodes is their moderate insertion potential relative to lithium. For instance, aluminum exhibits an insertion potential in the range of 0.3–0.4 V vs (W.-J. Zhang, 2011). Li/Li^+, which is sufficiently high to mitigate the risk of lithium dendrite formation a common issue with graphite anodes, which operate near 0.05 V vs. Li/Li^+ while remaining low enough to minimize energy losses during cycling (Bouguern et al., 2024).

However, the main limitation of metal anodes lies in their significant volume expansion upon lithiation. This phenomenon, which can exceed 200% in the case of elements such as bismuth, antimony, or tin, often leads to mechanical degradation of active particles and poor cycling stability (W.-J. Zhang, 2011). Moreover, during lithiation and delithiation cycles, significant mechanical and interfacial challenges arise, including the loss of electrode electrolyte contact and the rupture of the solid electrolyte interphase (SEI) layer (Ding et al., 2024; Du et al., 2024). These phenomena critically impair the cyclability and long-term stability of batteries employing high-capacity metal anodes (Qian et al., 2015). In contrast, graphite exhibits only modest volumetric expansion (~12%) during cycling, which contributes to its continued dominance in commercial lithium-ion batteries, despite its limited specific capacity.

To address these challenges, numerous studies have demonstrated that the miniaturization of active materials to the nanometric scale offers a promising strategy to enhance the mechanical integrity and cycling stability of alloy-based anodes (Cheng et al., 2023; Saleh & Hassan, 2023; Thakur et al., 2012; Z. Zhang et al., 2025). When nanoparticle agglomeration is effectively mitigated typically by embedding the particles within a composite matrix marked improvements in electrochemical performance have been observed. For instance, Yang et al. (J. Yang et al., 1999) reported that reducing the particle size of tin from 3 μm to 300 nm increased the number of stable cycles to achieved around 200 cycles under galvanostatic, charge-limited conditions, highlighting the crucial role of particle size in improving the durability of alloy-based anodes.

Beyond mechanical stability, the theoretical energy gain of metal anodes over graphite can be quantitatively assessed by considering both specific capacity and redox potential. Notably, metals such as magnesium (42%), tin (30%), and zinc (29%) exhibit substantial energy advantages, reinforcing their relevance as candidates for next-generation anode materials.

In parallel, metal–air batteries another class of systems utilizing metal anodes are garnering increasing attention due to their exceptionally high theoretical energy densities, which surpass those of conventional lithium-ion batteries (Barman et al., 2025). These batteries operate based on the oxidation of the metal anode ($M \rightarrow M^{n+} + n e^-$) and the reduction of atmospheric oxygen at the cathode ($O_2 + 2H_2O + 4e^- \rightarrow 4OH^-$), as illustrated in Figure 1 and detailed in Table 2. One of their most attractive features is the elimination of the need for a heavy cathode material, as oxygen is sourced directly from ambient air, contributing to lighter and more compact battery architectures (Li et al., 2023; Rani et al., 2024).

Table 2. Reaction mechanisms of same Metal–Air battery systems (Olabi et al., 2021). All values are cited directly from the original publications to ensure reproducibility.

Metal (Anode)	Electrolyte	Anode reaction	Cathode reaction	Overall reaction
Fer (Fe)	Aqueous alkaline	$Fe + 2OH^- \rightarrow Fe(OH)_2 + 2e^-$ $3Fe(OH)_2 + 2OH^- \rightarrow Fe_3Fe_4 + 4H_2O + 2e^-$	$O_2 + 2H_2O + 4e^- \rightarrow 4OH^-$	$2O_2 + 3Fe \rightarrow Fe_3O_4$
Aluminium (Al)		$Al + 4OH^- \rightarrow [Al(OH)_4]^- + 3e^-$		$3O_2 + 2Al \rightarrow Al_2O_3$
Zinc (Zn)		$Zn + 4OH^- \rightarrow [Zn(OH)_4]^{2-} + 2e^-$ $[Zn(OH)_4]^{2-} \rightarrow ZnO + 2OH^- + H_2O$		$O_2 + Zn \rightarrow ZnO$
Lithium (Li)	Non-aqueous	$Li \rightarrow Li^+ + e^-$	$O_2 + e^- \rightarrow O_2^-$ $O_2^- + Li^+ \rightarrow LiO_2$ $LiO_2 + Li^+ + e^- \rightarrow Li_2O_2$	$O_2 + 2Li \rightarrow Li_2O_2$

This simplified architecture results in a significant gain in specific energy, as shown in Figure 1.

Figure 1. Analysis of battery voltage, specific energy, and volumetric energy density in metal–air systems (Yaqoob et al., 2022) Data compiled from literature sources published.

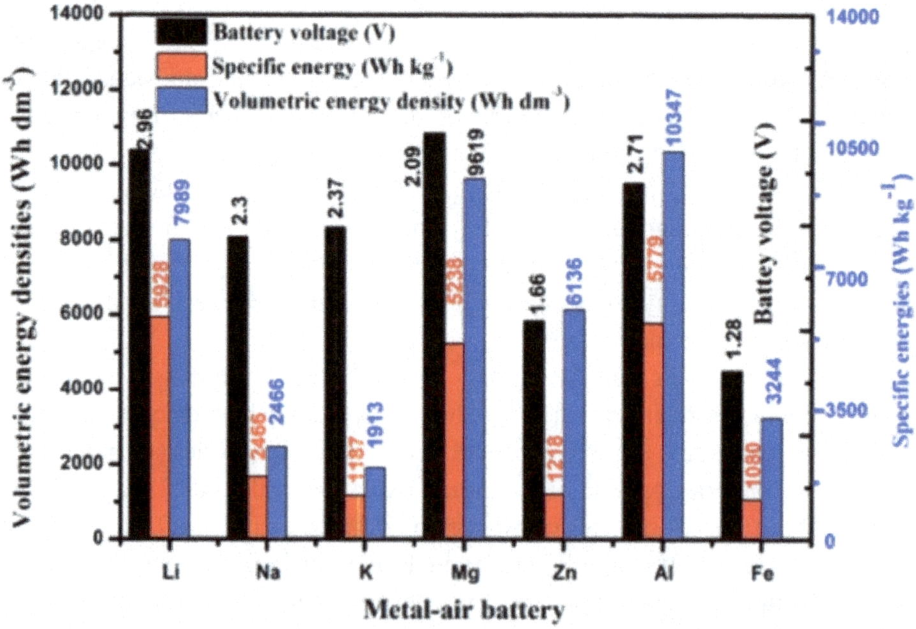

These observations underscore why lithium remains to be the material of choice for high-performance rechargeable batteries. Its exceptionally high specific capacity and extremely low electrochemical potential (Table 3) confer a distinct strategic advantage, even when compared to emerging technologies such as lithium–air batteries, which theoretically offer some of the highest energy densities among known systems (Li et al., 2023; Olabi et al., 2021; Yaqoob et al., 2022).

However, the practical deployment of lithium is hindered by its high chemical reactivity. Challenges such as dendritic growth during repeated charge–discharge cycles and pronounced instability in the presence of moisture and oxygen raise critical safety and durability concerns (Jia et al., 2024; Z. Zhang et al., 2024). These issues necessitate the implementation of stringent environmental and electrochemical control measures, which currently limit the commercial viability of lithium–air batteries.

Table 3. Physicochemical properties of selected metallic anodes. Data compiled from literature sources published (Li et al., 2023; Olabi et al., 2021; Yaqoob et al., 2022).

Metallic Anode	Redox Potential (V vs. SHE)	Abundance in Earth's Crust (ppm)	Ionic Radius (pm)
Li	–3.04	20	76
Na	–2.71	23,600	102
K	–2.90	20,900	138
Mg	–2.37	23,300	72
Ca	–2.87	41,500	100
Zn	–0.80	70	74
Al	-1.66	83000	53.5

Given the inherent limitations of lithium-based systems, current research is increasingly focused on alternative alkali metals (Hasan et al., 2025; Wulandari et al., 2023; Y. Zhang et al., 2025), particularly sodium (1,166 mAh g^{-1}) and potassium (685 mAh g^{-1}). These elements offer compelling advantages, including greater natural abundance and lower cost. For instance, a study by Zhao et al. (X. Zhao et al., 2020) reported that a potassium metal anode embedded in a porous carbon fiber scaffold coated with SnO_2 exhibited remarkable electrochemical stability for over 1,700 hours, effectivel-y suppressing dendrite formation.

Moreover, the larger ionic radii of Na^+ and K^+, combined with their more complex solvation behavior, hinder their efficient intercalation into conventional electrode hosts (Ml et al., 2021). This has prompted the development of new host materials tailored to accommodate these ions' unique physicochemical characteristics (Lakhnot et al., 2023).

Beyond monovalent systems, multivalent metal anodes including magnesium, zinc, aluminum, and iron present an attractive research direction due to their ability to transfer multiple electrons per ion, potentially enhancing the overall energy storage capacity (Olabi et al., 2021; Yaqoob et al., 2022). Nevertheless, these metals typically possess more positive standard reduction potentials than lithium, leading to lower cell voltages and thus reduced energy densities (Figure 1). Among these, zinc stands out for its compatibility with aqueous electrolytes, enabled by its high hydrogen overpotential (Ran et al., 2025). Despite the narrow electrochemical window of water (~1.23 V), aqueous Zn-based systems offer enhanced safety, environmental friendliness, and economic viability albeit at the cost of issues such as self-discharge and the generation of undesirable by-products (Cai et al., 2023). With a moderate specific capacity and a reduction potential of approximately –0.76 V vs SHE, zinc strikes a balance well suited to short-term commercial deployment.

Aluminum exhibits a high gravimetric capacity and a favorable reduction potential (–1.66 V vs SHE), yet its rechargeability in aqueous environments is severely hampered by the formation of a passivating oxide layer (Table 2), restricting its use predominantly to primary (non-rechargeable) battery systems (Rani et al., 2024). Magnesium offers an appealing compromise, combining a high theoretical capacity, suitable potential (–2.37 V vs SHE), intrinsic safety, and minimal dendrite formation (Ye et al., 2024). However, it suffers from limited ionic conductivity in conventional electrolytes and passivation issues, especially in aqueous media (Salado & Lizundia, 2022). Moreover, the oxygen reduction reaction in Mg-air batteries remains poorly catalyzed, representing a key technical bottleneck (Antony Jose et al., 2025).

Iron, although offering lower energy density, emerges as a compelling candidate for stationary and cost-sensitive applications due to its high natural abundance, low toxicity, low cost, and excellent electrochemical stability (He et al., 2021; Sandstrom & Ji, 2022).

Overall, no single metal meets all the desirable criteria for next-generation anodes (Nzereogu et al., 2022). Lithium offers unmatched specific capacity and low redox potential but faces serious safety and dendrite-related concerns (Hasan et al., 2025; Jia et al., 2024; Wulandari et al., 2023). Magnesium avoids dendrite formation and provides high energy density but suffers from sluggish kinetics and poor compatibility with standard electrolytes (Antony Jose et al., 2025; Ye et al., 2024). Heavier elements like tin and antimony exhibit superior volumetric capacities but introduce substantial mechanical strain upon cycling. The current scientific challenge thus lies in the design of advanced electrode architecture ranging from nanostructured materials and composites to hybrid systems that harness the strengths of these metals while mitigating their respective drawbacks.

II. HYDROGEN EVOLUTION AND ITS IMPACT ON DURABILITY OF METAL-AIR BATTERIES

As previously discussed, metal–air batteries, such as zinc–air, magnesium–air, and aluminum–air systems have garnered considerable interest due to their high theoretical energy densities and reliance on ambient oxygen as the cathodic reactant. These systems operate through redox processes involving a metal anode and oxygen from the surrounding environment. Despite their promising characteristics, the practical deployment of metal–air batteries is hindered by several technical obstacles, among which the hydrogen evolution reaction (HER) is particularly problematic (Gelman et al., 2016; Kube & Kopljar, 2023; J. Li et al., 2022).

In aqueous electrolytes, HER emerges as an undesirable side reaction at the metal anode, directly competing with the intended oxidation of the metal (W. Lu et al., 2025). This leads not only to the evolution of hydrogen gas but also to undesirable effects such as electrolyte depletion, fluctuations in pH, and structural deterioration of the electrode materials (Achanta et al., 2024; W. Lu et al., 2025; Nian et al., 2020). The issue is further exacerbated under alkaline conditions, often used to improve ionic conductivity and electrochemical kinetics, where HER tends to occur more aggressively (Leuaa & Chatzichristodoulou, 2023; Mahmood et al., 2018). Highly reactive metals like aluminum and magnesium exhibit particularly pronounced hydrogen generation, making the suppression of HER a key research focus in the development of metal–air technologies (Xu & Liu, 2019; Zhou et al., 2018, 2019).

Moreover, the accumulation of hydrogen gas in sealed systems can increase internal pressure, posing serious safety concerns (Le et al., 2023; Zhou et al., 2024). These risks highlight the urgent need for a comprehensive understanding of HER mechanisms and the development of effective mitigation strategies (Calabrese et al., 2024).

Fundamentally, HER is an electrochemical process in which protons, or water molecules in neutral or alkaline media, are reduced at the electrode surface to produce molecular hydrogen (H_2). During battery operation, particularly under low-current or idle conditions, water reduction can proceed concurrently with metal oxidation, resulting in hydrogen release (Y. Liu et al., 2024; Y. Lu et al., 2024; Zhou et al., 2018). Typically, HER manifests as a parasitic reaction at the metal anode, occurring during discharge or even under open-circuit conditions. The stoichiometry of HER varies depending on the pH of the electrolyte, reflecting differences in the electrochemical pathway in acidic, neutral, or alkaline environments (Gebremariam et al., 2023; Luo et al., 2023).

In acidic media $2H^+ + 2e^- \rightarrow H_2$

In neutral or alkaline media $2H_2O + 2e^- \rightarrow H_2 + 2OH^-$

The hydrogen evolution reaction (HER) typically proceeds via two fundamental steps: the Volmer step, which involves the electrochemical adsorption of hydrogen atoms onto the electrode surface, followed by either the Heyrovsky step (electrochemical desorption) or the Tafel step (chemical desorption) (Luo et al., 2023). The dominance of a particular pathway is influenced by various factors, including electrode material, electrolyte pH, operating temperature, and applied overpotential (Zhan et al., 2020; Zhao et al., 2023). In the context of metal–air batteries, the HER mechanism generally follows the Volmer–Heyrovsky, particularly due to the electrochemical environment at the anode (Gebremariam et al., 2023; Luo et al., 2023).

In these systems, the anode is usually composed of highly reactive metals such as zinc, aluminum, magnesium, or iron. These metals possess significantly negative standard electrode potentials, making them thermodynamically susceptible to both oxidation and HER in aqueous media (Ambat & Piotrowska, 2022; Ershov, 2024). In alkaline electrolytes, which are frequently used to enhance ion transport and reaction kinetics, the abundance of hydroxide ions (OH^-) further facilitates HER, often catalyzed directly by the metal surface (Ambat & Piotrowska, 2022; Kam et al., 2024).

For instance, in zinc–air batteries employing potassium hydroxide (KOH) as the electrolyte, zinc readily reacts with water and hydroxide ions to generate zincate ions ($Zn(OH)_4^{2-}$) and hydrogen gas (Al Bostami et al., 2025; Wittman et al., 2025). This reaction contributes to self-discharge, electrolyte consumption, and the formation of zinc dendrites, which pose a risk of short-circuiting the cell and reducing overall performance (Zuo et al., 2021). A typical parasitic reaction associated with HER is:

$$2H_2O + 2e^- \rightarrow H_2 + 2OH^-$$

This side reaction decreases the battery's energy efficiency and accelerates degradation through mechanisms such as electrode corrosion, local pH shifts, and the formation of insulating by-products on the anode surface (H. Du et al., 2024; Y. Liu et al., 2017; Zhang et al., 2025).

The thermodynamic favorability of HER can be understood by comparing standard electrode potentials: the standard potential for hydrogen evolution is 0 V versus the standard hydrogen electrode (SHE), which is often more positive than the oxidation potentials of many anode metals (Isse & Gennaro, 2010; W. Li et al., 2022; X. Liu et al., 2024). As a result, HER is thermodynamically allowed. However, the kinetics of HER, and thus the actual rate at which it proceeds, are governed by the overpotential and the catalytic nature of the anode surface (Bao et al., 2021; X. Liu et al., 2024). These factors depend not only on the intrinsic properties of the electrode material but also on the electrolyte composition, including pH and ionic strength (Gebremariam et al., 2023; Isse & Gennaro, 2010; Luo et al., 2023).

Additional variables such as temperature, electrolyte concentration, and surface morphology of the anode (e.g., roughness, porosity) also influence HER activity (Leuaa & Chatzichristodoulou, 2023). Together, these parameters determine the severity of HER-related losses in metal–air battery systems and are therefore critical considerations in the development of effective mitigation strategies.

Numerous studies have been dedicated to addressing the challenges associated with the hydrogen evolution reaction (HER) in metal–air batteries, particularly in systems utilizing reactive metals like aluminum. For example, K.–Y. Chan and R. Savinel (Chan & Savinell, 1991) were among the first to explore the impact of hydrogen-related side reactions on the performance of aluminum–air single cells operating with a caustic electrolyte. Their results revealed that battery efficiency could be improved by reducing the electrolyte flow rate, minimizing the distance between electrodes, and increasing electrolyte concentration.

Expanding on this issue, Hong Zhou et al. (Zhou et al., 2020) carried out a thorough investigation combining experimental observations and physics-based modeling to evaluate HER's influence on aqueous Al–air battery performance. Their analysis indicated that HER-induced bubble formation significantly impaired cell kinetics due to bubble shielding, which dominated over ohmic losses. Interestingly, HER also introduced localized micro-convection from bubble motion, slightly reducing mass transfer limitations. Based on these insights, the study recommended operating under mass transfer-limited conditions, particularly at high current densities, to mitigate HER. Additionally, high electrolyte velocities and lower OH^- concentrations were found to be effective in reducing gas bubble coverage under low-current discharge regimes.

Despite these advances, severe HER remains a major bottleneck for aluminum–air batteries (AABs), leading to a substantial drop in discharge voltage (below 0.9 V), capacity density (under 800 mAh/g), and anodic utilization efficiency (less than 30%) (Liu et al., 2024). These limitations significantly restrict the large-scale deployment of AABs.

Further research by Zihan Wang (Z. Wang et al., 2024) explored the dynamics of HER under high-concentration NaOH conditions. The study showed that higher current densities substantially suppressed HER, suggesting the suitability of concentrated NaOH for demanding discharge applications. However, a drawback was the accumulation of discharge byproducts on the aluminum surface, which altered its electrochemical behavior. Moreover, rapid hydrogen gas release in such systems can pose explosion hazards and serious safety concerns, as supported by prior works (Abd El Haleem et al., 2021; Cai et al., 2023; Calabrese et al., 2024). The release of hydrogen in alkaline media may also result in harmful alkaline mist, posing risks to human health by irritating skin and respiratory organs. Additionally, HER-induced thermal effects can raise electrolyte temperatures, deform battery casings, shorten lifespan, and jeopardize user safety (Calabrese et al., 2024; Le et al., 2023).

Given these issues, HER is widely recognized as a critical barrier to the practical development of these batteries. In recent years, research efforts have intensified globally to mitigate HER through modifications of both the anode materials and the electrolyte compositions.

In this context, the work of Thiruvenkatam Subramaniam et al. (Subramaniam et al., 2024) though focused on Zn–air systems, offers transferable strategies. They employed nanofluid electrolytes containing dispersed metal oxide nanoparticles such as SiOx and ZnO, which effectively delayed the HER onset and increased the overpotential. This not only reduced anode corrosion but also enhanced oxygen reduction and evolution reactions (ORR/OER), demonstrating the potential of nanoparticle-based electrolytes to simultaneously suppress HER and improve cathodic catalytic activity.

Likewise, Henning Weinrich et al. (Weinrich et al., 2022) explored HER mechanisms on iron electrodes in alkaline media using in situ gas chromatography. Although centered on Fe systems, their quantitative assessment of HER as a function of electrode potential provided valuable insight into optimizing battery performance. They showed that adjusting charge–discharge parameters, such as the cut-off potential, could significantly reduce HER, achieving charge efficiencies up to 96.7%. These findings offer a methodological foundation that could be adapted to aluminum–air and other metal–air batteries to enhance cycling stability and operational efficiency.

To date, a variety of approaches, ranging from alloy design and mechanical processing to advanced surface engineering, have been explored to mitigate HER (Nandi & Pumera, 2025; Shabeer et al., 2025). While these methods have shown promise, their cost and complexity often hinder large-scale implementation, especially in the context of metal-anode batteries (Fang et al., 2019; Z. Wang et al., 2025). Therefore, there is an increasing need for more practical, cost-effective, and scalable strategies for anode protection to support the broader adoption of these energy storage technologies (Yi et al., 2024).

Non-aqueous electrolytes and complex multi-electrolyte systems can effectively suppress HER; however, they often come at the cost of reduced energy density and heightened safety risks (F. Wang et al., 2018; Xue et al., 2025). Among the alternative solutions, the use of electrolyte additives has gained attention due to its simplicity and promising results (Lei et al., 2024; J. Li et al., 2025; S. Yang et al., 2025). Nonetheless, certain limitations persist, such as the continued dissolution of aluminum anodes, and the overall suppression of HER remains inadequate in many cases (Thatipamula et al., 2024; Z. Wang et al., 2025; Yi et al., 2024).

In this regard, the development and optimization of electrolyte additives have become central to improving the performance and longevity of metal–air batteries.

Recent investigations point to the effectiveness of hybrid additives, which combine inorganic and organic components to form a compact and homogeneous protective layer on the surface of aluminum anodes (A. Ahmed et al., 2024; Y. Du et al., 2024; Figueira et al., 2015; Raza et al., 2025). This film acts as a physical and chemical barrier that limits HER while preserving the anode's electrochemical activity.

Moreover, by carefully balancing the electron-donating and -withdrawing tendencies of organic molecules, it is possible to enhance HER inhibition (Raza et al., 2025). Certain nucleophilic functional groups can strongly adsorb onto the anode surface, displacing water molecules and thereby reducing their reactivity (Guo et al., 2025; Sun et al., 2025). Simultaneously, these additives can guide the uniform deposition of protective inorganic films, improving ion distribution and surface stability (Y. Du et al., 2024).

Such modifications not only help suppress HER but also reduce the gap between the anode and cathode, improving ionic conductivity and increasing the current output of the battery (Raza et al., 2025; H. Yang & Wu, 2022). Ultimately, the chemical structure, type, and concentration of organic functional groups in the additive play a critical role in determining the inhibition efficiency. These insights may also inform the design of tailored additives for other aqueous systems, including metal–air and metal-ion batteries.

III. PERFORMANCE OPTIMIZATION AND CHARGE–DISCHARGE ANALYSIS

Achieving optimal performance in metal-air batteries requires more than high theoretical energy densities it demands robust and stable cycling over extended periods. The electrochemical charge-discharge process, while central to energy storage, often induces structural and chemical transformations that degrade both electrodes and electrolytes. These include passivation layers on metal anodes, parasitic side reactions, and oxygen electrode instability, all of which reduce coulombic efficiency and shorten cycle life. The interplay between electrode design, electrolyte formulation, and operational protocols is therefore crucial to sustaining performance. This section explores current strategies aimed at enhancing cyclability, evaluating how innovations in materials engineering and system configuration influence key performance metrics. It also offers a comparative view of cycling behaviour across different metal-air chemistries, identifying both their limitations and pathways for improvement.

1. Impact of Electrode Composition and Surface Engineering

The performance and longevity of metal-air batteries are intrinsically tied to the physicochemical properties of their electrodes. In particular, the structural composition and surface characteristics of metal anodes play a decisive role in mitigating degradation processes such as corrosion, passivation, and hydrogen evolution. Recent advances have demonstrated that strategic engineering of electrode surfaces through alloying, electroplating, and nano-architectures can significantly enhance electrochemical performance across various metal-air systems.

Sandeep Srivastava et all (Srivastava et al., 2024). conducted a comprehensive study on aluminium-air batteries, focusing on the effect of surface treatments using nickel and nickel–copper electroplating. Their results showed that nickel-coated aluminium anodes outperformed the nickel–copper counterparts in both KOH and NaOH electrolytes. Electrochemical impedance spectroscopy and anodic polarization tests revealed that the nickel layer served as an effective barrier, reducing corrosion while maintaining active sites for charge transfer. Scanning Electron Microscopy (SEM) confirmed a uniform deposition layer that improved stability without inhibiting battery reactions, leading to superior discharge behaviour. In a separate investigation, Sang-Min Lee et all (Lee et al., 2013). addressed the issue of self-discharge in zinc-air batteries by introducing a uniform Al_2O_3 coating onto zinc anode particles. This surface modification significantly increased the overpotential for hydrogen evolution, thereby minimizing side reactions responsible for capacity loss. Morphological and structural analyses using SEM, EDS, XRD, and XPS confirmed the homogeneous distribution of Al_2O_3. Electrochemical tests showed a marked improvement in discharge time from 7 hours (pristine Zn) to over 10 hours at a current density of 25 mA cm^{-2} highlighting the effectiveness of the oxide layer in stabilising the anode electrolyte interface. For lithium-air systems, Keegan R. Adair et all (Adair et al., 2018). developed a three-dimensional (3D) copper nanowire scaffold to accommodate molten lithium, forming a Li@CuLi alloy upon infusion. This lithiophilic host structure mitigates dendrite formation and buffers volume changes during cycling. The engineered anode delivered over 200 cycles at 10 mA cm^{-2} in symmetric cell configurations and exhibited exceptional stability when paired with a $LiFePO_4$ cathode maintaining performance over 400 full charge-discharge cycles with minimal capacity fade. This approach illustrates the potential of nano-structured metallic hosts in extending the practical application of Li-air batteries. A schematic illustration of the different stages involved in the electrode fabrication process is presented in Fig. 2.

Figure 2. Schematic diagram of the different stages of electrode fabrication (Adair et al., 2018).

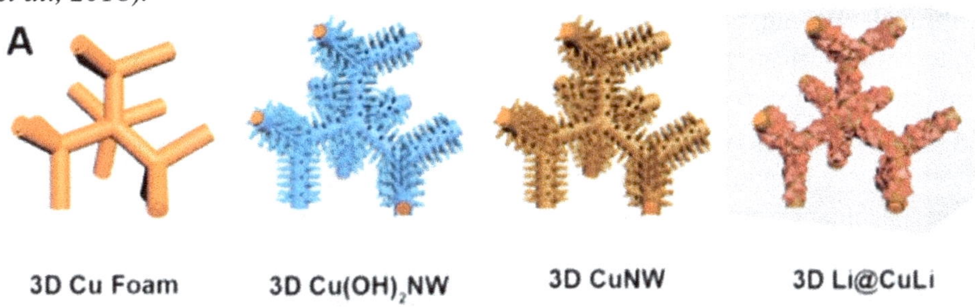

Further insights into the role of anode composition were provided by Y. Yu et al (Yu et al., 2021), who explored the effects of germanium modification on silicon anodes in Si-Ge air batteries using density functional theory (DFT). Their simulations revealed that replacing Si atoms with Ge increased the adsorption energies of GeO_2 units and reduced surface passivation, while also improving interfacial electronic conductivity. These results suggest that Si-Ge composite anodes offer promising electrochemical characteristics, driven by enhanced interfacial properties and higher current densities under operational conditions. J. Zhai et al (Zhai et al., 2021). further demonstrated the effectiveness of surface engineering by applying a conformal carbon coating via chemical vapor deposition (CVD) to manganese oxide (MnO) electrodes used in aqueous zinc-ion batteries. Unlike traditional wet-chemical methods, the CVD process produced a uniform and energy-compatible coating, enabling the electrode to achieve a high energy density of 351 Wh kg^{-1} at 625 W kg^{-1} with excellent rate capability. This structural stability was confirmed through a combination of XRD, XPS, FTIR, and GITT analyses. Their results show that precisely controlled surface coatings significantly improve charge storage mechanisms and highlight the potential of vapor-based modification for high-performance aqueous systems. Ö. Aslanbas et al (Aslanbas et al., 2018). examined the discharge behaviour and corrosion resistance of arc-melted Al/Si alloys used as anodes in metal-air batteries with $EMIm(HF)_{2.3}F$ ionic liquid electrolyte. Through systematic variation of silicon content across nine alloy compositions, they found that the discharge voltage decreased with higher Si content, with performance distinctions observed between hypoeutectic, hypereutectic, and high-silicon phases. SEM analysis post-discharge confirmed that aluminium oxidation dominated the electrochemical contribution. Potentiodynamic polarization and galvanic coupling analyses based on Evans diagrams revealed how microstructure and galvanic interactions shape corrosion behaviour. This study highlights how binary alloy composition directly governs anode activity and stability in air battery systems.

In summary, the composition and surface architecture of metal anodes play a crucial role in determining the stability and efficiency of metal-air batteries. Whether through alloying, protective coatings, or nanostructured host frameworks, these strategies significantly reduce parasitic reactions such as corrosion and dendrite formation. The reviewed studies clearly demonstrate that optimized electrode design not only improves cycle life but also enhances the overall energy delivery of the system. These insights form a critical foundation for understanding how electrode properties interact with electrolytes, as discussed in the following section.

2. Electrolyte Design and Its Role in Cycle Efficiency

The electrolyte plays a fundamental role in determining the coulombic efficiency, voltage stability, and overall reversibility of metal-air batteries. However, its function is deeply influenced by the interfacial properties of the electrode materials. Recent developments have focused on tuning this electrode-electrolyte interaction through surface modifications and compositional engineering, aiming to suppress parasitic reactions and enhance charge transport during cycling.

S. Moonngam et al (Moonngam et al., 2025). investigated an Al-3Zn alloy for aluminium-air batteries, using a modified friction stir process (FSP) to refine grain structure and control crystal orientation. Their post-thermal treatment optimized the microstructure to facilitate better electrochemical behaviour, with the modified anode achieving a discharge potential of 1.53 V at 20 mA/cm^2. This highlights the role of microstructure-electrolyte interaction in improving battery output, as optimized grain boundaries enhance charge transfer while reducing corrosion. In the case of zinc-air batteries, Y. Kim et al (Kim et al., 2019). demonstrated that CuO nanoparticles deposited on zinc surfaces can prevent direct exposure to alkaline electrolytes like KOH, thereby suppressing hydrogen evolution and minimizing self-discharge. The 0.1 wt% CuO-modified Zn showed the best cycling stability, maintaining performance over 24 hours of continuous discharge. XPS and SEM confirmed homogeneous coverage of CuO, emphasizing that surface passivation can act as a kinetic barrier, reducing unwanted side reactions at the electrolyte interface. Fig. 3 illustrates the synthesis pathway of CuO-modified Zn electrodes used in this work.

Figure 3. Sketch illustrating the synthesis process of CuO-modified Zn in Zn-air batteries (Kim et al., 2019).

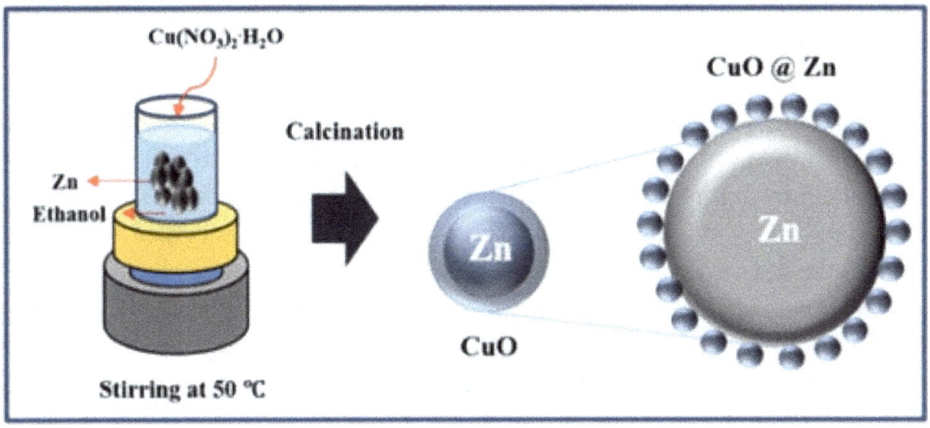

X. Cao et al (Cao et al., 2015). advanced a bifunctional electrocatalyst strategy by synthesizing $MnCo_2O_4$ coated with polypyrrole (MCO@PPy). This conductive polymer not only enhanced the electrical conductivity of the $MnCo_2O_4$ core but also facilitated faster electron transfer in ORR/OER reactions within the electrolyte environment. The result was a performance that matched or surpassed that of commercial Pt/C and RuO_2/C electrocatalysts, while also exhibiting higher stability. This illustrates how electrode-electrolyte synergy can be improved via intelligent material integration, especially in the air cathode domain. For silicon-air batteries, D. Chen et al (Chen et al., 2023). introduced metal-organic segment (MOS) coatings to mitigate the corrosion of silicon anodes in alkaline KOH environments. The MOS layer effectively lowered the critical KOH concentration required for stable operation (from 1.9 M to 1.3 M) and extended the discharge time of the battery from 250 h (bare Si) to 430 h (Si@MOS). This result underscores the importance of electrolyte-anode compatibility and the potential of surface engineering to reduce the aggressiveness of the required electrolyte formulation. Lastly, Y. Yu et al (Yu et al., 2021). employed DFT calculations to study how doping silicon anodes with Zn and Fe atoms can influence interfacial electrostatics and electronic conductivity in Si-air batteries. These dopants modified the Si/SiO_2 interface, reducing the bandgap and improving current flow across the junction. The findings indicate that even at the atomic level, modifying electrode surfaces can profoundly affect their interaction with electrolytes, ultimately increasing discharge capacity and overall system efficiency. Additional support for the electrolyte-centered enhancement of Al-air batteries was provided by M. El-Alouani et al (El-Alouani et al., 2025a,b), who investigated organic inhibitors including β(+) d-glucose (S1), Adonite (S2), and

2-n(4-nitrobenzylidene) malononitrile (Z4) in 4 M NaOH electrolytes. Their findings demonstrated a significant suppression of hydrogen evolution corrosion, with inhibition efficiencies above 65%, and corresponding increases in specific capacity and energy density reaching up to 1000 mAh g^{-1} and 1922 Wh kg^{-1} for treated systems. Surface analysis and DFT/MD simulations confirmed improved anode passivation and favourable molecular interactions, highlighting the importance of electrolyte formulation as a corrosion management strategy in alkaline Al-air batteries. Building on the growing interest in eco-friendly electrolyte modification, Y. Gao et al (Gao et al., 2025). introduced Toona Sinensis extract (TSE) as a green additive to suppress the self-corrosion of aluminium anodes in alkaline Al-air batteries. Rich in active components such as quercetin, kaempferol, and nicotinic acid, TSE was shown to form protective films through O-Al and N-Al bonding, as confirmed by DFT and MD simulations. With an optimized electrolyte containing 0.4 g L^{-1} TSE, the battery exhibited an anode utilization rate of 83.9% and a specific capacity of 2571.4 mAh g^{-1}, nearly doubling its original performance. These findings illustrate how natural molecular additives can play a critical role in enhancing both discharge efficiency and corrosion inhibition in aqueous Al-air systems.

Collectively, these studies demonstrate that improving the efficiency of charge-discharge cycling is not solely dependent on electrolyte composition itself, but also on how the electrolyte interfaces with the electrode material. Advances in surface modification techniques allow for better control over these interactions, reducing side reactions, extending battery life, and enhancing electrochemical stability across various metal-air battery chemistries.

3. Comparative Analysis of Charge-Discharge Cycling Behaviour

The cycling behaviour of metal-air batteries is a key performance metric that reflects their practical viability for long-term energy storage. It depends not only on the intrinsic chemistry of the metal and the air cathode, but also on the architecture of the cell, the design of the catalyst layers, the electrolyte environment, and the control of side reactions during repeated charge-discharge processes. Recent studies have revealed that strategic modifications to both materials and cell interfaces can significantly improve cycling stability, energy retention, and voltage efficiency across various metal-air battery chemistries.

(Salado et al., 2022). offered a comprehensive overview of current bottlenecks in metal-air batteries, placing a strong emphasis on the role of electrolyte systems in determining cycling behaviour. Their analysis spans aqueous, organic, gel-like, and solid-state electrolytes, highlighting how electrolyte engineering influences rechargeability, safety, and lifespan. They also incorporate a circular economy perspective, stressing the need for sustainable materials and formulations as next-generation batteries move towards commercial implementation. (Xu et al., 2019), explored a novel approach to improve the cycling performance of rechargeable Zn-air batteries (ZABs) using superwetting catalytic electrodes. By engineering the electrode surface to control O_2 adsorption and release, they significantly enhanced oxygen redox kinetics while minimizing gas bubble accumulation. Their superwetting tri-electrode system featuring superaerophilic Co-N-doped CNTs and superaerophobic NiFe-LDH arrays demonstrated an impressive peak power density of 245 mW cm^{-2} and sustained cycling stability beyond that of precious-metal-based reference cells. This work underscores the importance of interfacial engineering for robust charge-discharge reversibility. (Wang et al., 2022). focused on aqueous metal-air batteries such as Zn-air, Al-air, and Mg-air. They highlighted that despite the promise of high energy density and eco-compatibility, practical applications suffer from limited discharge voltages and poor rechargeability. The study examined electrolyte formulations and additives capable of stabilizing interfaces, buffering pH shifts, and suppressing parasitic reactions. This comparative review provides an essential framework for benchmarking charge-discharge performance in systems driven by aqueous electrolytes. Fig. 4 illustrates the operating mechanism of metal-air batteries and the critical interactions influencing cycling.

Figure 4. Illustration of metal-air battery.

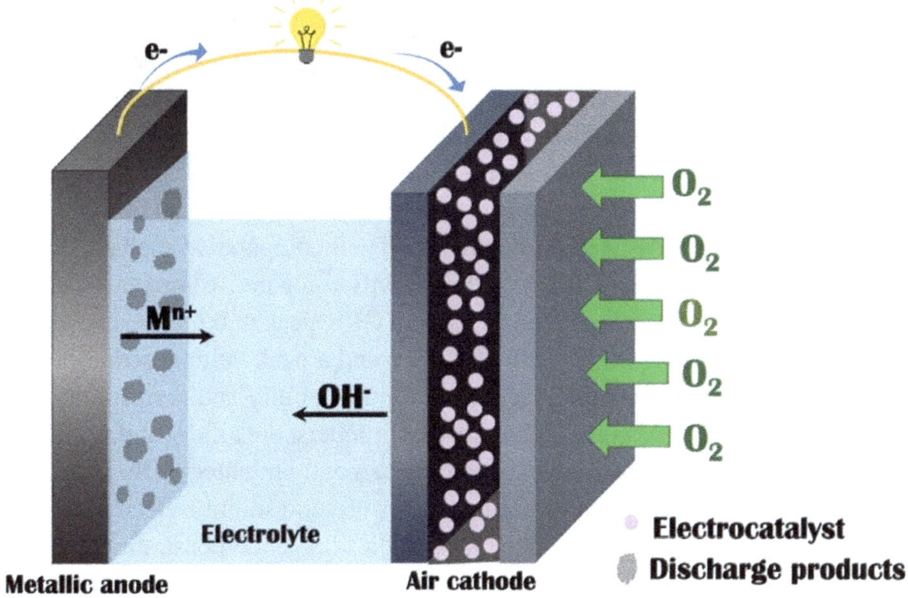

X. Huang et al (Huang et al., 2024). provided a detailed analysis of Mg-air batteries (MABs), noting that poor anodic efficiency and corrosion limit their cycling performance. They emphasized how anode microstructure including grain size, dislocations, and phase orientation affects hydrogen evolution and active material utilization. The authors propose that microstructure optimization and single-crystal design can significantly reduce capacity loss over multiple discharge cycles, offering a promising path toward durable MABs. P. Wu et al (Wu et al., 2022). addressed Al-air battery cycling limitations due to self-corrosion, which depletes active material and leads to unstable cycling. By introducing a sucrose-modified electrolyte, they successfully decreased water activity and hydrogen evolution. This allowed an increase in specific capacity from 356 mAh/g (standard electrolyte) to 2330 mAh/g, and extended discharge time from 6 to 22 hours under intermittent cycling. Their work highlights that modifying the electrolyte chemistry can profoundly influence battery runtime and recharge consistency. In terms of catalyst design, C. Huang et al (Huang et al., 2022). synthesized Ru-incorporated Co_3O_4 nanoparticles that dramatically improved both the oxygen reduction (ORR) and evolution reactions (OER). Used in a Zn-air battery, this material delivered a specific capacity of 788.1 mAh g^{-1} and power density of 101.2 mW cm^{-2}, outperforming commercial Pt/C systems and maintaining long-term cycling stability. The catalyst enhanced oxygen mobility and surface reactivity contributed to lower voltage gaps and increased round-trip efficiency. Lastly, Z. Chen et al (Chen et al., 2017). developed a sulfur/cobalt dual-modified electrolytic carbon (E-carbon) catalyst derived from CO_2. This catalyst exhibited electrocatalytic activity comparable to Pt/C, while maintaining higher methanol tolerance and stability during extended cycling. Their approach, which integrates waste-derived carbon and sustainable sulfidation chemistry, offers a dual benefit: high cycling efficiency and low environmental impact. Y. Zhou et al (Zhou et al., 2024). developed a superhydrophilic carbon fiber electrode (CF-O) functionalized with oxygen-containing groups, intended as a metal-free bifunctional air electrode for all-solid-state Zn-air batteries. Their comparative analysis showed that CF-O exhibited superior ORR and OER activity compared to untreated CF, with a low overpotential of 365.6 mV (OER) and an ORR peak at 0.683 V. The assembled batteries achieved a voltage gap of 1.07 V and a peak volume power density of 82.8 mW cm^{-3}, demonstrating excellent cycling stability and reversibility over 30 cycles. This low-cost and scalable approach underscores the viability of non-metallic electrode materials in advancing durable metal-air battery technologies. Z. Sun et al (Sun et al., 2025). investigated copper-zinc solid solution electrodes in the context of metal-air systems optimized for Ca-ion storage, demonstrating over 700 charge-discharge cycles at room temperature. This study provides a benchmark for comparing high-stability electrode systems, particularly in terms of cycle durability, voltage retention, and energy density reinforcing the relevance of multi-metal electrode integration in extending cycling lifetimes of air batteries.

These findings illustrate the multifactorial nature of charge-discharge performance in metal-air systems. Whether through electrolyte engineering, catalyst innovation, or interface stabilization, the optimization of each component contributes to a more reversible, stable, and high-performing cycling process.

CONCLUSION AND PERSPECTIVES

This chapter presents an in-depth critical review of metal-air battery technologies, focusing on the mechanisms governing their performance and the main obstacles limiting their large-scale deployment. Li-air, Zn-air, Na-air and Al-air batteries all promise exceptionally high energy densities, with Li-air systems reaching up to 3,500 Wh/kg and Al-air around 2,790 Wh/kg. However, their practical application is hampered by recurring problems such as corrosion, passivation and hydrogen evolution at the anode. Among the examined systems, Zn-air batteries currently offer the most balanced profile, with over 300 stable charge-discharge cycles and coulombic efficiencies in excess of 90%. Li-air batteries, on the other hand, remain limited by a short service life (<50 cycles), while Al-air systems suffer from rapid degradation due to intense hydrogen evolution, considerably reducing their efficiency and operational life.

Recent advances have proposed effective strategies to overcome these limitations. Surface-modified anodes, such as nickel-coated aluminum and Al_2O_3-modified zinc, have shown improved corrosion resistance and stability. Similarly, the use of specific electrolyte additives in alkaline solutions has reduced hydrogen evolution and improved anode utilization, particularly in aluminum-based systems. The improvements in catalytic interfaces, particularly in Zn-air batteries, have also improved cycle stability and voltage efficiency.

REFERENCES

Abd El Haleem, S., Wanees, S., & Farouk, A. (2021). Hydrogen Production on Aluminum in Alkaline Media. *Protection of Metals and Physical Chemistry of Surfaces, 57*(5), 906–916. DOI: 10.1134/S2070205121050099

Achanta, K. S. K., Prenzel, T., Winter, J., Waldvogel, S. R., & Tessonnier, J.-P. (2024). Mitigating the HER-Promoted Corrosion of Lead Cathodes during Electrosynthesis. ECS Meeting Abstracts, MA2024-01(56), 2985. https://doi.org/DOI: 10.1149/MA2024-01562985mtgabs

Adair, K. R., Iqbal, M., Wang, C., Zhao, Y., Banis, M. N., Li, R., Zhang, L., Yang, R., Lu, S., & Sun, X. (2018). Towards high performance Li metal batteries: Nanoscale surface modification of 3D metal hosts for pre-stored Li metal anodes. *Nano Energy, 54*, 375–382. DOI: 10.1016/j.nanoen.2018.10.002

Ahmed, A. M., Amin, S., & Mohamed, A. (2024). Current and emerging trends of inorganic, organic and eco-friendly corrosion inhibitors. *RSC Advances, 14*(43), 31877–31920. DOI: 10.1039/D4RA05662K

Al Bostami, R. D., Al Othman, A., Tawalbeh, M., & Olabi, A. G. (2025). Advancements in zinc-air battery technology and water-splitting. *Energy Nexus, 17*, 100387. DOI: 10.1016/j.nexus.2025.100387

Ambat, R., & Piotrowska, K. (2022). Basic theory of corrosion and important failure mechanisms connected to corrosion in electronics. In Ambat, R., & Piotrowska, K. (Eds.), *Humidity and Electronics* (pp. 19–91). Woodhead Publishing., DOI: 10.1016/B978-0-323-90853-5.00002-5

Antony Jose, S., Doering, E., Klein, N., Mena, E. I., Owens, C., Pronk, S., & Menezes, P. L. (2025). Magnesium–Air Batteries : Manufacturing, Processing, Performance, and Applications. *Processes (Basel, Switzerland), 13*(3), 607. Advance online publication. DOI: 10.3390/pr13030607

Areola, R. I., Adebiyi, A. A., Moloi, K. (2025). Integrated Energy Storage Systems for Enhanced Grid Efficiency: A Comprehensive Review of Technologies and Applications. *Energies (19961073), 18*(7). https://doi.org/DOI: 10.3390/en18071848

Aslanbas, Ö., Durmus, Y. E., Tempel, H., Hausen, F., Ein-Eli, Y., Eichel, R. A., & Kungl, H. (2018). Electrochemical analysis and mixed potentials theory of ionic liquid-based Metal–Air batteries with Al/Si alloy anodes. *Electrochimica Acta, 276*, 399–411. DOI: 10.1016/j.electacta.2018.04.176

Bai, A., Wang, L., Li, J., He, X., Wang, J., & Wang, J. (2015). Composite of graphite/phosphorus as anode for lithium-ion batteries. *Journal of Power Sources, 289*, 100–104. Advance online publication. DOI: 10.1016/j.jpowsour.2015.04.168

Bao, F., Kemppainen, E., Dorbandt, I., Bors, R., Xi, F., Schlatmann, R., van de Krol, R., & Calnan, S. (2021). Understanding the Hydrogen Evolution Reaction Kinetics of Electrodeposited Nickel-Molybdenum in Acidic, Near-Neutral, and Alkaline Conditions. *ChemElectroChem*, *8*(1), 195–208. DOI: 10.1002/celc.202001436

Barman, M., Pal, M., Biswas, R., & Dutta, A. (2025). A comprehensive review of metal-air batteries : Mechanistic aspects, advantages and challenges. *Catalysis Today*, *451*, 115229. DOI: 10.1016/j.cattod.2025.115229

Barman, M., Pal, M., Biswas, R., & Dutta, A. (2025). A comprehensive review of metal-air batteries: Mechanistic aspects, advantages and challenges. *Catalysis Today*, *115229*, 115229. Advance online publication. DOI: 10.1016/j.cattod.2025.115229

Behara, S. S., Thomas, J., & Van der Ven, A. (2024). Fundamental Thermodynamic, Kinetic, and Mechanical Properties of Lithium and Its Alloys. *Chemistry of Materials*, *36*(15), 7370–7387. DOI: 10.1021/acs.chemmater.4c01279

Bouguern, M. D., M R, A. K., & Zaghib, K. (2024). The critical role of interfaces in advanced Li-ion battery technology : A comprehensive review. *Journal of Power Sources*, *623*, 235457. DOI: 10.1016/j.jpowsour.2024.235457

Branco, H., Castro, R., & Lopes, A. S. (2018). Battery energy storage systems as a way to integrate renewable energy in small isolated power systems. *Energy for Sustainable Development*, *43*, 90–99. DOI: 10.1016/j.esd.2018.01.003

Bruce, P. G., Freunberger, S. A., Hardwick, L. J., & Tarascon, J. M. (2012). Li–O2 and Li–S batteries with high energy storage. *Nature Materials*, *11*(1), 19–29. DOI: 10.1038/nmat3191

Cai, P., Chen, K., Lu, Z., Mondal, R., Thotiyl, M. O., & Wen, Z. (2023). Aqueous OH−/H+ Dual-Ion Zn-Based Batteries. *ChemSusChem*, *16*(4), e202201034. DOI: 10.1002/cssc.202201034

Calabrese, M., Portarapillo, M., Di Nardo, A., Venezia, V., Turco, M., Luciani, G., & Di Benedetto, A. (2024). Hydrogen Safety Challenges : A Comprehensive Review on Production, Storage, Transport, Utilization, and CFD-Based Consequence and Risk Assessment. *Energies*, *17*(6), 6. Advance online publication. DOI: 10.3390/en17061350

Cao, X., Yan, W., Jin, C., Tian, J., Ke, K., & Yang, R. (2015). Surface modification of MnCo2O4 with conducting polypyrrole as a highly active bifunctional electrocatalyst for oxygen reduction and oxygen evolution reaction. *Electrochimica Acta*, *180*, 788–794. DOI: 10.1016/j.electacta.2015.08.160

Chan, K.-Y., & Savinell, R. F. (1991). Modeling Calculations of an Aluminum-Air Cell. *Journal of the Electrochemical Society*, *138*(7), 1976–1984. DOI: 10.1149/1.2085911

Che, S., Liu, H., Li, K., Li, B., & Li, X. (2025). Advances in high-performance aluminum-air batteries: A comprehensive review of electrode and electrolyte engineering strategies. *Journal of Electroanalytical Chemistry (Lausanne, Switzerland)*, *119400*, 119400. Advance online publication. DOI: 10.1016/j.jelechem.2025.119400

Chen, D., Zhang, X., Zhang, Y., Liu, Z., Deng, F., & Yu, Y. (2023). Si protected by metal-organic segments as anodes in Si-air batteries. *Surfaces and Interfaces*, *38*, 102777. DOI: 10.1016/j.surfin.2023.102777

Chen, J., Peng, X., Zhang, H., Kang, L., Jun, S. C., Yamauchi, Y., & Liu, S. (2025). Tailoring interfacial chemistry of aluminum alloy anodes for high-performance aqueous aluminum-ion batteries. *Progress in Materials Science*, *101611*. Advance online publication. DOI: 10.1016/j.pmatsci.2025.101611

Chen, Z., Gu, Y., Du, K., Wang, X., Xiao, W., Mao, X., & Wang, D. (2017). Enhanced electrocatalysis performance of amorphous electrolytic carbon from CO_2 for oxygen reduction by surface modification in molten salt. *Electrochimica Acta*, *253*, 248–256. DOI: 10.1016/j.electacta.2017.09.053

Cheng, X., Li, D., Jiang, Y., Huang, F., & Li, S. (2023). Nanostructure Engineering of Alloy-Based Anode Materials with Different Dimensions for Sodium/Potassium Storage. *Coatings*, *13*(12), 12. Advance online publication. DOI: 10.3390/coatings13122088

Di Somma, M., Graditi, G. (2022). Challenges and opportunities of the energy transition and the added value of energy systems integration. *Technologies for integrated energy systems and networks*, 1-14. https://doi.org/DOI: 10.1002/9783527833634.ch1

Ding, J., Li, X., Gong, L., & Tan, P. (2024). Investigating the failure mechanism of solid electrolyte interphase in silicon particles from an electrochemical-mechanical coupling perspective. *Advanced Powder Materials*, *3*(4), 100200. DOI: 10.1016/j.apmate.2024.100200

Du, H., Wang, Y., Kang, Y., Zhao, Y., Tian, Y., Wang, X., Tan, Y., Liang, Z., Wozny, J., Li, T., Ren, D., Wang, L., He, X., Xiao, P., Mao, E., Tavajohi, N., Kang, F., & Li, B. (2024). Side Reactions/Changes in Lithium-Ion Batteries : Mechanisms and Strategies for Creating Safer and Better Batteries. *Advanced Materials*, *36*(29), 2401482. DOI: 10.1002/adma.202401482

Du, Y., Li, R., Wang, T., Feng, Z., Dong, H., Fan, F., Deng, D., Zhu, J., Liu, Y., Dai, L., Wang, L., & He, Z. (2024). Dual effect of organic/inorganic artificial protective layer to construct high-performance zinc metal anode. *Chemical Engineering Journal*, *486*, 150139. DOI: 10.1016/j.cej.2024.150139

El-Alouani, M., Kharbouch, O., Dahmani, K., Errahmany, N., Saber, I., Galai, M., Benzekri, Z., Boukhris, S., Ebn Touhami, M., Al-Sadoon, M. K., & Al-Maswari, B. M. (2025). Enhancing Al–Air Battery Performance with Beta-d-Glucose and Adonite Additives: A Combined Electrochemical and Theoretical Study. *Langmuir*, *41*(1), 431–449. DOI: 10.1021/acs.langmuir.4c03720

El-Alouani, M., Kharbouch, O., Dahmani, K., Galai, M., Dkhireche, N., Benzekri, Z., Boukhris, S., Ebn Touhami, M., Almeer, R., Chafiq, M., & Chaouiki, A. (2025). Probing corrosion protective mechanism of 2- (4-nitrobenzylidene) malononitrile on anode for enhanced alkaline (4 M NaOH) Aluminum -air battery performance. *Journal of Alloys and Compounds*, *1010*, 178239. DOI: 10.1016/j.jallcom.2024.178239

Ershov, B. G. (2024). Electrode potentials involving atoms and ions of metals in intermediate oxidation states in aqueous solutions. *Journal of Molecular Liquids*, *402*, 124770. DOI: 10.1016/j.molliq.2024.124770

Fang, C., Wang, X., & Meng, Y. S. (2019). Key Issues Hindering a Practical Lithium-Metal Anode. *Trends in Chemistry*, *1*(2), 152–158. DOI: 10.1016/j.trechm.2019.02.015

Feyzi, E., M R, A. K., Li, X., Deng, S., Nanda, J., & Zaghib, K. (2024). A comprehensive review of silicon anodes for high-energy lithium-ion batteries : Challenges, latest developments, and perspectives. *Next Energy*, *5*, 100176. DOI: 10.1016/j.nxener.2024.100176

Figueira, R. B., Silva, C. J. R., & Pereira, E. V. (2015). Organic–inorganic hybrid sol–gel coatings for metal corrosion protection : A review of recent progress. *Journal of Coatings Technology and Research*, *12*(1), 1–35. DOI: 10.1007/s11998-014-9595-6

Gaele, M. F., & Di Palma, T. M. (2022). Polymer electrolytes for Al-air batteries: Current state and future perspectives. *Energy & Fuels*, *36*(21), 12875–12895. DOI: 10.1021/acs.energyfuels.2c02453

Gao, Y., Zhao, Q., Liu, W., Guo, F., Yin, Y., Zheng, X., Wu, J., Qiang, Y., & Wang, W. (2025). A green electrolyte additive based on Toona Sinensis extract for enhanced performance of alkaline Al-air battery. *Corrosion Science*, *252*, 112952. DOI: 10.1016/j.corsci.2025.112952

Gebremariam, G., Jovanović, A., & Pasti, I. (2023). The Effect of Electrolytes on the Kinetics of the Hydrogen Evolution Reaction. *Hydrogen*, *4*(4), 776–806. DOI: 10.3390/hydrogen4040049

Gelman, D., Shvartsev, B., & Ein-Eli, Y. (2016). Challenges and Prospect of Non-aqueous Non-alkali (NANA) Metal–Air Batteries. *Topics in Current Chemistry (Cham)*, *374*(6), 82. DOI: 10.1007/s41061-016-0080-9

Guo, Y., Zhao, R., Xu, Z., & Lai, C. Q. (2025). Enhancing Zn anode stability with bioderived electrolyte additive for aqueous Zn-ion batteries. *Journal of Power Sources*, *643*, 237071. DOI: 10.1016/j.jpowsour.2025.237071

Hasan, M. M., Haque, R., Jahirul, M. I., Rasul, M. G., Fattah, I. M. R., Hassan, N. M. S., & Mofijur, M. (2025). Advancing energy storage : The future trajectory of lithium-ion battery technologies. *Journal of Energy Storage*, *120*, 116511. DOI: 10.1016/j.est.2025.116511

He, Z., Xiong, F., Tan, S., Yao, X., Zhang, C., & An, Q. (2021). Iron metal anode for aqueous rechargeable batteries. *Materials Today. Advances*, *11*, 100156. DOI: 10.1016/j.mtadv.2021.100156

Heise, G. W. (1933). *U.S. Patent No. 1,899,615*. U.S. Patent and Trademark Office., https://patents.google.com/patent/US1899615A/

Hossain, M., Chowdhury, M. A., Hossain, N., Islam, M. A., & Mobarak, M. H. (2023). Advances of lithium-ion batteries anode materials—A review. *Chemical Engineering Journal Advances*, *16*, 100569. DOI: 10.1016/j.ceja.2023.100569

Huang, C., Ji, Q., Zhang, H., Wang, Y., Wang, S., Liu, X., Guo, Y., & Zhang, C. (2022). Ru- incorporated Co3O4 nanoparticles from self-sacrificial ZIF-67 template as efficient bifunctional electrocatalysts for rechargeable metal-air battery. *Journal of Colloid and Interface Science*, *606*, 654–665. DOI: 10.1016/j.jcis.2021.08.046

Huang, X., Dai, Q., Xiang, Q., Yang, N., Zhang, G., Shen, A., & Li, W. (2024). Microstructure design of advanced magnesium-air battery anodes. *Journal of Magnesium and Alloys*, *12*(2), 443–464. DOI: 10.1016/j.jma.2024.01.025

Imanishi, N., & Yamamoto, O. (2019). Perspectives and challenges of rechargeable lithium–air batteries. *Materials Today. Advances*, *4*, 100031. DOI: 10.1016/j.mtadv.2019.100031

Isse, A., & Gennaro, A. (2010). Absolute Potential of the Standard Hydrogen Electrode and the Problem of Interconversion of Potentials in Different Solvents. *The Journal of Physical Chemistry B*, *114*(23), 7894–7899. DOI: 10.1021/jp100402x

Ji, Y., Li, J., & Li, J. (2022). Recent development of electrolyte engineering for sodium metal batteries. *Batteries*, *8*(10), 157. DOI: 10.3390/batteries8100157

Jia, J., Zhang, H., Zhou, Y., Zhang, C., Sun, Y., Gao, J., Qin, Z., Tian, R., & Gao, Y. (2024). Stability and performance of lithium metal batteries using silicon and silver alloyed separator surfaces. *International Journal of Electrochemical Science*, *19*(9), 100742. DOI: 10.1016/j.ijoes.2024.100742

Kam, L., Maier, T., & Krischer, K. (2024). Electrolyte effects on the alkaline hydrogen evolution reaction : A mean-field approach. *Electrochimica Acta*, *497*, 144530. DOI: 10.1016/j.electacta.2024.144530

Khan, F. M. N. U., Rasul, M. G., Sayem, A. S. M., & Mandal, N. K. (2023). Design and optimization of lithium-ion battery as an efficient energy storage device for electric vehicles : A comprehensive review. *Journal of Energy Storage*, *71*, 108033. DOI: 10.1016/j.est.2023.108033

Khan, S. Y., Noor, T., Iqbal, N., & Ali, Z. (2025). A review on the development of perovskite based bifunctional electrocatalysts for oxygen electrodes in metal-air batteries. *Materials Research Bulletin*, *183*, 113189. DOI: 10.1016/j.materresbull.2024.113189

Kim, Y. J., & Ryu, K.-S. (2019). The surface-modified effects of Zn anode with CuO in Zn-air batteries. *Applied Surface Science*, *480*, 912–922. DOI: 10.1016/j.apsusc.2019.02.153

Kube, A., & Kopljar, D. (2023). Challenges in Metal-Air Batteries. In *Metal-Air Batteries* (pp. 361–374). CRC Press. DOI: 10.1201/9781003295761-25

Lakhnot, A. S., Panchal, R. A., Datta, J., Mahajani, V., Bhimani, K., Jain, R., Datta, D., & Koratkar, N. (2023). Intercalation Hosts for Multivalent-Ion Batteries. *Small Structures*, *4*(4), 2200290. DOI: 10.1002/sstr.202200290

Le, S. T., Nguyen, T. N., Linforth, S., & Ngo, T. D. (2023). Safety investigation of hydrogen energy storage systems using quantitative risk assessment. *International Journal of Hydrogen Energy*, *48*(7), 2861–2875. DOI: 10.1016/j.ijhydene.2022.10.082

Lee, D. U., Xu, P., Cano, Z. P., Kashkooli, A. G., Park, M. G., & Chen, Z. (2016). Recent progress and perspectives on bi-functional oxygen electrocatalysts for advanced rechargeable metal–air batteries. *Journal of Materials Chemistry. A, Materials for Energy and Sustainability*, *4*(19), 7107–7134. DOI: 10.1039/C6TA00173D

Lee, S. M., Kim, Y. J., Eom, S. W., Choi, N. S., Kim, K. W., & Cho, S.-B. (2013). Improvement in self-discharge of Zn anode by applying surface modification for Zn–air batteries with high energy density. *Journal of Power Sources*, *227*, 177–184. DOI: 10.1016/j.jpowsour.2012.11.046

Lei, Y., Wang, K., Jiang, S., Xu, X., Zheng, J., Yin, J., & Gao, Y. (2024). Recent Progress on Multifunctional Electrolyte Additives for High-Energy-Density Li Batteries – A Review. *ChemElectroChem*, *11*(14), e202300702. DOI: 10.1002/celc.202300702

Leuaa, P., & Chatzichristodoulou, C. (2023). HER Kinetics of Pt and Ni in Alkaline Medium at Elevated Temperatures (25-200 oC) and Pressures (1-50 bar). ECS Meeting Abstracts, MA2023-01(36), 2105. https://doi.org/DOI: 10.1149/MA2023-01362105mtgabs

Li, J., Zhang, K., Wang, B., & Peng, H. (2022). Light-Assisted Metal–Air Batteries : Progress, Challenges, and Perspectives. *Angewandte Chemie International Edition*, *61*(51), e202213026. DOI: 10.1002/anie.202213026

Li, J., Zhang, S., Yu, X., Li, J., Gu, H., Long, Y., Li, N., Han, J., & Yang, Z. (2025). Enhancing long-term cycling stability in aqueous zinc-ion batteries via effective control of bulk electrolyte with biomass-derived multifunctional additive. *Nano Energy*, *141*, 111049. DOI: 10.1016/j.nanoen.2025.111049

Li, T., Huang, M., Bai, X., & Wang, Y.-X. (2023). Metal–air batteries : A review on current status and future applications. *Progress in Natural Science*, *33*(2), 151–171. DOI: 10.1016/j.pnsc.2023.05.007

Li, W., Tian, H., Ma, L., Wang, Y., Liu, X., & Gao, X. (2022). Low-Temperature Water Electrolysis : Fundamentals, Progress, and New Strategies. *Materials Advances*, *3*(14), 5598–5644. DOI: 10.1039/D2MA00185C

Liu, H., Zhang, G., Zheng, X., Chen, F., & Duan, H. (2020). Emerging miniaturized energy storage devices for microsystem applications : From design to integration. *International Journal of Extreme Manufacturing*, *2*(4), 042001. DOI: 10.1088/2631-7990/abba12

Liu, X., Guo, Y., Ning, F., Liu, Y., Shi, S., Li, Q., Zhang, J., Lu, S., & Yi, J. (2024). Fundamental Understanding of Hydrogen Evolution Reaction on Zinc Anode Surface : A First-Principles Study. *Nano-Micro Letters*, *16*(1), 111. DOI: 10.1007/s40820-024-01337-0

Liu, Y., & Holze, R. (2022). Metal-Ion Batteries. *Encyclopedia*, *2*(3), 3. Advance online publication. DOI: 10.3390/encyclopedia2030110

Liu, Y., Wang, L., Cao, C., Shang, C., Wang, Z., Wang, H.-E., He, L., Yang, J., Cheng, H., Li, J., & Lu, Z. (2017). Understanding and Suppressing Side Reactions in Li-Air Batteries. *Materials Chemistry Frontiers, 1*(12), 2495–2510. Advance online publication. DOI: 10.1039/C7QM00353F

Liu, Y., Wu, Z., Qin, Z., Liu, Y., & Hu, W. (2024). Recent progress in inhibition of hydrogen evolution reaction in alkaline Al-air batteries. *National Science Open, 3*(6), 20240037. DOI: 10.1360/nso/20240037

Lu, W., Zheng, T., Zhang, X., He, T., Sun, Y., Li, S., Guan, B., Zhang, D., Wei, Z., Jiang, H., Fan, H. J., & Du, F. (2025). Band Engineering of Mn-P Alloy Enables HER-suppressed Aqueous Manganese Ion Batteries. *Angewandte Chemie International Edition, 64*(5), e202417171. DOI: 10.1002/anie.202417171

Lu, Y., Zhu, Y., Chen, Z., Chen, C., Li, X., Yu, H., Peng, K., & Tian, Z. (2024). Boosting the Performance of Aluminum–Air Batteries by Interface Modification. *ACS Applied Materials & Interfaces, 16*(29), 37818–37828. DOI: 10.1021/acsami.4c03156

Luo, M., Yang, J., Li, X., Eguchi, M., Yamauchi, Y., & Wang, Z.-L. (2023). Insights into the alloy/oxide or hydroxide interfaces in Ni-Mo-based electrocatalysts for hydrogen evolution under alkaline conditions. *Chemical Science (Cambridge), 14*(13), 3400–3414. Advance online publication. DOI: 10.1039/D2SC06298D

Lyu, L., Zheng, Y., Hua, Y., Li, J., Yi, Y., Sun, Y., & Xu, Z.-L. (2024). High-Energy Sodium Ion Batteries Enabled by Switching Sodiophobic Graphite into Sodiophilic and High-Capacity Anodes. *Angewandte Chemie International Edition, 63*(48), e202410253. DOI: 10.1002/anie.202410253

Mahmood, N., Yao, Y., Zhang, J.-W., Pan, L., Zhang, X., & Zou, J.-J. (2018). Electrocatalysts for Hydrogen Evolution in Alkaline Electrolytes : Mechanisms, Challenges, and Prospective Solutions. *Advanced Science (Weinheim, Baden-Wurttemberg, Germany), 5*(2), 1700464. DOI: 10.1002/advs.201700464

Ml, D., Lee, Y.-S., & Aravindan, V. (2021). Solvent Co-intercalation: An Emerging Mechanism in Li-, Na-, and K-Ion Capacitors. *ACS Energy Letters, 6*(12), 4228–4244. DOI: 10.1021/acsenergylett.1c01801

Moonngam, S., & Banjongprasert, C. (2025). Enhancing the electrochemical performance of aluminum-zinc alloy anodes by friction stir process for aluminum–air batteries. *Journal of Alloys and Compounds, 1018*, 178784. DOI: 10.1016/j.jallcom.2025.178784

Muhammad Saqib, Q., Mannan, A., Noman, M., Chougale, M. Y., Patil, C. S., Ko, Y., Kim, J., Patil, S. R., Yousuf, M., Ali Shaukat, R., Pyo Jeon, Y., Dubal, D., & Bae, J. (2024). Miniaturizing Power : Harnessing Micro-Supercapacitors for advanced micro-electronics. *Chemical Engineering Journal*, *490*, 151857. DOI: 10.1016/j.cej.2024.151857

Nagmani, P.D., Tyagi, A., & Puravankara, D. S. (2022). Lithium-Ion Battery Technologies for Electric Mobility – State-of-the-Art Scenario. *ARAI Journal of Mobility Technology, 2*(2). https://doi.org/DOI: 10.37285/ajmt.1.2.10

Nandi, S., & Pumera, M. (2025). Materials for aluminum batteries : Progress and challenges. *Chemical Engineering Journal*, *517*, 164283. DOI: 10.1016/j.cej.2025.164283

Nazir, G., Rehman, A., Lee, J.-H., Kim, C.-H., Gautam, J., Heo, K., Hussain, S., Ikram, M., AlObaid, A. A., Lee, S.-Y., & Park, S.-J. (2024). A Review of Rechargeable Zinc–Air Batteries : Recent Progress and Future Perspectives. *Nano-Micro Letters*, *16*(1), 138. DOI: 10.1007/s40820-024-01328-1

Nazir, G., Rehman, A., Lee, J. H., Kim, C. H., Gautam, J., Heo, K., Hussain, S., Ikram, M., AlObaid, A. A., Lee, S.-Y., & Park, S. J. (2024). A review of rechargeable zinc–air batteries: Recent progress and future perspectives. *Nano-Micro Letters*, *16*(1), 138. DOI: 10.1007/s40820-024-01328-1

Nian, Q., Zhang, X., Feng, Y., Liu, S., Sun, T., Ren, X., Tao, Z., Zhang, D., & Chen, J. (2020). Designing electrolyte structure to suppress hydrogen evolution reaction in aqueous batteries. *Research Square*. https://doi.org/DOI: 10.21203/rs.3.rs-95916/v1

Nzereogu, P. U., Omah, A. D., Ezema, F. I., Iwuoha, E. I., & Nwanya, A. C. (2022). Anode materials for lithium-ion batteries : A review. *Applied Surface Science Advances*, *9*, 100233. DOI: 10.1016/j.apsadv.2022.100233

Olabi, A. G., Sayed, E. T., Wilberforce, T., Jamal, A., Alami, A. H., Elsaid, K., Rahman, S. M. A., Shah, S. K., & Abdelkareem, M. A. (2021). Metal-air batteries—A review. *Energies*, *14*(21), 7373. DOI: 10.3390/en14217373

Østergaard, P. A., Duic, N., Noorollahi, Y., & Kalogirou, S. (2020). Latest progress in Sustainable Development using renewable energy technology. *Renewable Energy*, *162*, 1554–1562. DOI: 10.1016/j.renene.2020.09.124

Qian, J., Henderson, W., Xu, W., Bhattacharya, P., Engelhard, M., Borodin, O., & Zhang, J.-G. (2015). High Rate and Stable Cycling of Lithium Metal Anode. *ECS Meeting Abstracts, MA2015-01*, 1155-1155. https://doi.org/DOI: 10.1149/MA2015-01/15/1155

Ran, Y., Dong, F., Sun, S., & Lei, Y. (2025). Aqueous Zinc-Based Batteries : Active Materials, Device Design, and Future Perspectives. *Advanced Energy Materials, 2406139*, 2406139. Advance online publication. DOI: 10.1002/aenm.202406139

Rani, B., Yadav, J. K., Saini, P., Pandey, A. P., & Dixit, A. (2024). Aluminum–air batteries: Current advances and promises with future directions. *RSC Advances, 14*(25), 17628–17663. DOI: 10.1039/D4RA02219J

Raza, B., Shamraiz, U., Hu, M., Xu, X., Yu, B., Jing, S., Tao, Y., & Wang, J. (2025). Advances in Organic electrolytes for High-Performance zinc Batteries : Enhancing zinc anode Robustness and efficiency. *Chemical Engineering Journal, 515*, 163617. DOI: 10.1016/j.cej.2025.163617

Romio, M., Surace, Y., Eshraghi, N., Herzog, B., Eckmann, B., Cupid, D. M., Hoffmann, J., & Jahn, M. (2022). Advanced Investigation of the Electrolyte-Mg Anode Interphase for the Development of Mg-Ion Batteries. *ECS Meeting Abstracts, MA2022-02*(4), 465. https://doi.org/DOI: 10.1149/MA2022-024465mtgabs

Salado, M., & Lizundia, E. (2022). Advances, challenges, and environmental impacts in metal–air battery electrolytes. *Materials Today. Energy, 28*, 101064. DOI: 10.1016/j.mtener.2022.101064

Saleh, H. M., & Hassan, A. I. (2023). Synthesis and Characterization of Nanomaterials for Application in Cost-Effective Electrochemical Devices. *Sustainability (Basel), 15*(14), 14. Advance online publication. DOI: 10.3390/su151410891

Sandstrom, S. K., & Ji, X. (2022). Unlocking the Longevity of the Iron Metal Anode. *ACS Central Science, 8*(6), 686–688. DOI: 10.1021/acscentsci.2c00542

Shabeer, Y., Madani, S. S., Panchal, S., Mousavi, M., & Fowler, M. (2025). Different Metal–Air Batteries as Range Extenders for the Electric Vehicle Market: A Comparative Study. *Batteries, 11*(1), 35. DOI: 10.3390/batteries11010035

Shabeer, Y., Madani, S. S., Panchal, S., Mousavi, M., & Fowler, M. (2025). Different metal–air batteries as range extenders for the electric vehicle market: A comparative study. *Batteries, 11*(1), 35. DOI: 10.3390/batteries11010035

Shi, J., Jiang, K., Fan, Y., Zhao, L., Cheng, Z., Yu, P., Peng, J., & Wan, M. (2024). Advancing Metallic Lithium Anodes : A Review of Interface Design, Electrolyte Innovation, and Performance Enhancement Strategies. *Molecules (Basel, Switzerland), 29*(15), 15. Advance online publication. DOI: 10.3390/molecules29153624

Shinde, S. S., Wagh, N. K., Kim, S., & Lee, J. (2023). Li, Na, K, Mg, Zn, Al, and Ca Anode Interface Chemistries Developed by Solid-State Electrolytes. *Advanced Science (Weinheim, Baden-Wurttemberg, Germany)*, *10*(32), 2304235. DOI: 10.1002/advs.202304235

Shinde, S. S., Wagh, N. K., Kim, S. H., & Lee, J. H. (2023). Li, Na, K, Mg, Zn, Al, and Ca Anode Interface Chemistries Developed by Solid-State Electrolytes. *Advanced Science (Weinheim, Baden-Wurttemberg, Germany)*, *10*(32), 2304235. DOI: 10.1002/advs.202304235

Shittu, E., Suman, R., Ravikumar, M. K., Shukla, A. K., Zhao, G., Patil, S., & Baker, J. (2022). Life cycle assessment of soluble lead redox flow battery. *Journal of Cleaner Production*, *337*, 130503. DOI: 10.1016/j.jclepro.2022.130503

Srivastava, S., Ahuja, D., & Varshney, P. K. (2024). Impact of surface modification of electrode for aluminium air batteries. *Journal of Energy Storage*, *76*, 109588. DOI: 10.1016/j.est.2023.109588

Subramaniam, T., & Idris, M. B., K.s., S., K.s., R., & S., D. (2024). Mitigating hydrogen evolution reaction and corrosion of zinc in electrically rechargeable zinc-air batteries using nanofluid electrolytes. *Journal of Energy Storage*, *81*, 110457. DOI: 10.1016/j.est.2024.110457

Sun, M., Wan, K., Huang, Y., Yang, H., Zhou, X., Yan, C., Zhou, J., & Qian, T. (2025). Playing with Water Molecules : "Repulsing" or "Trapping" to Exclude Water-Induced Side Reactions on Zn Metal Anode. *Advanced Functional Materials*, *35*(13), 2417890. DOI: 10.1002/adfm.202417890

Sun, S.-Y., Zhang, X.-Q., Yan, X.-Y., Zheng, Z., Zhang, Q.-K., & Huang, J.-Q. (2025). *Advances in high-coulombic-efficiency lithium metal anodes under practical conditions in liquid electrolytes*. EES Batteries., DOI: 10.1039/D4EB00034J

Thakur, M., Isaacson, M., Sinsabaugh, S., Wong, M., Sibani, L., & Biswal, S. (2012). Gold-coated porous silicon films as anodes for lithium ion batteries (Jeu de données). In *Journal of Power Sources* (Vol. 205). https://doi.org/DOI: 10.1016/j.jpowsour.2012.01.058

Thatipamula, S., Malaarachchi, C., Alam, M. R., Khan, M. W., Babarao, R., & Mahmood, N. (2024). Non-aqueous rechargeable aluminum-ion batteries (RABs) : Recent progress and future perspectives. *Microstructures*, *4*(4), N/A, N/A. DOI: 10.20517/microstructures.2024.27

Uzun, E., Güler, M. O., Güler, A., Kuruahmet, D., Yıldırım Gültekin, S., Güngör, H., & Singil, M. M. (2025). Enhanced performance of lithium-ion battery anodes using CuO nanorod/graphene aerogel composites. *Diamond and Related Materials*, *155*, 112356. DOI: 10.1016/j.diamond.2025.112356

Wang, F., Borodin, O., Ding, M. S., Gobet, M., Vatamanu, J., Fan, X., Gao, T., Eidson, N., Liang, Y., Sun, W., Greenbaum, S., Xu, K., & Wang, C. (2018). Hybrid Aqueous/Non-aqueous Electrolyte for Safe and High-Energy Li-Ion Batteries. *Joule*, *2*(10), 2178. DOI: 10.1016/j.joule.2018.09.018

Wang, L., Snihirova, D., Deng, M., Vaghefinazari, B., Xu, W., Höche, D., Lamaka, S. V., & Zheludkevich, M. L. (2022). Sustainable aqueous metal-air batteries: An insight into electrolyte system. *Energy Storage Materials*, *52*, 573–597. DOI: 10.1016/j.ensm.2022.08.032

Wang, M., Zhang, F., Lee, C.-S., & Tang, Y. (2017). Low-Cost Metallic Anode Materials for High Performance Rechargeable Batteries. *Advanced Energy Materials*, *7*(23), 1700536. DOI: 10.1002/aenm.201700536

Wang, Z., Hu, S., Wang, D., Huang, J., Qi, J., Liu, H., Li, X., Han, C., & Cheng, H.-M. (2025). A HER-Inhibiting Layer Based on M-H Bond Regulation for Achieving Stable Zinc Anodes in Aqueous Zinc-Ion Batteries. *Advanced Functional Materials*, *2502186*(34), 2502186. Advance online publication. DOI: 10.1002/adfm.202502186

Wang, Z., Ni, Z., Chen, J., Dai, Y., Gao, Y., Zhang, Q., Dong, F., Xiong, S., Zhang, C., & Feng, J. (2025). Recent progress and challenges on emerging high-entropy materials for better Zn-Air and Zn-Ion batteries. *Energy Storage Materials*, *75*, 104064. DOI: 10.1016/j.ensm.2025.104064

Wang, Z., Wu, Z., Liu, Y., Zhang, J., Yang, Y., Xu, J., Zhao, T., Qin, Z., & Hu, W. (2024). Hydrogen Evolution and Electrochemical Behavior of Aluminum in Al-Air Batteries with Concentrated Naoh Electrolytes (SSRN Scholarly Paper 4869814). Social Science Research Network. https://doi.org/DOI: 10.2139/ssrn.4869814

Weinrich, H., Pleie, J., Schmid, B., Tempel, H., Kungl, H., & Eichel, R.-A. (2022). In Situ Hydrogen Evolution Monitoring During the Electrochemical Formation and Cycling of Pressed-Plate Carbonyl Iron Electrodes in Alkaline Electrolyte. *Batteries & Supercaps*, *5*(5), e202100415. DOI: 10.1002/batt.202100415

Wittman, R., Sacci, R., & Zawodzinski, T. (2025). Alkaline Zinc Passivation Mechanism is Controlled by Hydroxide Concentration. Journal of The Electrochemical Society, 172. https://doi.org/DOI: 10.1149/1945-7111/adc6c7

Wu, P., Zhao, Q., Yu, H., Tang, Z., Li, Y., Huang, D., Sun, D., Wang, H., & Tang, Y. (2022). Modification on water electrochemical environment for durable Al-Air Battery: Achieved by a Low-Cost sucrose additive. *Chemical Engineering Journal*, *438*, 135538. DOI: 10.1016/j.cej.2022.135538

Wulandari, T., Fawcett, D., Majumder, S. B., & Poinern, G. E. J. (2023). Lithium-based batteries, history, current status, challenges, and future perspectives. *Battery Energy*, *2*(6), 20230030. DOI: 10.1002/bte2.20230030

Xu, S., & Liu, J. (2019). Metal-based direct hydrogen generation as unconventional high density energy. *Frontiers in Energy*, *13*(1), 27–53. DOI: 10.1007/s11708-018-0603-x

Xu, W., Lu, Z., Zhang, T., Zhong, Y., Wu, Y., Zhang, G., Liu, J., Wang, H., & Sun, X. (2019). An advanced zinc air battery with nanostructured superwetting electrodes. *Energy Storage Materials*, *17*, 358–365. DOI: 10.1016/j.ensm.2018.06.029

Xu, W., Welty, C., Peterson, M. R., Read, J. A., & Stadie, N. P. (2022). Exploring the Limits of the Rapid-Charging Performance of Graphite as the Anode in Lithium-Ion Batteries. *Journal of the Electrochemical Society*, *169*(1), 010531. DOI: 10.1149/1945-7111/ac4b87

Xue, T., Mu, Y., Wei, X., Zhou, Z., Zhou, Y., Zhang, Z., Yang, C., Qiu, J., Zang, L., & Zeng, L. (2025). From Fundamentals to Practice : Electrolyte Strategies for Zinc-Ion Batteries in Extreme Temperature. *Carbon Neutralization*, *4*(1), e183. DOI: 10.1002/cnl2.183

Yanagisawa, A. (1999). Magnesium(II) and other alkali and alkaline earth metals. *ChemInform*, *30*(32), chin.199932321. Advance online publication. DOI: 10.1002/chin.199932321

Yang, C., Jiang, Z., Chen, X., Luo, W., Zhou, T., & Yang, J. (2024). Lithium Metal Based Battery Systems beyond 500 Wh Kg-1. *Chemical Communications*. Advance online publication. DOI: 10.1039/D4CC03177F

Yang, H., & Wu, N. (2022). Ionic conductivity and ion transport mechanisms of solid-state lithium-ion battery electrolytes : A review. *Energy Science & Engineering*, *10*(5), 1643–1671. DOI: 10.1002/ese3.1163

Yang, J., Wachtler, M., Winter, M., & Besenhard, J. O. (1999). Sub-Microcrystalline Sn and Sn - SnSb Powders as Lithium Storage Materials for Lithium-Ion Batteries. *Electrochemical and Solid-State Letters*, *2*(4), 161. DOI: 10.1149/1.1390769

Yang, S., Min, X., Fan, H., Xiao, J., Liu, Y., Mi, R., Wu, X., Huang, Z., Xi, K., & Fang, M. (2022). In situ characterization of lithium-metal anodes. *Journal of Materials Chemistry. A, Materials for Energy and Sustainability*, *10*(35), 17917–17947. DOI: 10.1039/D2TA04309B

Yang, S., Zhao, Y., & Zhi, C. (2025). Insights into the role of electrolyte additives for stable Zn anodes. *Energy Materials*, *5*(2), N/A, N/A. DOI: 10.20517/energymater.2024.169

Yaqoob, L., Noor, T., & Iqbal, N. (2022). An overview of metal-air batteries, current progress, and future perspectives. *Journal of Energy Storage*, *56*, 106075. DOI: 10.1016/j.est.2022.106075

Ye, J., Wang, C., Shao, Z., & Liao, K. (2024). From Li–O_2 to Li–Air Batteries: Challenges and Progress on Oxygen-Permeable Membranes. *Energy & Fuels*, *38*(16), 15001–15024. DOI: 10.1021/acs.energyfuels.4c02606

Ye, X., Xiao, X., Wu, Z., Zhan, Y., Wu, X., & Liu, S. (2024). Recent advances in rechargeable aqueous magnesium-ion batteries. *Journal of Materials Chemistry. A, Materials for Energy and Sustainability*, *12*(35), 23337–23363. DOI: 10.1039/D4TA03970J

Yi, X., Qi, G., Liu, X., Depcik, C., & Liu, L. (2024). Challenges and strategies toward anode materials with different lithium storage mechanisms for rechargeable lithium batteries. *Journal of Energy Storage*, *95*, 112480. DOI: 10.1016/j.est.2024.112480

Yu, Y., Gao, S., & Hu, S. (2021). Si modified by Zn and Fe as anodes in Si-air batteries with ameliorative properties. *Journal of Alloys and Compounds*, *883*, 160902. DOI: 10.1016/j.jallcom.2021.160902

Yu, Y., Hu, S., & Huang, J. (2021). Germanium-modified silicon as anodes in Si–Ge air batteries with enhanced properties. *Journal of Physics and Chemistry of Solids*, *157*, 110226. DOI: 10.1016/j.jpcs.2021.110226

Zahoor, A., Ghouri, Z. K., Hashmi, S., Raza, F., Ishtiaque, S., Nadeem, S., Ullah, I., & Nahm, K. S. (2019). Electrocatalysts for lithium–air batteries: Current status and challenges. *ACS Sustainable Chemistry & Engineering*, *7*(17), 14288–14320. DOI: 10.1021/acssuschemeng.8b06351

Zhai, J., Yan, J., Wang, G., Chen, S., Jin, D., Zhang, H., Zhao, W., Zhang, Z., & Huang, W. (2021). Surface modification of manganese monoxide through chemical vapor deposition to attain high energy storage performance for aqueous zinc–ion batteries. *Journal of Colloid and Interface Science*, *601*, 617–625. DOI: 10.1016/j.jcis.2021.05.142

Zhan, T., Kumar, A., Sevilla, M., Sridhar, A., & Zeng, X. (2020). Temperature Effects on CO 2 Electroreduction Pathways in an Imidazolium-Based Ionic Liquid on Pt Electrode. *The Journal of Physical Chemistry. C, Nanomaterials and Interfaces*, *124*(48), 26094–26105. DOI: 10.1021/acs.jpcc.0c06065

Zhang, H., Wang, L., Li, H., & He, X. (2021). Criterion for Identifying Anodes for Practically Accessible High-Energy-Density Lithium-Ion Batteries. *ACS Energy Letters*, *6*(10), 3719–3724. DOI: 10.1021/acsenergylett.1c01713

Zhang, J., Zhou, Q., Tang, Y., Zhang, L., & Li, Y. (2019). Zinc–air batteries: Are they ready for prime time. *Chemical Science (Cambridge)*, *10*(39), 8924–8929. DOI: 10.1039/C9SC04221K

Zhang, W.-J. (2011). A review of the electrochemical performance of alloy anodes for lithium-ion batteries. *Journal of Power Sources*, *196*(1), 13–24. DOI: 10.1016/j.jpowsour.2010.07.020

Zhang, Y., Cao, R., Ouyang, C., Jiang, L., Wang, Y., Yang, M., & Xia, H. (2025). Recent progress in alkali metal (Li/Na/K) hybrid-ion batteries : Pioneering the future of energy storage. *Journal of Materials Chemistry. A, Materials for Energy and Sustainability*, *13*(6), 3973–3990. DOI: 10.1039/D4TA06190J

Zhang, Z., Gao, L., Sun, X., Deng, N., Zhao, Y., & Kang, W. (2024). Matching strategy between sulfide solid electrolyte and various anodes : Electrolyte modification, interface engineering and electrode structure design. *Energy Storage Materials*, *69*, 103422. DOI: 10.1016/j.ensm.2024.103422

Zhang, Z., Li, Y., Yin, X., Li, S., Li, B., Zhao, N., Zhu, J., Dai, L., Wang, L., He, Z., & Feng, Z. (2025). Recent advances in cellulose-based separators for zinc-based batteries : Performances, mechanism and perspectives. *Green Energy & Environment*, *10*(10), 2029–2046. Advance online publication. DOI: 10.1016/j.gee.2025.03.010

Zhang, Z., Wu, Y., Mo, Z., Lei, X., Xie, X., Xue, X., Qin, H., & Jiang, H. (2025). Research progress of silicon-based anode materials for lithium-ion batteries. *RSC Advances*, *15*(14), 10731–10753. DOI: 10.1039/D5RA01268F

Zhao, C., Lu, Y., Yue, J., Pan, D., Qi, Y., Hu, Y.-S., & Chen, L. (2018). Advanced Na metal anodes. *Journal of Energy Chemistry*, *27*(6), 1584–1596. DOI: 10.1016/j.jechem.2018.03.004

Zhao, L., Li, Y., Yu, M., Peng, Y., & Ran, F. (2023). Electrolyte-Wettability Issues and Challenges of Electrode Materials in Electrochemical Energy Storage, Energy Conversion, and Beyond. *Advanced Science (Weinheim, Baden-Wurttemberg, Germany)*, *10*(17), 2300283. DOI: 10.1002/advs.202300283

Zhao, X., Chen, F., Liu, J., Cheng, M., Su, H., Liu, J., & Xu, Y. (2020). Enhanced surface binding energy regulates uniform potassium deposition for stable potassium metal anodes. *Journal of Materials Chemistry. A, Materials for Energy and Sustainability*, *8*(11), 5671–5678. DOI: 10.1039/C9TA14226F

Zhong, C., Weng, S., Wang, Z., Zhan, C., & Wang, X. (2023). Kinetic limits and enhancement of graphite anode for fast-charging lithium-ion batteries. *Nano Energy*, *117*, 108894. DOI: 10.1016/j.nanoen.2023.108894

Zhou, C., Bhonge, K., & Cho, K. T. (2018). Characterization of Effect of H2 Evolution in Aluminum Air Battery. ECS Meeting Abstracts, MA2018-02(1), 65. https://doi.org/DOI: 10.1149/MA2018-02/1/65

Zhou, C., Bhonge, K., & Cho, K. T. (2019). Effect of H2 Evolution in Aqueous Aluminum Air Battery. ECS Meeting Abstracts, MA2019-01(1), 1. https://doi.org/DOI: 10.1149/MA2019-01/1/1

Zhou, C., Bhonge, K., & Cho, K. T. (2020). Analysis of the effect of hydrogen-evolving side reaction in the aqueous aluminum-air battery. *Electrochimica Acta*, *330*, 135290. DOI: 10.1016/j.electacta.2019.135290

Zhou, C., Huang, Y., Zheng, Y., & Hua, Z. (2024). Hydrogen permeation behavior of rubber sealing materials for hydrogen infrastructure : Recent advances and perspectives. *International Journal of Hydrogen Energy*, *59*, 742–754. DOI: 10.1016/j.ijhydene.2024.02.042

Zhou, Y., Li, Z., Cheng, X., He, Y., Zhong, N., He, X., & Lan, Q. (2024). Carbon fiber with superhydrophilic interface as metal-free air electrode for all-solid-state Zn-air batteries. *Surfaces and Interfaces*, *55*, 105310. DOI: 10.1016/j.surfin.2024.105310

Zuo, Y., Wang, K., Pei, P., Wei, M., Liu, X., Xiao, Y., & Zhang, P. (2021). Zinc dendrite growth and inhibition strategies. *Materials Today. Energy*, *20*, 100692. DOI: 10.1016/j.mtener.2021.100692

Chapter 7
Advances in Plant-Based Materials for Sustainable Electrochemical Energy Devices

Khaoula Mzioud
https://orcid.org/0009-0007-8781-5024
Ibn Tofail University, Morocco

Zakia Aribou
https://orcid.org/0000-0002-5422-7062
Ibn Tofail University, Morocco

Sara Rached
Ibn Tofail University, Morocco

Amar Habsaoui
Ibn Tofail University, Morocco

Mohamed Ebn Touhami
Ibn Tofail University, Morocco

ABSTRACT

Transitions to circular and sustainable energy systems require the design of innovative material platforms that are safe for the environment and highly resource-use efficient. In this respect, plant-derived biomaterials are increasingly recognized as highly promising alternatives to conventional components in batteries, supercapacitors, and fuel cells. This chapter underlines the role that plant-based materials play in the development of greener and more sustainable energy infrastructures

DOI: 10.4018/979-8-3373-3156-0.ch007

by presenting their integration within the above-mentioned energy systems. Major issues such as electrical conductivity, durability, and scalability are comprehensively discussed. The most recent strategies in the realm of material hybridization, system integration, and bioinspiration are reviewed. The fast-developing intersection between natural material science and electrochemical engineering keeps driving innovation to the forefront in this interdisciplinary domain towards next-generation sustainable energy technologies.

1. INTRODUCTION

The transition towards sustainable and circular energy systems is among the major challenges of the 21st century (Asif, 2022), while fossil resources are depleting, demand from both environmental and social perspectives for clean technologies is increasing. In this context the scientific community develops new strategies to meet, in a more intensive way, the energy demand rising globally (Asif, 2024; Kandpal et al., 2024). It is electrochemical energy storage and conversion devices-particularly, batteries and supercapacitors-which are at the frontline of this effort because of their high efficiency and versatility in storing and delivering electrical energy (Shah et al., 2024). However, these systems conventionally have been made with materials mostly sourced from rare or non-renewable sources and usually produced through energy-intensive methods. Their use reveals several long-term sustainability and cost concerns with a high environmental impact (Aazem et al., 2024; Gallo et al., 2022).

In this respect, biobased materials and, mainly, those of plant origin represent a very fascinating option for being an alternative (Chowdhury et al., 2025; Saini et al., 2021; Vinodh et al., 2021), appreciated that with the advancement of material sciences, green chemistry, and biotechnology, they were able to confer unique properties on such materials, among which one could mention renewability, biodegradability, low environmental footprint, tunable physicochemical properties, and several others. For example, as reported by (Chinnappan et al., 2012; L. Zhang et al., 2021) works, the portfolio starting from cellulose, lignin, tannins, polysaccharides, plant-derived oils, and other biomolecules is quite broad in respect of components that can be valorized in an active or functional way within electrochemical systems.

These materials represent not only an ecological and ethical need but also provide technical properties competitive or superior to conventional alternatives quite frequently(Aslan & Altuntaş, 2023; J. Luo & Liu, 2023; Yhobu et al., 2022). Indeed, most biomass precursor-derived activated carbons have shown large specific surface area, hierarchical porosity, and very good charge storage capability, hence appearing as perfect candidates to act as electrodes in supercapacitors (Jain et al., 2021; Sarmah et al., 2024; Srivastava & Yadav, 2023).

Other examples concern biopolymers that, after chemical modification, could play the role of binders or solids. The renewed interest in bio-derived materials falls within the frame of a general paradigm change toward eco-design of energy technologies inspired by circular economics. Apart from material substitution, such an approach would mean a holistic rethinking of the value chain, starting from sourcing the raw material and life cycle assessment through recyclability and process sustainability to end-of-life impact (Lingaitiene & Burinskiene, 2024; Popescu et al., 2024; Ries et al., 2023). Furthermore, using agriculture residues, non-food biomass, and agro-industrial wastes as feedstocks for functional materials enhances economic and ecological viability (J. A. Kumar et al., 2023; Sarangi et al., 2022). However, translating these materials into practical electrochemical systems still remains one of the greatest challenges (M. M. S. Ahmed et al., 2024; Bertaglia et al., 2024). Besides, biomass composition variability, material stability, and compatibility with conventional fabrication processes, a number of challenges are involved in the process that need to be dealt with systematically(Kapoor et al., 2022; Schnepp, 2013; L. K. Shrestha et al., 2023). Other obstacles that need to be overcome involve biomass composition variability, material stability, and compatibility with conventional fabrication processes, most of which have to be approached systematically. Recently, many researchers have been exploring new strategies for overcoming these obstacles, including nanomaterial hybridization, chemical functionalization of biopolymers, and the design of bio-inspired architectures with controlled hierarchical structure(Aslam et al., 2025; Nadar et al., 2025). Such techniques are targeted toward the improvement of functional properties of bio-derived materials for next-generation energy storage devices offering high efficiency. Moreover, certain external factors such as government policies, international climate agreements, and socio-economic matters influence wide diffusion of plant-based energy materials. Appealing policy frameworks-that offer subsidy support, carbon pricing, and incentives for green innovation-enable research, commercialization, and deployment of biomass-based technologies (Mertens et al., 2024; Rajendran et al., 2025; W.-Y. Wu et al., 2024). Socioeconomic factors include public environmental awareness, resource renewability, and opportunities for rural development that quicken the rate of adoption. This includes the improvement in economic viability and reduction of dependence on critical raw materials(Mertens et al., 2024; Rajendran et al., 2025).

This chapter will review recent advances in the development and application of plant-based materials within energy storage and conversion. It defines their potential to become sustainable alternatives to conventional material by integrating performance, resource efficiency, and environmental responsibility. Based on interdisciplinary contributions in materials science, bioengineering, electrochemistry, and green chemistry, this work will outline opportunities and challenges associated with embedding the plant-based material into future landscapes of energy technologies.

2. ENERGY STORAGE DEVICES

Batteries and supercapacitors are the two most promising technologies that afford innovative solutions to meet ever-growing needs in energy storage and access (Yadlapalli et al., 2022; Q. Zhang et al., 2025). Therefore, this section discusses the design, operation, and materials constituting these two devices, in an attempt to understand their role within the frameworks of future energy systems.

2.1 Battery

Batteries represent indispensable energy storage devices owing to their high energy density and long-lasting ability of stored energy retention. Based on a wide range of operating modes of chemical-to-electrical energy conversion, a general battery usually consists of an anode, a cathode, and an electrolyte, which allows the migration of ions across both electrodes, thus producing an electric current (figure 1) (Guan et al., 2016; Mehta et al., 2020; Yue & Liang, 2018).

Figure 1. The electrochemical operating mechanism of a rechargeable battery (Singh et al., 2023)

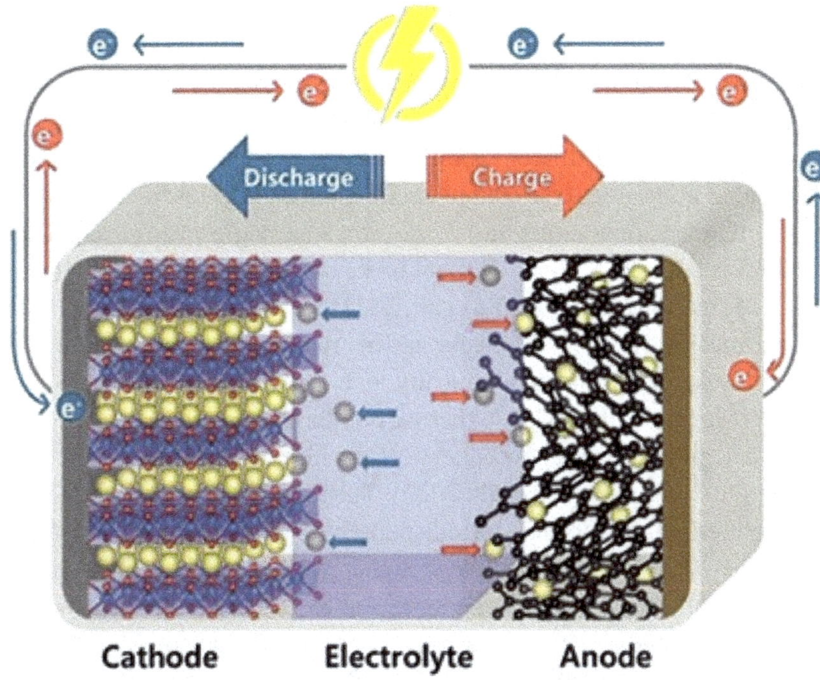

Batteries have been developed using many forms of mobile ions (Q. Chen, 2024; L. Li et al., 2021). Presently, LIBs, nickel-cadmium, and sodium-sulfur batteries control the entire market, but LIBs are the most widely spread because of their outstanding electrochemical properties such as high voltage, long cycle life, and low self-discharge rates, according to (C. Ma et al., 2023; Mehta et al., 2020). In the last years, lithium-ion accumulators have been increasingly becoming the technology of choice for very diverse applications, from low-power portable electronics to large-scale energy infrastructures (Hasan et al., 2025). Their applications, from electric vehicles to stationary storage systems, smart robotics to plug-in hybrid vehicles, are just but a few applications to add on their versatility and excellent energy performance (Abo-Khalil et al., 2023; Günther et al., 2025). LIBs work on the principle of Li^+ ion reversible motion across the two electrodes via an electrolyte. The ions migrate from the cathode to the anode during charging, and the reverse happens during discharge (Belmesov et al., 2023). This is conventionally realized by graphite on the anode because it presents an ideal interlayer spacing for Li^+ intercalation (Nzereogu et al., 2022),

However, graphite has a theoretical specific capacity of only 372 mAh.g-1 (Manyà & Gascó, 2020) Moreover, this makes the moderate efficiencies unsuitable in application fields that are increasingly exigent (Abdollahifar et al., 2022). While lithium-ion batteries outperform fossil fuels in terms of usable energy density, the dependence on limited resources raises growing concerns about their long-term viability (Kim et al., 2019; Lee et al., 2025). It is not too fanciful to consider the fact that global lithium reserves might become the critical resources of tomorrow, much like oil in the previous century. For example, a single electric vehicle battery, such as the Tesla Model S, requires about 64 kg of lithium-a quantity that can serve thousands of smartphones (J. Zhang et al., 2024). All this places enormous demand on strategic materials, mostly expensive and not renewable, hence the urgent need to design more sustainable and economically viable alternatives with a lower critical metal load (Madhu et al., 2023). Other carbon-based materials, including carbon nanotubes and graphene, do show better performance; however, these are expensive in mass production (Belmesov et al., 2023). These limitations are strong indicators that while batteries may currently be indispensable for long-term energy storage, the future depends on developing much safer, higher-performance, and more sustainable systems. Current research therefore focuses on the development of new electrode materials, more stable electrolytes, and hybrid concepts that combine the strengths of several technologies (Libber et al., 2024; Rashid Khan & Latif Ahmad, 2025; Reenu et al., 2024). In this regard, synergy between batteries and supercapacitors offers a promising route toward meeting future demands on power, energy, and sustainability.

2.2 Supercapacitors

Supercapacitors, which are also called electrochemical capacitors, fall between conventional capacitors and rechargeable batteries in providing intermediate storage systems for energy (Figure 2) (Abraham et al., 2023; J. Sun et al., 2022).

Figure 2. Performance Mapping of Energy Storage Devices on a Ragone Plot (S. K. Tiwari et al., 2020)

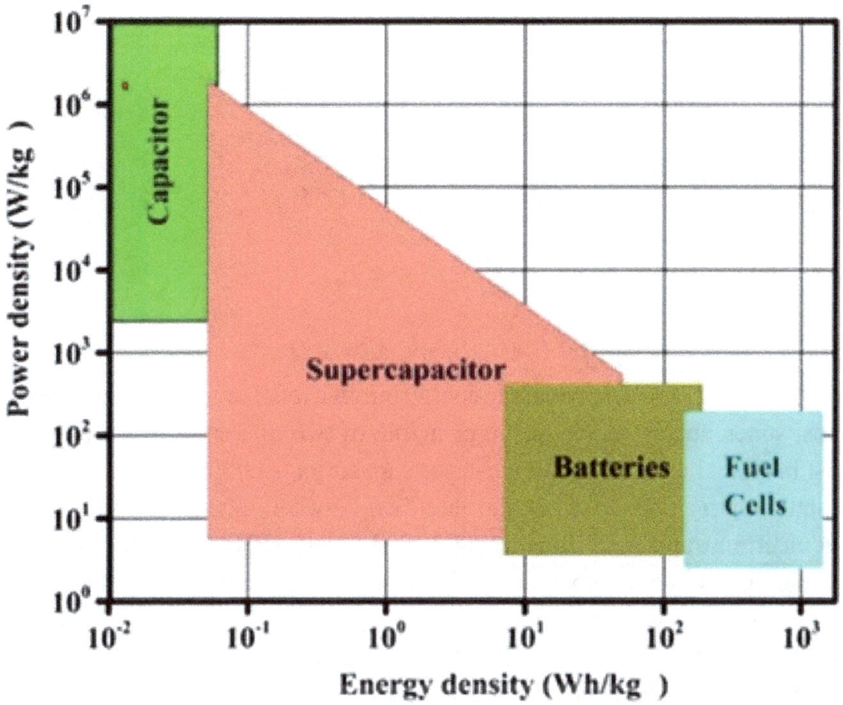

They have gained great attention in the energy storage area owing to their different capabilities. Their applications extend to a wide range from portable electronics and mobile devices to flexible displays, smart wearables, biomedical textiles, and antimicrobial fabric (Kou et al., 2014; Madhu et al., 2023; Monteiro et al., 2011; Palchoudhury et al., 2019; Yu et al., 2015). Among the prime strengths of supercapacitors is their ability to provide high power output in a very short period. This makes them suitable for applications requiring high amounts of power for short intervals, such as electrical systems with sudden surges (N. Kumar et al., 2025; L. Wang et al., 2025). This unique architecture (Figure 3) can support much higher power densities than traditional batteries. Apart from this, they also exhibit an excellent cycle life that generally surpasses hundreds of thousands of charge-discharge cycles to ensure long-term reliability (N. Liu & Gao, 2017; Qi et al., 2017; X. Zhang et al., 2021).

Figure 3. Design Architectures of Conventional and Flexible Supercapacitors (Lokhande et al., 2021)

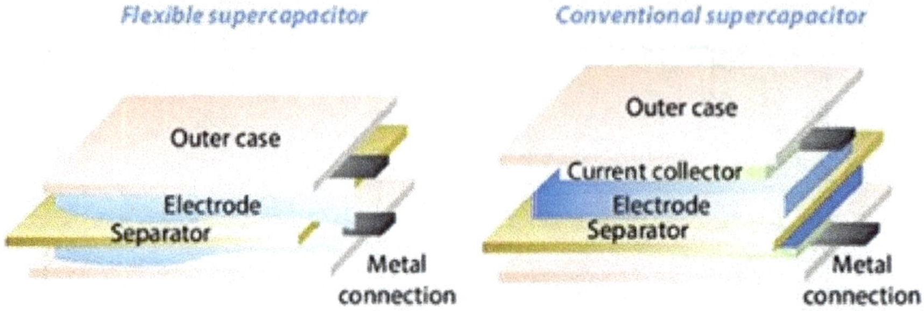

Unlike batteries, which accumulate energy through relatively slow electrochemical reactions, supercapacitors act on the principle of two major storage mechanisms. The first is typical of electric double-layer capacitors, EDLCs, and consists of the electrostatic charge separation at the interface between electrode and electrolyte without electron transfer (Shiraishi, 2003; Staaf et al., 2014). The second mechanism, occurring in pseudocapacitors (PCs), involves fast, surface-confined faradaic redox reactions. These are similar to battery processes but happen only on the surface of the electrode (Y. Jiang & Liu, 2019; Y. Liu et al., 2020). According to these principles of storage, supercapacitors are widely divided into three classes: EDLCs, pseudocapacitors, and hybrid capacitors (HCs). Each class shows peculiar characteristics that depend on the nature of the electrode materials and on the system design, as sketched in Figure 4. EDLCs generally use porous carbon materials like activated carbon, carbon nanotubes, carbon aerogels, and graphene, which ensure high reversibility and very small internal resistance and provide exceptional cycle stability (Sarno, 2020). These features make them show very good specific power. Energy storage, however, remains quite modest; thus, this type of supercapacitor does not enable long-term applications but only provides short periods of time with high power.

Figure 4. Schematic diagram representing EDLC, hybrid and pseudocapacitors (Kularatna et al., 2021)

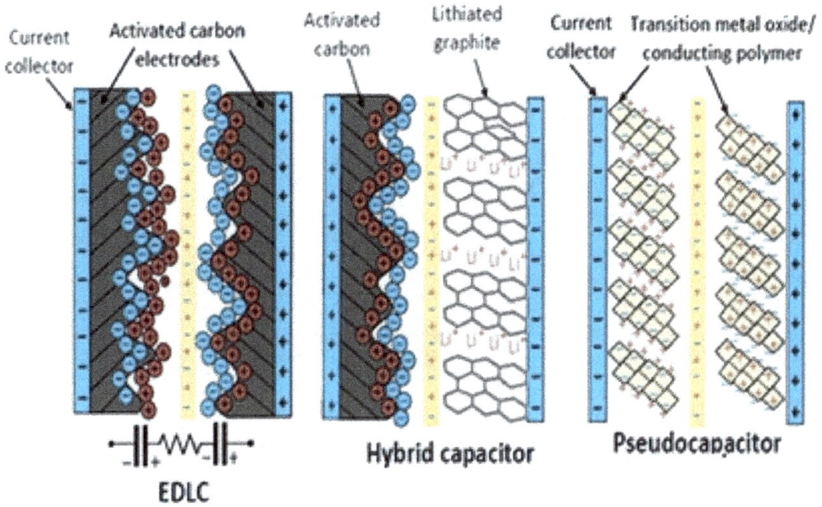

On the other hand, pseudocapacitors have much higher energy densities because of the use of redox-active materials that are capable of reversible electrochemical reactions (Gopi et al., 2024; Z. Wu et al., 2016; L. Zhou et al., 2018). In this respect, examples include transition metal oxides such as MnO_2, Co_3O_4, and NiO and conducting polymers like polyaniline and polypyrrole (Fu et al., 2019; Hoch, 2007).

Although those materials normally show good performance, further issues with these materials are poor electrical conductivity, limited cycle stability, or high costs, hence limiting their large-scale practical applications, as is the case with ruthenium oxide (Fu et al., 2019; Hoch, 2007; Sajedi-Moghaddam et al., 2017). Hybrid supercapacitors have been built based on the combination of the advantages of EDLCs and PCs. Such devices are constituted by two different kinds of electrodes, usually one presenting capacitive behavior and the other showing battery-like or pseudocapacitive properties (D. Gao et al., 2023; Patel et al., 2024). For an extended operating voltage, one may achieve better overall energy density, but generally, at the cost of worse charge-discharge kinetics, life cycles, or design complexities while preserving the electrochemical balance (Abraham et al., s. d.; Lakshmi & Vedhanarayanan, 2023). The main disadvantage of supercapacitors when compared with batteries seems to be their relatively low energy density. However, this disadvantage is redeemed by ultra-fast charging capability, usually in seconds, very good durability-up to 500,000 cycle-operating temperature ranges, from -40 °C up to 70 °C, and little maintenance. Table 1 summarizes the key operational characteristics that differentiate supercapacitors from batteries (Kularatna et al., 2021; Rafik et al., 2007). All these features make them especially suitable for niche applications like regenerative braking in electric vehicles, emergency backup systems, portable electronics, and power smoothing of solar generation systems (Dissanayake & Kularatna-Abeywardana, 2024; Sayed et al., 2025).

Table 1. Comparison of Key Performance Characteristics between Supercapacitors and Batteries (Kularatna et al., 2021)

Characteristics	Supercapacitors	Batteries
Power density (W/kg)	32	<1000
Energy density (Wh/kg)	1 to 20	10 to 100
Cycle life	50000 to 1000000	500 to 2000
Charge time (seconds)	1 to 60	3,600 to 18000
Discharge time (seconds)	6 to 1800	600 to 10800
Typical lifetime (years)	30	5
Operating temperature range (°C)	-40 to +70	-20 to +65

With such attributes, a rising strategy is the integration of supercapacitors with batteries within hybrid energy storage systems. This enables the use not only of the higher energy storage capacity of batteries but also of the very fast power delivery by supercapacitors, in order to provide a balanced and adaptable solution for more demanding applications, as seen in (Patel et al., 2024; Sayed et al., 2025). In terms of EDLC, pseudocapacitor, and hybrid types, selection among these supercapacitors is based on their particular target application. Currently, EDLCs are most commercially accessible for their simplicity, low cost, and strength, while on-going research in advanced pseudocapacitive and hybrid materials holds much promise in overcoming existing limitations, especially those related to energy density (Patel et al., 2024; Y. Zhang & Tang, 2020). Efficiency and cost are two major reasons for the development of next generation materials in making supercapacitors a core player in energy storage systems in the future (Appiah et al., 2022; Fathona et al., 2024; Mehta et al., 2020).

3. FUNDAMENTAL PARTS AND MATERIAL COMPOSITION OF ENERGY DEVICES

Most of the devices used for electrochemical energy storage, including primary and secondary batteries, supercapacitors, and hybrid systems, possess essentially similar architecture: two electrodes (positive and negative), an electrolyte, a separator, current collectors, and a protective casing. Each of these components contributes importantly to performance of the system in terms of energy density, power output, long-term stability, and operational safety (Benoy et al., 2022).

At the heart of these devices are electrodes, which are made from active materials capable of storing or taking up ions in the electrolyte (Staaf et al., 2014; Y. Zhang & Tang, 2020). Both the nature and structural properties of these materials influence the electrochemical behavior of the working system significantly (Abouelamaiem et al., 2018; Dissanayake & Kularatna-Abeywardana, 2024). Among them, the carbon-based materials occupy a special place due to their huge surface area, excellent electrical conductivity, and good chemical stability, including graphite, activated carbon, carbon nanotubes, graphene, and carbon black (Benoy et al., 2022; C. Song et al., 2024).

Among supercapacitors and rechargeable batteries, ACs are used owing to their affordability, technological maturity, and reasonable performance-to-cost ratio (Abdollahifar et al., 2022; Dissanayake & Kularatna-Abeywardana, 2024). However, conventional carbon materials synthesized from fossil-based precursors, such as graphite, ethylene, methane, benzene, and phenol, have several disadvantages: high energy consumption in the production process, unsatisfactory control of porosity, underdeveloped graphitic structure, and low degree of heteroatom doping (R. He, 2024; Mandal et al., 2023). These serious drawbacks limit the ionic transport and conductivity in next-generation energy storage systems(V.-T. Nguyen et al., 2024a).

Carbon black is obtained from petroleum and represents one of the most common additives in lithium-ion batteries (Powell et al., 2021). However, raises several issues regarding sustainability for alternative technologies, such as sodium-ion batteries. Therefore, there is growing interest in more value-added carbon nanostructures like aerogels, graphene, and carbon nanotubes, showing much better physicochemical properties(Nardecchia et al., 2012). Yet again, high production costs, problems in large-scale processing, and their aggregation during manufacturing hamper their industrial use (Pérez-Rodríguez et al., 2021). Research is then increasingly being focused on low-cost, abundant, and renewable alternatives furnishing optimized porosity, tailored graphitization, and effective heteroatom doping key features for sustainable, high-performance, and eco-friendly energy storage technologies (V.-T. Nguyen et al., 2024a).

In the parallel, pseudocapacitive and battery-type materials based on transition metal oxides-such as MnO_2, $LiCoO_2$, NiO, Fe_2O_3 are nowadays increasingly combined with either conventional carbon electrodes or functional materials to improve overall energy storage by faradaic redox mechanisms (Z. Wu et al., 2016). Hybrid systems made thereof allow both high energy density and high-power output. These functional materials still have open issues about cyclability, structural stability, and potential toxicity (Benoy et al., 2022; Raut et al., 2025).

Separators are commonly microporous polymer membranes that effectively prevent direct contact of the electrodes while allowing ionic transport (Raza et al., 2025; Z. Wang et al., 2020). It must offer adequate porosity, thermal resistance, chemical inertness, and mechanical strength. The electrolyte, which is usually liquid, gel, or solid, acts as an ionic conductor between the electrodes (X. Chen & Holze, 2024). Its composition has a great impact on the performance, safety, operating temperature, and other characteristics of the system(Raza et al., 2025). Nevertheless, most of these components are synthetic and fossil-based-polymer electrolytes and their binders such as PVDF-being hence potentially toxic, costly, and renewable sources (Lizundia et al., 2020; Schlemmer et al., 2021; M. Y. Zhang et al., 2017). Besides, most energy storage devices are designed with materials whose long-term availability is not predictable. Additionally, most of the manufacturing processes involved are energy-intensive and have hazardous by-products, which complicates end-of-life recycling and increases the cost (H. Chen et al., 2025; Hayagan et al., 2024; Machín et al., 2024). These factors highlight the need to develop electrode materials that are environmentally friendly, economically viable, and renewable. The goal is to achieve high electrochemical performance while being eco-friendly even at the levels of disposal and recycling (Madhu et al., 2023).

Within this context, bio-based materials are promising substitutes in the use of conventional synthetic components. Biomass-obtained carbon electrodes, metal oxides synthesized from plant extracts, and natural polymers are among the widely explored active groups in the development of electrochemical storage systems with high performance, long-term sustainability, and environmental compatibility(C. Ma et al., 2023; V.-T. Nguyen et al., 2024a; Rajput et al., 2024). The following sections focus specifically on these biodegradable, plant-derived materials, exploring their properties, functionalization, and applications in energy storage technologies

4. ADVANCES IN PLANT-BASED MATERIALS: FROM BIOCHEMISTRY TO ENERGY STORAGE

Plants represent a renewable resource, available in abundance and accessible, whose potential, both from the standpoint of scientific application and technological valorization, has been considered to grow steadily over recent years (Barbinta-Patrascu et al., 2024). The fact that, through photosynthesis, they are able to capture carbon dioxide in the atmosphere, produce biomass, and release oxygen places them at the center of ecological balance on Earth (Gayathri et al., 2021). Currently, more than 380000 plant species have been identified, from unicellular organisms to big trees, offering an inexhaustible reservoir of complex biological structures and exploitable chemical compounds (Wijerathne et al., 2025).

4.1 Historical use of plant- derived materials in energy-related fields

Historically, plants were used by humans as construction material, for nutrition, medicine, and for the preparation of utility resources. Nowadays, this rich biological heritage has become of great interest for obtaining functional materials suitable for a wide range of modern technologies, in particular energy-related applications (Barbinta-Patrascu et al., 2024; Fathona et al., 2024; Okoye et al., 2024a; Wijerathne et al., 2025; Y. Zhang & Tang, 2020). This contemporary interest builds upon a long history of plant-derived materials in energy devices, which has evolved gradually from basic applications to highly advanced technologies. Early uses of plant-based resources, such as oils, fats, and wood, served as fuels for lighting and heating (Ruiz-Gonzalez et al., 2023) with biomass remained the main energy source until the early 20th century, when fossil fuels became prominent in the field (Y. Li & Ding, 2024). The 20th century further witnessed the development of biodiesel and bioethanol as biofuel alternatives to petroleum-based fuels(Aakko-Saksa et al., 2018). together with biodiesel inventions such as the palletization of plant waste into solid biofuels (González et al., 2020). More specifically, as history unfolded, it went from the simple combustion of biomass to the sophisticated engineering of plant-derived compounds for high-performance energy storage and conversion technologies, reflecting scientific efforts increasing to utilize the abundance and unique physicochemical properties of plant material to address global energy and environmental challenges (M. Gao et al., 2018a; Ruiz-Gonzalez et al., 2023).

More recently, there has been the migration of plant-based materials from their current simple use as a source of energy to active material elements in batteries and supercapacitors and dye-sensitized solar cells(Biswal & Balasubramanian, 2026; V.-T. Nguyen et al., 2024b; Okoye et al., 2024b). This further stipulates sustainability, a circular economy, and the design of next-generation energy technologies that are reliable, appropriate and inexhaustible(Gupta et al., 2021; Halder et al., 2025). Among these, lignocellulosic biomass has gained momentum as one of the cornerstones of the abovementioned developments due to its renewable and ecofriendly nature for energy storage and catalysis applications (Okoye et al., 2024b; Ruiz-Gonzalez et al., 2023). Accordingly, prior to exploring the application of plant-derived materials in specific energy storage systems, the following section briefly examines the biochemical composition of these plants, focusing on the polymeric molecules that can be utilized in energy-related technologies.

4.2 Green biochemistry: Plant biopolymers

Plants are mainly built up from organic macromolecules like cellulose, hemicellulose, lignin, and pectin, which make up the skeleton of their cellular structure (Berglund et al., 2020; Kögel-Knabner, 2002). Such biomass resources are rich in biopolymers containing hydroxyl, carboxyl, and methoxyl functional groups; therefore, highly chemically active owing to transformation and functionalization (Barbinta-Patrascu et al., 2024; N. Zhang et al., 2015). Besides this, plants biosynthesize unique classes of secondary metabolites including flavonoids, alkaloids, phenolic acids, terpenes, and tannins that possess remarkable antioxidant, conductive or catalytic properties (Mzioud et al., 2023, 2024). This therefore, gives ways that such features may be designed into advanced materials such as carbon-based materials, natural polymers, organicinorganic hybrid systems among others may be designed with such features (Berglund et al., 2020; T. Li et al., 2021; Timell, 1967; Wijerathne et al., 2025; N. Zhang et al., 2015). In this context, the need to profound knowledge about plant biochemistry is necessary. The nature, structure, and reactivity of components issued from plants allow for a rational approach to valorization toward high added-value applications.

Thus, the valorization of plant-derived components or biomass is possible only by deeply understanding the biochemical mechanisms at the basis of plant systems, since this allows for identifying, extracting, and then transforming the most promising bioactive compounds into advanced technological applications. Indeed, the remarkable molecular diversity of plants represents a valuable reservoir for the development of high-performance function materials. As mentioned above, plant-based materials are one class of renewable, biodegradable, and low-cost resources that can be exploited to promote sustainability in developing electrochemical energy storage devices.

The plants' chemical richness resides especially within the structural polymers of cellulose, hemicellulose, and lignin and in the wide range of secondary metabolites, including flavonoids, tannins, phenolic acids, and alkaloids (Berglund et al., 2020; Mzioud et al., 2025). Basically, such biochemical complexity opens perspectives to designing materials with conductive, redox, or stabilizing properties used in the fabrication of electrodes, electrolytes, or separators (Bresser et al., 2018; Spirk, 2018; Subhani et al., 2024; T. Zhao et al., 2024). In this context, two main families of plant-derived materials can be distinguished:

- Natural polymers: cellulose and its derivatives may serve as matrices of solid electrolytes, as separators, or as binders.
- Carbon-based material from biomass: The pyrolysis of the residues of plants (leaves, husks, stems, etc.). The resultant carbon materials are highly porous

with good electrical conductivity and can be used for the fabrication of electrodes in supercapacitors and batteries.

Valorization of plants and their biochemistry therefore represents an innovative, low environmental impact approach to the development of next-generation energy storage systems (Bresser et al., 2018; Pérez-Rodríguez et al., 2021). In this context, the structural components of plant cell walls are composed of cellulose, hemicellulose, and lignin (Figure 5). These macromolecules impart structure and strength to plant tissues (W. Chen et al., 2022; Nishimura et al., 2018; Saldarriaga-Hernández et al., 2020; N. Zhang et al., 2015).

Figure 5. Lignocellulosic Structure and Main Biopolymer Components including Cellulose Hemicellulose and Lignin

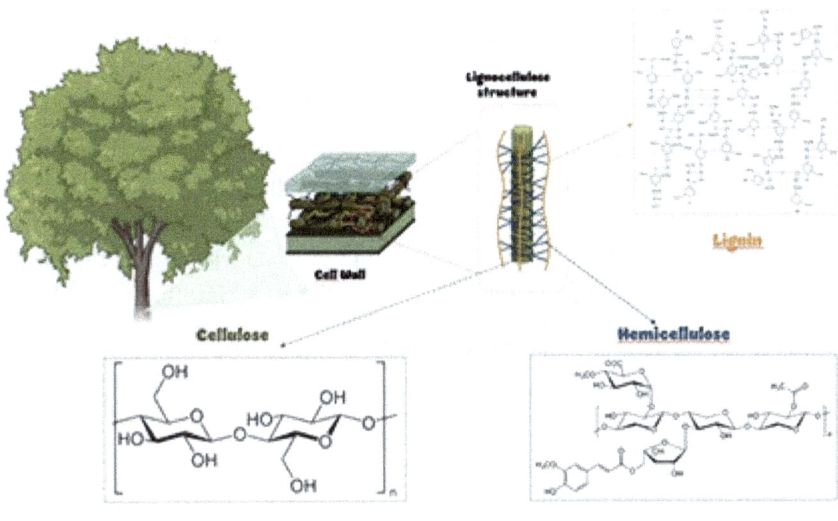

the relative proportion of these biopolymers can vary widely depending on the plant source and will directly affect both functional properties as well as the value of the plant biomass in energy storage applications (Dias et al., 2025; Dutta & Sit, 2024). Table 2 shows the relative contents of cellulose, hemicellulose, and lignin in various plant materials; those richest in cellulose include cotton and hemp fibers, while those with the highest contents of lignin include softwood and wheat straw (Beg et al., 2025).

Table 2. Cellulose, hemicellulose, and lignin content from various plant sources(Beg et al., 2025)

Sources	Lingnin (%)	Hemicellulose (%)	Cellulose (%)
Banana fibre	8	20	70
Cotton fibres	0.4	0.5	98
Corn stover	19	21	33
Jute	11	15	68
Hemp	7	11	76
Softwood	29	20	49
Soy hulls	18	13	56
Wheat straw	22	34	43

Apart from cellulose, hemicellulose, and lignin, other plant-based biopolymers like starch and pectin play an important role in the structure and energy storage of plants (Chelfouh et al., 2022; Harikumar & Batabyal, 2023; Thomas et al., 2025). Tubers, cereal grains, and legumes are high in starch content, while pectin occurs mainly in the cell walls of apples, citrus fruits, and beetroot. These biopolymers have typical chemical properties, conferring on them specific functionalities such as gelling, ionic complexation, and film formation, which make them useful in the manufacture of components like electrolytes, binders, and separator membranes (Das et al., 2023; T. D. Nguyen & Lee, 2025; Thrisha K & Saratha R, 2024).

4.3 Cellulose

Cellulose is the most abundant polysaccharide on Earth and forms a basic component of plant cell walls (Lerouxel et al., 2006). In plants, it is more concentrated in secondary cell walls, especially in tissues such as the bark, stems, and vascular tissues (Alberts et al., 2002). Such abundant presence in certain regions of plants is important for enabling plants to resist environmental stress, various climatic changes, and mechanical forces(Ahluwalia et al., 2021; Lerouxel et al., 2006).

Cellulose is a homopolysaccharide of D-glucose units interconnected by $\beta(1\rightarrow4)$ glycosidic bonds, giving it an orderly structure with the ability to form intra- and intermolecular hydrogen bonds for a long chain (Tayeb et al., 2018). It is semicrystalline in nature, containing crystalline and amorphous zones; hence, this gives it mechanical strength and reduced solubility in conventional solvents The main sources of cellulose are fibrous plants like cotton, flax, hemp, jute, and some lignocellulosic by-products such as straw, stems, and leaves. These latter products are usually considered secondary biomass (Ani et al., 2022; Elfaleh et al., 2023; D. Zhao et al., 2021). Because of its ability to make use of nature's materials, it has applications in several industries, especially for the preparation of biofriendly products (Aboul Hrouz et al., 2023; El Bourakadi et al., 2024; Lavanya et al., 2011; Rashid & Dutta, 2022). The usual route for the extraction of cellulose from plant biomass takes a complicated pathway and generally consists of methods for removing non-cellulosic substances such as lignin and hemicellulose (Abolore et al., 2024; Pinto et al., 2022). Most of the extraction treatments, like those performed with sodium hydroxide, dissolve the hemicellulose; afterward, treatments may follow with the aim of removing lignin by bleaching. Sophisticated work has also been carried out to improve yield and reduce the environmental impact by using organic solvents, enzymatic processes, and hydrothermal treatments (Abolore et al., 2024; L. Wang et al., 2024). Advanced techniques make the extraction of cellulose efficient and environment-friendly, putting this biomaterial among the core positions of materials that are being pursued for several industrial applications with an environmentally conscious scope.

4.4 Lignin

Lignin is considered one of three major components in lignocellulosic plants, the other two being cellulose and hemicellulose. Since it is situated in the interstitial areas of the cell wall, it naturally acts as a binder of the microfibrils of cellulose and chains of hemicellulose, participating in the structural cohesion of the plant matrix (Figure 5)(Alberts et al., 2002; Lerouxel et al., 2006). This configuration provides mechanical rigidity, resistance to moisture, and increases protection against biological agents, playing an important role in maintaining the stability of such lignocellulosic biomasses as wood, wheat straw, and sugarcane residues (Ojo, 2023; Sant'Ana Júnior et al., 2025).

The molecular-level structure of lignin is amorphous and three-dimensional, with high branching because of the polymerization of three major monolignols: p-coumaryl alcohol, coniferyl alcohol, and sinapyl alcohol (Feofilova & Mysyakina, 2016), forming H, G, and S units, respectively (figure 6). It is these respective ether and C–C linkages that form these various inter-unit linkages, leading to a complex heterogeneous architecture compared to the regularity observed in cellulose (Ali et al., 2024; Feofilova & Mysyakina, 2016).

Figure 6. Structural representation of lignin illustrating its three main monolignol units. p-coumaryl alcohol (unit G, in bleue), coniferyl alchohol (unit H, in purple), and sinapyl alcohol (unit S, in red). These monolignols differ in the number of methoxy substituent (-OCH_3) on the aromatic ring, wich determines the structtural diversity and properties of lignin macromolecules (Ali et al., 2024)

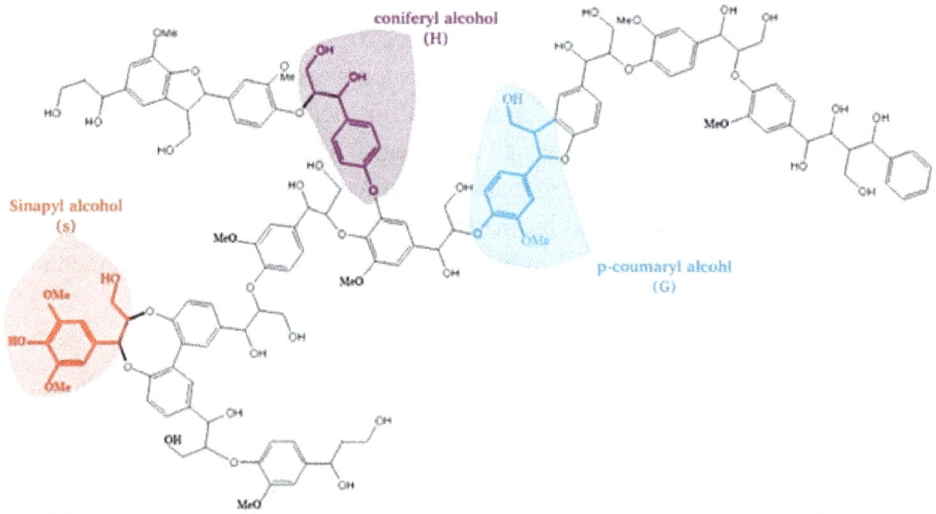

About half of the mass of lignin consists of condensed aromatic structures, conferring remarkable thermal stability and interesting electrochemical properties, mainly concerning electron transfer. It consists of several oxygenated functional groups such as methoxy, hydroxyl, and carbonyl primarily at α and β positions, which significantly increase its chemical reactivity and thus provide the possibility for various modifications of its structure. At the same time, such complexity restricts its solubility in ordinary solvents(Ali et al., 2024; Muddasar, Beaucamp, et al., 2024). On the other hand, partial solubilization could be advantageous in selective separations or conversion processes, as lignin is insoluble in water at acidic or neutral pH conditions but can be partly solubilized at alkaline conditions or in certain organic solvents (Lourenço et al., 2023; Saadan et al., 2024). Such characteristics are of particular relevance within the domain of energy storage, in so far as it is possible to transform lignin into functional carbon-based materials (Yao et al., 2022). Besides, the presence of this aromatic biopolymer in a wide variety of plant biomasses-such as wood, grass, herbaceous, aquatic plants, and even some algae (Ali et al., 2024), offers a strategic advantage toward transforming lignin into an attractive player for energy storage technologies. Its abundance guarantees a sustainable and economically viable supply of raw material, as required by the scaling-up of electrochemical materials derived from lignin.

4.5 Pectin

Pectin is a complex carbohydrate polymer that, with respect to structure, is included in the class of anionic heteropolysaccharides. This is one of the key contributors to cell wall structure in higher plants(Lerouxel et al., 2006; Thomas et al., 2025). From a molecular point of view, it is mainly characterized by α-D-galacturonic acid units linked together by α-(1→4) glycosidic bonds with carboxyl groups that might be methylated or acetylated. The chemical structure variability from its source and modification extent is affecting its physicochemical properties (Lara-Espinoza et al., 2018; X. Liu et al., 2021; Roman-Benn et al., 2023). Another point is that most of the structure of pectin is divided into three major regions, including homogalacturonan, composed of a continuous chain rich in galacturonic acid; rhamnogalacturonan I, composed of units of rhamnose and galacturonate, with side chains rich in arabinose and galactose; and rhamnogalacturonan II, which contains more than ten sugar types in its highly branched region, able to form dimers through borate-diol bonds(Humerez-Flores et al., 2021; Ropartz & Ralet, 2020).

Agro-industrial by-products are the main sources of pectin; in particular, the peels of citrus fruits, such as orange and lemon, but also apple pomace, which contain quite relevant homogalacturonic segments, represent promising feedstocks. Alternative sources of pectin include mango and banana peels or even cocoa shells, highly promising, especially for the higher content of rhamnogalacturonan I and rhamnogalacturonan II that could give peculiar functional characteristics(Chelfouh et al., 2022; Harikumar & Batabyal, 2024; Humerez-Flores et al., 2021).

Classically, pectin are extracted by acid hydrolysis under high temperature and acidic conditions (80-95°C, pH 1-2); this gives high extraction efficiency but with considerable limitations from an environmental point of view, very often leading to loss and/or deterioration of sensitive side-chain structures (Chandel et al., 2022; Obodo-Ovie et al., 2024). Other techniques developed are more environmentally friendly for the extraction of pectin: UAE and MAE improve efficiency and mildness during extraction and reduce thermal degradation (Haque et al., 2025). Enzyme treatments will attack only the bonds in the cell wall and enable much better preservation of its native structure.(Haque et al., 2025). Optimal efficiency of extraction, purity, functionality, and low ecological footprint is achievable by applying a combination of the proposed approaches. (Mellinas et al., 2020). Further development of such green technologies will allow sustainable exploitation of various plant raw materials and expansion of functional applications of pectin for new uses in food, pharmaceuticals, and electrochemistry.

4.6 Starch

Starch, one of the most abundant biopolymers, has been under wide investigation in renewable material fields, and its global production is estimated to be around 2.8 billion tons per year(Gopinath et al., 2023). It is synthesized by photosynthesis and stored as granules within cereals like corn, wheat, and rice, as well as tubers such as cassava and potatoes (Gopinath et al., 2023; Thomas et al., 2025). Structurally, starch contains two main fractions: amylose-a linear polymer made of $\alpha(1\rightarrow4)$ glycosidic bonds-and amylopectin, which is a branched polymer with $\alpha(1\rightarrow4)$ bonds along the main chain and $\alpha(1\rightarrow6)$ branching points (Robyt, 2008; Tong et al., 2023) (Figure 7).

Figure 7. Chemical Structure of Cellulose (Samir et al., 2022)

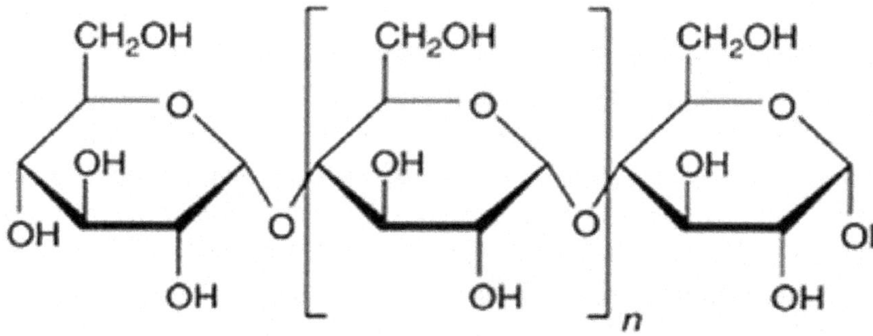

These two structures combine to impart unique physicochemical properties in starch, such as semi-crystalline regions, strong affinity for water, gelatinization under heat, and possible film or gel formation with the presence of water and heat (Cornejo-Ramírez et al., 2018; Gong et al., 2024). However, the high concentration of hydroxyl groups in it reduces the intrinsic hydrophilicity; therefore, it diminishes the performance of starch in high moisture conditions, reducing stability in some advanced applications.

Such chemical modification techniques include esterification, etherification, and oxidation, among others, with a view to improving the solubility, mechanical strength, thermal stability, and permeability of starch (Y. Fan & Picchioni, 2020; Mirzaaghaei et al., 2022). These modifications create wide application fields in construction materials, food processing, technical textiles, water treatment, and controlled drug-delivery systems, including organic electronics (Gopinath et al., 2023).

From this, it would be possible to draw a conclusion that natural polymers of plant and plant biomass, represented by cellulose, lignin, hemicellulose, and pectin, have a lot of structural-functional advantages. It is expected that, because of their abundance, biodegradability, and specific physicochemical properties, they may find a place in sustainable applications for energy storage, packaging, and bioengineering. In this respect, one can notice an increasing interest in such biopolymers for application as binders, separators, and electrolytes in batteries, making significant contributions to developing greener, long-lived electrochemical systems.

5. BIO-POLYMER-BASED BINDERS, SEPARATORS, AND ELECTROLYTES

The growing presence of biopolymers in energy storage technology underlining their capacity to meet the demands of sustainability and performance which electrochemical devices(Aziz et al., 2024; A. Tiwari & Kumar Singh, 2024). In nature, they are being investigated for major roles in three most important aspects: as binder, separator, and electrolyte functions of batteries and supercapacitors. They act as a binder by maintaining electrode integrity without affecting its electrochemical stability (Appiah et al., 2022). As separators, they provide physical separation of the electrodes with ion transport capability. Some of these biopolymers can even be tailored as electrolytes, hence further enhancing safety and efficiency in such systems (Appiah et al., 2022; Muddasar, Beaucamp, et al., 2024). This chapter intends to give a comprehensive description of those roles in detail and the precise properties required for the same in those applications.

5.1 Biopolymer-based separators for electrochemical applications

Separators play a vital role not only in batteries but also in supercapacitors, providing a separator for the electrodes while maintaining fast ion traffic for fast charging and discharging(J. Li et al., 2022; Raza et al., 2025). Natural polymers such as lignin and cellulose are increasingly used as separators due to ecological and functional properties (Y. He & Guo, 2024; H. Li et al., 2025; Silva et al., 2022; Y. Song et al., 2024). Among all the biopolymers, porosity, flexibility, and the possibility of chemical modification are the main characteristics that guarantee the performance of the separator with minimal impact on the environment.

5.1.1 Cellulose-based materials as electrochemical separators

Native cellulose extracted from plants can be transformed into various forms, including cellulose nanocrystals (CNC), cellulose nanofibers (CNF), regenerated cellulose, and cellulose acetate, depending on the treatments applied (Suresh Khurd & Kandasubramanian, 2022). These various forms possess a range of properties, such as flexibility, transparency, biocompatibility, or thermal conductivity, and are particularly adapted for energy storage applications, mainly as separators in batteries and supercapacitors (Muddasar et al., 2022; Saleh et al., 2025). Functional groups or conductive polymers can be attached by approaches like chemical grafting or layer assembly to endow cellulose with specific properties, such as electrical conductivity or hydrophobicity (Kang et al., 2015; Yazdani McCord et al., 2025). Modified cellulose, due to its chemical activity and the possibility of combining it with a wide range of materials, represents a promising starting material for the design of multifunctional composites dedicated to green electronics. Thanks to intrinsic properties, such as flexibility, light weight, and biodegradability, it is of particular interest for flexible and environmentally friendly electronic devices (Y. H. Jung et al., 2015). Such a combination of features has favored growing interest in its application for energy storage, where it can play several roles: flexible substrate, separator with tunable porosity, or solid electrolyte.

Several works have recognized that cellulose can enhance the performance, stability, and durability of both batteries and supercapacitors. A major area of research in recent years has focused on the investigation of cellulose-based materials in battery technologies. As a typical example, cellulose-based separators in LIBs are new alternatives to conventionally used separators made from PE and PP(P. Chen et al., 2024; K.-P. Yang et al., 2025), these separators guarantee enhanced mechanical strength, an increase in thermal stability, and non-flammability, hence overcoming the major safety issues that occur with conventional separators. Moreover, from the point of view of analytic studies, cellulose separators-if modified appropriately-remain or sometimes increase the conductivity compared to synthetic separators. In addition, the transport of ions is further optimized due to the high surface area and porous structure of cellulose; hence, it proves to be efficient for charge-discharge cycling cycles. Table 3 depicts a few studies conducted on the preparation of cellulose-based separators for batteries.

Table 3. Representative studies on cellulose-based battery separators

Cellulose type	Cellulose source	Separator thickness (µm)	Porosity (%)	Ionic conductivity (mS.cm^{-1})	Cycle stability	Ref.
CNF	Coffee waste	25	55	3.00	47.1% specific capacity retention	(Naveen Kumar et al., 2024)
CNF	Rice straw	30	51	3.40	100% after 5000 cycles at 0.5C	(Islam et al., 2021)
CNF	Bamboo pulp, hardwood, softwood, cotton, hemp	20-30	–	–	–	(Naveen Kumar et al., 2024)
RCS	Cotton pulp	19.74	61	1.25	72% after 100 cycles at 0.2C	(Y. Wang et al., 2021)
CSA	Cotton pulp	109	58.43	1.34	78.7% after 80 cycles at 0.5C	(T. Guo et al., 2019)

(Naveen Kumar et al., 2024) have developed a separator based on cellulose nanofibers extracted from coffee pulp by TEMPO oxidation. The resulting electrochemical parameter was illustrated in table 3. These results are remarkably compared with other bio-based materials, offering a combination of high porosity, electrolyte uptake and specific capacitance that could outperform commercial separators under high current density conditions. However, the capacitive retention is still low compared with synthetic polymer separators such as Celgard 2325, which show high retention values under low current density. A similar approch has been taken by (Islam et al., 2021), who used rice straw as a precursor for producing membranes by casting nanofibrillated cellulose solutions in the range of 0.5-0.8%. The optimal membrane, which had a thickness of 30 µm, showed a porosity of 51%, a specific capacity of 150.7 F.g^{-1}, and an energy density of 30.2 Wh.kg^{-1}. It demonstrated exceptional electrochemical stability with 100% retention after 5000 cycles outperforming the performance of the commercial separator used as a reference. This is an outstanding result in terms of durability which places rice straw as one of the candidate materials for long-term applications. However, the specific capacity remains lower than that observed for the membrane based on cellulose nanofibres from coffee pulp which may limit the performance for higher capacity applications. Otherwise (Sheng et al., 2019) investigated other vegetal sources such as bamboo, hardwood, softwood, cotton, and hemp for membrane preparation using vacuum filtration. The prepared membranes with thickness comparable to Celgard 2325 membranes (21-23 µm) presented excellent wettability (contact angles < 45°) and good thermal stability up to

160°C for 2 h. Samples coming from cotton and softwood showed high mechanical strength (>130 MPa). Even though these membranes have not been characterized so far, their physical and mechanical properties make them strong candidates as separators in lithium-ion batteries. However, the lack of electrochemical data prohibits their effective roles from being evaluated when compared with other biobased separators tested. A paradigm shift can be observed in work published by (Y. Wang et al., 2021) that has elaborated a separator from regenerated cellulose derived from cotton pulp using a nonsolvent phase inversion method. The membranes obtained have excellent thermal stability up to 180°C, 436% electrolyte absorption, an ionic conductivity of between 0.9 and 1.25 $mS.cm^{-1}$, and discharge capacities ranging between 110 and 117 $mAh.g^{-1}$, retaining 76-81% after 100 cycles, outperforming the Celgard separator. This approach therefore evidences competitive performance from the point of view of ionic conductivity and thermal stability, but the lower discharge capacity and the retention after cycles compared with nanocellulose-based separators make this material difficult to use in high specific capacity applications.(T. Guo et al., 2019) have also consolidated this approach through the incorporation of PVA into the regenerated cellulose. Thus, getting a composite separator 0.4 mm thick. The separator obtained presents good porosity and suitable wettability, mechanical strength of 53.7 MPa, and low thermal deformation at 200°C. With an ionic conductivity of 1.34 $mS.cm^{-1}$, this membrane represents a durable and safe solution for new generation batteries. Such a comparative analysis of cellulose-based separators, most of which were recently developed, puts forward the decisive role of both the nature of cellulose derivative used and of the manufacturing method chosen. Each form of cellulose confers specific properties on the membranes in terms of porosity, mechanical stability, and ionic conductivity. These characteristics themselves are modulated by the chemical or physical treatments applied upstream. On the other hand, the technique used for the shaping of the membranes plays an equally crucial role in the final performance of the separators. A variety of techniques has been used, going from molding and coating of commercial separators to more sophisticated techniques like electrospinning, vacuum filtration, and nonsolvent phase inversion methods(Babiker et al., 2023; Sheng et al., 2020; H. Zhu et al., 2016).

Both techniques have their advantages and limitations. Other works have pointed out an improvement of some properties of interest, like mechanical resistance or ionic conductivity, by the addition of additives or composting with other polymers (Bandyopadhyay & Nandan, 2025; Sakthi Balan & Aravind Raj, 2025; Shamsuri et al., 2024). Hybrid approaches enable the extension of application fields for biobased separators, while performance can be precisely tuned to meet the requests of modern energy storage systems. For this reason, the choice of cellulose type properly matched with the most suitable treatment strategy is expected to enable the development of sustainable, high-performance separators. Accordingly, the selection of the right type of cellulose and the right treatment strategy appear to be fundamental to developing high-performance and sustainable separators(Pajares et al., 2024; Turossi et al., 2025).

5.1.2 Lignin-based materials as electrochemical separators

Besides cellulose, lignin has also emerged as a promising functional material in the design of separator membranes for rechargeable batteries(H. Y. Jung et al., 2022). Interest in lignin is also related to its properties of being able to modulate porosity, wettability, and interaction with electrolytes-properties fundamental to the optimization of both safety and electrochemical performance.(M.-J. Uddin et al., 2017) reported an electrospun membrane based on lignin and PVA. This electrospinning from environmentally benign aqueous solution resulted in a non-woven membrane characterized by an ultra-fine, highly porous three-dimensionally interconnected structure. PVA can provide a supporting fibrous matrix, while the presence of lignin significantly enhances wettability, porosity, and thermal stability. This novel membrane exhibited good compatibility with different electrolytes and outperformed the commercial Celgard separator at high C-rates. Compared with (M. Zhao et al., 2015) explore an alternative strategy by formulating lignin with polyacrylonitrile (PAN), a polymer of good reputation for its thermal robustness. The electrospun L-PAN composite had a porosity of 74%, much higher than that of the PP separator. It demonstrated outstanding dimensional stability at an elevated temperature up to 150°C, showing excellent capacitance retention of 95% after 50 cycles at 0.2C in Li/$LiFePO_4$ cells. Different from what was conducted by Uddin, here, the lignin seems to function more in enhancing thermal and mechanical stability synergistically with PAN. These two works represent converging strategies to incorporate lignin into polymer separators but with somehow different performances targeted: While the lignin/PVA combination focuses on ionic compatibility and wettability, the lignin/PAN system chases thermal resistance and mechanical strength. It suggests that the choice of the host polymer is extremely important to the functional expression of lignin in the final material. In the work by (Jeong et al., 2020), the functionalization of lignin was extended to the preparation of lignin-based graft polymers via CRP. These UV-crosslinked materials are under development for SPEs in all-solid-state lithium metal batteries. In addition to their high mechanical stability, they offer good ionic conductivity and effectively inhibit dendrite formation. This work evidence that lignin has moved from a passive structural role to an active one in ion conduction, thus enabling its application in new-generation all-solid systems. In this respect, the research of (Z. Zhang, Yi, et al., 2019) follows a different strategy, adopting LNPs as a surface coating for Celgard membranes in Li-S batteries. Such treatment aims at reducing the shuttle effect of soluble polysulphides, one of the most relevant problems of these batteries. Indeed, by virtue of the donor groups present on the LNP surface, a chemical interaction with polysulphides is induced, limiting their diffusion and improving cyclic stability over 500 cycles. Here, lignin plays the role of a chemical trap, exploiting its surface chemistry instead of morphology. Several natural polymers have been combined with different shaping techniques, to fulfill specific performance requirements. Therefore, the works in this frame show that lignin-based separators, when incorporated into polymer matrices such as PVA or PAN, take advantage of its intrinsic properties to improve wettability, porosity, thermal stability, and electrolytic adhesion. The most important role is played by the host polymer choice, since depending on the system, either ionic conduction and electrochemical compatibility, or alternatively, mechanical strength and high-temperature resistance (B. Aziz et al., 2018; Coviello et al., 2025).

At the same time, cellulose-based membranes, or their derivative NaNF, regenerated cellulose, showed high porosity, excellent electrolyte absorption, good ionic conductivity, and notable thermal stability. Electrochemical performance varied according to the source of cellulose, the eventual chemical treatments applied, and the manufacturing method. Processes such as electrospinning, molding, vacuum filtration, or phase inversion directly affect the membrane internal structure, which influences the retention capabilities, cyclic stability, and durability of the device(Grzybek et al., 2024; J. Wang et al., 2024). Hybrid methods, using biobased polymers with additives or synthetic polymers, can optimize targeted properties, inclusive of mechanical strength and dendrite elimination, thus enabling these separators to widen their application fields within advanced systems for energy storage.

5.2 Biopolymer-Based Electrolytes for Energy Storage Systems

Coupled with the progress made in bio-based separators, increasing interest is also being focused on electrolytes derived from natural polymers, which favor the performance of electrochemical devices. Besides their contribution to ion transport between electrodes, these electrolytes are indeed active protagonists of electrochemical stability, system safety, and durability under extreme operating conditions(B. Aziz et al., 2018). In such a context, interest in biosourced polymers is originally based on their biodegradability and non-toxicity(Aziz et al., 2024). Biopolymers like cellulose, lignin, starch, or pectin are under evaluation as electrolyte matrices, pure or formulated with conductive salts or green solvents(Coviello et al., 2025; Mellinas et al., 2020; Schlemmer et al., 2021; Turossi et al., 2025). Thanks to particular preparation methods such as gelation, cross-linking, or impregnation, these materials enable the realization of solid or quasi-solid electrolytes suitable for a safer, more durable battery architecture(X. Chen & Holze, 2024; Ding et al., 2024; Vikraman et al., 2025).

5.2.1 Cellulose-based materials as an electrolyte for electrochemical systems

The development of cellulose-based polymer electrolytes represents an emerging and strategic application in the field of energy storage devices, particularly for solid-state batteries and gel electrolytes(Hadad et al., 2022a; Shi et al., 2025). Conventional polymer electrolytes usually suffer from poor ionic conductivity at ambient temperature and their effectiveness in next-generation systems is compromised. These limitations can be overcome through the incorporation of cellulose into the polymer matrix. Cellulose gives electrolytes better mechanical integrity due to its three-dimensional structure and the ability to form dense networks; it thus facilitates ion transport(Saeed et al., 2025; C. Tang et al., 2024; L. Xu et al., 2024). Recent research has revealed that the application of cellulose in its nanocellulose form, or regenerated and modified forms, leads to enhancement in both mechanical stability and ionic conductivity of solid or semi-solid electrolytes(Du et al., 2025; Lizundia et al., 2020). The addition of functional components such as ionic liquids or inorganic fillers like metal oxides or conductive salts to these hybrid systems enhances overall electrochemical performance by weakening lithium dendrite formation and prolonging the life cycle of batteries(J. Huang et al., 2025; Manfredy Marquina et al., 2023; X. Zhang et al., 2024). Besides, the capability for cellulose to form hydrogels and aerogels with high porosity and high electrolyte retention opens the way to new architectures that are flexible and highly conductive. Such features make cellulose a key material in the design of durable, safe and high-performance electrolytes for next-generation energy storage systems. Table 4 below summarizes some representative studies on cellulose-based polymer electrolytes, highlighting their structures, physicochemical properties and electrochemical performance.

Table 4. Selected Studies on Cellulose-Derived Polymer Electrolytes for Energy Storage Applications

Cellulose type	Complex compound	Ionic conductivity ($S.cm^{-1}$)	Applications	Notes	Ref
Cellulose acetate (CA)	Zinc nitrate ($Zn(NO_3)_2$)	2.7×10^{-3}	Zinc-ion Batteries	Amorphous structure, good electrochemical stability, OCV of 1.72 V	(Bhuvaneswari et al., 2022)
	Ammonium nitrate (NH_4NO_3)	1.02×10^{-3}	Proton batteries	Potential to replace Nafion, needs performance improvements	(Monisha et al., 2017)
2-hydroxyethylcellulose (2-HEC)	Ammonium nitrate	$\sim 10^{-3}$	Supercapacitor	Plasticization improves flexibility and ionic mobility	(Hafiza & Isa, 2020)
Methylcellulose (MC)	$Mg(NO_3)_2$ $6H_2O$	1.02×10^{-4}	-	Structure amorphe, faible conductivité ionique.	(Koliyoor et al., 2022)
Methylcellulose (MC)	($LiClO_4$)	3.66×10^{-5}	Lithium-ion batteries	Good mechanical strength and thermal stability	(Ndruru et al., 2020)
	(NH_4Br)	1.89×10^{-4} à 1.67×10^{-3}	Double-layer supercapacitors	Plasticisation with glycerol improves conductivity	(Kadir et al., 2018)
	Ammonium carbonate	7.71×10^{-6}	-	Ionic conductivity still too low ($\sim 10^{-6}$ $S.cm^{-1}$) for some applications	(Sohaimy & Isa, 2020)

Table 4 clearly demonstrates that cellulose derivatives could work as polymer electrolytes due to their ability to form amorphous structures with higher conduction properties. Besides cellulose derivatives, other biopolymers from biomass material have also drawn interest regarding electrolyte applications. Among such biomaterials, lignin has also been in application to limit lithium dendrite growth, thus ensuring good battery cyclability(Cheng et al., 2016; Jeong et al., 2020). By removing the one-sidedness, in continuation, an overview on the possibility of lignin as a bio-based electrolyte material and the associated enhancement in electrochemical performance is given.

5.2.2 Lignin-based materials as bio-derived electrolytes

The potential of lignin as an electrolyte has been indicated by several studies. Cao et al. (Cao et al., 2019)have prepared a composite electrolyte based on sulphonated lignin and exfoliated boron nitride that realizes 92% capacity retention after 1800 cycles at 4 mA/cm^2 and inhibits dendrite formation effectively. Shi Wang et al developed (S. Wang et al., 2018) a self-supporting film consisting of a hydrophilic copolymer and lignin. The high ionic conductivity of the separator is 10^{-3} S/cm, and the lithium-ion transfer number was determined to be 0.63, which outperformed a commercial separator. Other convincing results come from Bo Liu et al. (B. Liu et al., 2018), who developed a lignin/PVP composite membrane with improved thermal stability, an extended electrochemical window (4.8 V) and an enhanced mechanical capacity of 670%, with a conductivity of 2.52×10^{-3} S/cm. In comparison, although some synthetic polymer electrolytes such as PETEA gel achieve slightly better performance (601.2 mAh/g at 1C; conductivity of 1.13×10^{-2} S/cm), lignin-based electrolytes offer a more sustainable alternative, while providing an interesting balance between efficiency, safety and respect for the environment (M. Liu et al., 2016). Taken together, the results suggest that lignin, traditionally regarded as a by-product, can be used as a multifunctional material in the development of safer, more efficient and environmentally sustainable lithium-ion batteries.

5.2.3 Pectin-based materials as solid electrolytes

Like lignin and cellulose, pectin is gradually establishing itself as a biopolymer of interest in the field of biobased electrolytes(J. Huang et al., 2025). Thanks to its macromolecular structure rich in functional groups and its wide availability, it offers interesting prospects for the development of energy storage devices that are both efficient and environmentally friendly. Thanks to its carboxyl-rich structure, it can efficiently complex metal cations-Mg^{2+}, Li^{+}, Zn^{2+} and promises a potential application in solid polymer electrolytes(S. Ahmed et al., 2023). Recently, a magnesium nitrate-doped pectin film has proved capable of an ionic conductivity of the order of 7.7×10^{-4} $S.cm^{-1}$ at room temperature, an Mg^{2+} cation transfer number close to 0.29, and electrochemical stability up to 3.8 V. In this respect, it represents a potential opportunity for pectin in the development of eco-compatible energy storage devices such as magneto-ionic batteries, supercapacitors, or biodegradable sensors (Patra et al., 2024). Furthermore, its biopolymer (chitosan, alginate) or nanomaterials compatibility opens prospects in designing high-performance, multi-functional composite materials. Pectin today is positioned at the crossing point of green chemistry, electrochemical technologies, and material engineering, offering a fascinating example of residual biomass conversion into high added-value applications.

Results reported in the recent literature highlight the important role that is played by the anionic backbone of pectin in the creation of active coordination sites with metal ions, a crucial mechanism through which ionic dissociation is favored, hence considerably enhancing the conductivity of biobased polymer electrolytes (Harikumar & Batabyal, 2024). Such a structural characteristic has been exploited in the formulation of flexible electrolyte films, such as those from pectin/PEG/$CaCl_2$ blends, where calcium chloride acts as a cross-linking agent.

The application of these films in symmetrical supercapacitors has achieved a remarkable specific capacitance (879 mF cm^{-2}) and excellent cycling stability (82.92% retention after 5000 cycles), demonstrating their relevance for portable, flexible and biocompatible energy devices.

Furthermore, the versatility of pectin in different ionic environments is well represented by the study by Kiruthika et al. (Kiruthika et al., 2020), who combined this biopolymer with $MgCl_2$ for magnesium battery electrolytes. Of the different tested compositions, the 30% pectin: 70% $MgCl_2$ ratio offered the highest ionic conductivity (1.14×10^{-3} S cm^{-1}), suggesting a favorable synergy between the functional groups of pectin and the magnesium ions. Other works, such as those by Perumal et al. (Perumal et al., 2018), confirmed the performance of pectin in association with LiCl, whose conductivity reached up to 2.08×10^{-3} S.cm^{-1}. This performance is well correlated to a mainly amorphous structure, propitious for ionic mobility owing to the flexibility of the polymer chain, which favors segmental displacement and hopping-type conduction mechanisms. Lately, Perumal (Perumal et al., 2019) explored an electrolyte based on pectin and $LiClO_4$, featuring a conductivity of 5.15 $\times 10^{-5}$ S.cm^{-1} at room temperature. Although that value is much lower compared to what was achieved with LiCl, it is enough to explain how lithium ions will remain the main charge carriers in these systems, hence exploring the ability of pectin to host different ionic salts with respect to the specific applications. Indeed, in addition to plasticizers, the presence of dissociative salts such as LiTFSI, together with ethylene carbonate, enabled the attainment of a conductivity of 2.43×10^{-4} S.cm^{-1} within thermally stable amorphous matrices(Rajendran & Saratha, 2021). In fact, these results underpin how formulation represents a key step in the optimization of the properties of pectin-based solid electrolytes.

Recently, Eswaragomathy et al. (Eswaragomathy et al., 2023) prepared a zinc ion-conducting membrane from pectin and $ZnCl_2$ by solution casting. XRD and FTIR analysis confirmed the amorphous character and ionic complexation of the membrane, respectively. The optimal formulation (40% pectin: 60% $ZnCl_2$) showed an excellent ionic conductivity of 4.49×10-3 S.cm^{-1} and an electrochemical window of 2.36 V. The zinc-ion battery based on this membrane reached a 1.84 V open-circuit voltage, showing that this solution was viable in electrochemical systems. Meanwhile, Vahin et al. (Vahini et al., 2019) investigated pectin-based electrolytes doped with sodium nitrate ($NaNO_3$), for which the conductivity increased from 2.7 $\times 10^{-9}$ S/cm of pure pectin to 4.2 $\times 10^{-6}$ S/cm of the best formulation, corresponding to 9% $NaNO_3$. The dielectric properties and thermally activated behavior of these electrolytes prove their suitability for applications involving low-power devices.

Mishra et al. (Mishra et al., 2009) also prepared amidated pectin-based electrolyte membranes, which were cross-linked with glutaraldehyde. Different spectroscopic, thermal, and mechanical characterizations showed increased crystallinity, while thermal stability presented values inferior to unmodified pectin.

A maximum conductivity of 1.098 $\times 10^{-3}$ S.cm^{-1} was obtained for the AP-3 sample. This work, therefore, brings into focus the applicability of these membranes in low-cost fuel cell applications. All in all, the present work illustrates that pectin is a promising platform in the fabrication of sustainable and eco-friendly energy storage devices, considering its tunable chemical structure, electrochemical performance, and compatibility with a wide range of metal ions.

5.2.4 Starch-based materials as bio-derived electrolytes

Within this frame, other natural polysaccharides, like starch, are also attracting increased interest in the study of functional biopolymers. For instance, polymer electrolytes based on Manihot esculenta starch have been developed and integrated into supercapacitor-type devices(Zahid et al., 2018). These materials have exhibited good ionic conductivity, besides satisfactory electrochemical capacity and accelerated biodegradation behavior in a natural environment. All these reinforce their interest within the frame of a circular economy and low-carbon technologies(Dalwadi et al., 2023). The use of such biopolymers will enable a reduction in production costs while providing safety of use and environmental compatibility, especially at the end of their life cycle(S. Ahmed et al., 2023). (T. Tiwari et al., 2011)pointed out the interest in using a solid electrolyte based on potato starch reinforced by a cross-linking agent (glutaraldehyde) and a plasticizer (PEG). The NaI-doped system showed remarkable ionic conductivity (~10^{-4} S/cm) and high humidity independence, up to 80% RH, which gives it potential in devices operating in open environments. Such performance results from the synergy created between chemical cross-linking and plasticization, facilitating ionic transport by the formation of extended channels favorable to charge carrier migration. IR analysis confirmed the formation of a starch-salt ion complex, which underlines the key role played by the molecular interactions in the optimization of the conduction. Similarly, (Jothi et al., 2022) have prepared solid proton electrolytes, based on a corn starch-polyvinylpyrrolidone mixture, doped with ammonium acetate salt. The optimal system, AA-60, reached a conductivity of 1.09×10^{-6} S/cm. An Overlapping Large Polaron Tunneling (OLPT) was identified as the main conduction mechanism. This kind of conduction is based on coherent carrier hopping across long distances provided by the semi-crystalline structure of the polymer. A high proportion of free ions, combined with high cation mobility, characteristic of a good balance between ion conduction and structural stability, is also found in the material. An open circuit voltage of 0.89 V and an excellent autonomy of 63 h were delivered when integrated into Liew et al. (Liew & Ramesh, 2015) incorporated an ionic liquid, 1-butyl-3-methylimidazolium hexafluorophosphate ($BmImPF_6$), into a starch and $LiPF_6$ matrix; this significantly enhanced the ionic conductivity (~2×10^{-4} S/cm) and electrochemical properties of a double-layer supercapacitor. This ionic liquid increased the amorphous phase and decreased the glass transition temperature (Tg), supporting ion mobility, which enlarged the electrochemical stability window up to 2.9 V.

The presented work shows how small additions of ionic compounds in biopolymer matrices can finely tune the thermomechanical and electrochemical properties of systems. In the work by (Hadad et al., 2022b), carboxymethylated starch (CMS) and acetate starch (SA) were used as binders in solid and gel electrolytes for LIB applications. By crossing these starches with polyvinyl alcohol, a stable polymer network with up to 1.13×10^{-2} S/cm conductivity was obtained in gelled form. Batteries designed this way showed an electrochemical capacity over 210 mAh/g, as well as charging efficiencies close to 100% in successive cycles. Such performances reflect the effectiveness of the starch chemical structuring in improving the electrode/electrolyte interface-a central element for LIBs. (Hapuarachchi et al., 2020) modified starch with PEG chains to optimize the mechanical properties of silicon-based electrodes. This biopolymer reached an initial capacity of 3486 mAh/g with 98% capacity retention from the second cycle onward. These characteristics can be ascribed to starch hydroxyl groups, which enable good adhesion to Si nanoparticles, while PEG groups increase lithium-ion diffusion. The duality of such an adhesive and conductive role makes this system of particular interest for Si-based anodes, which are known to have problems in volume stability. A fluorinated starch-based binder modified with dopamine, forming a cross-linked network, has been reported by Biyu Jin et al.(Jin et al., 2021), contributing to the structural integrity of electrodes. This binder reached an initial discharge capacity of 3801 mAh/g, with 88% capacity retention after 100 cycles.

Reversible interactions between polymer functional groups and silicon reinforce electrode cohesion, preventing mechanical degradation. The influence of pH on mechanical and electrochemical properties of cassava starch-based films was investigated by Almario(Almario et al., 2019). Films elaborated at moderate pH (5-9) presented enhanced flexibility and stability, while films obtained at acid pH were brittle. This difference is related to the influence of pH on starch crystallinity, which conditions the proportion of amorphous phase favorable to ionic conduction. Electrochemical measurements by cyclic voltammetry indicated higher peak intensities at acid pH but less stability, underlining that for optimum performance, a compromise between reactivity and stability must be found. Most recently, in the domain of supercapacitors, Mario Rapisarda et al.(Rapisarda et al., 2021) have fabricated an electrode based on graphene oxide gel and starch, which showed a specific capacity of 173.8 F/g with exceptional stability over more than 17000 cycles.

Molecular interactions between starch and GO favored structuring that was propitious to charge storage, underlining the interest of biopolymers for ecological and sustainable storage devices. (Zuo et al., 2022) also developed a Manihot esculenta starch-based gel electrolyte for flexible zinc-air batteries. The gel provided high conductivity, good flexibility, and remarkable stability over 200 cycles, with a power density of 84 mW/cm^2 and a specific capacity of 752 mAh/g. This result validates the application of starch biogels in portable systems. Starch-based materials offer, therefore, considerable advantages whether used as electrolytes or matrices in energy storage devices, in terms of durability, efficiency, and environmental compatibility. Chemical modification of starch, through the insertion of functional groups such as PEGs or fluorinated groups, enables the modification of its properties to meet the demands of advanced energy storage technologies (table 5). These developments, in addition to their encouraging electrochemical performance, turn starch-based biopolymers into one of the most promising materials for future applications in energy.

Table 5. Starch-Based Electrolytic Materials for Energy Storage

Materials	Application	Key findings	Observations	Ref.
Potato starch + NaI + glutaraldehyde + PEG	Solid electrolyte for batteries	Conductivity ~10^{-4} S/cm, good humidity stability (~80% RH)	Synergy between cross-linking and plasticization; starch-Na^+ ionic interactions promoting conduction	(T. Tiwari et al., 2011)
Corn starch + PVP + ammonium acetate	Proton battery	Conductivity 1.09×10^{-6} S/cm; operating time 63 h; open-circuit voltage 0.89 V	OLPT conduction mechanism; high proportion of free ions and good cation mobility	(Jothi et al., 2022)
Starch + $LiPF_6$ + ionic liquid ($BmImPF_6$)	Supercapacitor	Conductivity ~2×10^{-4} S/cm; electrochemical stability up to 2.9 V	Increase amorphous phase, reduce Tg, improve ionic mobility	(Liew & Ramesh, 2015)

continued on following page

Table 5. Continued

Materials	Application	Key findings	Observations	Ref.
PEG-modified starch	LIB	Initial capacity 3486 mAh/g; 98% retention from second cycle onwards	OH (starch) and PEG groups promote adhesion and diffusion of Li^+ ions.	(Hapuarachchi et al., 2020)
Cassava starch	Solid electrolyte for LIB	Greater flexibility at pH 5-9; high electrochemical peaks at acidic pH	pH influences crystallinity/ amorphism, and hence flexibility and electrochemical stability	(Almario et al., 2019)
GO gel + starch	Supercapacitor	Capacity 173.8 F/g; > 17000 cycles	Starch-GO interactions favor structuring for charge storage	(Rapisarda et al., 2021)
Cassava (Manihot esculenta) starch gel	Flexible zinc-air battery	Power density 84 mW/cm^2; specific capacity 752 mAh/g; > 200 cycles	Flexibility and conductivity for portable devices; gel structure promotes ion diffusion	(Zuo et al., 2022)
Starch + DMAc + different concentrations of LiCl	Solid ionic conductors, sensitive membranes	Increased conductivity, high water absorption, reduced crystallinity and enhanced thermal stability	LiCl increases hydrogen interactions with starch, reduces crystallinity, promotes plasticity; water also improves conductivity	(Ning et al., 2009)
Starch/PVA-based hydrogel + glycerol + $CaCl_2$	Supercapacitorand flexible strain sensors	Improved conductivity, extensibility (>790%), specific capacitance of 107.2 mF/cm^2, high sensitivity (gauge factor 3.422)	Multiple hydrogen bonds between components, improved PVA/starch compatibility, anti-freeze effect of $CaCl_2$ and enhanced plasticity thanks to glycerol	(J. Lu et al., 2021)
TSP (Tamarind Seed Polysaccharide) + LiBr	Electrolytic membranes for lithium-ion batteries	Ionic conductivity of 4.83×10^{-4} S/cm, open circuit voltage of 1.63 V, amorphous structure favorable to ionic mobility	Interaction between TSP functional groups and Li^+ ions confirmed by FTIR; amorphization detected by XRD facilitating ion transport	(Premalatha et al., 2017)

5.3 Bottom of Form Bio-Based Polymeric Binders for Electrochemical Devices

While the role of biopolymers in the design of more durable separators and electrolytes has been highlighted, there is also increasing attention being paid to their application as binder materials in electrodes. These binders show great importance in maintaining the adherence of active materials with the current collector while contributing toward the mechanical and electrochemical stability of electrodes(Hu et al., 2023; Y. Wang et al., s. d.; Yoon et al., 2024). Traditionally, fluoropolymers such as PVDF, PTFE or Nafion have been widely used in this context(L. Fan et al., 2020; Shyu & Wang, 2021; Teng et al., 2013). However, their non-biodegradable nature, potential toxicity and unsuitable behavior in certain electrolytic environments are encouraging the search for bio-based solutions(Hu et al., 2023; Jin et al., 2021). In this respect, binders derived from renewable resources, such as cellulose, are promising solutions, combining performance and respect for the environment. To use cellulose as a binder in supercapacitors, the first step is to transform it into a slurry. The various types of cellulose are first dissolved in a solvent, such as water or an ionic liquid, then stirred for about an hour at room temperature, followed by the addition of the active and conductive materials.

The mixture is homogenized for an hour with a magnetic stirrer. Finally, after application to the current collector(Böckenfeld et al., 2013; Krause et al., 2011; Kühnel et al., 2011).This method enables the homogeneous dispersion of electrode components and takes advantage of the intrinsic hydrogen-bonding ability of cellulose to ensure mechanical cohesion and flexibility(Akhlaq et al., 2023). Various recent works have been focused on the use of cellulose-based binders in electrodes, underlining how their structure, morphology, and processing conditions influence electrochemical performance. Mian et al (Mian et al., 2023) demonstrated, for the first time, the capability of CNFs to function as green binders for electrodes made from activated biochar, which exhibited a 1.4-fold improvement in specific capacitance, yielding 268.4 F/g at 5 A/g, over PVDF. Improved wettability and better access to the electrolyte resulted in reduced internal resistance of 0.76 Ω, leading to excellent cycling stability of 96.2% after 10000 cycles. These findings established that CNFs can be used not just as structural binders but as active performers in such electrochemical systems.

In a parallel, (Fone et al., 2024) explored ethyl cellulose as a binder for aqueous systems. The hydrophobic nature of ethyl cellulose provided a stable electrode structure with strikingly improved capacitance compared to the conventional binders. This must be considered important, as even the nonhydrophilic cellulose derivatives can be optimized for high-performance applications, provided they interact appropriately with the electrode matrix and electrolyte. Interesting results show that not all studies report enhancements in performance due to CNFs. (Andres et al., 2018) have recorded that CNFs yielded lower capacitance and higher resistance than MFC. This was attributed to the dense film-forming tendency of CNFs, which may impede ion mobility and further decrease porosity. In contrast, MFC showed better dispersion and conductivity, indicating that morphological properties and processing methods are more dominant over the performance of binders. Another great example that demonstrated the structural diversity of CNFs was the work of (R. Chen et al., 2021), who used CNFs as a scaffold in a three-dimensional graphene-carbon aerogel composite. In this respect, CNFs acted as both binder and structure-directing agent to create a lightweight aerogel which featured outstanding mechanical integrity at 240 kPa and high areal capacitance of 1625 mF cm^{-2}. This multifunctionality of CNFs should be considered a breakthrough in designing binders that links mechanical robustness with electrochemical activity. More recently, (Q. Yuan & Ma, 2021) incorporated CNFs into a MoS_2-polypyrrole film and improved the conductivity from 9.70 S/cm up to mechanical strength of 21.3 MPa. The as-prepared film delivered a specific capacitance of 0.734 F.cm^{-2} with a retention of 84.5% after 2000 cycles. Such synergy points out the capability of CNFs to enhance ionic and electronic pathways while bringing flexibility, so crucial in wearable or flexible energy devices. Further, this work by (Ghodhbane et al., 2023) on nanocellulose as a binder for flexible supercapacitors contributes meaningfully to the ever-growing literature of cellulose-based binders in energy storage applications. In fact, the results of cellulose, especially in its nanocellulose form, will contribute in many respects: from flexibility and biocompatibility to providing excellent electrochemical performance. This work has explained the ways in which the 1D fibrous hierarchical structure of cellulose improves the electrochemical performance, mentioning that 4-6 wt.% cellulose is the best proportion considering the great rise in specific capacitance and cycling stability. This result is according to the work carried out by (Liao et al., 2022) who also illustrated the importance of cellulose nanofibers in composite aerogels. In Liao's study, it was indicated that a small amount of CNF effectively improved the mechanical properties of the aerogel with no sacrifice in the specific surface area, hence contributing to high capacitance and long-term stability. Both studies focused on cellulose's role in ensuring high porosity and conductivity, quintessential features of high-performance supercapacitors.

These results are in agreement with the work done by (Qu et al., 2021), which indicated that nanocellulose can be added to a composite material for improved electrochemical performance. Qu's work on carbon nanotubes in porous Co_3O_4 proved that nanocellulose not only improved the mechanical properties of the composite but also resulted in the enhancement of electrochemical performance for such composite materials, especially symmetric supercapacitors. The synergy that nanocellulose confers on conductive materials in various forms, from carbon nanotubes to metal oxides, underlines the flexibility of cellulose as a binder for various configurations and its role in improving overall performance. These studies show that cellulose-based binders, either nanocellulose or cellulose nanofibres, improve supercapacitor performance due to their effect on porosity, conductivity, and mechanical stability-all factors relevant to high capacity and durability.

Nowadays, other biopolymers such as lignin have gained new attention for their aromatic structure and functional diversity, made readily available as by-products from the paper industry. Unlike previous trends where the main applications of lignin were thermal or as a carbon precursor(Muddasar, Culebras, et al., 2024), its repositioning in recent research as an electroactive component underlines its multifunctionality in rechargeable batteries. As a binder, the interactions of lignin with active material surfaces can happen via hydroxyl, methoxy, or carboxyl groups, directly impacting the SEI formation and stability(Bryntesen et al., 2022). In return, these interactions, as discussed by Bryntesen et al. (2022), modulate the local electrochemical environment and participate not only in cycling stability but also in improved ionic conductivity and reduced long-term degradation phenomena(Bryntesen et al., 2022; Muddasar, Culebras, et al., 2024).

In addition, its natural viscoelastic behavior can help absorb mechanical stresses induced by volume variations during charge and discharge cycles, a critical factor for electrode materials with high expansion rates, such as silicon or sulfur(Qin et al., 2024; Tenhaeff et al., 2014). In this context, several studies have explored the use of lignin as an electrochemical binder (Table 6), among the study by (Domínguez-Robles et al., 2017), which demonstrated the feasibility of using different lignin extracted from the same biomass (wheat straw) as binders in graphite-based anodes. Whatever the extraction method (kraft, soda, organosolv), all the lignin achieved stable cycling over 50 cycles, with a specific capacity of around 310 $mAh.g^{-1}$. The better thermal performance of the organosolv lignin did not induce a significant difference on the electrochemical scale, suggesting that lignin can tolerate some structural variability without compromising its binder function. also, (C. Luo et al., 2018) designed a biobased hybrid binder by grafting sodium polyacrylic acid (PAA-Na) onto an alkaline lignin. This branched copolymer, used in silicon anodes, demonstrated a remarkable capacity of 1457 $mAh.g^{-1}$ after 100 cycles at 0.2 C, far surpassing the CMC used as a reference.

This behavior is attributed to the binder's increased absorption capacity and structural flexibility, able to compensate for large volumetric variations of silicon. Other approaches consist of the incorporation of lignin into multi-functional composite matrices. (T. Chen et al., 2016) have prepared a heat-treated Si-lignin composite in which lignin acts as binder and conductive carbon precursor. Such material exhibited outstanding specific capacitance, up to 2106 $mAh.g^{-1}$ after 100 cycles at 0.2 $A.g^{-1}$, demonstrating the interest of lignin in the conception of integrated carbon structures. Regarding less mature sodium-ion batteries, which are promising for stationary applications. (Gond et al., 2021) tested LgSA as a binder in a hard carbon-based anode. Indeed, compared to CMC, it allowed the formation of a finer and more stable SEI, while capacity remained stable over more than 400 cycles, underlining an improved control of the electrode/electrolyte interface. Increasing the mechanical properties of the film with the addition of SBR-a synthetic copolymer confirms the relevance of hybrid systems. On the other hand, (Y. Ma et al., 2019) extended the field of application of lignin to high voltage cathodes, such as $LiNi_{0.5}Mn_{1.5}O_4$, and showed its capability in limiting oxidative degradation through its antioxidant properties. This stabilization resulted in a capacity retention of 80% after 1000 cycles at 1 C, that is, twice the value obtained with PVDF. (H. Lu et al., 2016) underlined the universality of lignin as a binder by applying it for either a graphite anode and a $LiFePO_4$ cathode, showing results comparable to those obtained with conventional binders and confirming the versatility of lignin for two-sided Li-ion systems. As a result, this work demonstrates that lignin, in its natural or modified form, constitutes a credible alternative to petro-based binders in rechargeable batteries. Its biobased origin, chemical modularity, and multifunctionality make it a prime candidate for future generations of sustainable electrochemical materials. In the same period, cellulose, thanks to its solubility in water, its compatibility with different active materials, and its ability to create homogeneous networks that ensure excellent mechanical cohesion and electrochemical stability, also established itself as a high-performance binder. With respect to cellulose, it has better processability in aqueous media and some real experience in devices needing high formulation flexibility, such as supercapacitors and some metal oxide batteries. On the contrary, lignin can be distinguished for its aromatic interactions and its capability to stabilize electrochemical interfaces, thus providing further added value to SEI management and electrodes with a high volume of change. These two biopolymers could open new frontiers in the design of multifunctional binders that meld electrochemical performance with environmental compatibility and low cost.

Table 6. Lignin as a Binder in Energy Storage Systems

Lignin type	Electrochemical system	Electrochemical performance	Ref
Raw lignins (kraft, soda, organosolv)	LIB (anode)	Stable capacity over 50 cycles	(Domínguez-Robles et al., 2017)
Natural lignin	LIB (cathode 5 V)	94.1% retention after 1000 cycles	(Y. Ma et al., 2019)
Purified lignin (leached)	LIB (anode/ cathode)	305 / 148 $mAh.g^{-1}$ respectively	(H. Lu et al., 2016)
Sodium lignosulfonate (LgSA)	SIB (anode)	>95% retained capacity (40 cycles)	(Gond et al., 2021)
Sodium polyacrylate-grafted lignin (PAL-NaPAA)	LIB (anode)	1914 $mAh.g^{-1}$ after 100 cycles at 0.84 $A.g^{-1}$	(C. Luo et al., 2018)

continued on following page

Table 6. Continued

Lignin type	Electrochemical system	Electrochemical performance	Ref
Lignin + Si, pyrolysed at 600°C	LIB (anode)	3086 mAh.g^{-1} (initial), 2378 mAh.g^{-1} after 100 cycles	(T. Chen et al., 2016)
Lignin (BP-based separator host matrix)	Zinc-ion battery	Ionic conductivity: 9.1 × 10^{-3} S.cm^{-1} (in 2M $ZnSO_4$), mechanical strength: 250 MPa; capacity: 236 mAh.g^{-1} at 0.1C; improved H^+/cation conduction and electrolyte absorption	(D. Yuan et al., 2019)
Lignin + copolymer (composite film)	Lithium-ion battery	Ionic conductivity: 0.63, specific capacity: 150 mAh g^{-1} at 1 C, over 450 cycles, inhibition of lithium dendrite growth	(S. Wang et al., 2018)
Lignin + polyvinylpyrrolidone (PVP)	Lithium-ion battery	Ionic conductivity: 2.52 × 10^{-3} S.cm^{-1}, capacity: 110 mAh.g^{-1} at 10 C, excellent capacity retention after 400 cycles, dendrite inhibition	(B. Liu et al., 2018)
Lignin	Lithium-ion battery	Improved ionic conductivity, ion transport between cathode and anode, suppression of dendrite formation	(Cheng et al., 2016)
Lignin + polymers (composite)	Lithium-sulfur battery	Ionic conductivity: 1.13 × 10^{-2} S.cm^{-1}, capacity at 1 C: 601.2 mAh.g^{-1}, capacity retention: 81.9% after 400 cycles, suppression of polysulfide shuttling effect	(Y. Liu et al., 2016)
Lignin + LCP	Lithium-ion battery	Mechanical strength: 1.6 MPa, elongation: 16%, thermal stability up to 300°C; ionic conductivity: 0.63 × 10^{-3} S.cm^{-1}; Li^+ transfer no.: 0.63; electrochemical window: 4.7 V; capacity: 150 mAh.g^{-1} (1C, 450 cycles), 110 mAh g^{-1} (10C).	(S. Wang et al., 2018)

In this frame of the transition to non-polluting materials, the interest in biopolymers as binders, separators, or electrolytes is just one element in a larger eco-design strategy of the next-generation electrochemical components. Nevertheless, the total performance of electrochemical storage devices depends largely on nature and the architecture of active materials. In this sense, carbon-based materials hold a predominant position among other materials, basically because of their high electrical conductivity, chemical stability, easily scalable specific surface area and compatibility with a great number of electrochemical configurations. Therefore, the investigation of carbon-based electrodes becomes a necessary phase to the performance optimization of the new generation of batteries and Bio-based polymers used as binders, separators, and electrolytes in supercapacitors and batteries are to be reckoned as the next-generation energy storage devices that are not only durable but also exhibit superior electrochemical performance. By addressing the main issues such as porosity, conductivity, and mechanical integrity, these biopolymers become indispensable in the ultimate performance optimization of such systems. This move is in line with the ongoing transition to renewable and sustainable materials, thus giving the planet-friendly alternatives to the usual synthetic polymers. In addition, the use of biomass-derived carbons in the electrode materials to couple with natural polymer binders is like a rocket fuel effect in the performance parameters of energy density, cycle life, and eco-friendliness of future supercapacitors and batteries, which can reach very high.

6. BIOMASS-DERIVED CARBON-BASED ELECTRODES

In the past, carbon-based materials like activated carbon, carbon nanotubes (CNTs), and graphene sheets have been the standard and have shown impressive electrochemical properties, thus have been widely used in energy storage and conversion devices, mostly supercapacitors and batteries. There are many studies which confirm the statements such as (Antiohos et al., 2013; Inagaki et al., 2010; Tatrari et al., 2021). Nevertheless, the manufacturing of these materials is still largely dependent on fossil fuel resources, which brings about worries of price, geopolitical issues, and the ecologic side effect(Kamgang Maffo et al., 2024).

In this context, plant biomass represents a valuable alternative. Mainly made up of carbohydrates, lignin and other biopolymers, it is abundant, inexpensive and naturally porous, making it an ideal raw material for developing high-performance carbon-based materials(Elyasi et al., 2024; Z. Zhang, Yang, et al., 2019). The world's biomass is around 4.5×10^{11} tonnes a year, mostly of plant origin, offering enormous potential for energy recovery, while also forming part of a circular economy by recovering organic waste(Casau et al., 2022; R. Chen et al., 2021). Various types of biomasses have been used to produce high-performance carbon materials (table 7), including coconut shells, cotton, green tea waste, etc.

Table 7. Various Biomasses Used to Produce High-Performance Carbon Materials

Biomaterial Precursor	Specific Surface Area (m^2/g)	Electrolyte	Specific Capacitance ($F.g^{-1}$)	Cycles	Ref.
***Borassus flabellifer* flower**	633.43	1 M KOH	238.2	1000	(Sivachidambaram et al., 2017)
Soybean root	2690	6 M KOH	328	10000	(N. Guo et al., 2020)
Pomelo peel	2167	1 M TEA BF4-PC	163	10000	(F. Sun et al., 2018)

continued on following page

Table 7. Continued

Biomaterial Precursor	Specific Surface Area (m^2/g)	Electrolyte	Specific Capacitance ($F.g^{-1}$)	Cycles	Ref.
Cotton	1508	6 M KOH	278	10000	(W. Jiang et al., 2019)
Green tea waste	1057	1 M H2SO4	162	5000	(Sankar et al., 2019)
Bean dregs	3104	6 M KOH	268	-	(Y. Yang et al., 2017)
Kapok flower	1904.1	6 M KOH	286.8	5000	(Zheng et al., 2021)
Coconut shell	1506	1 M TEMABF4/ propylene carbonate (PC)	92	5000	(Xia et al., 2018)
Shell of broad beans	655	6 M KOH	202	3000	(G. Xu et al., 2015)
Pleurotus eryngii	3255	6 M KOH	236	15000	(Y. Yuan et al., 2018)
Eucommia ulmoides wood	1456	1 M H2SO4	158	5000	(G. Sun et al., 2018)
Osmanthus flower	1463	6 M KOH	255	10000	(Zou et al., 2018)
Tea leaves	911	6 M KOH	167	16	(X. Song et al., 2019)
Sakura	1433.8	6 M KOH	265.8	2000	(F. Ma et al., 2019)
Tassel tree flower	3310	6 M KOH	394	5000	(T. Ma et al., 2019)
Tasmanian Blue Gum	971	1 M KOH	212	5000	(Prasankumar et al., 2022)

Each biomass contributes specific properties depending on its composition and intrinsic structure. For example, (Prasankumar et al., 2022) prepared activated carbons from Tasmanian Blue Gum bark, showing a specific surface area of 971 m^2/g and an excellent 212 F/g capacity at 1 A/g, as well as a high CO_2 capture capacity (4.5 mmol/g). (Chu et al., 2020) developed a hierarchical porous carbon from Franchet groundcherry pericarp, obtaining a specific capacity of 335 F/g and an energy density of 22.3 Wh/kg. Similarly, (Vijayakumar et al., 2019) have used industrial cotton waste to obtain electrodes with a very high energy density (30.94 Wh/L) and a lifetime in excess of 15000 cycles. (Zheng et al., 2021), using kapok flowers, produced a high-performance mesoporous carbon with a capacity of 286.8 F/g and excellent stability (97.4% after 5000 cycles).

Other notable works include those by (H. Liu et al., 2021), using camphor leaves to obtain a porous graphene naturally doped with O/N (397 F/g), and those by (W. Jiang et al., 2019), who developed a hollow tubular carbon optimizing ionic transport and mechanical stability. Thesecases highlight that the selection of precursors and the exact control of the activation process are the main factors for achieving optimal electrochemical performance. Advanced strategies have also been explored to further improve the properties of biomass-derived carbons. (Y. Yang et al., 2017) have transformed bean residues into porous carbons with a very high specific surface area (up to 3104 m^2/g) and a capacity of 268 F/g. (Y. Yuan et al., 2018), using fungi (*Pleurotus eryngii*), prepared carbons with a specific surface area of 3255 m^2/g and excellent cyclic stability (>93% after 15000 cycles). (T. Ma et al., 2019) have developed a simple pyrolysis-activation method to produce 3D honeycomb carbons, reaching 394 F/g and enabling the production of flexible supercapacitors.

Other innovative approaches include those of the work on Osmanthus flowers by Zou et al., on Borassus flabellifer flowers by (Sivachidambaram et al., 2017), and that of Fei Ma et al. (F. Ma et al., 2019) using sakura petals, all of which have shown high electrochemical performances resulting from controlled porous structuring. In the same way, (C. Wang et al., 2019) created a sulphur-doped carbon-silica hybrid from bamboo leaves, thus proving the stability of the material for more than 30000 cycles. Besides that, (Z. Liu et al., 2018) converted reed leaves into a three-dimensional Si-C composite for lithium-ion batteries, thus illustrating the adaptability of using plant biomass for applications other than supercapacitors.

The entire body of work demonstrates that plant biomass stands out as a highly valuable source in the creation of eco-friendly carbon . The wide variety of precursors available, combined with the flexibility of conversion processes, means that the porous structure, doping level, graphitization and electrical conductivity can be finely tuned. The results obtained (table 7) indicate that by optimizing the nature of the precursor, the activation conditions and the post-treatment methods, it is possible to rival or even surpass the performance of conventional commercial carbons.

What's more, using these approaches to recycle agricultural and industrial waste not only helps to reduce environmental impact, but also contributes to the emergence of new local economic sectors based on renewable resources. So, to encourage the development of this eco-responsible approach, it is essential to continue research into the precise engineering of the porous structure, and to gain a better understanding of the activation and transformation mechanisms to obtain a carbon with optimized performance.

In this respect, the transformation of biomass into carbons is mainly carried out by thermochemical methods, which consist of breaking down the main components of plants in an inert atmosphere at high temperature(Ashok Kumar et al., 2023; Kariim et al., 2022). The four main approaches are: high-temperature pyrolysis, hydrothermal carbonization and molten-salt carbonization. In addition, physical or chemical activation can be used to modulate morphology, graphitization and the incorporation of heteroatoms, thereby optimizing electrode performance.

7. CONVERSION OF BIOMASS INTO CARBON MATERIALS: PYROLYSIS AND ACTIVATION STRATEGIES

7.1. Hydrothermal carbonization

The hydrothermal process is a low-impact method that involves transforming plant precursors under high pressure and high temperature, using water as a solvent (Yoshimura & Byrappa, 2008). Hydrolysis, dehydration, decarboxylation, polymerization, and arylation are the five steps involved in this process. HMF and furfural are some of the resultant products emanating from these steps, followed by transformation to cyclic aromatic frameworks (Antonetti et al., 2015; Pavlovič et al., 2013). The product may be spherical or with interpenetrating porosity, higher reactivity due to lower relative density and lower oxygen surface content (H. He et al., 2023).

However, directly carbonized hydrothermal products have a low specific surface area and limited porosity, which reduces their effectiveness, particularly as electrode materials. Further activation is therefore required to develop their porous structure, increase their specific surface area and improve their electrochemical properties, making these materials more suitable for advanced applications.

7.2. High-temperature pyrolysis

Pyrolysis represents a fundamental thermochemical process for converting lignocellulosic biomass into carbonaceous materials(K n et al., 2022, 2022). Carried out in an anaerobic atmosphere, this thermal degradation triggers a series of complex reactions, leading to the release of volatile compounds (H_2O, CO, CO_2, CH_4) and the formation of a solid carbonaceous matrix (figure 8).

Figure 8. Thermal decomposition pathways of cellulose showing the formation of bio-based and charcoal at different temperatures. Blue arrows indicate decomposition and volatilization processes, while red arrows represent polymerization and condensation reactions.

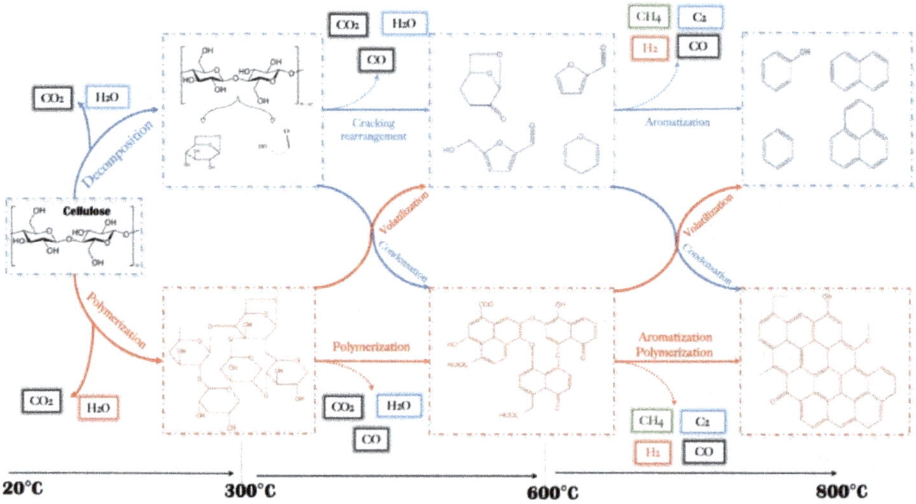

Decomposition profiles vary according to the nature of the plant macromolecules: hemicelluloses depolymerize at temperatures between 220 and 315°C, followed by cellulose between 315 and 400°C, while lignin, due to its more heterogeneous structure, degrades over a wider thermal range (200-550°C)(W.-H. Chen et al., 2023; Ong et al., 2021). Although this method is robust and suited to the recovery of a wide variety of agro-industrial residues (husks, straw, lignocellulosic waste), the carbons obtained by pyrolysis generally have a porosity that is not very hierarchical, dominated by macropores, and a relatively limited specific surface area(Q. Wang et al., 2023). These textural properties restrict their effectiveness in highly demanding applications, such as electrochemical devices. Consequently, additional treatments, in particular chemical or physical activation, are often required to optimize their functional characteristics.

7.3. Molten salt carbonization

Molten salt carbonization is a controlled pyrolysis method for preparing porous carbon from plant precursors. Theprocess is done by mixing the salts with the precursors, and then the salts are melted at high temperatures to generate a specific reaction environment (X. Zhu et al., 2023). When the heating starts, the precursors get dehydrated and decarboxylated. After a while, the salts melt and, by distributing themselves evenly within the precursors, facilitate the formation of porous structures (X. Liu & Antonietti, 2014). The activated carbon structure, which is the result of the molten salts' action, has been further developed in terms of porosity, and thus a microporous coal with a well-developed pore system has been obtained. By comparison with the hydrothermal method, molten-salt carbonization allows a higher level of carbonization to be attained and new morphologies, such as graphene-like structures, to be introduced (Plavniece et al., 2022; Q. Wu et al., 2025). Moreover, the porous carbon can be washed off the salts, and the salts can be recycled (Q. Wu et al., 2025). The method is very promising for the production of high-performance carbon materials; however, the precise mechanism is still a mystery and needs a lot more research to be able to determine the factors that influence the process (Plavniece et al., 2022).

7.4. Activation

Electrochemical devices such as supercapacitors and batteries require biochar of highly porous nature which can be derived by activation of biochar. Normally straight-channel plant-derived biochar becomes porous after its surface has been partially oxidized by the interaction with an activating agent in an inert atmosphere at high temperature (T. Zhang et al., 2023). There are two different ways of activating biochar; physical activation which uses oxidizing agents such as steam or CO_2 and chemical activation which reacts with compounds such as KOH, $ZnCl_2$ or H_3PO_4 (I. Yang et al., 2021). Even though a physically activated sample is relatively cheap and easy to prepare, the oxidized surface is limited and only slightly porous structures are obtained (specific surface area <1100 m^2/g), leading to a mismatch of the latest electrochemical devices' requirements (Shahcheragh et al., 2023; I. Yang et al., 2021).In contrast, chemical activation offers far superior performance. KOH induces the formation of complex microporous networks via a series of synergistic mechanisms, including the attack of carbon by potassium by-products, the insertion of metal vapor at high temperature, and the expansion of the graphitic network (Król et al., 2011; I. Yang et al., 2021). These effects give biochar a specific surface area more than 1500 m^2/g, increased electronic conductivity and optimal pore hierarchy (micropores for ionic storage, mesopores for electrolytic transport) (Król et al., 2011). In comparison, H_3PO_4 and $ZnCl_2$ favor the development of mesopores via catalytic dehydration or acid gasification mechanisms, particularly suited to the rapid diffusion of ions in Li-ion or Na-ion batteries(Y. Guo et al., 2025).

The processing and activation method plays a decisive role in the final porous architecture of biobased carbons. Many researchers have explored various activation strategies, adapted to the nature of the precursor and the electrochemical performance targets (Haider et al., 2025; Kwiatkowski & Broniek, 2017; Siipola et al., 2018).

Carbon-based porous materials from plant biomass are still a viable source of high potential, but there are still issues related to the trade-off between microporosity and mesoporousness and the optimization of specific capacity. A less aggressive activation method and the use of chemical doping, for instance with nitrogen, might be a way to overcome these barriers, thus improving the resistance of the materials and their performance in electrochemical applications (Temesgen et al., 2023; Y. Wang et al., 2022).

8. CHALLENGES, LIMITATIONS, AND KNOWLEDGE GAPS

The incorporation of biobased components in electrochemical units is a radical change in the energy sector to be more eco-friendly, as these are materials of a plant origin, which are still quite unexploited, and have very attractive properties from the environmental standpoint (Escobar et al., 2021; R. Guo et al., 2020; Molahalli et al., 2024a; Okoye et al., 2024c). It is a part of a sustainability approach and focuses on lower the environmental burden while minimizing reliance on non-renewable sources as depicted (Rosen, 2009; Ruiz Serrano et al., 2025). Plant biomass, being one of the most plentiful sources on the planet, is currently considered one of the main candidates to be the origin of new-generation environmentally friendly electrode materials, which are mainly used in fuel cells, supercapacitors, and batteries (Escobar et al., 2021; Z. Gao et al., 2017; X. Tang et al., 2021a). In which, lignin is the one that is highlighted because of its cheapness, renewable character, and a structure full of functional groups, which are able to stimulate electrochemical reactions(Jia et al., 2022a). Besides that, nanocellulose (NC) is becoming more and more popular, which can be explained by its good mechanical characteristics, being light, and having the ability to be changed by doping or graphitization to make the conductivity better (R. Guo et al., 2020; Molahalli et al., 2024b; X. Tang et al., 2021b).

As showing in the previous section carbon-based materials from biomass usually have a hierarchical porous structure and can be doped with heteroatoms (N, O, S, P) to facilitate ion diffusion and increase electrochemical activity (Molahalli et al., 2024a). Due to these features, they can be used as cathode or anode electrodes or in supercapacitors. The different natural sources such as coconut, cotton, sugarcane, pomelo, tamarind seeds, have been utilized by pyrolysis, hydrothermal synthesis, microwave irradiation, or laser treatment to obtain biogenic carbon nanomaterials (Sharma, 2024). These substances have the potential to be used in various spheres like energy storage, water purification, and electromagnetic shielding (Sharma, 2024).

Howeverseveral major gaps remain in the research and development of plant-based materials for electrochemical devices despite these promising

(i) Standardization and scalability of processes: While a wide range of methods (hydrothermal, pyrolytic, microwave) have been cited (Sharma, 2024), the massive production of uniform, stable, and materials with high performance is still open to question because of the exorbitant energy costs and the negative impact of these processes on the environment (Butera et al., 2020; I. Uddin et al., 2022).

(ii) Understanding structure property relationships: several studies point out the positive effects of doping and porosity (Molahalli et al., 2024a), however, the exact correlation for botanical composition (cellulose, hemicellulose, lignin), treatment, and electrochemical performance is still not clear. One example is the material like NH_4HCO_2-doped tamarind seed polysaccharide that is a proton-conducting biopolymer electrolyte (Premalatha et al., 2018), which reveals the enormous functional potential of plant resources that is still largely untapped.

(iii) Long-term durability and stability: although specific capacities and power densities are frequently reported(Xiao et al., 2024), studies on degradation mechanisms, cycle life, and the effect of humidity on bio-based electrolytes are still limited.

(iv) Characterizationand modeling tools: The hybrid nature of these materials calls for sophisticated methods to reveal their porosity, morphology, and surface chemistry (Jia et al., 2022b), along with predictive modeling to both structure and performance .

(v) Life cycle assessment (LCA): Although sustainability is one of the major points of consideration (Escobar et al., 2021; M. Gao et al., 2018b; Molahalli et al., 2024a), there are not many studies that have looked at a comprehensive supply chain assessment from growing to end of life(Albawab et al., 2020; Ghobakhloo et al., 2024; Jia et al., 2022b). A comprehensive approach using multi-criteria performance indices would allow for a realistic estimation of environmental and economic benefits.

Consequently, the area remains in need of a comprehensive and organized study of synthesis pathways, electrochemical processes, and the overall environmental friendliness of biomaterials, in addition to the fact that great potential has been revealed.It is a necessary step to fill these gaps with combined, interdisciplinary research to plant-based materials a viable way of being used in the next generation of electrochemical devices to make a considerable contribution to a sustainable energy future (Péter, 2020; Sá & Helena, 2024; P. Shrestha & Bazylak, 2023). Moreover, their massive implementation is affected by numerous factors. One of the main industrial trends is the increasing need for sustainable and high-performance materials, primarily in the energy sector (Varghese et al., 2025)This demand has led to innovation in material science, which has in turn encouraged advances in the use of biomass and carbon-rich materials for electrochemical energy storage systems (Varghese et al., 2025; Z. Zhu et al., 2024). While material performance enhancement has been acknowledged as progress, it is still one of the major challenges for large-scale applications (Z. Zhu et al., 2024). Carbon material from biomass sourced from agro-forestry residues or municipal wastes is being developed for better performance in both batteries and supercapacitors. Some examples are porous carbons obtained from lignin and nanocarbons obtained from cotton that can be used as cheap, multi-functional, and environmentally friendly electrode materials. However, as a central issue, consistent performance, scalability, and long-term stability should be faced to ensure the industrial feasibility of plant-based materials in the next generation of sustainable energy .

9. CONCLUSION & OUTLOOK

The chapter bio-derived materials to the energy storage technologies have been substantially increased and has been reviewed by the chapter bio-derived materials to the energy storage technologies has been substantially increased. The discussion has been mainly around the utilization of biopolymers as binders, separators, or electrolytes and carbons derived from biomass for the manufacturing of electrodes. These natural substitutes for standard synthetic components lead to the evolution of supercapacitors and batteries with enhanced porosity, ion transport, mechanical strength, and overall sustainability. Their use is in line with the current trend of greener and more resource-conscious energy systems. However, a few issues remain to be addressed.

The diversity of biomass sources can lead to the variation of the properties of the materials. In addition, the process of transforming biomass into high-performance carbon materials may require energy-consuming steps or the use of non-environmentally friendly chemicals. One of the issues in the matter is that of the best interaction between bio-based polymers and other parts of the device, particularly in terms of stability, conductivity, and resistance to harsh operational conditions. Besides, this area is the future to be very bright. The advancements in green processing methods and the molecular tailoring of bio-based materials will facilitate the production of such materials that are not only electrochemically superior but also have a lower environmental footprint.

The development of hybrids, where functionalized biocarbon materials are combined with smart or responsive polymers, may provide the next level of synergy, efficiency, and sustainability. Besides, the use of these materials in upcoming technologies such as biodegradable electronics, wearable energy devices, or grid-level storage is a deliberate move towards solving the escalating global demand for clean and circular energy solutions.

REFERENCES

Aakko-Saksa, P. T., Cook, C., Kiviaho, J., & Repo, T. (2018). Liquid organic hydrogen carriers for transportation and storing of renewable energy – Review and discussion. *Journal of Power Sources, 396*, 803–823. DOI: 10.1016/j.jpowsour.2018.04.011

Aazem, I., Walden, R., Jelmy, E. J., Radhkrishnan, S., John, H., & Pillai, S. C. (2024). Materials for energy-efficient systems and environmental remediation. In Kulkarni, N. V., & Kharissov, B. I. (Eds.), *Handbook of Emerging Materials for Sustainable Energy* (pp. 741–777). Elsevier., DOI: 10.1016/B978-0-323-96125-7.00029-0

Abdollahifar, M., Molaiyan, P., Lassi, U., Wu, N. L., & Kwade, A. (2022). Multifunctional behaviour of graphite in lithium–sulfur batteries. *Renewable & Sustainable Energy Reviews, 169*, 112948. DOI: 10.1016/j.rser.2022.112948

Abo-Khalil, A. G., Sobhy, A., Abdelkareem, M. A., & Olabi, A. G. (2023). Advancements and challenges in hybrid energy storage systems : Components, control strategies, and future directions. *International Journal of Thermofluids, 20*, 100477. DOI: 10.1016/j.ijft.2023.100477

Abolore, R. S., Jaiswal, S., & Jaiswal, A. K. (2024). Green and sustainable pretreatment methods for cellulose extraction from lignocellulosic biomass and its applications : A review. *Carbohydrate Polymer Technologies and Applications, 7*, 100396. DOI: 10.1016/j.carpta.2023.100396

Abouelamaiem, D., Mostazo-López, M. J., He, G., Patel, D., Neville, T., Parkin, I., Lozano-Castello, D., Morallon, E., Cazorla-Amorós, D., Jorge Sobrido, A. B., Wang, R., Titirici, M., Shearing, P., & Brett, D. (2018). New insights into the electrochemical behaviour of porous carbon electrodes for supercapacitors. *Journal of Energy Storage, 19*, 19. DOI: 10.1016/j.est.2018.08.014

Aboul Hrouz, S., Amadine, O., Sair, S., Dânoun, K., Essemlali, Y., & Zahouily, M. (2023). Nanocellulose : Extraction, Mechanical Properties, and Applications. In H. Abdellaoui, S. M. R., & S. Siengchin (Éds.), *Mechanics of Nanomaterials and Polymer Nanocomposites* (p. 105-128). Springer Nature. DOI: 10.1007/978-981-99-2352-6_6

Abraham, T. R., Sunil, K., Shah, M., Ashok, N., & Thomas, S. (2023). Energy storage devices: batteries and supercapacitors. Nanobiohybrids for Advanced Wastewater Treatment and Energy Recovery, 61.

Ahluwalia, O., Singh, P. C., & Bhatia, R. (2021). A review on drought stress in plants : Implications, mitigation and the role of plant growth promoting rhizobacteria. *Resources. Environmental Sustainability*, *5*, 100032. DOI: 10.1016/j.resenv.2021.100032

Ahmed, M. M. S., Hasan, M., Chowdhury, M. S., Rahman, M. K., Islam, M. S., Hossain, M. S., Islam, M. A., Hossain, N., & Mobarak, M. H. (2024). Prospects and challenges of energy storage materials : A comprehensive review. *Chemical Engineering Journal Advances*, *20*, 100657. DOI: 10.1016/j.ceja.2024.100657

Ahmed, S., Sharma, P., Bairagi, S., Rumjit, N. P., Garg, S., Ali, A., Lai, C. W., Mousavi, S. M., Hashemi, S. A., & Hussain, C. M. (2023). Nature-derived polymers and their composites for energy depository applications in batteries and supercapacitors : Advances, prospects and sustainability. *Journal of Energy Storage*, *66*, 107391. DOI: 10.1016/j.est.2023.107391

Akhlaq, M., Mushtaq, U., Naz, S., & Uroos, M. (2023). Carboxymethyl cellulose-based materials as an alternative source for sustainable electrochemical devices : A review. *RSC Advances*, *13*(9), 5723–5743. DOI: 10.1039/D2RA08244F

Albawab, M., Ghenai, C., Bettayeb, M., & Janajreh, I. (2020). Sustainability Performance Index for Ranking Energy Storage Technologies using Multi-Criteria Decision-Making Model and Hybrid Computational Method. *Journal of Energy Storage*, *32*, 101820. DOI: 10.1016/j.est.2020.101820

Alberts, B., Johnson, A., Lewis, J., Raff, M., Roberts, K., & Walter, P. (2002). The Plant Cell Wall. In *Molecular Biology of the Cell* (4th ed.). Garland Science., https://www.ncbi.nlm.nih.gov/books/NBK26928/

Ali, S., Rani, A., Dar, M. A., Qaisrani, M. M., Noman, M., Yoganathan, K., Asad, M., Berhanu, A., Barwant, M., & Zhu, D. (2024). Recent Advances in Characterization and Valorization of Lignin and Its Value-Added Products : Challenges and Future Perspectives. *Biomass*, *4*(3), 3. Advance online publication. DOI: 10.3390/biomass4030053

Almario, A. A., Mogollón, C. G., & Caballero, E. C. (2019). Effect of elaboration pH on the electroactivity of cassava starch solid biopolymer electrolyte films. *Rasayan Journal of Chemistry*, *12*(4), 1766–1773. DOI: 10.31788/RJC.2019.1245302

Andres, B., Dahlström, C., Blomquist, N., Norgren, M., & Olin, H. (2018). Cellulose binders for electric double-layer capacitor electrodes : The influence of cellulose quality on electrical properties. *Materials & Design*, *141*, 342–349. DOI: 10.1016/j.matdes.2017.12.041

Ani, P. C., Nzereogu, P. U., Agbogu, A. C., Ezema, F. I., & Nwanya, A. C. (2022). Cellulose from waste materials for electrochemical energy storage applications : A review. *Applied Surface Science Advances, 11*, 100298. DOI: 10.1016/j.apsadv.2022.100298

Antiohos, D., Romano, M., Chen, J., Razal, J. M., Antiohos, D., Romano, M., Chen, J., & Razal, J. M. (2013). Carbon Nanotubes for Energy Applications. In *Syntheses and Applications of Carbon Nanotubes and Their Composites*. IntechOpen., DOI: 10.5772/51784

Antonetti, C., Bonari, E., Licursi, D., Nassi o Di Nasso, N., & Raspolli Galletti, A. (2015). Hydrothermal Conversion of Giant Reed to Furfural and Levulinic Acid : Optimization of the Process under Microwave Irradiation and Investigation of Distinctive Agronomic Parameters. *Molecules (Basel, Switzerland), 20*(12), 12. Advance online publication. DOI: 10.3390/molecules201219760

Appiah, E. S., Dzikunu, P., Mahadeen, N., Ampong, D. N., Mensah-Darkwa, K., Kumar, A., Gupta, R. K., & Adom-Asamoah, M. (2022). Biopolymers-Derived Materials for Supercapacitors : Recent Trends, Challenges, and Future Prospects. *Molecules (Basel, Switzerland), 27*(19), 19. Advance online publication. DOI: 10.3390/molecules27196556

Ashok Kumar, S. S., Bashir, S., Pershaanaa, M., Kamarulazam, F., Saidi, N. M., Goh, Z. L., Ma, I. A. W., Kunjunee, V., Jamaluddin, A., Ramesh, K., Ramesh, S., Ramesh, S., & Manikam, R. (2023). A review on the recent progress of the plant-based porous carbon materials as electrodes for high-performance supercapacitors. *Journal of Materials Science, 58*(15), 6516–6555. DOI: 10.1007/s10853-023-08413-7

Asif, M. (2022). Sustainable energy transition in the 21st century. In Asif, M. (Ed.), *Handbook of Energy and Environmental Security* (pp. 27–38). Academic Press., DOI: 10.1016/B978-0-12-824084-7.00023-0

Asif, M. (2024). Sustainable Energy Transition : Technological Perspective. In *Handbook of Energy and Environment in the 21st Century*. CRC Press. DOI: 10.1201/9781032715438-1

Aslam, M. A., Ahsen, R., Usha, Z. R., Iqbal, O., Ali, S., Li, N., & Wang, Z. (2025). Dimensionality-driven sustainable biocarbon-based microwave absorbers : From bio-waste to functional materials. *Next Materials, 8*, 100584. DOI: 10.1016/j.nxmate.2025.100584

Aslan, S., & Altuntaş, D. B. (2023). Biocomposites and Bioplastics in Electrochemical Applications. In *Handbook of Bioplastics and Biocomposites Engineering Applications* (pp. 491–512). John Wiley & Sons, Ltd., DOI: 10.1002/9781119160182.ch23

Aziz, B., Woo, T. J., Kadir, M. F. Z., & Ahmed, H. M. (2018). A Conceptual Review on Polymer Electrolytes and Ion Transport Models. *Journal of Science. Advanced Materials and Devices*, *3*(1), 1–17. Advance online publication. DOI: 10.1016/j.jsamd.2018.01.002

Aziz, S. B., Abdulwahid, R. T., Mohammed, S. J., Aziz, D. M., Hamsan, M. H., Abdul Halim, N., Al-Saeedi, S. I., Karim, W. O., Woo, H. J., & Kadir, M. F. Z. (2024). Clean energy storage device derived from biopolymers with moderate charge-discharge cycles : Structural and electrochemical properties. *Journal of Industrial and Engineering Chemistry*, *130*, 673–687. DOI: 10.1016/j.jiec.2023.11.044

Babiker, D. M. D., Usha, Z. R., Wan, C., Hassaan, M. M., & El, . (2023). Recent progress of composite polyethylene separators for lithium/sodium batteries. *Journal of Power Sources*, *564*, 232853. DOI: 10.1016/j.jpowsour.2023.232853

Bandyopadhyay, S., & Nandan, B. (2025). Strategies for improving ionic conductivity and mechanical stability of solid polymer electrolytes for lithium batteries *via* physical and chemical interlocking. *Energy Storage Materials*, *78*, 104233. DOI: 10.1016/j.ensm.2025.104233

Barbinta-Patrascu, M.-E., Bita, B., & Negut, I. (2024). From Nature to Technology : Exploring the Potential of Plant-Based Materials and Modified Plants in Biomimetics, Bionics, and Green Innovations. *Biomimetics (Basel, Switzerland)*, *9*(7), 7. Advance online publication. DOI: 10.3390/biomimetics9070390

Beg, M., Saju, J., Alcock, M., Mavelil, A. T., Markapudi, P. R., Yu, H., & Manjakkal, L. (2025). Biodegradable biopolymers for electrochemical energy storage devices in a circular economy. *RSC Sustainability*, *3*(1), 37–63. DOI: 10.1039/D4SU00468J

Belmesov, A. A., Glukhov, A. A., Kayumov, R. R., Podlesniy, D. N., Latkovskaya, E. M., Repina, M. A., Ivanov, N. P., Tsvetkov, M. V., & Shichalin, O. O. (2023). Using Aquatic Plant-Derived Biochars as Carbon Materials for the Negative Electrodes of Li-Ion Batteries. *Coatings*, *13*(12), 12. Advance online publication. DOI: 10.3390/coatings13122075

Benoy, S. M., Pandey, M., Bhattacharjya, D., & Saikia, B. K. (2022). Recent trends in supercapacitor-battery hybrid energy storage devices based on carbon materials. *Journal of Energy Storage*, *52*, 104938. DOI: 10.1016/j.est.2022.104938

Berglund, J., Mikkelsen, D., Flanagan, B. M., Dhital, S., Gaunitz, S., Henriksson, G., Lindström, M. E., Yakubov, G. E., Gidley, M. J., & Vilaplana, F. (2020). Wood hemicelluloses exert distinct biomechanical contributions to cellulose fibrillar networks. *Nature Communications*, *11*(1), 4692. DOI: 10.1038/s41467-020-18390-z

Bertaglia, T., Costa, M., Lanceros-Méndez, S., & Crespilho, F. N. (2024). Eco-friendly, sustainable, and safe energy storage : A nature-inspired materials paradigm shift. *Materials Advances*, *5*(19), 7534–7547. DOI: 10.1039/D4MA00363B

Bhuvaneswari, B., Sivabharathy, M., Prasad, L. G., & Selvasekarapandian, S. (2022). Structural, thermal and electrochemical characterization of cellulose acetate–based solid biopolymer electrolyte for zinc ion batteries. *Ionics*, *28*(8), 3865–3875. DOI: 10.1007/s11581-022-04616-1

Biswal, B. K., & Balasubramanian, R. (2026). Biomass-derived carbon as sustainable materials for application in clean energy storage technologies : Progress and outlook. *Carbon*, *246*, 120931. DOI: 10.1016/j.carbon.2025.120931

Böckenfeld, N., Jeong, S. S., Winter, M., Passerini, S., & Balducci, A. (2013). Natural, cheap and environmentally friendly binder for supercapacitors. *Journal of Power Sources*, *221*, 14–20. DOI: 10.1016/j.jpowsour.2012.07.076

Bresser, D., Buchholz, D., Moretti, A., Varzi, A., & Passerini, S. (2018). Alternative binders for sustainable electrochemical energy storage – the transition to aqueous electrode processing and bio-derived polymers. *Energy & Environmental Science*, *11*(11), 3096–3127. DOI: 10.1039/C8EE00640G

Bryntesen, S. N., Tolstorebrov, I., Svensson, A., Shearing, P., Lamb, J., & Burheim, O. (2022). Introducing Lignin as a Binder Material for the Aqueous Production of NMC Cathodes for Li-ion Batteries. *Materials Advances*, *4*(2), 523–541. Advance online publication. DOI: 10.1039/D2MA00850E

Butera, G., Fendt, S., Jensen, S. H., Ahrenfeldt, J., & Clausen, L. R. (2020). Flexible methanol production units coupling solid oxide cells and thermochemical biomass conversion via different gasification technologies. *Energy*, *208*, 118432. DOI: 10.1016/j.energy.2020.118432

Cao, D., Zhang, Q., Hafez, A. M., Jiao, Y., Ma, Y., Li, H., Cheng, Z., Niu, C., & Zhu, H. (2019). Lignin-Derived Holey, Layered, and Thermally Conductive 3D Scaffold for Lithium Dendrite Suppression. *Small Methods*, *3*(5), 1800539. DOI: 10.1002/smtd.201800539

Casau, M., Dias, M. F., Matias, J. C. O., & Nunes, L. J. R. (2022). Residual Biomass : A Comprehensive Review on the Importance, Uses and Potential in a Circular Bioeconomy Approach. *Resources*, *11*(4), 4. Advance online publication. DOI: 10.3390/resources11040035

Chandel, V., Biswas, D., Roy, S., Vaidya, D., Verma, A., & Gupta, A. (2022). Current Advancements in Pectin : Extraction, Properties and Multifunctional Applications. *Foods*, *11*(17), 2683. DOI: 10.3390/foods11172683

Chelfouh, N., Gaël, C., Rousselot, S., Foran, G., Briqueleur, E., Shoghi, F., Caradant, L., & Dollé, M. (2022). Apple Pectin-Based Hydrogel Electrolyte for Energy Storage Applications. *ACS Sustainable Chemistry & Engineering*, *10*(48), 15802–15812. Advance online publication. DOI: 10.1021/acssuschemeng.2c04600

Chen, H., Yang, Y., & Dong, Z. (2025). The effect of product design on recycling efficiency of lithium-ion batteries through structural equation modeling and life cycle assessment. *Scientific Reports*, *15*(1), 12352. DOI: 10.1038/s41598-025-87663-8

Chen, P., Lin, X., Yang, B., Gao, Y., Xiao, Y., Li, L., Zhang, H., Li, L., Zheng, Z., Wang, J., & Chou, S. (2024). Cellulose Separators for Rechargeable Batteries with High Safety : Advantages, Strategies, and Perspectives. *Advanced Functional Materials*, *34*(49), 2409368. DOI: 10.1002/adfm.202409368

Chen, Q. (2024). Types of Batteries for New Energy Electric Vehicles. *Science and Technology of Engineering. Chemistry and Environmental Protection*, *1*(5), 5. Advance online publication. DOI: 10.61173/1e54eb37

Chen, R., Li, X., Huang, Q., Ling, H., Yang, Y., & Wang, X. (2021). Self-assembled porous biomass carbon/RGO/nanocellulose hybrid aerogels for self-supporting supercapacitor electrodes. *Chemical Engineering Journal*, *412*, 128755. DOI: 10.1016/j.cej.2021.128755

Chen, T., Zhang, Q., Pan, J., Xu, J., Liu, Y., Al-Shroofy, M., & Cheng, Y.-T. (2016). Low-Temperature Treated Lignin as Both Binder and Conductive Additive for Silicon Nanoparticle Composite Electrodes in Lithium-Ion Batteries. *ACS Applied Materials & Interfaces*, *8*(47), 32341–32348. DOI: 10.1021/acsami.6b11500

Chen, W., Dong, T., Bai, F., Wang, J., & Li, X. (2022). Lignin–carbohydrate complexes, their fractionation, and application to healthcare materials : A review. *International Journal of Biological Macromolecules*, *203*, 29–39. DOI: 10.1016/j.ijbiomac.2022.01.132

Chen, W.-H., Naveen, C., Ghodke, P. K., Sharma, A. K., & Bobde, P. (2023). Co-pyrolysis of lignocellulosic biomass with other carbonaceous materials : A review on advance technologies, synergistic effect, and future prospectus. *Fuel*, *345*, 128177. DOI: 10.1016/j.fuel.2023.128177

Chen, X., & Holze, R. (2024). Polymer Electrolytes for Supercapacitors. *Polymers*, *16*(22), 22. Advance online publication. DOI: 10.3390/polym16223164

Cheng, X.-B., Hou, T.-Z., Zhang, R., Peng, H.-J., Zhao, C.-Z., Huang, J.-Q., & Zhang, Q. (2016). Dendrite-Free Lithium Deposition Induced by Uniformly Distributed Lithium Ions for Efficient Lithium Metal Batteries. *Advanced Materials*, *28*(15), 2888–2895. DOI: 10.1002/adma.201506124

Chinnappan, B., Baskar, D., Dhillon, R., & Editors. (2012). *Biomass Conversion : The interface of biotechnology, chemistry and materials science.*

Chowdhury, M., Oliullah, M. S., Islam, R. T., Hurayra, M. A., Mahmud, M. Z. A., Hasan, N., Rahman, M. K., Islam, M. S., Khan, J., & Hossain, N. (2025). Biomaterials for energy storage : Synthesis, properties, and performance. *Green Technologies and Sustainability*, *3*(2), 100152. DOI: 10.1016/j.grets.2024.100152

Chu, D., Nashalian, A., Zhou, Y., Zhao, X., Guo, D., He, R., Chen, W., & Chen, J. (2020). Low-Cost and Nature-Friendly Hierarchical Porous Carbon for Enhanced Capacitive Electrochemical Energy Storage. *ACS Applied Energy Materials*, *3*(8), 7246–7250. DOI: 10.1021/acsaem.0c01289

Cornejo-Ramírez, Y. I., Martínez-Cruz, O., Del Toro-Sánchez, C. L., Wong-Corral, F. J., Borboa-Flores, J., & Cinco-Moroyoqui, F. J. (2018). The structural characteristics of starches and their functional properties. *CYTA: Journal of Food*, *16*(1), 1003–1017. DOI: 10.1080/19476337.2018.1518343

Coviello, L., Montalbano, G., Piovano, A., Izaguirre, N., Vitale-Brovarone, C., Gerbaldi, C., & Fiorilli, S. (2025). The Impact of Recovered Lignin on Solid-State PEO-Based Electrolyte Produced via Electrospinning : Manufacturing and Characterisation. *Polymers*, *17*(7), 7. Advance online publication. DOI: 10.3390/polym17070982

Dalwadi, S., Goel, A., Kapetanakis, C., Salas-de la Cruz, D., & Hu, X. (2023). The Integration of Biopolymer-Based Materials for Energy Storage Applications : A Review. *International Journal of Molecular Sciences*, *24*(4), 4. Advance online publication. DOI: 10.3390/ijms24043975

Das, A., Ringu, T., Ghosh, S., & Pramanik, N. (2023). A comprehensive review on recent advances in preparation, physicochemical characterization, and bioengineering applications of biopolymers. *Polymer Bulletin*, *80*(7), 7247–7312. DOI: 10.1007/s00289-022-04443-4

Dias, J. C., Marques, S., Branco, P. C., Rodrigues, T., Torres, C. A. V., Freitas, F., Evtyugin, D. V., & Silva, C. J. (2025). Biopolymers Derived from Forest Biomass for the Sustainable Textile Industry. *Forests*, *16*(1), 1. Advance online publication. DOI: 10.3390/f16010163

Ding, J., Yang, Y., Poisson, J., He, Y., Zhang, H., Zhang, Y., Bao, Y., Chen, S., Chen, Y. M., & Zhang, K. (2024). Recent Advances in Biopolymer-Based Hydrogel Electrolytes for Flexible Supercapacitors. *ACS Energy Letters*, *9*(4), 1803–1825. DOI: 10.1021/acsenergylett.3c02567

Dissanayake, K., & Kularatna-Abeywardana, D. (2024). A review of supercapacitors : Materials, technology, challenges, and renewable energy applications. *Journal of Energy Storage*, *96*, 112563. DOI: 10.1016/j.est.2024.112563

Domínguez-Robles, J., Sánchez, R., Díaz-Carrasco, P., Espinosa, E., García-Domínguez, M. T., & Rodríguez, A. (2017). Isolation and characterization of lignins from wheat straw : Application as binder in lithium batteries. *International Journal of Biological Macromolecules*, *104*, 909–918. DOI: 10.1016/j.ijbiomac.2017.07.015

Du, Z., Wang, N., & Du, J. (2025). Recent Progress in Cellulose-Based Conductive Hydrogels. *Polymers*, *17*(8), 8. Advance online publication. DOI: 10.3390/polym17081089

Dutta, D., & Sit, N. (2024). A comprehensive review on types and properties of biopolymers as sustainable bio-based alternatives for packaging. *Food Biomacromolecules*, *1*(2), 58–87. DOI: 10.1002/fob2.12019

El Bourakadi, K., Semlali, F.-Z., Hammi, M., & El Achaby, M. (2024). A review on natural cellulose fiber applications : Empowering industry with sustainable solutions. *International Journal of Biological Macromolecules*, *281*, 135773. DOI: 10.1016/j.ijbiomac.2024.135773

Elfaleh, I., Abbassi, F., Habibi, M., Ahmad, F., Guedri, M., Nasri, M., & Garnier, C. (2023). A comprehensive review of natural fibers and their composites : An eco-friendly alternative to conventional materials. *Results in Engineering*, *19*, 101271. DOI: 10.1016/j.rineng.2023.101271

Elyasi, S., Saha, S., Hameed, N., Mahon, P. J., Juodkazis, S., & Salim, N. (2024). Emerging trends in biomass-derived porous carbon materials for hydrogen storage. *International Journal of Hydrogen Energy*, *62*, 272–306. DOI: 10.1016/j.ijhydene.2024.02.337

Escobar, B., Martínez-Casillas, D. C., Pérez-Salcedo, K. Y., Rosas, D., Morales, L., Liao, S. J., Huang, L. L., & Shi, X. (2021). Research progress on biomass-derived carbon electrode materials for electrochemical energy storage and conversion technologies. *International Journal of Hydrogen Energy*, *46*(51), 26053–26073. DOI: 10.1016/j.ijhydene.2021.02.017

Eswaragomathy, S., Selvanayagam, S., Selvasekarapandian, S., Muniraj Vignesh, N., Aafrin Hazaana, S., & Meera Naachiyar, R. (2023). Preparation of pectin biopolymer electrolyte for zinc-ion battery application. *Ionics*, *29*(6), 2329–2340. DOI: 10.1007/s11581-023-05005-y

Fan, L., Shi, J., & Xi, Y. (2020). PVDF-Modified Nafion Membrane for Improved Performance of MFC. *Membranes (Basel)*, *10*(8), 185. DOI: 10.3390/membranes10080185

Fan, Y., & Picchioni, F. (2020). Modification of starch : A review on the application of "green" solvents and controlled functionalization. *Carbohydrate Polymers*, *241*, 116350. DOI: 10.1016/j.carbpol.2020.116350

Fathona, I. W., Ismardi, A., Supartiwi, H. A., & Yabuki, A. (2024). Use of plant extracts in fabricating manganese-based film electrode for capacitor. *Electrochimica Acta*, *484*, 144060. DOI: 10.1016/j.electacta.2024.144060

Feofilova, E. P., & Mysyakina, I. S. (2016). Lignin : Chemical structure, biodegradation, and practical application (a review). *Applied Biochemistry and Microbiology*, *52*(6), 573–581. DOI: 10.1134/S0003683816060053

Fone, T. M., Jones, D. R., Phillips, C. O., Chasri, A., & Deganello, D. (2024). Ethyl Cellulose As a Sustainable Binder for Activated Carbon Electrodes in Supercapacitors. *ECS Meeting Abstracts*, *MA2024-02*(6), 711. DOI: 10.1149/MA2024-026711mtgabs

Fu, L., Qu, Q., Holze, R., Kondratiev, V. V., & Wu, Y. (2019). Composites of metal oxides and intrinsically conducting polymers as supercapacitor electrode materials : The best of both worlds? *Journal of Materials Chemistry. A, Materials for Energy and Sustainability*, *7*(25), 14937–14970. DOI: 10.1039/C8TA10587A

Gallo, M., Moreschi, L., & Borghi, A. D. (2022). A CRITICAL ENVIRONMENTAL ANALYSIS OF STRATEGIC MATERIALS TOWARDS ENERGY TRANSITION. *Detritus*, *20*(20), 3–12. DOI: 10.31025/2611-4135/2022.15223

Gao, D., Luo, Z., Liu, C., & Fan, S. (2023). A survey of hybrid energy devices based on supercapacitors. *Green Energy & Environment*, *8*(4), 972–988. DOI: 10.1016/j.gee.2022.02.002

Gao, M., Pan, S.-Y., Chen, W.-C., & Chiang, P.-C. (2018a). A cross-disciplinary overview of naturally derived materials for electrochemical energy storage. *Materials Today. Energy*, *7*, 58–79. DOI: 10.1016/j.mtener.2017.12.005

Gao, M., Pan, S.-Y., Chen, W.-C., & Chiang, P.-C. (2018b). A cross-disciplinary overview of naturally derived materials for electrochemical energy storage. *Materials Today. Energy*, *7*, 58–79. DOI: 10.1016/j.mtener.2017.12.005

Gao, Z., Zhang, Y., Song, N., & Li, X. (2017). Biomass-derived renewable carbon materials for electrochemical energy storage. *Materials Research Letters*, *5*(2), 69–88. DOI: 10.1080/21663831.2016.1250834

Gayathri, R., Mahboob, S., Govindarajan, M., Al-Ghanim, K. A., Ahmed, Z., Al-Mulhm, N., Vodovnik, M., & Vijayalakshmi, S. (2021). A review on biological carbon sequestration : A sustainable solution for a cleaner air environment, less pollution and lower health risks. *Journal of King Saud University. Science*, *33*(2), 101282. DOI: 10.1016/j.jksus.2020.101282

Ghobakhloo, M., Mahdiraji, H. A., Iranmanesh, M., & Jafari-Sadeghi, V. (2024). From Industry 4.0 Digital Manufacturing to Industry 5.0 Digital Society : A Road-map Toward Human-Centric, Sustainable, and Resilient Production. *Information Systems Frontiers*. Advance online publication. DOI: 10.1007/s10796-024-10476-z

Ghodhbane, M., Ashraf, J. M., Karam, Z., Lonkar, S., Alshaya, A., & Busà, C. (2023). Cellulose nanofibers as a green binder for symmetric carbon nanotubes-based supercapacitors. *Electrochimica Acta*, *461*, 142584. DOI: 10.1016/j.electacta.2023.142584

Gond, R., Asfaw, H. D., Hosseinaei, O., Edström, K., Younesi, R., & Naylor, A. J. (2021). A Lignosulfonate Binder for Hard Carbon Anodes in Sodium-Ion Batteries : A Comparative Study. *ACS Sustainable Chemistry & Engineering*, *9*(37), 12708–12717. DOI: 10.1021/acssuschemeng.1c05263

Gong, Y., Xiao, S., Yao, Z., Deng, H., Chen, X., & Yang, T. (2024). Factors and modification techniques enhancing starch gel structure and their applications in foods:A review. *Food Chemistry: X*, *24*, 102045. DOI: 10.1016/j.fochx.2024.102045

González, W. A., López, D., & Pérez, J. F. (2020). Biofuel quality analysis of fallen leaf pellets : Effect of moisture and glycerol contents as binders. *Renewable Energy*, *147*, 1139–1150. DOI: 10.1016/j.renene.2019.09.094

Gopi, C. V. V. M., Alzahmi, S., Al-Haik, M. Y., Kumar, Y. A., Hamed, F., Haik, Y., & Obaidat, I. M. (2024). Recent advances in pseudocapacitive electrode materials for high energy density aqueous supercapacitors : Combining transition metal oxides with carbon nanomaterials. *Materials Today Sustainability*, *28*, 100981. DOI: 10.1016/j.mtsust.2024.100981

Gopinath, G., Ayyasamy, S., Shanmugaraj, P., Swaminathan, R., Subbiah, K., & Kandasamy, S. (2023). Effects of biopolymers in energy storage applications : A state-of-the-art review. *Journal of Energy Storage*, *70*, 108065. DOI: 10.1016/j.est.2023.108065

Grzybek, P., Dudek, G., & van der Bruggen, B. (2024). Cellulose-based films and membranes : A comprehensive review on preparation and applications. *Chemical Engineering Journal, 495*, 153500. DOI: 10.1016/j.cej.2024.153500

Guan, L., Yu, L., & Chen, G. Z. (2016). Capacitive and non-capacitive faradaic charge storage. *Electrochimica Acta, 206*, 464–478. DOI: 10.1016/j.electacta.2016.01.213

Günther, S., Bensmann, A., & Hanke-Rauschenbach, R. (2025). Representative energy management strategies for hybrid energy storage systems derived from a meta-review. *Renewable & Sustainable Energy Reviews, 216*, 115610. DOI: 10.1016/j.rser.2025.115610

Guo, N., Luo, W., Guo, R., Qiu, D., Zhao, Z., Wang, L., Jia, D., & Guo, J. (2020). Interconnected and hierarchical porous carbon derived from soybean root for ultra-high rate supercapacitors. *Journal of Alloys and Compounds, 834*, 155115. DOI: 10.1016/j.jallcom.2020.155115

Guo, R., Zhang, L., Lu, Y., Zhang, X., & Yang, D. (2020). Research progress of nanocellulose for electrochemical energy storage : A review. *Journal of Energy Chemistry, 51*, 342–361. DOI: 10.1016/j.jechem.2020.04.029

Guo, T., Song, J., Jin, Y., Sun, Z., & Li, L. (2019). Thermally stable and green cellulose-based composites strengthened by styrene-co-acrylate latex for lithium-ion battery separators. *Carbohydrate Polymers, 206*, 801–810. DOI: 10.1016/j.carbpol.2018.11.025

Guo, Y., Ji, S., Liu, F., Zhu, Z., Xiao, J., Liu, K., Zhang, Y., Liao, S., & Zeng, X. (2025). A review of the preparation and characterization techniques for closed pores in hard carbon and their functions in sodium-ion batteries. *Energy Materials, 5*(3), N/A, N/A. DOI: 10.20517/energymater.2024.63

Gupta, H., Singh, S. K., Srivastava, N., Meghnani, D., Tiwari, R. K., Mishra, R., Patel, A., Tiwari, A., Saroj, A. L., & Singh, R. K. (2021). Improved High Voltage Performance of Li-ion Conducting Coated Ni-rich NMC Cathode Materials for Rechargeable Li Battery. *ACS Applied Energy Materials, 4*(12), 13878–13889. DOI: 10.1021/acsaem.1c02681

Hadad, S., Hamrahjoo, M., Dehghani, E., Salami-Kalajahi, M., Eliseeva, S. N., Moghaddam, A. R., & Roghani-Mamaqani, H. (2022a). Cellulose-based solid and gel polymer electrolytes with super high ionic conductivity and charge capacity for high performance lithium ion batteries. *Sustainable Materials and Technologies, 33*, e00503. DOI: 10.1016/j.susmat.2022.e00503

Hadad, S., Hamrahjoo, M., Dehghani, E., Salami-Kalajahi, M., Eliseeva, S. N., Moghaddam, A. R., & Roghani-Mamaqani, H. (2022b). Starch acetate and carboxymethyl starch as green and sustainable polymer electrolytes for high performance lithium ion batteries. *Applied Energy*, *324*, 119767. DOI: 10.1016/j.apenergy.2022.119767

Hafiza, M. N., & Isa, M. I. N. (2020). Correlation between structural, ion transport and ionic conductivity of plasticized 2-hydroxyethyl cellulose based solid biopolymer electrolyte. *Journal of Membrane Science*, *597*, 117176. DOI: 10.1016/j.memsci.2019.117176

Haider, R., Sagadevan, S., Cameron, N. R., & Johan, M. R. (2025). Biomass-derived activated carbon for high-performance energy storage devices. *Journal of Power Sources*, *633*, 236404. DOI: 10.1016/j.jpowsour.2025.236404

Halder, S., Bhattacharya, D., Maind, R., & Chakraborty, S. (2025). Embracing the Transition from Conventional Energy to Nonconventional Energy for a Better Future and Sustainability : Innovative Energy Storage Materials. *ChemistrySelect*, *10*(40), e04465. DOI: 10.1002/slct.202504465

Hapuarachchi, S. N. S., Wasalathilake, K. C., Nerkar, J. Y., Jaatinen, E., O'Mullane, A. P., & Yan, C. (2020). Mechanically Robust Tapioca Starch Composite Binder with Improved Ionic Conductivity for Sustainable Lithium-Ion Batteries. *ACS Sustainable Chemistry & Engineering*, *8*(26), 9857–9865. DOI: 10.1021/acssuschemeng.0c02843

Haque, S. M., Kabir, A., Ratemi, E., Elzagheid, M., Appu, S. P., Ghani, S. S., & Sarief, A. (2025). Greener Pectin Extraction Techniques : Applications and Challenges. *Separations*, *12*(3), 3. Advance online publication. DOI: 10.3390/separations12030065

Harikumar, M. E., & Batabyal, S. K. (2023). Fabrication of pectin biopolymer-based biocompatible freestanding electrodes for supercapacitor applications. *Polymers for Advanced Technologies*, *34*(9), 2890–2902. DOI: 10.1002/pat.6106

Harikumar, M. E., & Batabyal, S. K. (2024). Biopolymer pectin with calcium ion crosslinker as biocompatiable electrolyte for energy storage applications. *Electrochimica Acta*, *489*, 144150. DOI: 10.1016/j.electacta.2024.144150

Hasan, M. M., Haque, R., Jahirul, M. I., Rasul, M. G., Fattah, I. M. R., Hassan, N. M. S., & Mofijur, M. (2025). Advancing energy storage : The future trajectory of lithium-ion battery technologies. *Journal of Energy Storage*, *120*, 116511. DOI: 10.1016/j.est.2025.116511

Hayagan, N., Gaalich, I., Loubet, P., Croguennec, L., Aymonier, C., Philippot, G., & Olchowka, J. (2024). Challenges and Perspectives for Direct Recycling of Electrode Scraps and End-of-Life Lithium-ion Batteries. *Batteries & Supercaps*, *7*(6), e202400120. DOI: 10.1002/batt.202400120

He, H., Zhang, R., Zhang, P., Wang, P., Chen, N., Qian, B., Zhang, L., Yu, J., & Dai, B. (2023). Functional Carbon from Nature : Biomass-Derived Carbon Materials and the Recent Progress of Their Applications. *Advanced Science (Weinheim, Baden-Wurttemberg, Germany)*, *10*(16), 2205557. DOI: 10.1002/advs.202205557

He, R. (2024). Research progress of carbon-based anode materials for lithium-ion batteries. *Advances in Engineering Technology Research*, *12*(1), 1. Advance online publication. DOI: 10.56028/aetr.12.1.776.2024

He, Y., & Guo, Z. (2024). Natural polymers-based separation membrane for high-efficient separation of oil water mixture. *Nano Today*, *57*, 102367. DOI: 10.1016/j.nantod.2024.102367

Hoch, M. J. R., Senin, H. B., & Idris, N. H. (2007). The Intriguing Properties of Transition Metal Oxides. *AIP Conference Proceedings*, *909*(1), 3–8. DOI: 10.1063/1.2739818

Hu, Z., Zhao, R., Yang, J., Wu, C., & Bai, Y. (2023). Binders for Si based electrodes : Current status, modification strategies and perspective. *Energy Storage Materials*, *59*, 102776. DOI: 10.1016/j.ensm.2023.102776

Huang, J., Li, C., Jiang, D., Gao, J., Cheng, L., Li, G., Luo, H., Xu, Z.-L., Shin, D.-M., Wang, Y., Lu, Y., & Kim, Y. (2025). Solid-State Electrolytes for Lithium Metal Batteries : State-of-the-Art and Perspectives. *Advanced Functional Materials*, *35*(1), 2411171. DOI: 10.1002/adfm.202411171

Humerez-Flores, J. N., Verkempinck, S. H. E., Kyomugasho, C., Moldenaers, P., Van Loey, A. M., & Hendrickx, M. E. (2021). Modified Rhamnogalacturonan-Rich Apple Pectin-Derived Structures : The Relation between Their Structural Characteristics and Emulsifying and Emulsion-Stabilizing Properties. *Foods*, *10*(7), 7. Advance online publication. DOI: 10.3390/foods10071586

Inagaki, M., Konno, H., & Tanaike, O. (2010). Carbon Materials for Electrochemical Capacitors. *Journal of Power Sources*, *195*(24), 7880–7903. DOI: 10.1016/j.jpowsour.2010.06.036

Islam, Md. A., Ong, H. L., Villagracia, A. R., A. Halim, K. A., Ganganboina, A. B., & Doong, R.-A. (. (2021). Biomass–derived cellulose nanofibrils membrane from rice straw as sustainable separator for high performance supercapacitor. *Industrial Crops and Products*, *170*, 113694. DOI: 10.1016/j.indcrop.2021.113694

Jain, A., Ghosh, M., Krajewski, M., Kurungot, S., & Michalska, M. (2021). Biomass-derived activated carbon material from native European deciduous trees as an inexpensive and sustainable energy material for supercapacitor application. *Journal of Energy Storage*, *34*, 102178. DOI: 10.1016/j.est.2020.102178

Jeong, D., Shim, J., Shin, H., & Lee, J.-C. (2020). Sustainable Lignin-Derived Cross-Linked Graft Polymers as Electrolyte and Binder Materials for Lithium Metal Batteries. *ChemSusChem*, *13*(10), 2642–2649. DOI: 10.1002/cssc.201903466

Jia, R., He, C., Li, Q., Liu, S.-Y., & Liao, G. (2022a). Renewable plant-derived lignin for electrochemical energy systems. *Trends in Biotechnology*, *40*(12), 1425–1438. DOI: 10.1016/j.tibtech.2022.07.017

Jia, R., He, C., Li, Q., Liu, S.-Y., & Liao, G. (2022b). Renewable plant-derived lignin for electrochemical energy systems. *Trends in Biotechnology*, *40*(12), 1425–1438. DOI: 10.1016/j.tibtech.2022.07.017

Jiang, W., Li, L., Pan, J., Senthil, R. A., Jin, X., Cai, J., Wang, J., & Liu, X. (2019). Hollow-tubular porous carbon derived from cotton with high productivity for enhanced performance supercapacitor. *Journal of Power Sources*, *438*, 226936. DOI: 10.1016/j.jpowsour.2019.226936

Jiang, Y., & Liu, J. (2019). Definitions of Pseudocapacitive Materials : A Brief Review. *Energy & Environmental Materials*, *2*(1), 30–37. DOI: 10.1002/eem2.12028

Jin, B., Wang, D., Song, L., Cai, Y., Ali, A., Hou, Y., Chen, J., Zhang, Q., & Zhan, X. (2021). Biomass-derived fluorinated corn starch emulsion as binder for silicon and silicon oxide based anodes in lithium-ion batteries. *Electrochimica Acta*, *365*, 137359. DOI: 10.1016/j.electacta.2020.137359

Jothi, M. A., Vanitha, D., Sundaramahalingam, K., & Nallamuthu, N. (2022). Utilisation of corn starch in production of 'eco friendly' polymer electrolytes for proton battery applications. *International Journal of Hydrogen Energy*, *47*(67), 28763–28772. DOI: 10.1016/j.ijhydene.2022.06.192

Jung, H. Y., Lee, J. S., Han, H. T., Jung, J., Eom, K., & Lee, J. T. (2022). Lignin-Based Materials for Sustainable Rechargeable Batteries. *Polymers*, *14*(4), 4. Advance online publication. DOI: 10.3390/polym14040673

Jung, Y. H., Chang, T.-H., Zhang, H., Yao, C., Zheng, Q., Yang, V. W., Mi, H., Kim, M., Cho, S. J., Park, D.-W., Jiang, H., Lee, J., Qiu, Y., Zhou, W., Cai, Z., Gong, S., & Ma, Z. (2015). High-performance green flexible electronics based on biodegradable cellulose nanofibril paper. *Nature Communications*, *6*(1), 7170. DOI: 10.1038/ncomms8170

K n, Y., T, P. D., P, S., S, K., R, Y. K., Varjani, S., AdishKumar, S., Kumar, G., & J, R. B. (2022). Lignocellulosic biomass-based pyrolysis : A comprehensive review. *Chemosphere, 286*, 131824. DOI: 10.1016/j.chemosphere.2021.131824

Kadir, M. F. Z., Salleh, N. S., Hamsan, M. H., Aspanut, Z., Majid, N. A., & Shukur, M. F. (2018). Biopolymeric electrolyte based on glycerolized methyl cellulose with NH4Br as proton source and potential application in EDLC. *Ionics*, *24*(6), 1651–1662. DOI: 10.1007/s11581-017-2330-4

Kamgang Maffo, E., Clementine-Christelle, M.-E., & Kombou, V. (2024). *Production of Activated Carbons : Processes, Applications, and Environmental Considerations* (SSRN Scholarly Paper 5025307). Social Science Research Network. DOI: 10.2139/ssrn.5025307

Kandpal, V., Jaswal, A., Santibanez Gonzalez, E. D. R., & Agarwal, N. (2024). Challenges and Opportunities for Sustainable Energy Transition and Circular Economy. In Kandpal, V., Jaswal, A., Santibanez Gonzalez, E. D. R., & Agarwal, N. (Eds.), *Sustainable Energy Transition : Circular Economy and Sustainable Financing for Environmental, Social and Governance (ESG) Practices* (pp. 307–324). Springer Nature Switzerland., DOI: 10.1007/978-3-031-52943-6_11

Kang, H., Liu, R., & Huang, Y. (2015). Graft modification of cellulose : Methods, properties and applications. *Polymer*, *70*, A1–A16. DOI: 10.1016/j.polymer.2015.05.041

Kapoor, R. T., Rafatullah, M., Qamar, M., Qutob, M., Alosaimi, A. M., Alorfi, H. S., & Hussein, M. A. (2022). Review on Recent Developments in Bioinspired-Materials for Sustainable Energy and Environmental Applications. *Sustainability (Basel)*, *14*(24), 24. Advance online publication. DOI: 10.3390/su142416931

Kariim, I., Swai, H., & Kivevele, T. (2022). Recent advances in thermochemical conversion of biomass into drop-in fuel:a review. *Scientific African*, *17*, e01352. DOI: 10.1016/j.sciaf.2022.e01352

Kim, U.-H., Nam, G. W., Yoon, D. R., & Sun, Y.-K. (2019). Pushing the Limit of Layered Cathodes for High-Energy Density, Long-Term Cycling, and High Thermal Stability Rechargeable Li Ion Batteries. *ECS Meeting Abstracts, MA2019-03*(2), 131. DOI: 10.1149/MA2019-03/2/131

Kiruthika, S., Malathi, M., Selvasekarapandian, S., Tamilarasan, K., & Maheshwari, T. (2020). Conducting biopolymer electrolyte based on pectin with magnesium chloride salt for magnesium battery application. *Polymer Bulletin*, *77*(12), 6299–6317. DOI: 10.1007/s00289-019-03071-9

Kögel-Knabner, I. (2002). The macromolecular organic composition of plant and microbial residues as inputs to soil organic matter. *Soil Biology & Biochemistry*, *34*(2), 139–162. DOI: 10.1016/S0038-0717(01)00158-4

Koliyoor, J., Ismayil, , Hegde, S., Vasachar, R., & Sanjeev, G. (2022). Novel solid biopolymer electrolyte based on methyl cellulose with enhanced ion transport properties. *Journal of Applied Polymer Science*, *139*(12), 51826. DOI: 10.1002/app.51826

Kou, L., Huang, T., Zheng, B., Han, Y., Zhao, X., Gopalsamy, K., Sun, H., & Gao, C. (2014). Coaxial wet-spun yarn supercapacitors for high-energy density and safe wearable electronics. *Nature Communications*, *5*(1), 3754. DOI: 10.1038/ncomms4754

Krause, A., Kossyrev, P., Oljaca, M., Passerini, S., Winter, M., & Balducci, A. (2011). Electrochemical double layer capacitor and lithium-ion capacitor based on carbon black. *Journal of Power Sources*, *196*(20), 8836–8842. DOI: 10.1016/j.jpowsour.2011.06.019

Król, M., Gryglewicz, G., & Machnikowski, J. (2011). KOH activation of pitch-derived carbonaceous materials—Effect of carbonization degree. *Fuel Processing Technology*, *92*(1), 158–165. DOI: 10.1016/j.fuproc.2010.09.019

Kühnel, R.-S., Böckenfeld, N., Passerini, S., Winter, M., & Balducci, A. (2011). Mixtures of ionic liquid and organic carbonate as electrolyte with improved safety and performance for rechargeable lithium batteries. *Electrochimica Acta*, *56*(11), 4092–4099. DOI: 10.1016/j.electacta.2011.01.116

Kularatna, N., Subasinghage, K., Gunawardane, K., Jayananda, D., & Ariyarathna, T. (2021). Supercapacitor-Assisted Techniques and Supercapacitor-Assisted Loss Management Concept : New Design Approaches to Change the Roadmap of Power Conversion Systems. *Electronics (Basel)*, *10*(14), 14. Advance online publication. DOI: 10.3390/electronics10141697

Kumar, J. A., Sathish, S., Prabu, D., Renita, A. A., Saravanan, A., Deivayanai, V. C., Anish, M., Jayaprabakar, J., Baigenzhenov, O., & Hosseini-Bandegharaei, A. (2023). Agricultural waste biomass for sustainable bioenergy production : Feedstock, characterization and pre-treatment methodologies. *Chemosphere*, *331*, 138680. DOI: 10.1016/j.chemosphere.2023.138680

Kumar, N., Aepuru, R., Lee, S.-Y., & Park, S.-J. (2025). Recent advances in phosphorene : A promising material for supercapacitor applications. *Materials Science and Engineering R Reports*, *163*, 100932. DOI: 10.1016/j.mser.2025.100932

Kwiatkowski, M., & Broniek, E. (2017). An analysis of the porous structure of activated carbons obtained from hazelnut shells by various physical and chemical methods of activation. *Colloids and Surfaces. A, Physicochemical and Engineering Aspects*, *529*, 443–453. Advance online publication. DOI: 10.1016/j.colsurfa.2017.06.028

Lakshmi, K. C. S., & Vedhanarayanan, B. (2023). High-Performance Supercapacitors : A Comprehensive Review on Paradigm Shift of Conventional Energy Storage Devices. *Batteries*, *9*(4), 4. Advance online publication. DOI: 10.3390/batteries9040202

Lara-Espinoza, C., Carvajal-Millán, E., Balandrán-Quintana, R., López-Franco, Y., & Rascón-Chu, A. (2018). Pectin and Pectin-Based Composite Materials : Beyond Food Texture. *Molecules : A Journal of Synthetic Chemistry and Natural Product Chemistry, 23*(4), 942. DOI: 10.3390/molecules23040942

Lavanya, D., Kulkarni, P., Dixit, M., Raavi, P. K., & Krishna, L. N. V. (2011). Sources of cellulose and their applications- A review. *International Journal of Drug Formulation and Research*, *2*, 19–38.

Lee, M., Park, J.-H., Park, D.-J., & Paek, S.-M. (2025). *A Ge/Geo2/Titanate Nanocomposite with High Energy Density and Enhanced Long-Term Stability for Lithium-Ion Batteries* (SSRN Scholarly Paper 5087635). Social Science Research Network. DOI: 10.1016/j.jssc.2025.125256

Lerouxel, O., Cavalier, D. M., Liepman, A. H., & Keegstra, K. (2006). Biosynthesis of plant cell wall polysaccharides—A complex process. *Current Opinion in Plant Biology*, *9*(6), 621–630. DOI: 10.1016/j.pbi.2006.09.009

Li, H., Askari, S., Kulachenko, A., Ek, M., & Sevastyanova, O. (2025). Eco-friendly and strong lignin-containing microfibrillated cellulose films for high-performance separators of aqueous zinc batteries. *International Journal of Biological Macromolecules*, *290*, 138711. DOI: 10.1016/j.ijbiomac.2024.138711

Li, J., Jia, H., Ma, S., Xie, L., Wei, X.-X., Dai, L., Wang, H., Su, F., & Chen, C.-M. (2022). Separator Design for High-Performance Supercapacitors : Requirements, Challenges, Strategies, and Prospects. *ACS Energy Letters*, *8*(1), 56–78. DOI: 10.1021/acsenergylett.2c01853

Li, L., Zhang, W., Pan, W., Wang, M., Zhang, H., Zhang, D., & Zhang, D. (2021). Application of expanded graphite-based materials for rechargeable batteries beyond lithium-ions. *Nanoscale, 13*(46), 19291–19305. DOI: 10.1039/D1NR05873H

Li, T., Chen, C., Brozena, A. H., Zhu, J. Y., Xu, L., Driemeier, C., Dai, J., Rojas, O. J., Isogai, A., Wågberg, L., & Hu, L. (2021). Developing fibrillated cellulose as a sustainable technological material. *Nature, 590*(7844), 47–56. DOI: 10.1038/s41586-020-03167-7

Li, Y., & Ding, R. (2024). Perovskite fluorides for electrochemical energy storage and conversion : Structure, performance and mechanisms. *Nano Energy, 124*, 109430. DOI: 10.1016/j.nanoen.2024.109430

Liao, Z., Cheng, J., Yu, J.-H., Tian, X.-L., & Zhu, M.-Q. (2022). Graphene aerogel with excellent property prepared by doping activated carbon and CNF for free-binder supercapacitor. *Carbohydrate Polymers, 286*, 119287. DOI: 10.1016/j.carbpol.2022.119287

Libber, M., Gariya, N., & Kumar, M. (2024). A comprehensive analysis of supercapacitors with current limitations and emerging trends in research. *Journal of Solid State Electrochemistry, 29*(2), 513–527. DOI: 10.1007/s10008-024-06107-x

Liew, C.-W., & Ramesh, S. (2015). Electrical, structural, thermal and electrochemical properties of corn starch-based biopolymer electrolytes. *Carbohydrate Polymers, 124*, 222–228. DOI: 10.1016/j.carbpol.2015.02.024

Lingaitiene, O., & Burinskiene, A. (2024). Development of Trade in Recyclable Raw Materials : Transition to a Circular Economy. *Economies, 12*(2), 2. Advance online publication. DOI: 10.3390/economies12020048

Liu, B., Huang, Y., Cao, H., Song, A., Lin, Y., Wang, M., & Li, X. (2018). A high-performance and environment-friendly gel polymer electrolyte for lithium ion battery based on composited lignin membrane. *Journal of Solid State Electrochemistry, 22*(3), 807–816. DOI: 10.1007/s10008-017-3814-x

Liu, H., Chen, W., Zhang, R., Xu, C., Huang, X., Peng, H., Huo, C., Xu, M., & Miao, Z. (2021). Bioinspired *in situ* self-catalyzing strategy towards graphene nanosheets with hierarchical structure derived from biomass for advanced supercapacitors. *Applied Surface Science, 566*, 150692. DOI: 10.1016/j.apsusc.2021.150692

Liu, M., Zhou, D., He, Y.-B., Fu, Y., Qin, X., Miao, C., Du, H., Li, B., Yang, Q.-H., Lin, Z., Zhao, T. S., & Kang, F. (2016). Novel gel polymer electrolyte for high-performance lithium–sulfur batteries. *Nano Energy, 22*, 278–289. DOI: 10.1016/j.nanoen.2016.02.008

Liu, N., & Gao, Y. (2017). Recent Progress in Micro-Supercapacitors with In-Plane Interdigital Electrode Architecture. *Small, 13*(45), 1701989. DOI: 10.1002/smll.201701989

Liu, X., & Antonietti, M. (2014). Molten salt activation for synthesis of porous carbon nanostructures and carbon sheets. *Carbon, 69*, 460–466. DOI: 10.1016/j.carbon.2013.12.049

Liu, X., Renard, C. M. G. C., Rolland-Sabaté, A., & Le Bourvellec, C. (2021). Exploring interactions between pectins and procyanidins : Structure-function relationships. *Food Hydrocolloids, 113*, 106498. DOI: 10.1016/j.foodhyd.2020.106498

Liu, Y., Jiang, S. P., & Shao, Z. (2020). Intercalation pseudocapacitance in electrochemical energy storage : Recent advances in fundamental understanding and materials development. *Materials Today. Advances, 7*, 100072. DOI: 10.1016/j.mtadv.2020.100072

Liu, Y., Lin, D., Liang, Z., Zhao, J., Yan, K., & Cui, Y. (2016). Lithium-coated polymeric matrix as a minimum volume-change and dendrite-free lithium metal anode. *Nature Communications, 7*(1), 10992. DOI: 10.1038/ncomms10992

Liu, Z., Zhu, Z., Dai, J., & Yan, Y. (2018). Waste Biomass Based-Activated Carbons Derived from Soybean Pods as Electrode Materials for High-Performance Supercapacitors. *ChemistrySelect, 3*(21), 5726–5732. DOI: 10.1002/slct.201800609

Lizundia, E., Costa, C. M., Alves, R., & Lanceros-Méndez, S. (2020). Cellulose and its derivatives for lithium ion battery separators : A review on the processing methods and properties. *Carbohydrate Polymer Technologies and Applications, 1*, 100001. DOI: 10.1016/j.carpta.2020.100001

Lokhande, P. e., Chavan, U. s., Bhosale, S., Kalam, A., & Deokar, S. (2021). New-Generation Materials for Flexible Supercapacitors. In *Flexible Supercapacitor Nanoarchitectonics* (pp. 277–313). John Wiley & Sons, Ltd., DOI: 10.1002/9781119711469.ch11

Lourenço, A., Gominho, J., Lourenço, A., & Gominho, J. (2023). Lignin as Feedstock for Nanoparticles Production. In *Lignin—Chemistry, Structure, and Application*. IntechOpen. DOI: 10.5772/intechopen.109267

Lu, H., Cornell, A., Alvarado, F., Behm, M., Leijonmarck, S., Li, J., Tomani, P., & Lindbergh, G. (2016). Lignin as a Binder Material for Eco-Friendly Li-Ion Batteries. *Materials (Basel), 9*(3), 3. Advance online publication. DOI: 10.3390/ma9030127

Lu, J., Gu, J., Hu, O., Fu, Y., Ye, D., Zhang, X., Zheng, Y., Hou, L., Liu, H., & Jiang, X. (2021). Highly tough, freezing-tolerant, healable and thermoplastic starch/poly(vinyl alcohol) organohydrogels for flexible electronic devices. *Journal of Materials Chemistry. A, Materials for Energy and Sustainability*, *9*(34), 18406–18420. DOI: 10.1039/D1TA04336F

Luo, C., Du, L., Wu, W., Xu, H., Zhang, G., Li, S., Wang, C., Lu, Z., & Deng, Y. (2018). Novel Lignin-Derived Water-Soluble Binder for Micro Silicon Anode in Lithium-Ion Batteries. *ACS Sustainable Chemistry & Engineering*, *6*(10), 12621–12629. DOI: 10.1021/acssuschemeng.8b01161

Luo, J., & Liu, T. L. (2023). Electrochemical valorization of lignin : Status, challenges, and prospects. *Journal of Bioresources and Bioproducts*, *8*(1), 1–14. DOI: 10.1016/j.jobab.2022.11.003

Ma, C., Zhang, M., Ding, Y., Xue, Y., Wang, H., Li, P., & Wu, D. (2023). Green Production of Biomass-Derived Carbon Materials for High-Performance Lithium–Sulfur Batteries. *Nanomaterials (Basel, Switzerland)*, *13*(11), 11. Advance online publication. DOI: 10.3390/nano13111768

Ma, F., Ding, S., Ren, H., & Liu, Y. (2019). Sakura-based activated carbon preparation and its performance in supercapacitor applications. *RSC Advances*, *9*(5), 2474–2483. DOI: 10.1039/C8RA09685F

Ma, T., Zhang, X., & Wang, Y. (2019). Tassel tree flowers-derived hierarchically porous carbons with high surface area for high-performance flexible all-solid-state symmetric supercapacitors. *Journal of Energy Storage*, *26*, 101014. DOI: 10.1016/j.est.2019.101014

Ma, Y., Chen, K., Ma, J., Xu, G., Dong, S., Chen, B., Li, J., Chen, Z., Zhou, X., & Cui, G. (2019). A biomass based free radical scavenger binder endowing a compatible cathode interface for 5 V lithium-ion batteries. *Energy & Environmental Science*, *12*(1), 273–280. DOI: 10.1039/C8EE02555J

Machín, A., Cotto, M. C., Díaz, F., Duconge, J., Morant, C., & Márquez, F. (2024). Environmental Aspects and Recycling of Solid-State Batteries : A Comprehensive Review. *Batteries*, *10*(7), 7. Advance online publication. DOI: 10.3390/batteries10070255

Madhu, R., Periasamy, A. P., Schlee, P., Hérou, S., & Titirici, M.-M. (2023). Lignin : A sustainable precursor for nanostructured carbon materials for supercapacitors. *Carbon*, *207*, 172–197. DOI: 10.1016/j.carbon.2023.03.001

Mandal, M., Chattopadhyay, K., Jain, A., & Bhattacharya, S. K. (2023). Porous Carbon Materials for Supercapacitor Applications. In Grace, A. N., Sonar, P., Bhardwaj, P., & Chakravorty, A. (Eds.), *Handbook of Porous Carbon Materials* (pp. 117–146). Springer Nature., DOI: 10.1007/978-981-19-7188-4_5

Manfredy Marquina, L., La Torre Riveros, L., Jauja Ccana, V., Muedas-Taipe, G., Isaacs, M., & La Rosa-Toro, A. (2023). Recent studies on polymer electrolytes containing ionic liquids and their applications in lithium-ion batteries. *Journal of Electroanalytical Chemistry (Lausanne, Switzerland), 948*, 117819. DOI: 10.1016/j.jelechem.2023.117819

Manyà, J. J., & Gascó, G. (2020). Biochar as a Sustainable Resource to Drive Innovative Green Technologies. In *Biochar as a Renewable-Based Material* (p. 1-33). WORLD SCIENTIFIC (EUROPE). DOI: 10.1142/9781786348975_0001

Mehta, S., Jha, S., & Liang, H. (2020). Lignocellulose materials for supercapacitor and battery electrodes : A review. *Renewable & Sustainable Energy Reviews, 134*, 110345. DOI: 10.1016/j.rser.2020.110345

Mellinas, C., Ramos, M., Jiménez, A., & Garrigós, M. C. (2020). Recent Trends in the Use of Pectin from Agro-Waste Residues as a Natural-Based Biopolymer for Food Packaging Applications. *Materials (Basel), 13*(3), 3. Advance online publication. DOI: 10.3390/ma13030673

Mertens, J., Dewulf, J., Breyer, C., Belmans, R., Gendron, C., Geoffron, P., Goossens, L., Fischer, C., Du Fornel, E., Hayhoe, K., Hirose, K., Le Cadre-Loret, E., Lester, R., Maigné, F., Maitournam, H., de Miranda, P. E. V., Verwee, P., Sala, O., Webber, M., & Debackere, K. (2024). From emissions to resources : Mitigating the critical raw material supply chain vulnerability of renewable energy technologies. *Mineral Economics, 37*(3), 669–676. DOI: 10.1007/s13563-024-00425-2

Mian, M. M., Kamana, I. M. L., An, X., Abbas, S. C., Ahommed, M. S., He, Z., & Ni, Y. (2023). Cellulose nanofibers as effective binders for activated biochar-derived high-performance supercapacitors. *Carbohydrate Polymers, 301*, 120353. DOI: 10.1016/j.carbpol.2022.120353

Mirzaaghaei, M., Nasirpour, A., Keramat, J., Goli, S. A. H., Dinari, M., Desobry, S., & Durand, A. (2022). Chemical modification of waxy maize starch by esterification with saturated fatty acid chlorides : Synthesis, physicochemical and emulsifying properties. *Food Chemistry, 393*, 133293. DOI: 10.1016/j.foodchem.2022.133293

Mishra, R., Anis, A., Mondal, S., Dutt, M., & Banthia, A. (2009). Reparation and characterization of amidated pectin based polymer electrolyte membranes. *Chinese Journal of Polymer Science, 27*(5), 639–646. DOI: 10.1142/S0256767909004333

Molahalli, V., Sharma, A., Bijapur, K., Soman, G., Chattham, N., & Hegde, G. (2024a). Low-cost bio-waste carbon nanocomposites for sustainable electrochemical devices : A systematic review. *Materials Today. Communications*, *38*, 108034. DOI: 10.1016/j.mtcomm.2024.108034

Molahalli, V., Sharma, A., Bijapur, K., Soman, G., Chattham, N., & Hegde, G. (2024b). Low-cost bio-waste carbon nanocomposites for sustainable electrochemical devices : A systematic review. *Materials Today. Communications*, *38*, 108034. DOI: 10.1016/j.mtcomm.2024.108034

Monisha, S., Mathavan, T., Selvasekarapandian, S., Milton Franklin Benial, A., Aristatil, G., Mani, N., Premalatha, M., & Vinoth pandi, D. (2017). Investigation of bio polymer electrolyte based on cellulose acetate-ammonium nitrate for potential use in electrochemical devices. *Carbohydrate Polymers*, *157*, 38–47. DOI: 10.1016/j.carbpol.2016.09.026

Monteiro, J., Garrido, N., & Fonseca, R. (2011). Efficient supercapacitor energy usage in mobile phones. *2011 IEEE International Conference on Consumer Electronics -Berlin (ICCE-Berlin)*, 318-321. DOI: 10.1109/ICCE-Berlin.2011.6031796

Muddasar, M., Beaucamp, A., Culebras, M., & Collins, M. (2022). Cellulose : Characteristics and applications for rechargeable batteries. *International Journal of Biological Macromolecules*, *219*, 788–803. Advance online publication. DOI: 10.1016/j.ijbiomac.2022.08.026

Muddasar, M., Beaucamp, A., Culebras, M., & Collins, M. N. (2024). High performance all lignin derived supercapacitors for energy storage applications. *Materials Today Sustainability*, *26*, 100767. DOI: 10.1016/j.mtsust.2024.100767

Muddasar, M., Culebras, M., & Collins, M. N. (2024). Lignin and its carbon derivatives : Synthesis techniques and their energy storage applications. *Materials Today Sustainability*, *28*, 100990. DOI: 10.1016/j.mtsust.2024.100990

Mzioud, K., Ferraa, N., Aribou, Z., Rached, S., Ouakki, M., Habsaoui, A., & Touhami, M. (2025). Green Corrosion Protection : Plant Extracts as Sustainable Solutions. (p. 215-254). DOI: 10.4018/979-8-3693-6341-6.ch008

Mzioud, K., Habsaoui, A., Imatra, H., Haida, S., Rached, S., Msairi, S., Douira, A., Alqahtani, A. S., Noman, O. M., Tarayrah, M., & Touhami, M. E. (2023). Physicochemical characterization, antioxidant and antifungal activities of essential oils of Urginea maritima and Allium sativum. *Open Chemistry, 21*(1). *Open Chemistry*, *21*(1), 20230149. Advance online publication. DOI: 10.1515/chem-2023-0149

Mzioud, K., Habsaoui, A., Rached, S., Ech-chihbi, E., Ouakki, M., Salghi, R., & Touhami, M. E. (2024). Experimental investigation and theoretical modeling of copper corrosion inhibition by Urginea maritima essential oil. *Materials Today Sustainability, 27. Scopus., 27*, 100906. Advance online publication. DOI: 10.1016/j.mtsust.2024.100906

Nadar, N. R., Akkinepally, B., Harisha, B. S., Ibrahim, E. H., Rao, H. J., Dash, T., Sharma, S. C., Hussain, I., & Shim, J. (2025). Nature-inspired materials as sustainable electrodes for energy storage devices : Recent trends and future aspects. *Journal of Energy Storage, 106*, 114779. DOI: 10.1016/j.est.2024.114779

Nardecchia, S., Carriazo, D., Ferrer, M. L., Gutiérrez, M. C., & del Monte, F. (2012). Three dimensional macroporous architectures and aerogels built of carbon nanotubes and/or graphene : Synthesis and applications. *Chemical Society Reviews, 42*(2), 794–830. DOI: 10.1039/C2CS35353A

Naveen Kumar, S., Johnsi, M., Asha Shalini, V. J., Kavitha, N. P., & Balasubramanian, N. (2024). Cellulose nanofibril separator from *Coffea arabica* waste for supercapacitor applications. *Industrial Crops and Products, 214*, 118459. DOI: 10.1016/j.indcrop.2024.118459

Ndruru, S., Wahyuningrum, D., Bundjali, B., & Arcana, I. M. (2020). Preparation and characterization of biopolymer electrolyte membranes based on liclo4-complexed methyl cellulose as lithium-ion battery separator. *J. Eng. Technol. Sci, 52*(1), 28–50. DOI: 10.5614/j.eng.technol.sci.2020.52.1.3

Nguyen, T. D., & Lee, J. S. (2025). Dynamic Bonds in Biopolymers : Enhancing Performance and Properties. *Polymers, 17*(4), 4. Advance online publication. DOI: 10.3390/polym17040457

Nguyen, V.-T., Cho, K., Choi, Y., Hwang, B., Park, Y.-K., Nam, H., & Lee, D. (2024a). Biomass-derived materials for energy storage and electrocatalysis : Recent advances and future perspectives. *Biochar, 6*(1), 96. DOI: 10.1007/s42773-024-00388-1

Nguyen, V.-T., Cho, K., Choi, Y., Hwang, B., Park, Y.-K., Nam, H., & Lee, D. (2024b). Biomass-derived materials for energy storage and electrocatalysis : Recent advances and future perspectives. *Biochar, 6*(1), 96. DOI: 10.1007/s42773-024-00388-1

Ning, W., Xingxiang, Z., Haihui, L., & Jianping, W. (2009). *N, N*-dimethylacetamide/lithium chloride plasticized starch as solid biopolymer electrolytes. *Carbohydrate Polymers, 77*(3), 607–611. DOI: 10.1016/j.carbpol.2009.02.002

Nishimura, H., Kamiya, A., Nagata, T., Katahira, M., & Watanabe, T. (2018). Direct evidence for α ether linkage between lignin and carbohydrates in wood cell walls. *Scientific Reports*, *8*(1), 6538. DOI: 10.1038/s41598-018-24328-9

Nzereogu, P. U., Omah, A. D., Ezema, F. I., Iwuoha, E. I., & Nwanya, A. C. (2022). Anode materials for lithium-ion batteries : A review. *Applied Surface Science Advances*, *9*, 100233. DOI: 10.1016/j.apsadv.2022.100233

Obodo-Ovie, O., Alyassin, M., Smith, A., & Morris, G. (2024). The Effect of Different Extraction Conditions on the Physicochemical Properties of Novel High Methoxyl Pectin-like Polysaccharides from Green Bell Pepper (GBP). *Macromol*, *4*(2), 420–436. DOI: 10.3390/macromol4020024

Ojo, E. (2023). An Overview of Lignocellulose and Its Biotechnological Importance in High-Value Product Production. *Fermentation (Basel, Switzerland)*, *9*(11), 990. DOI: 10.3390/fermentation9110990

Okoye, P. U., Lobato-Peralta, D. R., Alemán-Ramirez, J. L., Duque-Brito, E., Arias, D. M., Okolie, J. A., & Sebastian, P. J. (2024a). Plant-based Materials for Energy Application. In *Plant Biomass Derived Materials* (pp. 441–464). John Wiley & Sons, Ltd., DOI: 10.1002/9783527839032.ch17

Okoye, P. U., Lobato-Peralta, D. R., Alemán-Ramirez, J. L., Duque-Brito, E., Arias, D. M., Okolie, J. A., & Sebastian, P. J. (2024b). Plant-based Materials for Energy Application. In *Plant Biomass Derived Materials* (pp. 441–464). John Wiley & Sons, Ltd., DOI: 10.1002/9783527839032.ch17

Okoye, P. U., Lobato-Peralta, D. R., Alemán-Ramirez, J. L., Duque-Brito, E., Arias, D. M., Okolie, J. A., & Sebastian, P. J. (2024c). Plant-based Materials for Energy Application. In *Plant Biomass Derived Materials* (pp. 441–464). John Wiley & Sons, Ltd., DOI: 10.1002/9783527839032.ch17

Ong, H. C., Yu, K. L., Chen, W.-H., Pillejera, M. K., Bi, X., Tran, K.-Q., Pétrissans, A., & Pétrissans, M. (2021). Variation of lignocellulosic biomass structure from torrefaction : A critical review. *Renewable & Sustainable Energy Reviews*, *152*, 111698. DOI: 10.1016/j.rser.2021.111698

Pajares, E., Maestu, J. F., Fernandez-de-Mendiola, I., Silvan, U., Costa, P., Agirrezabal-Telleria, I., Tubio, C. R., Corona-Galván, S., & Lanceros-Mendez, S. (2024). Strategies for Improving Sustainability in the Development of High-Performance Styrenic Block Copolymers by Developing Blends with Cellulose Derivatives. *Polymers*, *16*(6), 6. Advance online publication. DOI: 10.3390/polym16060856

Palchoudhury, S., Ramasamy, K., Gupta, R. K., & Gupta, A. (2019). Flexible Supercapacitors : A Materials Perspective. *Frontiers in Materials*, *5*, 83. Advance online publication. DOI: 10.3389/fmats.2018.00083

Patel, A., Patel, S. K., Singh, R. S., & Patel, R. P. (2024). Review on recent advancements in the role of electrolytes and electrode materials on supercapacitor performances. *Discover Nano*, *19*(1), 188. DOI: 10.1186/s11671-024-04053-1

Patra, N., Ramesh, P., Donthu, V., & Ahmad, A. (2024). Biopolymer-based composites for sustainable energy storage : Recent developments and future outlook. *Journal of Materials Science: Materials in Engineering*, *19*(1), 34. DOI: 10.1186/s40712-024-00181-9

Pavlovič, I., Knez, Ž., & Škerget, M. (2013). Hydrothermal Reactions of Agricultural and Food Processing Wastes in Sub- and Supercritical Water : A Review of Fundamentals, Mechanisms, and State of Research. *Journal of Agricultural and Food Chemistry*, *61*(34), 8003–8025. Advance online publication. DOI: 10.1021/jf401008a

Pérez-Rodríguez, S., Alegre, C., Sebastián, D., & Lázaro, M. J. (2021). Emerging carbon nanostructures in electrochemical processes. In Sadjadi, S. (Ed.), *Emerging Carbon Materials for Catalysis* (pp. 353–388). Elsevier., DOI: 10.1016/B978-0-12-817561-3.00010-X

Perumal, P., Christopher Selvin, P., & Selvasekarapandian, S. (2018). Characterization of biopolymer pectin with lithium chloride and its applications to electrochemical devices. *Ionics*, *24*(10), 3259–3270. DOI: 10.1007/s11581-018-2507-5

Perumal, P., Christopher Selvin, P., Selvasekarapandian, S., & Sivaraj, P. (2019). Structural and Electrical Properties of Bio-polymer Pectin with LiClO4 Solid Electrolytes for Lithium Ion Polymer Batteries. *Materials Today: Proceedings*, *8*, 196–202. DOI: 10.1016/j.matpr.2019.02.100

Péter, L. (2020). Materials science aspects of the research directions of electrochemistry. *Journal of Solid State Electrochemistry*, *24*(9), 2161–2162. DOI: 10.1007/s10008-020-04551-z

Pinto, E., Aggrey, W. N., Boakye, P., Amenuvor, G., Sokama-Neuyam, Y. A., Fokuo, M. K., Karimaie, H., Sarkodie, K., Adenutsi, C. D., Erzuah, S., & Rockson, M. A. D. (2022). Cellulose processing from biomass and its derivatization into carboxymethylcellulose : A review. *Scientific African*, *15*, e01078. DOI: 10.1016/j.sciaf.2021.e01078

Plavniece, A., Dobele, G., Volperts, A., & Zhurinsh, A. (2022). Hydrothermal Carbonization vs. Pyrolysis : Effect on the Porosity of the Activated Carbon Materials. *Sustainability (Basel)*, *14*(23), 23. Advance online publication. DOI: 10.3390/su142315982

Popescu, C., Dissanayake, H., Mansi, E., & Stancu, A. (2024). Eco Breakthroughs : Sustainable Materials Transforming the Future of Our Planet. *Sustainability (Basel)*, *16*(23), 23. Advance online publication. DOI: 10.3390/su162310790

Powell, M. D., LaCoste, J. D., Fetrow, C. J., Fei, L., & Wei, S. (2021). Bio-derived nanomaterials for energy storage and conversion. *Nano Select*, *2*(9), 1682–1706. DOI: 10.1002/nano.202100001

Prasankumar, T., Salpekar, D., Bhattacharyya, S., Manoharan, K., Yadav, R. M., Campos Mata, M. A., Miller, K. A., Vajtai, R., Jose, S., Roy, S., & Ajayan, P. M. (2022). Biomass derived hierarchical porous carbon for supercapacitor application and dilute stream CO2 capture. *Carbon*, *199*, 249–257. DOI: 10.1016/j.carbon.2022.07.057

Premalatha, M., Mathavan, T., Selvasekarapandian, S., Monisha, S., Selvalakshmi, S., & Vinoth Pandi, D. (2017). Tamarind seed polysaccharide (TSP)-based Li-ion conducting membranes. *Ionics*, *23*(10), 2677–2684. DOI: 10.1007/s11581-017-1989-x

Premalatha, M., Mathavan, T., Selvasekarapandian, S., & Selvalakshmi, S. (2018). Structural and electrical characterization of tamarind seed polysaccharide (TSP) doped with NH4HCO2. *AIP Conference Proceedings*, *1942*(1), 070005. DOI: 10.1063/1.5028803

Qi, D., Liu, Y., Liu, Z., Zhang, L., & Chen, X. (2017). Design of Architectures and Materials in In-Plane Micro-supercapacitors : Current Status and Future Challenges. *Advanced Materials*, *29*(5), 1602802. DOI: 10.1002/adma.201602802

Qin, T., Yang, H., Li, Q., Yu, X., & Li, H. (2024). Design of functional binders for high-specific-energy lithium-ion batteries : From molecular structure to electrode properties. *Industrial Chemistry & Materials*, *2*(2), 191–225. DOI: 10.1039/D3IM00089C

Qu, K., Sun, Z., Shi, C., Wang, W., Xiao, L., Tian, J., Huang, Z., & Guo, Z. (2021). Dual-acting cellulose nanocomposites filled with carbon nanotubes and zeolitic imidazolate framework-67 (ZIF-67)–derived polyhedral porous Co3O4 for symmetric supercapacitors. *Advanced Composites and Hybrid Materials*, *4*(3), 670–683. DOI: 10.1007/s42114-021-00293-2

Rafik, F., Gualous, H., Gallay, R., Crausaz, A., & Berthon, A. (2007). Frequency, thermal and voltage supercapacitor characterization and modeling. *Journal of Power Sources*, *165*(2), 928–934. DOI: 10.1016/j.jpowsour.2006.12.021

Rajendran, S., Al-Samydai, A., Palani, G., Trilaksana, H., Sathish, T., Giri, J., Saravanan, R., Lalvani, J. I. J., & Nasri, F. (2025). Replacement of Petroleum Based Products With Plant-Based Materials, Green and Sustainable Energy—A Review. *Engineering Reports*, *7*(4), e70108. DOI: 10.1002/eng2.70108

Rajendran, S., & Saratha, R. (2021). An evaluation of solid-state electrolyte based on pectin and lithium bis (trifluoromethanesulphonyl)imide for lithium-ion batteries. *Materials Today: Proceedings*, *47*, 819–824. DOI: 10.1016/j.matpr.2020.12.557

Rajput, V., Naik, B., Kumar, V., Kumar, V., Bhatt, S. C., & Rustagi, S. (2024). Advancements in energy storage applications : Harnessing the potential of fish industry waste. *Discover Materials*, *4*(1), 95. DOI: 10.1007/s43939-024-00161-y

Rapisarda, M., Marken, F., & Meo, M. (2021). Graphene oxide and starch gel as a hybrid binder for environmentally friendly high-performance supercapacitors. *Communications Chemistry*, *4*(1), 1–11. DOI: 10.1038/s42004-021-00604-0

Rashid, S., & Dutta, H. (2022). Industrial Applications of Cellulose Extracted from Agricultural and Food Industry Wastes. In *Handbook of Biomass Valorization for Industrial Applications* (pp. 417–443). John Wiley & Sons, Ltd., DOI: 10.1002/9781119818816.ch18

Rashid Khan, H., & Latif Ahmad, A. (2025). Supercapacitors : Overcoming current limitations and charting the course for next-generation energy storage. *Journal of Industrial and Engineering Chemistry*, *141*, 46–66. DOI: 10.1016/j.jiec.2024.07.014

Raut, B., Ahmed, M., Kim, H.-Y., Rahman Khan, M. M., & Rafiqul Bari, A. K. M. (2025). Battery-Type Transition Metal Oxides in Hybrid Supercapacitors : Synthesis and Applications. *Batteries*, *11*(2), 60. DOI: 10.3390/batteries11020060

Raza, N., Aziz, K., Javed, M. S., Tariq, R., Kanwal, N., Raza, W., Sarfraz, M., Khan, S., Ismail, M. A., Akkinepally, B., & Wang, X. (2025). Advancements in separator design for supercapacitor technology : A review of characteristics, manufacturing processes, limitations, and emerging trends. *Journal of Energy Storage*, *111*, 115328. DOI: 10.1016/j.est.2025.115328

Reenu, S., Sonia, , Phor, L., Kumar, A., & Chahal, S. (2024). Electrode materials for supercapacitors : A comprehensive review of advancements and performance. *Journal of Energy Storage*, *84*, 110698. DOI: 10.1016/j.est.2024.110698

Ries, L., Wartzack, S., & Zipse, O. (2023). Creating Sustainable Products. In Zipse, O., Hornegger, J., Becker, T., Beckmann, M., Bengsch, M., Feige, I., & Schober, M. (Eds.), *Road to Net Zero : Strategic Pathways for Sustainability-Driven Business Transformation* (pp. 123–157). Springer International Publishing., DOI: 10.1007/978-3-031-42224-9_5

Robyt, J. F. (2008). Starch : Structure, Properties, Chemistry, and Enzymology. In Fraser-Reid, B. O., Tatsuta, K., & Thiem, J. (Eds.), *Glycoscience : Chemistry and Chemical Biology* (pp. 1437–1472). Springer., DOI: 10.1007/978-3-540-30429-6_35

Roman-Benn, A., Contador, C. A., Li, M.-W., Lam, H.-M., Ah-Hen, K., Ulloa, P. E., & Ravanal, M. C. (2023). Pectin : An overview of sources, extraction and applications in food products, biomedical, pharmaceutical and environmental issues. *Food Chemistry Advances*, *2*, 100192. DOI: 10.1016/j.focha.2023.100192

Ropartz, D., & Ralet, M.-C. (2020). Pectin Structure. In Kontogiorgos, V. (Ed.), *Pectin : Technological and Physiological Properties* (pp. 17–36). Springer International Publishing., DOI: 10.1007/978-3-030-53421-9_2

Rosen, M. A. (2009). Energy Sustainability : A Pragmatic Approach and Illustrations. *Sustainability (Basel)*, *1*(1), 55–80. DOI: 10.3390/su1010055

Ruiz-Gonzalez, A., Wang, M., & Haseloff, J. (2023). Sustainable Approaches to Incorporate Plant-Based Biomaterials in Power Generation. *Solids*, *4*(2), 133–155. DOI: 10.3390/solids4020009

Ruiz Serrano, A., Musumeci, A., Li, J. J., Ruiz Serrano, M., & Serrano Barquin, C. (2025). Rationality and the exploitation of natural resources : A psychobiological conceptual model for sustainability. *Environment, Development and Sustainability*, *27*(6), 13167–13189. DOI: 10.1007/s10668-024-04470-3

Sá, D., & Helena, M. (2024). Electrochemical Devices to Power a Sustainable Energy Transition—An Overview of Green Hydrogen Contribution. *Applied Sciences (Basel, Switzerland)*, *14*(5), 2168. DOI: 10.3390/app14052168

Saadan, R., Hachimi Alaoui, C., Ihammi, A., Chigr, M., & Fatimi, A. (2024). A Brief Overview of Lignin Extraction and Isolation Processes : From Lignocellulosic Biomass to Added-Value Biomaterials. *Environmental and Earth Sciences Proceedings*, *31*(1), 1. Advance online publication. DOI: 10.3390/eesp2024031003

Saeed, N., Liaquat, R., Iqbal, N., & Khan, T. A. (2025). Synthesis of cellulose-based hydrogel electrolyte from sawdust waste for enhanced ionic conductivity in ssZIBs. *Materials Science and Engineering B*, *313*, 117951. DOI: 10.1016/j.mseb.2024.117951

Saini, S., Chand, P., & Joshi, A. (2021). Biomass derived carbon for supercapacitor applications [Review]. *Journal of Energy Storage*, *39*, 102646. DOI: 10.1016/j.est.2021.102646

Sajedi-Moghaddam, A., Saievar-Iranizad, E., & Pumera, M. (2017). Two-dimensional transition metal dichalcogenide/conducting polymer composites : Synthesis and applications. *Nanoscale*, *9*(24), 8052–8065. DOI: 10.1039/C7NR02022H

Sakthi Balan, G., & Aravind Raj, S. (2025). Effect of additives on the performance of extruded polymer composites : A comprehensive review. *Journal of Thermoplastic Composite Materials*, *38*(2), 810–853. DOI: 10.1177/08927057241261314

Saldarriaga-Hernández, S., Velasco-Ayala, C., Leal-Isla Flores, P., de Jesús Rostro-Alanis, M., Parra-Saldivar, R., Iqbal, H. M. N., & Carrillo-Nieves, D. (2020). Biotransformation of lignocellulosic biomass into industrially relevant products with the aid of fungi-derived lignocellulolytic enzymes. *International Journal of Biological Macromolecules*, *161*, 1099–1116. DOI: 10.1016/j.ijbiomac.2020.06.047

Saleh, A. K., El-Sayed, M. H., El-Sakhawy, M. A., Alshareef, S. A., Omer, N., Abdelaziz, M. A., Jame, R., Zheng, H., Gao, M., & Du, H. (2025). Cellulose-based Conductive Materials for Bioelectronics. *ChemSusChem*, *18*(6), e202401762. DOI: 10.1002/cssc.202401762

Samir, A., Ashour, F. H., Hakim, A. A. A., & Bassyouni, M. (2022). Recent advances in biodegradable polymers for sustainable applications. *NPJ Materials Degradation*, *6*(1), 1–28. DOI: 10.1038/s41529-022-00277-7

Sankar, S., Ahmed, A. T. A., Inamdar, A. I., Im, H., Im, Y. B., Lee, Y., Kim, D. Y., & Lee, S. (2019). Biomass-derived ultrathin mesoporous graphitic carbon nanoflakes as stable electrode material for high-performance supercapacitors. *Materials & Design*, *169*, 107688. DOI: 10.1016/j.matdes.2019.107688

Sant'Ana Júnior, D. B., Kelbert, M., Hermes de Araújo, P. H., & de Andrade, C. J. (2025). Physical Pretreatments of Lignocellulosic Biomass for Fermentable Sugar Production. *Sustainable Chemistry*, *6*(2), 2. Advance online publication. DOI: 10.3390/suschem6020013

Sarangi, P., & Subudhi, S. (2022). Utilization of agricultural waste biomass and recycling toward circular bioeconomy. *Environmental Science and Pollution Research International*, *30*(4), 8526–8539. Advance online publication. DOI: 10.1007/s11356-022-20669-1

Sarmah, S., Kakati, B. K., Kucernak, A. R., & Deka, D. (2024). Fabrication of hierarchically structured supercapacitor using N and S co-doped activated carbons derived from Samanea saman biomass. *Energy Storage*, *6*(2), e616. DOI: 10.1002/est2.616

Sarno, M. (2020). Nanotechnology in energy storage : The supercapacitors. In Basile, A., Centi, G., Falco, M. D., & Iaquaniello, G. (Eds.), *Studies in Surface Science and Catalysis* (Vol. 179, pp. 431–458). Elsevier., DOI: 10.1016/B978-0-444-64337-7.00022-7

Sayed, K., Aref, M., Abokhalil, A. G., Mossa, M. A., Sayed, K., Aref, M., Abokhalil, A. G., & Mossa, M. A. (2025). *Hybrid and Advanced Energy Storage Systems : Integration, Applications, and Future Trends*. IntechOpen., DOI: 10.5772/intechopen.1010067

Schlemmer, W., Selinger, J., Hobisch, M. A., & Spirk, S. (2021). Polysaccharides for sustainable energy storage – A review. *Carbohydrate Polymers*, *265*, 118063. DOI: 10.1016/j.carbpol.2021.118063

Schnepp, Z. (2013). ChemInform Abstract : Biopolymers as a Flexible Resource for Nanochemistry. *ChemInform*, *44*(21), chin.201321266. Advance online publication. DOI: 10.1002/chin.201321266

Shah, S. S., Aziz, M. A., & Oyama, M. (2024). Advancing Electrochemical Energy Storage : The Synergistic Impact of Sulfur Nano-Confinement in Hierarchically Porous Jute-Derived Activated Carbon for Enhanced Supercapacitor Performance. *ECS Meeting Abstracts, MA2024-01*(1), 11. DOI: 10.1149/MA2024-01111mtgabs

Shahcheragh, S., Mohagheghi, M., & Shirpay, A. (2023). Effect of physical and chemical activation methods on the structure, optical absorbance, band gap and urbach energy of porous activated carbon. *SN Applied Sciences*, *5*(12), 313. Advance online publication. DOI: 10.1007/s42452-023-05559-6

Shamsuri, A. A., Md. Jamil, S. N. A., Yusoff, M. Z. M., & Abdan, K. (2024). Polymer Composites Containing Ionic Liquids : A Study of Electrical Conductivity. *Electronic Materials*, *5*(4), 4. Advance online publication. DOI: 10.3390/electronicmat5040013

Sharma, G. (2024). Biogenic carbon nanostructured materials for detection of cancer and medical applications : A mini review. *Hybrid Advances*, *5*, 100166. DOI: 10.1016/j.hybadv.2024.100166

Sheng, J., Chen, T., Wang, R., Zhang, Z., Hua, F., & Yang, R. (2020). Ultra-light cellulose nanofibril membrane for lithium-ion batteries. *Journal of Membrane Science*, *595*, 117550. DOI: 10.1016/j.memsci.2019.117550

Sheng, J., Wang, R., & Yang, R. (2019). Physicochemical Properties of Cellulose Separators for Lithium Ion Battery : Comparison with Celgard2325. *Materials (Basel)*, *12*(1), 1. Advance online publication. DOI: 10.3390/ma12010002

Shi, X., Dong, C., Ma, L., Yang, M., Chen, X., Zhang, Y., Liu, L., & Long, Z. (2025). A sustainable and recyclable cellulose gel electrolyte enables stable zinc metal anode for green aqueous batteries. *Chemical Engineering Journal*, *504*, 158659. DOI: 10.1016/j.cej.2024.158659

Shiraishi, S. (2003). Electric Double Layer Capacitors. In Yasuda, E., Inagaki, M., Kaneko, K., Endo, M., Oya, A., & Tanabe, Y. (Eds.), *Carbon Alloys* (pp. 447–457). Elsevier Science., DOI: 10.1016/B978-008044163-4/50027-9

Shrestha, L. K., Shrestha, R. G., Shahi, S., Gnawali, C. L., Adhikari, M. P., Bhadra, B. N., & Ariga, K. (2023). Biomass Nanoarchitectonics for Supercapacitor Applications. *Journal of Oleo Science*, *72*(1), 11–32. DOI: 10.5650/jos.ess22377

Shrestha, P., & Bazylak, A. (2023). Pursuit of next-generation electrochemical energy devices. *Electrochimica Acta*, *464*, 142810. DOI: 10.1016/j.electacta.2023.142810

Shyu, J.-C., & Wang, Y.-L. (2021). Effects of Nafion ionomer content on the performance of membraneless direct formate fuel cells. *Energy Science & Engineering*, *9*(11), 2139–2150. DOI: 10.1002/ese3.971

Siipola, V., Tamminen, T., Källi, A., Kupila, R., Romar, H., Rasa, K., Keskinen, R., Hyväluoma, J., Hannula, M., & Wikberg, H. (2018). Effects of biomass type, carbonization process, and activation method on the properties of bio-based activated carbons. *BioResources*, *13*(3), 5976–6002. Advance online publication. DOI: 10.15376/biores.13.3.5976-6002

Silva, A. C. Q., Silvestre, A. J. D., Vilela, C., & Freire, C. S. R. (2022). Natural Polymers-Based Materials : A Contribution to a Greener Future. *Molecules (Basel, Switzerland)*, *27*(1), 1. Advance online publication. DOI: 10.3390/molecules27010094

Singh, A. N., Islam, M., Meena, A., Faizan, M., Han, D., Bathula, C., Hajibabaei, A., Anand, R., & Nam, K.-W. (2023). Unleashing the Potential of Sodium-Ion Batteries : Current State and Future Directions for Sustainable Energy Storage. *Advanced Functional Materials*, *33*(46), 2304617. DOI: 10.1002/adfm.202304617

Sivachidambaram, M., Vijaya, J. J., Kennedy, L. J., Jothiramalingam, R., Al-Lohedan, H. A., Munusamy, M. A., Elanthamilan, E., & Merlin, J. P. (2017). Preparation and characterization of activated carbon derived from the Borassus flabellifer flower as an electrode material for supercapacitor applications. *New Journal of Chemistry*, *41*(10), 3939–3949. DOI: 10.1039/C6NJ03867K

Sohaimy, M. I. H. A., & Isa, M. I. N. M. (2020). Natural Inspired Carboxymethyl Cellulose (CMC) Doped with Ammonium Carbonate (AC) as Biopolymer Electrolyte. *Polymers, 12*(11), 11. Advance online publication. DOI: 10.3390/polym12112487

Song, C., Han, K., Teng, Z., Wang, M., Pei, Y., & Liu, J. (2024). Low metal content and hierarchically porous structure of "Fabaceae-Plants" derived activated carbon for high-performance supercapacitors. *Journal of Energy Storage*, *87*, 111430. DOI: 10.1016/j.est.2024.111430

Song, X., Ma, X., Li, Y., Ding, L., & Jiang, R. (2019). Tea waste derived microporous active carbon with enhanced double-layer supercapacitor behaviors. *Applied Surface Science*, *487*, 189–197. DOI: 10.1016/j.apsusc.2019.04.277

Song, Y., Zhao, G., Zhang, S., Xie, C., Yang, R., & Li, X. (2024). Chitosan nanofiber paper used as separator for high performance and sustainable lithium-ion batteries. *Carbohydrate Polymers*, *329*, 121530. DOI: 10.1016/j.carbpol.2023.121530

Spirk, S. (2018). Polysaccharides in Batteries. In Spirk, S. (Ed.), *Polysaccharides as Battery Components* (pp. 9–57). Springer International Publishing., DOI: 10.1007/978-3-319-65969-5_2

Srivastava, N., & Yadav, D. (2023). Polysaccharide Biomaterials for Electrochemical Applications. In *Sustainable Materials for Electrochemical Capacitors* (pp. 391–414). John Wiley & Sons, Ltd., DOI: 10.1002/9781394167104.ch16

Staaf, L. G. H., Lundgren, P., & Enoksson, P. (2014). Present and future supercapacitor carbon electrode materials for improved energy storage used in intelligent wireless sensor systems. *Nano Energy*, *9*, 128–141. DOI: 10.1016/j.nanoen.2014.06.028

Subhani, T., Khademolqorani, S., Banitaba, S. N., Ramadan, M., Khaliq, A., Chaudhry, I. A., & Osman, A. I. (2024). Advancements in Battery Materials : Bio-Based and Mineral Fillers for Next-Generation Solid Polymer Electrolytes. *ACS Applied Materials & Interfaces*, *16*(46), 63089–63108. Advance online publication. DOI: 10.1021/acsami.4c11214

Sun, F., Wang, L., Peng, Y., Gao, J., Pi, X., Qu, Z., Zhao, G., & Qin, Y. (2018). Converting biomass waste into microporous carbon with simultaneously high surface area and carbon purity as advanced electrochemical energy storage materials. *Applied Surface Science*, *436*, 486–494. DOI: 10.1016/j.apsusc.2017.12.067

Sun, G., Qiu, L., Zhu, M., Kang, K., & Guo, X. (2018). Activated carbons prepared by hydrothermal pretreatment and chemical activation of *Eucommia ulmoides* wood for supercapacitors application. *Industrial Crops and Products*, *125*, 41–49. DOI: 10.1016/j.indcrop.2018.08.082

Sun, J., Luo, B., & Li, H. (2022). A Review on the Conventional Capacitors, Supercapacitors, and Emerging Hybrid Ion Capacitors : Past, Present, and Future. *Advanced Energy and Sustainability Research*, *3*(6), 2100191. DOI: 10.1002/aesr.202100191

Suresh Khurd, A., & Kandasubramanian, B. (2022). A systematic review of cellulosic material for green electronics devices. *Carbohydrate Polymer Technologies and Applications*, *4*, 100234. DOI: 10.1016/j.carpta.2022.100234

Tang, C., Yu, L., Jiang, Q., Gu, R., Zhang, Y., Liu, Z., Cai, W., Wu, H., Zhang, Y., & Yao, M. (2024). Polymer-Based Electrolyte for Lithium-Based High-Energy-Density and Safe Energy Storages Devices : Strategy and Mechanisms. *Renewables*, *2*(5), 297–340. DOI: 10.31635/renewables.024.202400068

Tang, X., Liu, D., Wang, Y.-J., Cui, L., Ignaszak, A., Yu, Y., & Zhang, J. (2021a). Research advances in biomass-derived nanostructured carbons and their composite materials for electrochemical energy technologies. *Progress in Materials Science*, *118*, 100770. DOI: 10.1016/j.pmatsci.2020.100770

Tang, X., Liu, D., Wang, Y.-J., Cui, L., Ignaszak, A., Yu, Y., & Zhang, J. (2021b). Research advances in biomass-derived nanostructured carbons and their composite materials for electrochemical energy technologies. *Progress in Materials Science*, *118*, 100770. DOI: 10.1016/j.pmatsci.2020.100770

Tatrari, G., Karakoti, M., Tewari, C., Pandey, S., Singh Bohra, B., Dandapat, A., & Gopal Sahoo, N. (2021). Solid waste-derived carbon nanomaterials for supercapacitor applications : A recent overview. *Materials Advances*, *2*(5), 1454–1484. DOI: 10.1039/D0MA00871K

Tayeb, A., Amini, E., Ghasemi, S., & Tajvidi, M. (2018). Cellulose Nanomaterials—Binding Properties and Applications : A Review. *Molecules (Basel, Switzerland)*, *23*(10), 2684. DOI: 10.3390/molecules23102684

Temesgen, T., Tilahun, E., Gonfa, B., Tufa, L. T., Sabir, F., Tadesse, S., & Dessie, Y. (2023). Advancements in biomass derived porous carbon materials and their surface influence effect on electrode electrochemical performance for sustainable supercapacitors : A review. *Journal of Energy Storage*, *73*, 109293. DOI: 10.1016/j.est.2023.109293

Teng, X., Sun, C., Dai, J., Liu, H., Su, J., & Li, F. (2013). Solution casting Nafion/polytetrafluoroethylene membrane for vanadium redox flow battery application. *Electrochimica Acta*, *88*, 725–734. DOI: 10.1016/j.electacta.2012.10.093

Tenhaeff, W., Rios, O., More, K., & McGuire, M. (2014). Highly Robust Lithium Ion Battery Anodes from Lignin : An Abundant, Renewable, and Low-Cost Material. *Advanced Functional Materials*, *24*(1), 86–94. Advance online publication. DOI: 10.1002/adfm.201301420

Thomas, S. M., Gómez-Romero, P., & González-Gil, R. M. (2025). Polysaccharides : The Sustainable Foreground in Energy Storage Systems. *Polysaccharides*, *6*(1), 1. Advance online publication. DOI: 10.3390/polysaccharides6010005

Thrisha, K., & Saratha, R. (2024). Natural polymer-based electrolytes for energy storage devices—An overview. *Ionics*, *30*(3), 1245–1266. DOI: 10.1007/s11581-023-05315-1

Timell, T. E. (1967). Recent progress in the chemistry of wood hemicelluloses. *Wood Science and Technology*, *1*(1), 45–70. DOI: 10.1007/BF00592255

Tiwari, A., & Kumar Singh, R. (2024). Biopolymers for Energy Applications. In *Bio-Based Polymers : Farm to Industry. Volume 3 : Emerging Trends and Applications* (Vol. 1487, p. 1-12). American Chemical Society. DOI: 10.1021/bk-2024-1487.ch001

Tiwari, S. K., Thakur, A. K., Adhikari, A. D., Zhu, Y., & Wang, N. (2020). Current Research of Graphene-Based Nanocomposites and Their Application for Supercapacitors. *Nanomaterials (Basel, Switzerland)*, *10*(10), 10. Advance online publication. DOI: 10.3390/nano10102046

Tiwari, T., Srivastava, N., & Srivastava, P. C. (2011). Electrical transport study of potato starch-based electrolyte system. *Ionics*, *17*(4), 353–360. DOI: 10.1007/s11581-010-0516-0

Tong, C., Ma, Z., Chen, H., & Gao, H. (2023). Toward an understanding of potato starch structure, function, biosynthesis, and applications. *Food Frontiers*, *4*(3), 980–1000. DOI: 10.1002/fft2.223

Turossi, T. C., Júnior, H. L. O., Monticeli, F. M., Dias, O. T., & Zattera, A. J. (2025). Cellulose-Derived Battery Separators : A Minireview on Advances Towards Environmental Sustainability. *Polymers*, *17*(4), 4. Advance online publication. DOI: 10.3390/polym17040456

Uddin, I., Ali, H., Al-Sehemi, A. G., Muhammad, S., Hamad, N., Taha, T. A., AlSalem, H. S., Alenad, A. M., Palamanit, A., Hayat, A., & Sohail, M. (2022). Efficient pyrolysis process of lignin over dual catalyst bed for the production of Phenols and Aromatics. *South African Journal of Botany*, *149*, 109–116. DOI: 10.1016/j.sajb.2022.05.061

Uddin, M.-J., Alaboina, P. K., Zhang, L., & Cho, S.-J. (2017). A low-cost, environment-friendly lignin-polyvinyl alcohol nanofiber separator using a water-based method for safer and faster lithium-ion batteries. *Materials Science and Engineering B, 223*, 84–90. DOI: 10.1016/j.mseb.2017.05.004

Vahini, M., Muthuvinayagam, M., & Isa, M. I. N. (2019). Preparation and Characterization of Biopolymer Electrolytes Based on Pectin and NaNO3 for Battery Applications. *Polymer Science, Series A, 61*(6), 823–831. DOI: 10.1134/S0965545X19060129

Varghese, S., Ralph, P., & Kuzhiumparambil, U. (2025). Sustainable carbon-rich materials from algae and biomass : Synthesis, characterisation, and applications. *Results in Engineering, 27*, 106439. DOI: 10.1016/j.rineng.2025.106439

Vijayakumar, M., Bharathi Sankar, A., Sri Rohita, D., Rao, T. N., & Karthik, M. (2019). Conversion of Biomass Waste into High Performance Supercapacitor Electrodes for Real-Time Supercapacitor Applications. *ACS Sustainable Chemistry & Engineering, 7*(20), 17175–17185. DOI: 10.1021/acssuschemeng.9b03568

Vikraman, V. K., Kumar, D. P., Boopathi, G., Komalabharathi, P., & Subramanian, P. (2025). Bio-Based Polymer Electrolytes for Supercapacitor Applications. In Baskar, C., Ramakrishna, S., & Rosa, A. D. L. (Eds.), *Encyclopedia of Green Materials* (pp. 159–165). Springer Nature., DOI: 10.1007/978-981-97-4618-7_207

Vinodh, R., Sasikumar, Y., Kim, H.-J., Atchudan, R., & Yi, M. (2021). Chitin and chitosan based biopolymer derived electrode materials for supercapacitor applications : A critical review. *Journal of Industrial and Engineering Chemistry, 104*, 155–171. DOI: 10.1016/j.jiec.2021.08.019

Wang, C., Zhang, S.-H., Zhang, L., Xi, R., Jiang, D.-P., Chen, Z.-Y., Huang, H., Ding, L.-Y., & Pan, G.-B. (2019). Natural bamboo leaves derived sulphur-doped mesoporous heteroatom enriched carbon for high-performance supercapacitors and gas sensors. *Journal of Power Sources, 443*, 227183. DOI: 10.1016/j.jpowsour.2019.227183

Wang, J., Abbas, S. C., Li, L., Walker, C. C., Ni, Y., & Cai, Z. (2024). Cellulose Membranes : Synthesis and Applications for Water and Gas Separation and Purification. *Membranes (Basel), 14*(7), 7. Advance online publication. DOI: 10.3390/membranes14070148

Wang, L., Gao, X., Wang, D., Shang, H., Zhao, Y., & Zhang, B. (2025). Nickel-carbon composites toward supercapacitor and self-charging systems : A review. *Fuel, 381*, 133639. DOI: 10.1016/j.fuel.2024.133639

Wang, L., Li, G., Chen, X., Yang, Y., Liew, R. K., Abo-Dief, H. M., Lam, S. S., Sellami, R., Peng, W., & Li, H. (2024). Extraction strategies for lignin, cellulose, and hemicellulose to obtain valuable products from biomass. *Advanced Composites and Hybrid Materials*, *7*(6), 219. DOI: 10.1007/s42114-024-01009-y

Wang, Q., Li, Y., Yu, Z., Li, X., Yin, S., Ji, W., Hu, Y., Cai, W., & Wang, X. (2023). Highly porous carbon derived from hydrothermal-pyrolysis synergistic carbonization of biomass for enhanced CO2 capture. *Colloids and Surfaces. A, Physicochemical and Engineering Aspects*, *673*, 131787. DOI: 10.1016/j.colsurfa.2023.131787

Wang, S., Zhang, L., Wang, A., Liu, X., Chen, J., Wang, Z., Zeng, Q., Zhou, H., Jiang, X., & Zhang, L. (2018). Polymer-Laden Composite Lignin-Based Electrolyte Membrane for High-Performance Lithium Batteries. *ACS Sustainable Chemistry & Engineering*, *6*(11), 14460–14469. DOI: 10.1021/acssuschemeng.8b03117

Wang, Y., Chen, Y., Zhao, H., Li, L., Ju, D., Wang, C., & An, B. (2022). Biomass-Derived Porous Carbon with a Good Balance between High Specific Surface Area and Mesopore Volume for Supercapacitors. *Nanomaterials (Basel, Switzerland)*, *12*(21), 21. Advance online publication. DOI: 10.3390/nano12213804

Wang, Y., Liu, X., Sheng, J., Zhu, H., & Yang, R. (2021). Nanoporous Regenerated Cellulose Separator for High-Performance Lithium Ion Batteries Prepared by Nonsolvent-Induced Phase Separation. *ACS Sustainable Chemistry & Engineering*, *9*(44), 14756–14765. DOI: 10.1021/acssuschemeng.1c04350

Wang, Y., Ma, X., Yang, X., Zhang, R., Hong, H., Wang, S., Li, Q., Chen, Z., Huang, Z., Lv, H., & Zhi, C. (s. d.). A Multifunctional Binder for Current-Collector-Free Zn Powder Anodes. *Advanced Materials, n/a*(n/a), 2419702. DOI: 10.1002/adma.202419702

Wang, Z., Zhu, M., Pei, Z., Xue, Q., Li, H., Huang, Y., & Zhi, C. (2020). Polymers for supercapacitors : Boosting the development of the flexible and wearable energy storage. *Materials Science and Engineering R Reports*, *139*, 100520. DOI: 10.1016/j.mser.2019.100520

Wijerathne, B., Liao, T., Jiang, X., Zhou, J., & Sun, Z. (2025). Plant-inspired surfaces and interfaces for sustainable technologies. *Materials Futures*, *4*(1), 012301. DOI: 10.1088/2752-5724/ad93ea

Wu, Q., Gong, F., Liu, S., Yuan, X., & Xiao, R. (2025). Porous carbon derived from molten salt method for electrochemical energy storage. *Journal of Power Sources*, *644*, 237091. DOI: 10.1016/j.jpowsour.2025.237091

Wu, W.-Y., Gao, M., Yeo Jueyuan, R., Lin, M., Wang, S., Thitsartarn, W., Zhang, X., Kai, D., Wang, P., Qu, Z., Xu, J., Jun Loh, X., & Zhu, Q. (2024). Plant oil-based phase change materials for sustainable thermal energy storage : A review. *Fuel, 378*, 132940. DOI: 10.1016/j.fuel.2024.132940

Wu, Z., Zhu, Y., Ji, X., & Banks, C. E. (2016). Transition Metal Oxides as Supercapacitor Materials. In Ozoemena, K. I., & Chen, S. (Eds.), *Nanomaterials in Advanced Batteries and Supercapacitors* (pp. 317–344). Springer International Publishing., DOI: 10.1007/978-3-319-26082-2_9

Xia, J., Zhang, N., Chong, S., Li, D., Chen, Y., & Sun, C. (2018). Three-dimensional porous graphene-like sheets synthesized from biocarbon via low-temperature graphitization for a supercapacitor. *Green Chemistry*, *20*(3), 694–700. DOI: 10.1039/C7GC03426A

Xiao, X., Meng, Q., Zhu, J., Zhang, Y., Ma, Y., Huo, H., Yin, G., & Lou, S. (2024). Mechanics and electrochemistry in nature-inspired functional batteries : Fundamentals, configurations and devices. *Energy & Environmental Science*, *17*(3), 974–1006. DOI: 10.1039/D3EE03392A

Xu, G., Han, J., Ding, B., Nie, P., Pan, J., Dou, H., Li, H., & Zhang, X. (2015). Biomass-derived porous carbon materials with sulfur and nitrogen dual-doping for energy storage. *Green Chemistry*, *17*(3), 1668–1674. DOI: 10.1039/C4GC02185A

Xu, L., Li, Y., Fu, J., Shi, L., Li, C., & Ma, R. (2024). Advances in Functional Cellulose Hydrogels as Electrolytes for Flexible Zinc-Ion Batteries. *Nanomaterials (Basel, Switzerland)*, *14*(20), 20. Advance online publication. DOI: 10.3390/nano14201645

Yadlapalli, R. T., Alla, R. R., Kandipati, R., & Kotapati, A. (2022). Super capacitors for energy storage : Progress, applications and challenges. *Journal of Energy Storage*, *49*, 104194. DOI: 10.1016/j.est.2022.104194

Yang, I., Jung, M., Kim, M.-S., Choi, D., & Jung, J. (2021). Physical and chemical activation mechanisms of carbon materials based on the microdomain model. *Journal of Materials Chemistry. A, Materials for Energy and Sustainability*, *9*(15), 9815–9825. Advance online publication. DOI: 10.1039/D1TA00765C

Yang, K.-P., Xie, S., Han, Z.-M., Liu, H., Yin, C.-H., Sun, W.-B., Ling, Z.-C., Yang, H.-B., Li, D.-H., Guan, Q.-F., & Yu, S.-H. (2025). A cellulose-based lithium-ion battery separator with regulated ionic transport and high thermal stability for extreme environments. *Nano Research*, *18*(1), 94906994–949069941. DOI: 10.26599/NR.2025.94906994

Yang, Y., Zha, J., Liu, L., & Yuan, N. (2017). Facile Preparation of Highly Porous Carbon from Bean Dregs for Enhanced Electrochemical Performance. *International Journal of Electrochemical Science*, *12*(8), 7326–7340. DOI: 10.20964/2017.08.36

Yao, M., Bi, X., Wang, Z., Yu, P., Dufresne, A., & Jiang, C. (2022). Recent advances in lignin-based carbon materials and their applications : A review. *International Journal of Biological Macromolecules*, *223*, 980–1014. DOI: 10.1016/j.ijbiomac.2022.11.070

Yazdani McCord, M. R., Seppälä, A., Pourakbari-Kasmaei, M., Zimmerman, J. B., & Rojas, O. J. (2025). From low conductivity to high energy efficiency : The role of conductive polymers in phase change materials. *Chemical Engineering Journal*, *508*, 160804. DOI: 10.1016/j.cej.2025.160804

Yhobu, Z., Siddiqa, A., Padaki, M., & Budagumpi, S., & D. H., N. (. (2022). Lignocellulose Biopolymers and Electronically Conducting Polymers : Toward Sustainable Energy Storage Applications. *Energy & Fuels*, *36*(24), 14625–14656. DOI: 10.1021/acs.energyfuels.2c03101

Yoon, J., Lee, J., Kim, H., Kim, J., & Jin, H.-J. (2024). Polymeric Binder Design for Sustainable Lithium-Ion Battery Chemistry. *Polymers*, *16*(2), 2. Advance online publication. DOI: 10.3390/polym16020254

Yoshimura, M., & Byrappa, K. (2008). Hydrothermal processing of materials : Past, present and future. *Journal of Materials Science*, *43*(7), 2085–2103. DOI: 10.1007/s10853-007-1853-x

Yu, D., Qian, Q., Wei, L., Jiang, W., Goh, K., Wei, J., Zhang, J., & Chen, Y. (2015). Emergence of fiber supercapacitors. *Chemical Society Reviews*, *44*(3), 647–662. DOI: 10.1039/C4CS00286E

Yuan, D., Manalastas, W.Jr, Zhang, L., Chan, J. J., Meng, S., Chen, Y., & Srinivasan, M. (2019). Lignin@Nafion Membranes Forming Zn Solid–Electrolyte Interfaces Enhance the Cycle Life for Rechargeable Zinc-Ion Batteries. *ChemSusChem*, *12*(21), 4889–4900. DOI: 10.1002/cssc.201901409

Yuan, Q., & Ma, M.-G. (2021). Conductive polypyrrole incorporated nanocellulose/MoS2 film for preparing flexible supercapacitor electrodes. *Frontiers of Materials Science*, *15*(2), 227–240. DOI: 10.1007/s11706-021-0549-5

Yuan, Y., Yi, R., Sun, Y., Zeng, J., Li, J., Hu, J., Zhao, Y., Sun, W., Zhao, C., Yang, L., & Zhao, C. (2018). Porous Activated Carbons Derived from Pleurotus eryngii for Supercapacitor Applications. *Journal of Nanomaterials*, *2018*(1), 7539509. DOI: 10.1155/2018/7539509

Yue, Y., & Liang, H. (2018). Anatase TiO2 Nanoparticles Supported By Cu/Ni Micro-Channeled Current Collector as a Novel Electrode for Lithium-Ion Batteries. *ECS Meeting Abstracts, MA2018-01*(3), 431. DOI: 10.1149/MA2018-01/3/431

Zahid, A., Masri, M. N., Hussin, M. H., & Abu Bakar, M. B. (2018). The preliminary study on cassava (Manihot Esculenta) as gel polymer electrolyte for zinc-air battery. In *AIP Conference Proceedings* (Vol. 2030). DOI: 10.1063/1.5066919

Zhang, J., Song, C., & Xiang, J. (2024). A simulation-driven prediction model for state of charge estimation of electric vehicle lithium battery. *Heliyon*, *10*(10), e30988. Advance online publication. DOI: 10.1016/j.heliyon.2024.e30988

Zhang, L., Guo, L., & Wei, G. (2021). Recent Advances in the Fabrication and Environmental Science Applications of Cellulose Nanofibril-Based Functional Materials. *Materials (Basel)*, *14*(18), 18. Advance online publication. DOI: 10.3390/ma14185390

Zhang, M. Y., Li, M. X., Chang, Z., Wang, Y. F., Gao, J., Wu, Y. P., & Huang, W. (2017). A Sandwich PVDF/HEC/PVDF Gel Polymer Electrolyte for Lithium Ion Battery. *Electrochimica Acta*, *245*, 752–759. Advance online publication. DOI: 10.1016/j.electacta.2017.05.154

Zhang, N., Li, S., Xiong, L., Hong, Y., & Chen, Y. (2015). Cellulose-hemicellulose interaction in wood secondary cell-wall. *Modelling and Simulation in Materials Science and Engineering*, *23*(8), 085010. DOI: 10.1088/0965-0393/23/8/085010

Zhang, Q., Soham, D., Liang, Z., & Wan, J. (2025). Advances in wearable energy storage and harvesting systems. *Med-X*, *3*(1), 3. DOI: 10.1007/s44258-024-00048-w

Zhang, T., Zhang, T., Wang, F., & Ran, F. (2023). Pretreatment Process Before Heat Pyrolysis of Plant-based Precursors Paving Way for Fabricating High-Performance Hard Carbon for Sodium-Ion Batteries. *ChemElectroChem*, *10*(24), e202300442. DOI: 10.1002/celc.202300442

Zhang, X., Cheng, S., Fu, C., Yin, G., Wang, L., Wu, Y., & Huo, H. (2024). Advancements and Challenges in Organic–Inorganic Composite Solid Electrolytes for All-Solid-State Lithium Batteries. *Nano-Micro Letters*, *17*(1), 2. DOI: 10.1007/s40820-024-01498-y

Zhang, X., Jiang, C., Liang, J., & Wu, W. (2021). Electrode materials and device architecture strategies for flexible supercapacitors in wearable energy storage. *Journal of Materials Chemistry. A, Materials for Energy and Sustainability*, *9*(13), 8099–8128. DOI: 10.1039/D0TA12299H

Zhang, Y., & Tang, Z. (2020). Porous carbon derived from herbal plant waste for supercapacitor electrodes with ultrahigh specific capacitance and excellent energy density. *Waste Management (New York, N.Y.)*, *106*, 250–260. DOI: 10.1016/j.wasman.2020.03.032

Zhang, Z., Yang, S., Li, H., Zan, Y., Li, X., Zhu, Y., Dou, M., & Wang, F. (2019). Sustainable Carbonaceous Materials Derived from Biomass as Metal-Free Electrocatalysts. *Advanced Materials*, *31*(13), 1805718. DOI: 10.1002/adma.201805718

Zhang, Z., Yi, S., Wei, Y., Bian, H., Wang, R., & Min, Y. (2019). Lignin Nanoparticle-Coated Celgard Separator for High-Performance Lithium–Sulfur Batteries. *Polymers*, *11*(12), 12. Advance online publication. DOI: 10.3390/polym11121946

Zhao, D., Zhu, Y., Cheng, W., Chen, W., Wu, Y., & Yu, H. (2021). Cellulose-Based Flexible Functional Materials for Emerging Intelligent Electronics. *Advanced Materials*, *33*(28), e2000619. DOI: 10.1002/adma.202000619

Zhao, M., Wang, J., Chong, C., Yu, X., Wang, L., & Shi, Z. (2015). An electrospun lignin/polyacrylonitrile nonwoven composite separator with high porosity and thermal stability for lithium-ion batteries. *RSC Advances*, *5*(122), 101115–101120. DOI: 10.1039/C5RA19371K

Zhao, T., Xiao, P., Luo, M., Nie, S., Li, F., & Liu, Y. (2024). Eco-Friendly Lithium Separators : A Frontier Exploration of Cellulose-Based Materials. *International Journal of Molecular Sciences*, *25*(13), 13. Advance online publication. DOI: 10.3390/ijms25136822

Zheng, L.-H., Chen, M.-H., Liang, S.-X., & Lü, Q.-F. (2021). Oxygen-rich hierarchical porous carbon derived from biomass waste-kapok flower for supercapacitor electrode. *Diamond and Related Materials*, *113*, 108267. DOI: 10.1016/j.diamond.2021.108267

Zhou, L., Li, C., Liu, X., Zhu, Y., Wu, Y., & van Ree, T. (2018). 7—Metal oxides in supercapacitors. In Wu, Y. (Ed.), *Metal Oxides in Energy Technologies* (pp. 169–203). Elsevier., DOI: 10.1016/B978-0-12-811167-3.00007-9

Zhu, H., Luo, W., Ciesielski, P. N., Fang, Z., Zhu, J. Y., Henriksson, G., Himmel, M. E., & Hu, L. (2016). Wood-Derived Materials for Green Electronics, Biological Devices, and Energy Applications. *Chemical Reviews*, *116*(16), 9305–9374. DOI: 10.1021/acs.chemrev.6b00225

Zhu, X., Sun, M., Zhu, X., Guo, W., Luo, Z., Cai, W., & Zhu, X. (2023). Molten salt shielded pyrolysis of biomass waste : Development of hierarchical biochar, salt recovery, CO2 adsorption. *Fuel*, *334*, 126565. DOI: 10.1016/j.fuel.2022.126565

Zhu, Z., Men, Y., Zhang, W., Yang, W., Wang, F., Zhang, Y., Zhang, Y., Zeng, X., Xiao, J., Tang, C., Li, X., & Zhang, Y. (2024). Versatile carbon-based materials from biomass for advanced electrochemical energy storage systems. *eScience, 4*(5), 100249. DOI: 10.1016/j.esci.2024.100249

Zou, R., Quan, H., Wang, W., Gao, W., Dong, Y., & Chen, D. (2018). Porous carbon with interpenetrating framework from *Osmanthus* flower as electrode materials for high-performance supercapacitor. *Journal of Environmental Chemical Engineering*, *6*(1), 258–265. DOI: 10.1016/j.jece.2017.11.080

Zuo, Y., Wang, K., Wei, M., Zhao, S., Zhang, P., & Pei, P. (2022). Starch gel for flexible rechargeable zinc-air batteries. *Cell Reports. Physical Science*, *3*(1), 100744. Advance online publication. DOI: 10.1016/j.xcrp.2022.100744

Chapter 8
Exploring the Electrochemical Potential of Calcium Phosphates:
From Corrosion Inhibition to Energy Storage

Nouhaila Ferraa
https://orcid.org/0000-0003-1817-7798
Ibn Tofail University, Morocco

Sara Rached
https://orcid.org/0009-0009-0163-1553
Ibn Tofail University, Morocco

Zakia Aribou
https://orcid.org/0000-0002-5422-7062
Ibn Tofail University, Morocco

Moussa Ouakki
https://orcid.org/0000-0002-6265-4734
Ibn Tofail University, Morocco

Mohammed Cherkaoui
Ibn Tofail University, Morocco

Mounia Bennani Ziatni
Ibn Tofail University, Morocco

ABSTRACT

Calcium phosphates are inorganic materials. They have good electrochemical properties due to their stability, low cost, and compatibility with biological systems. These features make them suitable for many applications. This chapter looks at their role in corrosion protection and explores opportunities for their use in energy storage. it includes an experimental study that compares the anticorrosive effectiveness of different calcium phosphates on carbon steel in harsh neutral and acidic environments. The results show significant inhibition rates, ranging from 87% to 92.5% in a 3% NaCl solution and exceeding 95% in acidic conditions. SEM/EDX analyses confirmed the formation of protective layers on the carbon steel surface, support-

DOI: 10.4018/979-8-3373-3156-0.ch008

ing the effectiveness of these inhibitors. Lastly, it addresses some key challenges of calcium phosphates as electrode materials for energy storage and looks at future opportunities to improve performance.

1. INTRODUCTION

Calcium phosphates are one of inorganic minerals family that has been deeply studied, due to their exceptional physicochemical features. They naturally occur in biological tissues, such as teeth and bones, and for their outstanding biocompatibility, bioactivity, and ability to promote bone regeneration, they are particularly appreciated (Abouricha et al., 2024; Crespilho et al., 2025; Sane, 2017). For these features, calcium phosphates are considered reference biomaterials in the medical field and find application in controlled-release drug delivery systems, prosthesis coatings, bone substitutes, and orthopedic implants(Ballouze et al., 2021; Habraken et al., 2016; Huang et al., 2024; Hwang et al., 2012; Shadanbaz & Dias, 2012). Besides the biological applications, calcium phosphates present several interesting physico-chemical advantages, such as a modular crystal structure, good chemical stability, and possibility of ion doping(El Hallaoui et al., 2024). These features are opening the way to exciting electrochemical applications and, in particular, to chosen materials for energy storage and corrosion inhibitors(Abouricha et al., 2024; Ferraa et al., 2024; Ferraa, et al., 2025) (Figure 1). Among the innovators in the space, calcium phosphates such as hydroxyapatite ($Ca_{10}(PO_4)_6(OH)_2$), tricalcium phosphate ($Ca_9(HPO_4)(PO_4)_5(OH)$, and octacalcium phosphate $Ca_8(PO_4)_{3.5}(HPO_4)_{2.5}(OH)_{0.5}$. These compounds have the major advantage of being non-toxicity, besides the additional characteristic of forming dense adherent barriers on metallic surfaces which enable them to withstand better ionic exchange, and guard against corrosive damage in saline and acidic environments (Ferraa et al., 2024; Ouakki et al., 2018).

Figure 1. Electrochemical Application of calcium phosphates

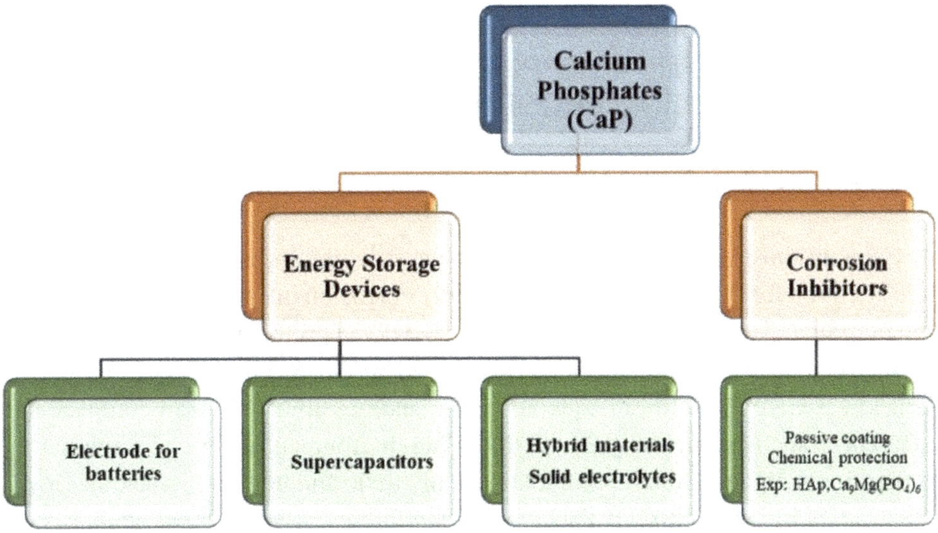

Calcium phosphate compounds also appear promising for applications in sustainable and environmental chemistry (Crespilho et al., 2025; Yan et al., 2023). Moreover, functional cations such as Mg^{2+}, Zn^{2+}, or Co^{2+}, can be doped that seem to greatly improve these materials regarding not only their efficiency(Li et al., 2017, 2018), but ionic conductivity as well(Suzuki, 2010).Meanwhile, there's great potential regarding the inclusion of these materials into the electrodes and supercapacitors of sodium-ion batteries (El Aggadi et al., 2023). Their apatite structure enables reversible ion intercalation/deintercalation, which is important for charging and discharging a battery(Iqbal et al., 2025; Jain & Payal, 2020). Hybrid composites with conductive materials, such as carbon, graphene, and metal oxides, can also significantly enhance their electrochemical performance(Islam et al., 2023; Rabecca et al., 2025). This chapter documents the role of calcium phosphates in corrosion inhibition as the first application and later in energy storage systems as a more recent application, followed by other electrochemical uses, in order to provide a systematic account of the corpus of literature on the subject. Then, we provide a case study on three calcium phosphate compounds to test how efficient they are as corrosion inhibitors in two aggressive media. Lastly, we address the obvious shortcomings of CaP as a complex system with integrated functions: low intrinsic conductivity, long-term stability, and compatibility with other technologies. These drawbacks, alongside the most advanced energy applications, present new perspectives to foster the evolution of hybrid calcium phosphates.

2. ELECTROCHEMICAL APPLICATION REVIEW OF CALCIUM PHOSPHATES

2.1. Function as Corrosion Inhibitor

Calcium phosphates and other phosphate-based inhibitors have received considerable attention for their promise as eco-friendly corrosion inhibitors on a wide range of materials and in various environments, especially in protecting steel in aggressive media, such as in acidic or chloride-rich media(Šimková et al., 2019). Because of their low toxicity, stability in aqueous media and possibility to form compact and adherent protective films, they are considered an environmentally friendly alternative to traditional organic inhibitors that are usually toxic and nonbiodegradable(Goyal et al., 2018; Lai et al., 2021). This is in line with the growing global emphasis on green inhibitors in corrosion studies a rapidly developing field of research dealing with environmentally responsible solutions without efficiency compromise(Aribou et al., 2025; Ferraa et al., 2025; Mzioud et al., 2025). Green inhibitors, whether inorganic or naturally based, aim at replacing harmful chemicals with ones considered biodegradable, nontoxic, and with fewer polluting properties(Ferraa et al., 2025).

In this regard arises because they are chemically similar to the mineral phase of bone and can provide passive layers on metal surfaces, acting as a barrier to corrosive agents(Ferraa et al., 2024; Rhilassi et al., 2024). Th electrochemical methods such as electrochemical impedance spectroscopy (EIS) and potentiodynamic polarization (PDP) indicating that an inhibition efficiencies greater than 90% have been shown for calcium phosphates, dependent among others on crystal morphology, Ca/P ratio and dopant incorporation(El Bakri et al., 2024; Matić et al., 2022). The presence of cations such as magnesium or cobalt in the inhibitor material improves the compactness, stability and adhesion of the protective layer (Ferraa et al., 2025; Mansoori et al., 2018). Hence, calcium phosphates are representative of a step toward green, more efficient and ecologically safer methods of corrosion protection concerned with both industrial and ecological interests. A summary of performance and inhibition efficiencies of several phosphate-based inhibitors on different metallic substrates is reported in the literature, as represented in Table 1.

Table 1. Corrosion inhibition performance of various calcium phosphate compounds in different corrosive media.

Phosphate Inhibitors	Conc.	Medium/ Metal	EI %	References
$CaZr(PO_4)_2$,$4H_2O$	-	NaCl 0.1 M/ Zinc	25% and 77%	(Bouali et al., 2017)
$MCuP_2O_7$ With M = Ba, Ca and Zn	10^{-3} M	1.0 M HCl/ Mild steel	89%	(Harmouchi et al., 2023)
Octacalcium phosphate (OCP_a)	50 ppm	3% NaCl/ Carbon steel	92%	(Ferraa et al., 2023)
Hydroxyapatite (HAp) $Ca_{10}(PO_4)_6(OH)_2$	10^{-3} M	HCl 1 M/ Mild steel	90%	(Chafki et al., 2015)
	100 ppm	3% NaCl/ Carbon steel	91%	(Ferraa et al., 2024)
Mg/Na-doped Apatite	100 ppm	H_3PO_4 2.0 M/ Mild steel	89%	(Ouakki et al., 2018)
$CaPO_3F.2H_2O$	1.0 mM	3% NaCl/ Nickel	96%	(Laamari et al., 2004)
Apatitic Tricalcium (PTCa) $Ca_9(PO_4)_5(HPO_4)(OH)_2$	200 ppm	H_2SO_4 0.5M/ Carbon steel	95%	(Ferraa et al., 2022)
	200 ppm	3% NaCl/ Carbon steel	87%	(Ferraa et al., 2022)
	50 ppm	HCl 1 M/ Carbon steel	93%	(Ferraa et al., 2022)
$Ca(H_2PO_4)_2$	0.6 g/L	3.5 NaCl/ AZ91D magnesium	89.29% after 48 h immersion	(Xu & Huang, 2023)
Anhydrous Tricalcium Phosphate	10^{-4} M	1.0 M HCl/ Mild steel	93.79%	(Chafki et al., 2015)
Amorphous tricalcium phosphate (TCPam)	5.10^{-6} M	1 M HCl/ Mild steel	87%	(Elaidi et al., 2014)
Mg^{2+}@PTC_a Mg^{2+}@HAp	100 ppm 50 ppm	3% NaCl/ Carbon steel	91.2% 92.9%	(Ferraa et al., 2025)

Calcium phosphates inhibitors are widely used in different corrosive environments for the protection of metals against corrosion. Bouali et al. used both polarization curve analysis with charge transfer resistance measurements to assess the anticorrosion performance of $CaZr(PO_4)_2$,$4H_2O$ for zinc protection in a 0.1 M NaCl solution. Compared to the unprotected sample, the study revealed a significant increase in R_{ct} for the Ca^{2+}-ZrP-treated sample, reaching a value of 4621 $\Omega.cm^2$, which clearly indicated a marked enhancement in the electrochemical barrier against corrosion. The inhibitory efficiency was improved by the insertion of Ca with ZrP (Ca^{2+}-ZrP) with a percentage of 77% compared to ZrP alone, which has an efficacy reaching just 25% after 33 hours of immersion, underlining the reinforcing and sustaining action of calcium on the protective layer. These results showed that the proposed strategy works effectively in such an aggressive environment, but longer exposure time tests are needed to establish completely the long-term resistance of the suggested system(Bouali et al., 2017).

The work by El Harmouchi et al. aimed to investigate the anticorrosive performances of copper-based pyrophosphates, $MCuP_2O_7$ (M = Ba, Ca, Zn), in protecting mild steel in 1.0 M HCl solution. These materials showed an inhibition efficiency up to 89% at a concentration of 10^{-3} M due to their ability to form a compact and adherent protective film demonstrated on SEM-EDX analyses. In addition, this work showed that the nature of the divalent cation is of critical importance on the overall protective behavior of the material: in fact, highest efficiency has been shown by the phases containing the Ba cation(Harmouchi et al., 2023). Ferraa et al. carried out a series of works evaluating in great detail the efficiency of calcium phosphate-based compounds as corrosion inhibitors of carbon steel in aggressive media environments.They tested apatitic tricalcium phosphate (PTC_a) showing an inhibition efficiency of 95% in 0.5 M H_2SO_4, 93% in 1.0 M HCl, and 87% in 3% NaCl(Ferraa et al., 2022; Ferraa et al., 2022). Such work was followed by a promising evaluation of the OCP, which obtained an efficiency of 92% in 3% NaCl, underlining the versatility of the calcium phosphate phases in protecting against the saline corrosion phenomena(Ferraa et al., 2023). Their work about hydroxyapatite, HAp, further confirmed these findings: in saline solutions, a corrosion inhibition efficiency of 91% was recorded(Ferraa et al., 2024). Such conclusive findings with all these tested materials unveil their prospects as ecofriendly corrosion inhibitors in different corrosive media.

To enhance the efficiency of the calcium phosphate, Ouakki et al. studied Mg^{2+} and Na^+ doped apatite for the inhibition of mild steel corrosion in a 2.0 M H_3PO_4 solution. An inhibition efficiency of 89% was recorded at a concentration of 100 ppm (Ouakki et al., 2018). The doping enhances the adhesion and compactness of the protective film. Chafki et al. have shown that anhydrous tricalcium phosphate at 10^{-4} M exerts an inhibition efficiency of 93.79% for the corrosion of mild steel in 1.0 M HCl. This result confirms the potentiality of apatite compounds to retard corrosion kinetics(Chafki et al., 2015). Elaidi et al. studied an amorphous phase of tricalcium phosphate (TCPam) at a very low concentration, 5×10^{-6} M, which achieved an inhibition efficiency of 87% for mild steel in 1 M HCl. This result emphasizes the role played by the amorphous structure in promoting a homogenous coverage of the metal surface(Elaidi et al., 2014). Calcium phosphates have also been studied for different metals such as nickel and magnesium, which gives good inhibitive action. Laamari et al. have reported that calcium fluorophosphate, $CaPO_3F.2H_2O$, at 1.0 mM, furnishes an excellent corrosion inhibition efficiency of 96% for nickel in a 3% NaCl solution. This result underlines the interest in incorporating fluorine anions in the molecule to strengthen the protective barrier(Laamari et al., 2004). Xu & Huang showed that $Ca(H_2PO_4)_2$ at 0.6 g/L reaches an inhibition efficiency of 89.29% for AZ91D magnesium alloy in a 3.5% NaCl solution. This work opens promising avenues for lightweight alloys, though further investigation of the kinetics of phosphate film formation would be necessary to confirm these results in the long term(Xu & Huang, 2023). The performed modifications bring important compactness to the protective film and improve the chemical stability of the deposit. In fact, several studies as we have already seen, manage to reach high inhibition efficiencies even at low concentrations, 50-200 ppm; this will make them advantageous for industrial applications by reducing both economic and environmental impacts. Additionally, the nature of the inhibitors whether amorphous or crystalline affects their performance: amorphous TCP offers rapid and effective inhibition at low concentrations, while more ordered structures like hydroxyapatite and tricalcium phosphate form more durable protective films over time.

Besides their efficiency as inhibitors in solution, phosphate-based materials in general have shown great potential as protective coatings, forming stable and durable barriers that shield metal surfaces from corrosive environments. A summary of previous investigations focusing on phosphate-based coatings and their inhibition mechanism in preventing corrosion are presented in Table 2. For instance, calcium aluminum polyphosphate silicate hydrate showed that this compound provided effective cathodic inhibition on hot-dip galvanized steel in a 0.05 M NaCl medium. The protective performance of such a compound improved remarkably from 33.8% after 1.5 hours to 74.7% after 24 hours, mainly due to a significant increase in the polarization resistance of the metal surface(Meeusen et al., 2019).

A distinct synergistic effect has been observed in the case of mixing various corrosion-inhibiting pigments. The combined use of a modified zinc phosphate pigment with a calcium-based ion-exchange pigment showed improved protective action for galvanized steel exposed to simulated acid rain. The synergy enhanced the compactness and adhesion of the film formed, consisting of mixed calcium and zinc phosphates, thus offering mixed anodic–cathodic inhibition. In such a system, the charge-transfer resistance of the metal increased 10–20 times compared to single-pigment formulations and reached a level observed for chromate-based inhibitors(Pokhmurs'kyi et al., 2003).

Regarding reinforcing steel in concrete, phosphate ions are known to be very efficient inhibitors in chloride-aggressive conditions. The addition of sodium phosphate significantly increased the critical chloride-to-hydroxide ratio required to initiate pitting corrosion-from 0.6 to about 15-through the promotion of a stable protective layer of iron phosphate ($Fe_3(PO_4)_2.8H_2O$) on the steel surface(Nahali et al., 2015). Further studies indeed confirmed that phosphates are mixed-type inhibitors and yield a two-layered protective film that maintains its efficiency against pitting even after 90 days of exposure to chloride-contaminated environments (Yohai et al., 2013).

Table 2. Corrosion inhibition mechanisms of various phosphate-based coatings

Inhibitors/ Coating	Medium/Metal	Inhibition Mechanism	Ref
Calcium aluminum polyphosphate silicate hydrate	0.05 M NaCl /Galvanized steel	Cathodic inhibition	(Meeusen et al., 2019)
Modified Zinc Phosphate + Ca-containing pigment	Artificial acid rain / Galvanized steel	Mixed cathodic-anodic control	(Pokhmurs'kyi et al., 2003)
Mixture of $CaHPO_4$ and $CaMoO_4$	NaCl solution / Low-carbon steel	Protective layer formation	(Przywecka et al., 2019)
Zn-modified CaP coating ($CaZn_2(PO_4)_2.2H_2O$)	Simulated Body Fluid (SBF) /Mg-Zn-Ca alloy	Barrier protection, compact structure	(J. Wang et al., 2021)
Na_3PO_4 (Sodium Phosphate)	Saturated $Ca(OH)_2$ with NaCl / Steel reinforcement	Anodic inhibition, passivation	(Nahali et al., 2015)
Phosphate ions (Na_3PO_4)	Pore Simulating Solution (PSS)+ 0.3 M Cl^- / Steel reinforcement	Mixed-type inhibition, duplex film	(Yohai et al., 2013)
Electrodeposited CaP coatings (HAP and Ca-def HAP)	Physiological solutions (SBF, DMEM, Ringer's) / Ti_6Al_4V alloy	Protective layer, pore filling	(Drevet et al., 2012)

Overall, the body of literature shows that calcium phosphate compounds, whether apatitic, amorphous or doped, exhibit excellent efficiency as corrosion inhibitors in a wide range of environments, from acidic to neutral and saline media. However, beyond the impressively high numerical efficiencies reported often over 90% a further mechanistic understanding has to be developed in order to explain in detail the basic processes ruling their protective action.

Based on comparative electrochemical results, adsorption parameters and surface analyses, it can be proposed that calcium phosphates act by a dual inhibition mechanism: (i) chemical adsorption of phosphate anions on anodic sites, forming insoluble iron or metal phosphates, and (ii) physical barrier formation where the calcium rich deposits provide a compact and adherent film that blocks the transport of corrosive species (Cl^-, H^+, SO_4^{2-}). The reinforcement effect of Ca^{2+} observed by Bouali et al. and other authors is very likely due to the ionic cross linking of phosphate networks, enhancing film cohesion and its resistance to dissolution in aggressive media. However, the long-term performance of these inhibitors and their interfacial mechanisms have not been sufficiently studied, since most works rely on short term tests. Advanced in situ techniques are needed for a better understanding of the structural evolution of the protective layers. The incorporation of calcium phosphates in hybrid systems coatings or mixed inhibitors is the most promising approach for achieving synergistic anodic and cathodic protection, although much more optimization is required for homogeneous and stable film formation.

From both environmental and industrial perspectives, low effective concentrations, typically below 200 ppm and nontoxicity are advantages of calcium phosphate inhibitors over traditional chromate-based systems. However, future research should be oriented toward the association of their corrosion inhibition efficiency with other emerging functionalities, such as electrochemical energy storage or photoelectrocatalytic activity, in order to obtain really dual function materials that bridge corrosion protection and energy applications. In the following, the focus will be put on some recent works in which calcium phosphate-based compounds have been studied not only as passive corrosion inhibitors but also as active electrochemical materials with an inherent capability of storing charge.

2.2. FUNCTION IN ENERGY STORAGE SYSTEMS

Calcium phosphates have gained huge interest in the energy storage as well(Arul et al., 2018; Asghar et al., 2023; Panneerselvam et al., 2024; Tuncer et al., 2019). Their high chemical stability, along with reversible ion exchange, positions these materials as excellent electrode candidates for lithium and sodium-ion batteries (Zhang et al., 2025) and supercapacitors (Eagambaram et al., 2025). Hence, CaPs represent an eco-friendly and efficient approach for responding to the energy demands of the future. Their intrinsic thermal and chemical robustness, combined with proven biocompatibility and a porous crystalline framework, enables good hosting of metal ions, which promotes effective ionic conductivity. Such properties render CaPs perfect for the design of durable, high-performance and eco-friendly energy storage systems (Sane, 2017; Shrivastava et al., 2025). Among them, magnesium phosphate has been particularly explored as a supercapacitor, either in its pure state or incorporated into composites with conductive materials like carbon or graphene, which help to overcome the limited electronic conductivity of the material(Ma & Zhang, 2019). In these composites, charge transfer processes were considerably enhanced, while the overall electrochemical efficiency of the system was improved. Figure 2 provide a summary of major findings from existing studies that highlight the potential of calcium phosphate-based compounds as active materials in energy storage devices.

Figure 2. Summary of major findings demonstrating the potential of calcium phosphate-based compounds as active materials for energy storage applications. (A) Galvanostatic charge–discharge (GCD) curves of CdS-CaP nanocomposites with ratios (a) 1:1, (b) 1:2, and (c) 2:1; (d) variation of specific capacitance with current density; (e) dependence of specific capacitance on scan rate; and (f) Nyquist plots illustrating charge transfer resistance and ion diffusion behavior(Saleem et al., 2025). (B) GCD curves of (a) Co^{2+}-doped HAp and (b) pure HAp nanoparticles recorded at various current densities; (c) variation of specific capacitance (Csp) and (d) energy density as a function of current density(Safari-Gezaz et al., 2025). (C) Charge–discharge profiles of the $Na_4CrFe(PO_4)_3$//PEO-LiTFSI-15%HAp//Na cell (Abouricha et al., 2024). (D) Galvanostatic charge-discharge curves of (a) HAp, (b) 0.01Mg-HAp, and (c) 0.1Mg-HAp obtained at different applied current densities (Arul et al., 2018)

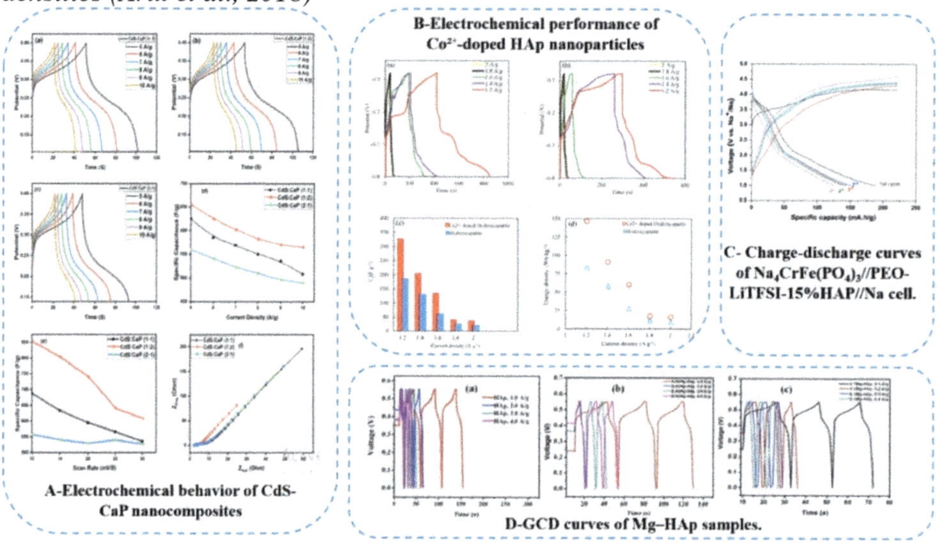

Abouricha et al. prepared hydroxyapatite (HAp) by using co-precipitation technique and added it as an inorganic filler to a solid polymer electrolyte of sodium salt (NaTFSI) and polyethylene oxide (PEO). The mechanical properties of the polymer were significantly enhanced by the addition of HAp without sacrificing its ionic conductivity. The resultant composite electrolyte was tested in sodium-ion battery using a sodium metal anode and a newly developed $Na_4CrFe(PO_4)_3$-based cathode. The electrochemical test showed that as early as the fifth cycle, a specific discharge capacity of 124 mAh/g could be achieved at a current of C/20. The obtained results point to the feasibility of calcium phosphates for the development of new composite electrolytes in the frame of sodium-based energy storage and show interest in HAp as a functional filler in solid polymer electrolytes(Abouricha et al., 2024).

Doped calcium phosphates, in which the calcium ion is partially replaced by magnesium, zinc or other transition metals, present better electrochemical performances, characterized by higher specific capacities with good cycle stability and satisfactory performances at high current rates(Masese & Kanyolo, 2024). Their structural flexibility and the possibility of reversible redox during the insertion/removal of ions allow them to be particularly appealing for advanced applications. In this context, Arul et al. prepared magnesium-doped calcium phosphate (Mg-HAp) by a wet-chemical route followed by microwave-assisted processing. With a constant capacitive value of about 0.55V and a longer discharge time, the sample with the highest magnesium concentration (0.1 Mg-HAp) exhibited the best behavior as an electrode material for supercapacitors (Arul et al., 2018). These materials also exhibited a prolonged release of the active principle. Therefore, suggesting another possible application of these materials as controlled drug delivery systems.

Beyond modifying the specific capacity, the addition of Mg^{2+} ions to the apatite structure enhances the multiple functionalities of the material. Magnesium-doped calcium phosphates thus appear as interesting multi-purpose materials for energy storage and biomedical uses(Habraken et al., 2016; Huang et al., 2024).On the same trend, Safari-Gezaz and coworkers have explored pure hydroxyapatite HAp -doped with cobalt ions Co^{2+}, as a suitable electrode material for supercapacitors. The nanoparticles were elaborated using the sol-gel method. Cyclic voltammetry (CV) and galvanostatic charge-discharge (GCD) results from electrochemical methods demonstrated that the electrode with Co^{2+}-doped HAp nanoparticles provided a high specific capacitance of 324 F/g and an energy density of 145.8 Wh/kg at a current density of 1.2 A/g. The effect of cobalt doping is responsible for this notable improvement, indicating that Co^{2+}-doped HAp is a viable substitute for the creation of high-performance electrodes for upcoming energy storage technologies(Safari-Gezaz et al., 2025). Moreover, their low toxicity, natural abundance and minimal environmental impact position calcium phosphates as key materials for safer and greener energy storage systems. As research continues to advance, calcium phosphate-based materials are expected to play an increasingly significant role in the design of next-generation electrochemical storage devices(Attia et al., 2021; El Aggadi et al., 2023; Shrivastava et al., 2025). More recently, research on the use of calcium phosphate-based composites for supercapacitor applications has also been presented. For example, CdS-CaP composites synthesized through the hydrothermal route have been investigated as hybrid dielectric-electrochemical materials. This composite exhibited pseudocapacitive charge storage with a maximum specific capacitance of 851.87 F. g^{-1} and a specific energy of 14.5 Wh.kg^{-1}, which was ascribed to its porous and rough surface morphology that enhances ion transport, while further assisted by its high polarizability(Saleem et al., 2025). Despite extensive investigations of calcium phosphates, their practical use as active energy storage materials is largely unexplored.

These compounds have huge potential for applications in electrochemistry, especially supercapacitors or hybrid energy storage devices, due to their intrinsic structural stability, ionic conductivity and surface reactivity. Mechanistically, the presence of Ca^{2+} and PO_4^{3-} ions should enable charge transfer and reversible ion intercalation, and doping or chemical modification may further enhance electronic conductivity and surface activity. However, in reality, very few thorough investigations into their electrochemical performance have been conducted. The effects of crystallinity versus amorphous phases, pathways for efficient ion diffusion, and long-term cycling stability are poorly understood. Future research efforts should aim at more in-depth in situ electrochemical characterizations and the engineering of hybrid or modified calcium phosphate systems to optimize charge storage capacity and cycling durability. Such work may establish calcium phosphate-based materials as green, multi-functional options for advanced energy storage technologies.

3. ELECTROCHEMICAL STUDY OF CALCIUM PHOSPHATES AS CORROSION INHIBITORS

Based on the comprehensive review of calcium phosphates and their broad scope of applications, in particular, electrochemical sectors, the present part focused more on their application as corrosion inhibitors. In this context, the section here reports a comparative study of calcium phosphate compounds: Apatite tricalcium phosphate (PTC_a), hydroxyapatite (HAp), and Apatitic Octacalcium Phosphate (OCP_a), with regard to their performance in protecting Carbon steel from corrosion by two complementary electrochemical techniques: potentiodynamic polarization (PDP) and electrochemical impedance spectroscopy (EIS). The experiments were performed in two important corrosive media: 3% NaCl solution, simulating a marine environment and 1 M HCl solution representing a very acidic industrial medium. Obtained results enable the comparison in direct terms of the performance of phosphate compounds against corrosion and enlighten their mechanism. Besides, based on electrochemical data, a thermodynamic analysis has been conducted to gain deeper insight into the process of inhibitor adsorption onto the surface of the carbon steel.

3.1. Operational Conditions

3.1.1. Materials and Sample Preparation

Two aggressive electrolytic solutions were used to evaluate the anticorrosive efficiency of various phosphate-based supports on carbon steel: hydrochloric acid (HCl 1.0 M), prepared from concentrated HCl (37%), and sodium chloride (NaCl 3%) which was prepared by dissolving 30 g of NaCl in one liter of distilled water. The apatite calcium phosphates used as inhibitors were prepared by the double decomposition method(Bakri et al., 2025; Heughebaert & Montel, 1970; Rhilassi et al., 2024). These materials were subsequently added to the electrolytic media at concentrations between 25 and 400 ppm. The substrates used for the electrochemical tests were carbon steel plates, each of area 1 cm^2, and of the following chemical composition: Mn: 0.47%, Si: 0.24%, C: 0.11%, Cu: 0.14%, Al: 0.03%, Mo: 0.02%, Cr: 0.01%, Ni: 0.1%, W: 0.06%, V: <0.003%, Co: <0.0012%, the rest being iron. The samples were pre-treated prior to the experiments with due care in order to reproduce the results accurately. The carbon steel surfaces were abraded using silicon carbide sandpaper of progressively finer grit sizes (from 50 to 2000), followed by polishing with papers of grain size smaller than 1 μm, yielding a smooth homogenous, mirror-like surface with no scratches. After this, the samples were rinsed thoroughly with distilled water, degreased with acetone, and air-dried before immersion in the electrolytic solutions.

3.1.2. Experimental Electrochemical Conditions

The electrochemical behavior of the phosphate supports was studied using a steel electrode. Two electrochemical techniques were used: potentiodynamic polarization (PDP) and electrochemical impedance spectroscopy (EIS). Potentiodynamic polarization curves for all the inhibitors studied were recorded at a scan rate of 1 mV/s, with the potential ranging from -1200 mV to 300 mV for the NaCl 3% medium, and from -900 mV to -100 mV for the acidic media. The Inhibitory efficacy ($\eta_{pp}\%$) was calculated using the following relationship:

$$_{pp}(\%) = \left(\frac{I_{corr,0} - I_{corr}}{I_{corr.0}}\right) \times 100 \text{ Eq.1}$$

Where $I_{corr,0}$ and I_{corr} denote corrosion current densities in the absence and presence of phosphate inhibitors respectively for each medium studied. Electrochemical impedance measurements (EIS) were conducted over a frequency range [100 kHz-100 mHz] for the acidic media and from [100 kHz-10 mHz] for NaCl the 3%

solution. These measurements were performed at open circuit potential (OCP), with a 10mV amplitude sinusoidal potential perturbation. The inhibition efficiency ($\eta_{EIS}\%$) calculated, following relationship (Eq.2).

$$\eta_{EIS\,\%} = \frac{R_p^{inh} - R_p^0}{R_p^{inh}} \times 100 \text{ Eq.2}$$

With R_p^{inh} and R_p^0 are respectively the polarization resistance in the presence and the absence of the inhibitor. Prior to each electrochemical experiment, the working electrode was immersed in the aggressive solutions for 30 minutes, both in the absence and presence of the inhibitors. Electrochemical measurements were carried out using a potentiostat/galvanostat/PGZ100, controlled by VoltaMaster4 analysis software.

3.2. Electrochemical Investigation

3.2.1. Performance in Neutral NaCl 3% Electrolyte

Marine environments are particularly aggressive, mainly due to the chloride ions, which strongly accelerate the degradation of metallic materials including bronze and steel. Among the research works devoted to the search for protective solutions, the use of corrosion inhibitors of natural origin has been considered for several years, such as plant extracts (Benzidia et al., 2022; Maurya et al., 2021; Zakir Hossain et al., 2020), organic compounds(Dagdag et al., 2015; Velrani et al., 2014), and other substances (Tristijanto et al., 2020). The present section compares the corrosion inhibition potential of PTC_a, HAp and OCP_a materials, based on their electrochemical performance in a neutral environment.

A. Electrochemical Impedance Spectroscopy

EIS is a non-destructive and versatile technique that is now widely used for the investigation of electrochemical properties of materials and evaluations of their corrosion resistance(Ferraa et al., 2025).The technique essentially involves the application of a sinusoidal alternating current perturbation and measurement of the system's response, which would give information on the charge transfer processes across the interface between the electrode and solution. EIS was performed after 30 min of immersion in 3% NaCl solutions without and with phosphate inhibitors. The impedance spectra obtained as NYQUIST plots over a frequency range of 100 kHz to 10 mHz. A sinusoidal perturbation with an amplitude of 10 mV is applied to record the system's response for the evaluation of the efficiency of corrosion inhibition. Figure 3 shows NYQUIST plots recorded for each phosphate at optimum concentrations (50 ppm for OCP_a, 100 ppm for HAp and 200 ppm for PTC_a).

Figure 3. NYQUIST impedance diagrams of carbon steel in 3% NaCl solution without and with the optimum concentration of the three inhibitors (PTC_a, HAp and OCP_a)

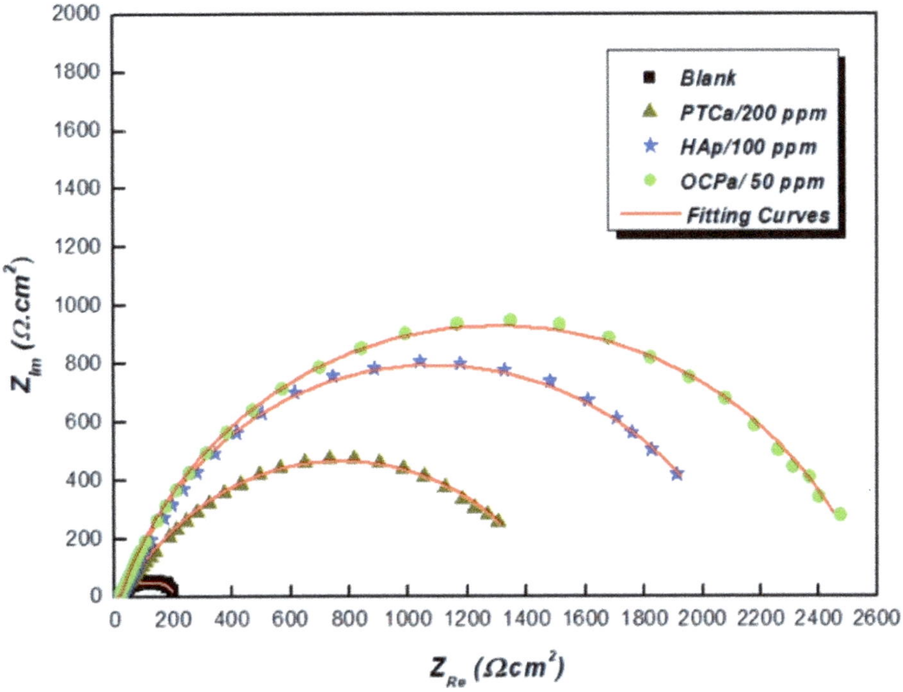

From these NYQUIST plots, it was seen that the curves are made up of two loops, which to the unaided eye are not clearly different: a high-frequency loop, usually associated with the formation on the steel surface of a film by inhibitor molecules, iron oxides, and hydroxides; and a low-frequency loop associated with a phenomena of charge transfer (Dahmani et al., 2017; Ghenimi et al., 2025). In addition, all inhibitors presented approximately the same shape in plots, which means the mechanism process of corrosion has not changed. The diameter of the loop in the presence of inhibitors increases compared to the blank, with the largest diameter recorded at 50 ppm of OCP_a, followed by 100 ppm of HAp and 200 ppm of PTC_a(Ferraa et al., 2023; Ferraa et al., 2022).

Figure 4. BODE diagrams of carbon steel in 3% NaCl with and without PTC$_a$, HAp and OCP$_a$ at optimum concentrations.

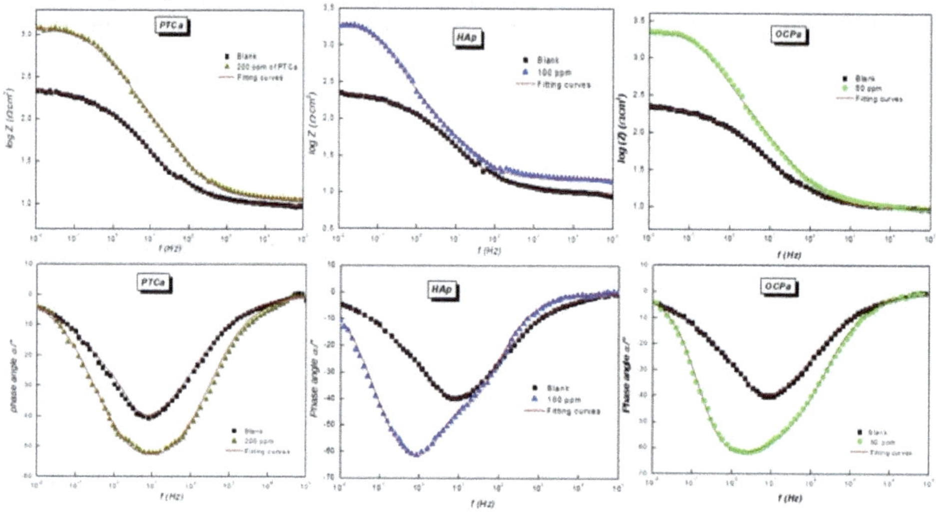

The BODE plots, recorded in the absence and presence of the optimal concentrations of the inhibitors PTC$_a$, HAp, and OCP$_a$, are shown in figure 4. These plots provide valuable insight into the electrochemical behavior of the system and support the interpretation of the NYQUIST diagrams, thus justifying the choice of the equivalent electrical circuit used for modeling. It can be seen from figure 4 that a remarkable phase broadening is observed at the intermediary and low frequencies for all three phosphate-based inhibitors. This behavior suggests the overlapping of two time constants, one of which may be ascribed to the formation of a protective film on the carbon steel surface.

Such an interpretation is further supported by two well-defined relaxation processes associated with the adsorption of PTC$_a$, HAp, and OCP$_a$ on the carbon steel surface as obtained by several works(Tan et al., 2018; Tristijanto et al., 2020).

In order to analyze the results obtained by EIS and achieve more accurate modeling of electrochemical behavior, two different equivalent electrical circuits were considered for the three phosphate-based inhibitors. The simulation results at their optimal concentrations are given in Table 3.

Table 3. Modeling impedance curves with two different electrical circuits.

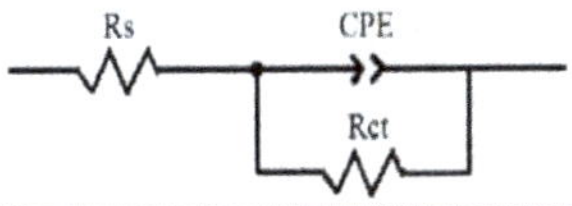

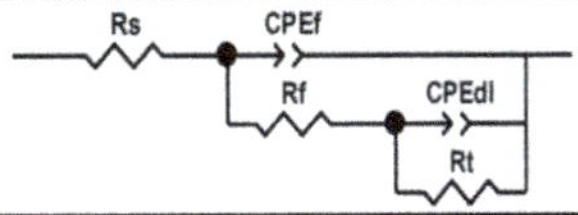

PTCa

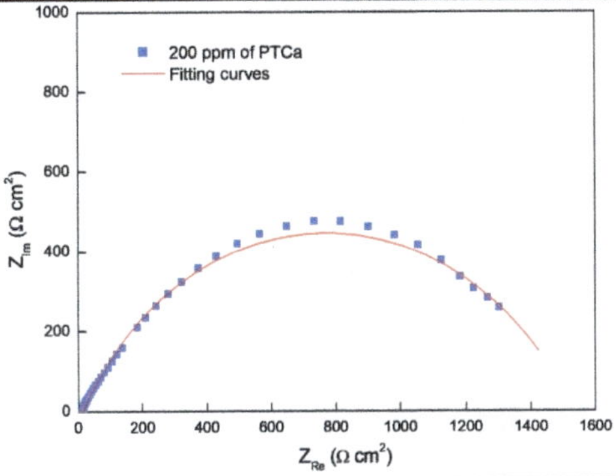

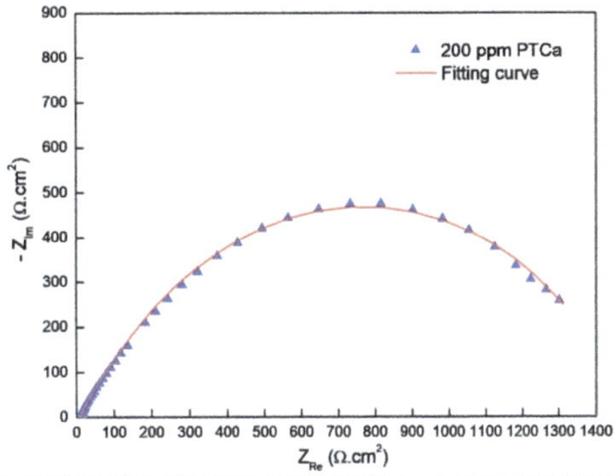

HAp

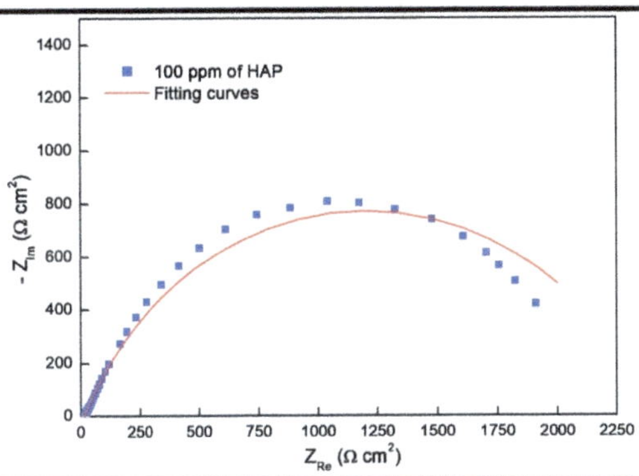

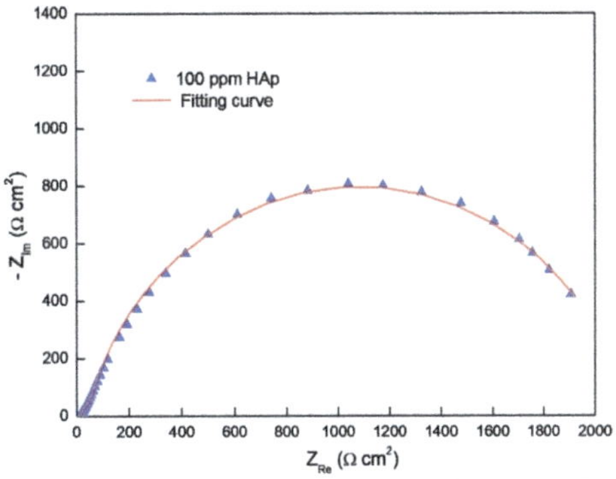

OCPa

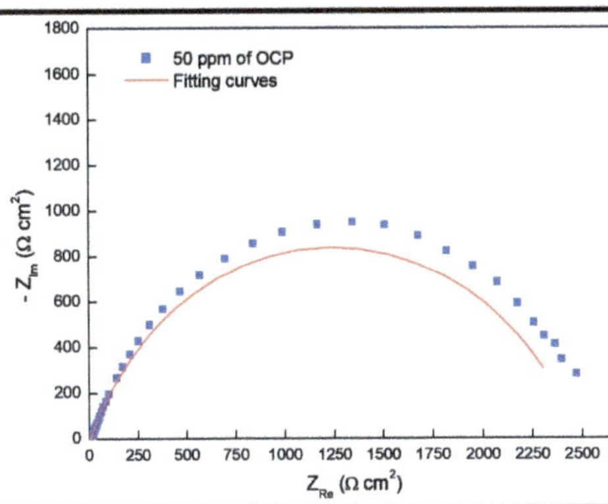

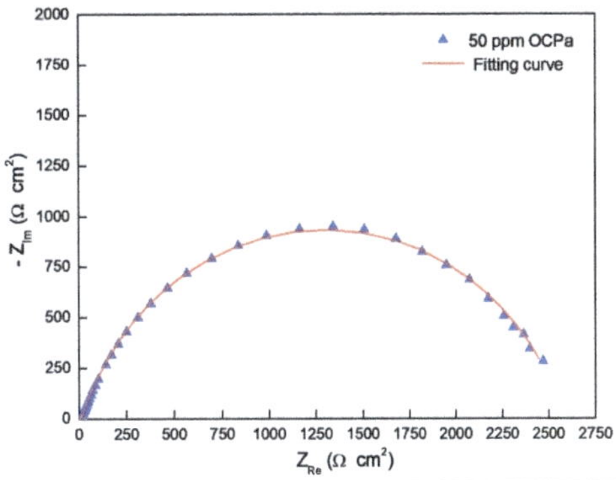

The model in Figure 5, among those proposed, best fitted the experimental impedance spectra; thus, it is the one most suitable for describing the system. The fitted electrochemical parameters include: R_s, the solution resistance; R_{ct}, the charge transfer resistance; R_f, the resistance associated with the protective film; and Q_f and Q_{tc}, constant phase elements (CPE) used in place of the ideal film capacitance (C_f) and double-layer capacitance (C_{dl}), respectively (Zhou et al., 2012).

Figure 5. Equivalent circuit representing carbon steel EIS behavior.

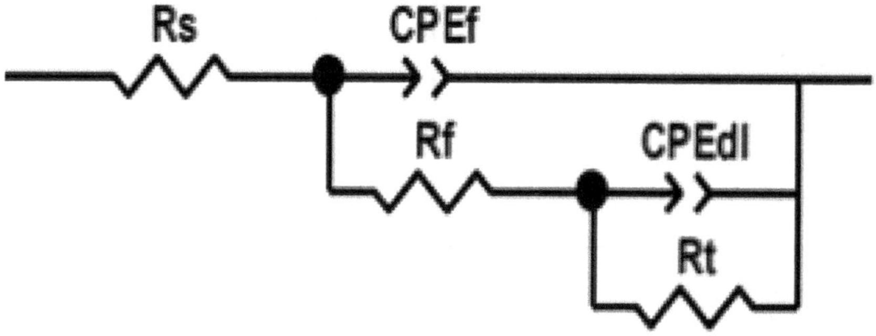

Table 4 presents the electrochemical parameters and the inhibition efficiency (η_{EIS}(%)) of apatite phosphates at optimum concentration.

Table 4. Electrochemical impedance parameters of the carbon steel interface in the absence and presence of inhibitors at 298 K.

	[ ] (ppm)	R_s (Ω.cm²)	R_f (Ω.cm²)	n_f	Q_f (μF.S^{n-1})	R_{ct} (Ω.cm²)	n_{ct}	Q_{ct} (μF.S^{n-1})	R_p (Ω.cm²)	θ	EIS %
NaCl 3%	--	10.2	38.0	0.705	1399	161.0	1	1451	199.0	-	-
PTC$_a$	200	12.0	312.7	0.743	339	1163.0	0.711	270	1475.7	0.865	86.5
HAp	100	15.0	147.5	0.771	651	1993.0	0.952	288	2140.5	0.907	90.7
OCP$_a$	50	9.3	196.5	0.782	324.4	2450	0.813	139.5	2646.5	0.925	92.5

The inhibition efficiency reached 86.5%, 90.5% and 92.5% for PTC_a, HAp and OCP_a, respectively, values that are significantly higher than those reported in previous studies exploring various concentrations of other inhibitors(Umoren et al., 2022). This is evidenced by the value for the charge transfer resistance R_{ct} reaching 2450 Ω.cm² for OCP_a. Also, a remarkably increase in film resistance R_f was observed, combined with a significant decrease in Q_f values (Loto et al., 2016;

Öğretir et al., 1999). The low values of Q_f suggest a dense, stable and protective film on the surface of carbon steel(Ghenimi et al., 2025; Kardas, 2005). This infers a homogeneous structure that may be effectively protective for the underlying steel against the corrosion process(Al-Sabagh et al., 2012; Choe et al., 2015; Rahmouni et al., 2005). Additionally, polarization resistance R_p increased from 199.0 $\Omega.cm^2$ (blank) to 2646.5 $\Omega.cm^2$ by the addition of OCP_a, reflecting the higher inhibition performance of OCP_a phosphate compared with other compounds(Ferraa et al., 2024; Tazi et al., 2016).

B. Potentiodynamic Polarization Measurement

This electrochemical method is based on the plotting of I = f (E) curves in a semi-logarithmic scale at a slow potential sweep rate, giving direct determination to the value of the corrosion current density. Potentiodynamic polarization curves (Figure 6) were recorded after stabilization of the electrode potential for 30 minutes in 3% NaCl solution, both without and with 200 ppm, 100 ppm and 50 ppm of PTCa, HAp, and OCPa, respectively. The scan rate was 1 mV/s, while the potential range was from -1200 mV to +300 mV.

Figure 6. Effect of Optimum concentrations of PTCa, HAp, and OCPa on carbon steel polarization in 3% NaCl at 298 K

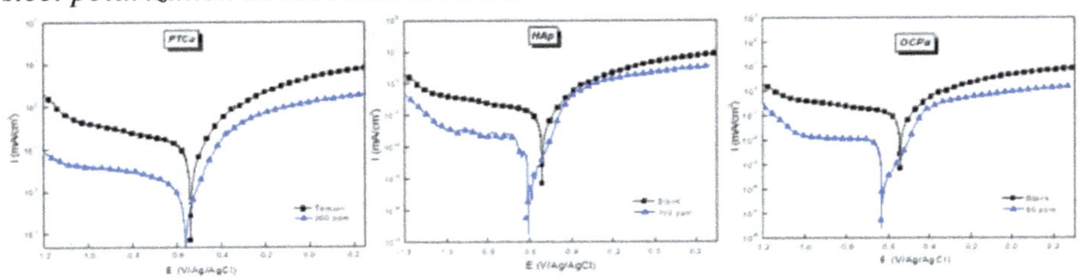

The addition of the inhibitors to the aggressive medium caused a small displacement of the corrosion potential from that of the free sample. Therefore, a significant reduction in cathodic current density was observed, while current density was also substantially lowered in the anodic branch. The addition of PTC_a, HAp and OCP_a inhibitors reduces the corrosion rate of carbon steel, which is evident from the lower current densities. At the same time, the overall shape of both anodic and cathodic polarization curves remains consistent. This shows that the inhibitors mainly work by slowing down the metal dissolution process without changing the basic cathodic or anodic reaction mechanisms. In other words, these phosphate materials offer effective protection without altering the basic corrosion process of the carbon steel(Dagdag et al., 2015).The electrochemical parameters of carbon steel in a 3%

NaCl solution namely the corrosion potential (E_{corr}), anodic (β_a) and cathodic (β_c) Tafel slopes, and the corrosion current density (I_{corr}) were determined using the Tafel extrapolation method (Table 5).

Table 5. PDP electrochemical parameters in the presence and the absence of the optimum concentration of phosphate materials (PTC_a, HAp, and OCP_a).

	Conc. [ ] ppm	- E_{corr} mV/Ag/AgCl	i_{corr} µA cm^{-2}	- β_c mV dec^{-1}	β_a mV dec^{-1}	η_{PP} %
NaCl 3%	--	538	464	141	149	-
PTCa	200	565	61,6	132	125	86.7
HAp	100	609	40,4	139	128	91.3
OCPa	50	632	31,9	125	132	93.1

From table 5, a shift in corrosion potential of less than ±85 mV was observed for the three calcium phosphate inhibitors studied, suggesting a mixed inhibition mechanism with a predominant cathodic influence (Al-Sabagh et al., 2012; Oguzie et al., 2007; Tasić et al., 2018). In addition, a substantial decrease in corrosion current density was recorded compared to the uninhibited system. The values of inhibition efficiencies were found to be 86.7%, 91.3%, and 93.1% for PTCa, HAp, and OCP_a, respectively. Figure 7 presents the variation in inhibition efficiency with the addition of different concentrations of phosphate-based inhibitors. A strong correlation is observed between the results obtained using the two methods, EIS and PDP, for the electrochemical analysis of the three inhibitors. Notably, OCP_a demonstrates the excellent inhibition efficiency. The higher presence of phosphorus atoms in the form of HPO_4^{2-} within the chemical composition likely influences the reactivity and performance of these inhibitors in material protection.

Figure 7. Evolution of inhibitory efficacy as a function of PTC_a, HAp and OCP_a concentrations in 3% NaCl by both EIS and PDP methods.

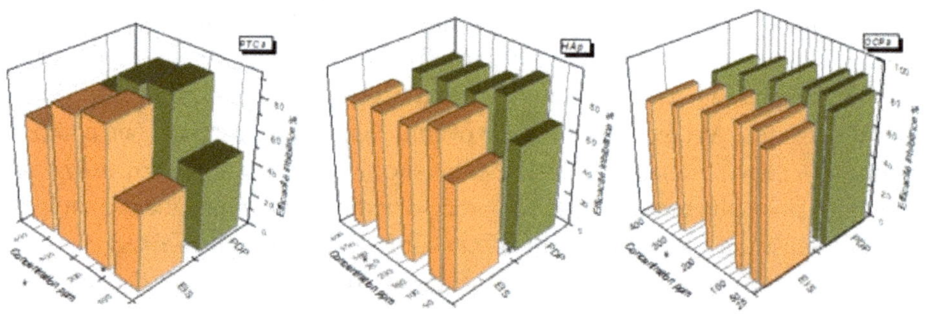

C. Temperatures Effect and Activation Thermodynamic Parameters

The polarization curves of carbon steel were recorded in the absence and presence of the optimal concentrations of phosphates in a 3% NaCl medium, at a temperature varying from 298 K to 328. Figure 8 presents curves derived from the parameters extracted from polarization curves(Ferraa et al., 2024). Accordingly, with the rise in temperature, the corrosion current density increases both in the absence and presence of phosphates, proving a physical adsorption of the three inhibitors on the carbon steel surface(Essaadaoui et al., 2018). This increase can be related to the appearance of desorption or the detachment of the film formed on the carbon steel substrate(Tazi et al., 2016). Besides, the inhibition efficiency in the presence of studied phosphate decreases slightly as the temperature rises(Anupama et al., 2016; Cao et al., 2021).

Figure 8. Effect of temperature on corrosion current density and inhibition efficiency of carbon steel, in 3% NaCl medium, with and without the optimum concentrations of phosphate inhibitors.

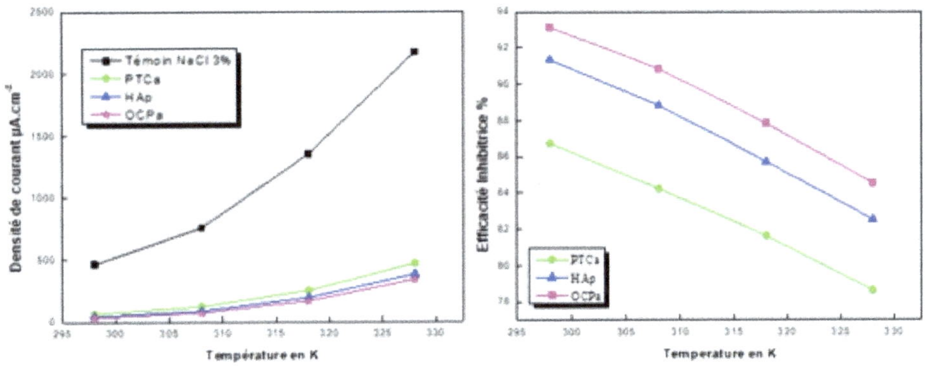

ARRHENIUS plots were developed to obtain a deeper understanding of the adsorption mechanism of the inhibitors. The following procedure was used in determining the activation energy (E_a) for the corrosion process: the linear relationship between ln i_{corr} and 1/T for a 3% NaCl solution was considered both with and without the inhibitors. Enthalpy ΔH_a, and entropy, ΔSa, of adsorption were also calculated from the plot of $\ln(i_{corr}/T)$ against 1/T. The slope of the straight line obtained gave ΔHa/R, and the intercept was [ln(R/Nh) + ΔSa/R], where R is the universal gas constant, N is the Avogadro's number and h is the Planck's constant (Figure 9). The calculated thermodynamic activation parameters are presented in Table 6.

Figure 9. ARRHENIUS lines for carbon steel in 3% NaCl in the absence and presence of PTC_a, HAp and OCP_a.

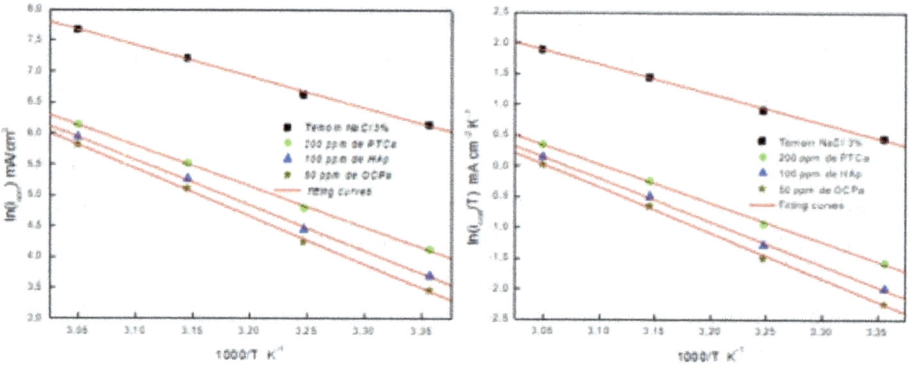

Table 6. Activation parameters for carbon steel in 3% NaCl in the absence and presence of phosphate inhibitors

Compounds		E_a *(KJ/mol)*	ΔH_a *(KJ/mol)*	ΔS_a *(J/mol.K)*
NaCl 3%	**Blank**	42.3	39.7	-60.8
	PTCa	55.1	52.5	-34.6
	HAp	61.2	58.6	-17.6
	OCPa	64.4	61.8	-8.9

The analysis of the data presented in the table 6 indicates an increase in activation energy (E_a) in the presence of the three inhibitors, following the order: $E_a(PTC_a)$ < E_a(HAp) < $E_a(OCP_a)$, compared to the uninhibited medium(Boughoues et al., 2020). This increase in E_a indicates that the main interaction of inhibitors with the metal surface corresponds to physisorption(Mzioud et al., 2024). The positive values of ΔH_a support the endothermic nature of the dissolution process for carbon steel in the presence of phosphate-based materials. Under these conditions, the increase in entropy ΔS_a upon the addition of inhibitors implies a greater disorder, likely due to the formation of an activated complex(Boughoues et al., 2020). Such behavior reflects an associative mechanism in the rate-determining step, it points out that the transition state is more ordered than the initial reactants, in agreement with a complexation rather than dissociation process(Velrani et al., 2014).

3.2.2. Performance in Acidic HCL 1.0 M Electrolyte

Acidic environments are highly corrosive to materials(Oyekunle et al., 2019; Umoren et al., 2022). Carbon steel, a material used in many industries, has been found to show the most significant deterioration under acidic conditions (Oyekunle et al., 2019). It is exposed to an acidic environment during such processes as pickling, industrial descaling, acid cleaning and oil well acidification (Hossain et al., 2020; Saji, 2010) and undergoes rapid deterioration of its properties(Abdallah et al., 2016). In order to avoid this problem, several authors have proposed the use of non-toxic corrosion inhibitors, whose action consists in adsorption on the surface of the material and protecting it from corrosion. In the second part of these section of the chapter, the calcium phosphate compounds PTC_a and OCP_a was tested as corrosion inhibitors for carbon steel in 1.0 M HCl solution. Both stationary and transient electrochemical methods were used to evaluate the carbon steel corrosion behavior.

A. Electrochemical Impedance Spectroscopy Investigation

EIS tests were performed after 30 min of exposure in a 1.0 M HCl solution with and without corrosion inhibitors. Impedance spectra were measured at OCP in a frequency range from 100 kHz to 100 mHz with a sinusoidal AC signal of 10 mV amplitude. Figure 10 depicts NYQUIST plots and corresponding BODE plots for different inhibitor concentrations from 25 to 300 ppm. The presence of one capacitive semicircle from the NYQUIST plots for the uninhibited solution, as well as those containing various concentrations of phosphates, suggests that the corrosion of carbon steel is a pure charge transfer process (El Harmouchi et al., 2024). This indicates that addition of inhibitors does not significantly change the fundamental corrosion process(Corrales-Luna et al., 2019; Daoud et al., 2014). In addition, for phosphate inhibitors, capacitive loop diameters increased and reached a maximum for 50 ppm for both inhibitors. BODE plots reflect one time constant in the presence of inhibitors, which supports the appropriateness of an equivalent electrical circuit model that has been used to fit the impedance data shown in Figure 11. A single phase angle maximum is observed, reaching -78° for PTC_a and -69° for OCP_a, which indicates that the electrochemical behavior is primarily controlled by charge transfer kinetics(Shao et al., 2022).

Figure 10. Electrochemical impedance diagrams (NYQUIST and BODE) of carbon steel in the absence and the presence of various concentrations of PTC_a and OCP_a in HCl 1.0 M

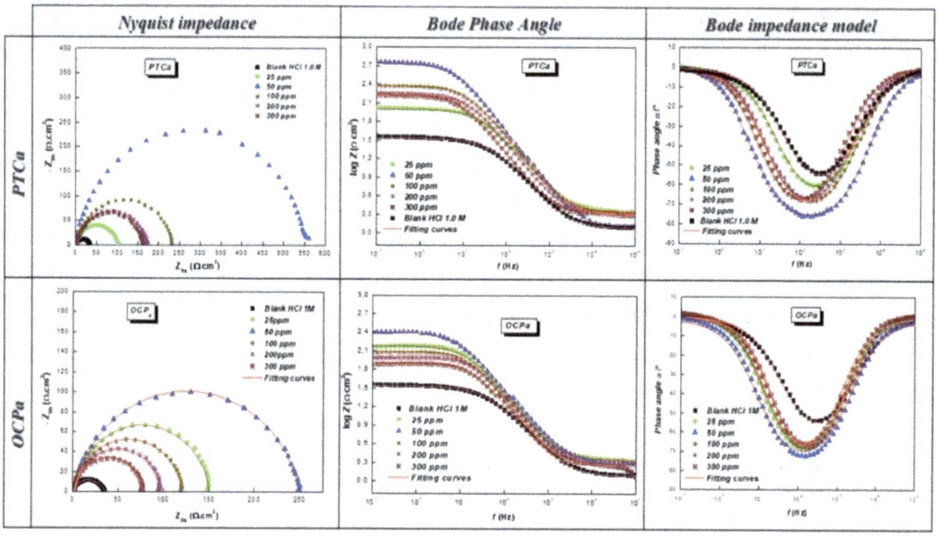

A simple equivalent electrical circuit (Figure 11) was proposed to model the experimental data. This circuit provided an excellent parametric fit to the experimental impedance spectra and allowed for the identification of key electrochemical parameters, including solution resistance (R_s), charge transfer resistance (R_{ct}), constant phase element (CPE).

Figure 11. Equivalent circuit designed to match the experimental impedance data.

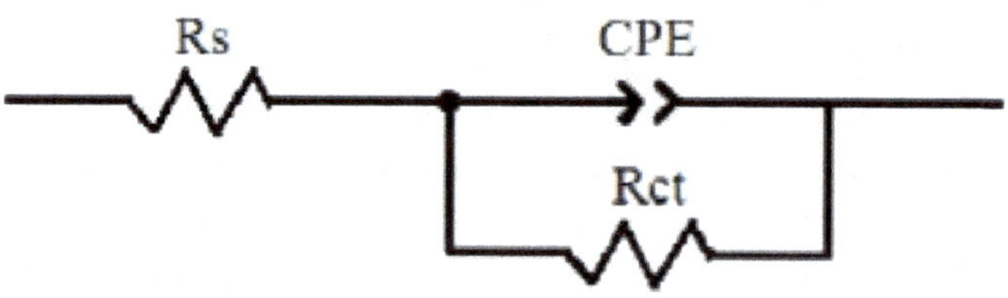

The extracted parameters are summarized in Table 7 for various concentrations of inhibitors corrosive media studied. Since the double layer at the interface does not act as an ideal capacitor, a constant phase element was employed to yield a better fit. The double-layer capacitance C_{dl} is related to the thickness of the double layer by the following equation

$$C_{dl} = \frac{\varepsilon \varepsilon_0}{d} S \quad \text{Eq.3}$$

Where: ε is the local dielectric constant, ε_0 is the permittivity of free space, d is the film thickness and S is the surface area.

Table 7. PDP electrochemical parameters for carbon steel without and with the different concentrations of inhibitors.

Medium	*[ ] ppm*	R_s $\Omega.cm^2$	R_{ct} $\Omega.cm^2$	C_{dl} $\mu F.cm^{-2}$	n_{dl}	Q $\mu F.Sn^{-1}$		$_{imp}$ %
HCl 1.0 M	--	1.12	35.00	121.00	0.773	419.00	-	-
	25	2.20	99.45	62.40	0.844	137.50	0.648	64.8
	50	1.93	549.70	49.15	0.900	70.43	0.936	93.6
TCP_a	100	2.60	231.00	56.12	0.876	95.81	0.848	84.8
	200	2.41	170.40	68.98	0.884	115.30	0.794	79.4
	300	2.28	159.60	105.50	0.893	163.20	0.781	78.1
	25	2.523	147.8	79.4	0.928	119.7	0.763	76.3
	50	2.076	247.4	73.71	0.867	109.3	0.858	85.8
OCP_a	100	1.858	119.1	88.27	0.913	125.1	0.706	70.6
	200	1.861	95.38	81.22	0.920	131.2	0.633	63.3
	300	1.732	75.87	91.88	0.913	164.2	0.539	53.9

The parameters presented in Table 6 show that the addition of the inhibitor in the corrosive environments leads to a significant increase in charge transfer resistance (R_{ct}), indicating the formation of a protective film on the steel surface(Behpour et al., 2010; Mehta et al., 2022). Simultaneously, the observed decrease in double-layer capacitance (C_{dl}) and the CPE constant (Q) reflects the effective adsorption of inhibitor molecules onto the metal surface. This reduction in C_{dl} may also be attributed to a decrease in the dielectric constant (Heakal et al., 2017) and/or an increase in the thickness of the adsorbed inhibitor layer on the steel surface(Gerengi & Sahin, 2012; Ghenimi et al., 2025). The parameters in Table 7 show that adding the inhibitor in corrosive environments increases the charge transfer resistance (R_{ct}). This behavior is typical of how a protective film forms on the carbon steel surface. A decrease in the double-layer capacitance (C_{dl}) and the CPE constant (Q) indicates effective adsorption of the inhibitor molecule onto the metal surface. The drop in C_{dl} could be explained by a lower local dielectric constant (Heakal et al., 2017) or by a thicker layer of adsorbed inhibitor on the carbon surface(Gerengi & Sahin, 2012; Ghenimi et al., 2025). An increase in the inhibitor concentration beyond 50 ppm decreases the charge transfer resistance and increases the double layer capacitance.

Furthermore, the obtained exponent n deviates slightly from the unity value, indicating a deviation from the behavior of an ideal capacitor that is expected from the surface heterogeneity of the used metal substrate (M. et al., 2017). Based on this result, PTCa had superior efficiency for corrosion inhibition of carbon steel in 1M HCl, with a maximum efficiency of 93.6% at 50 ppm. At the same concentration, OCP_a showed 85.8%.

B. Potentiodynamic Polarization Measurement

Figure 12 presents the polarization curves of carbon steel in 1.0 M HCl solution, recorded at 298 K in the presence and absence of the investigated phosphates (PTC_a and OCP_a). The measurements were performed over a potential range of -900 mV to -100 mV, using a scan rate of 1 mV/s.

Figure 12. Polarization curves for carbon steel in HCl 1.0 M in the absence and presence of inhibitors at 298 k.

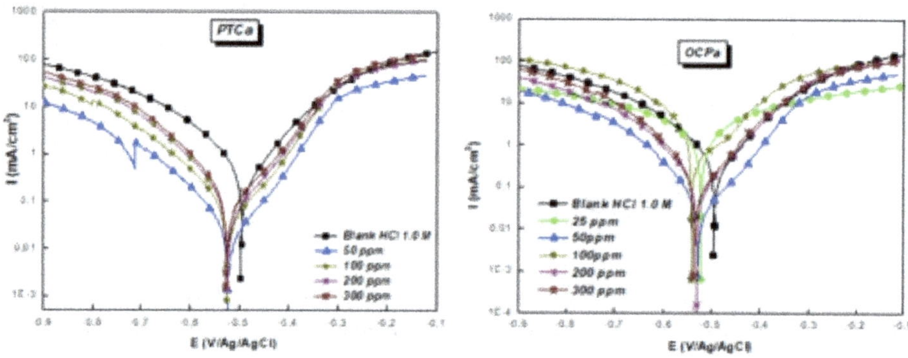

It can be seen that the hydrogen reduction mechanism in the cathodic part and the iron oxidation mechanism in the anodic part are controlled by charge transfer for the blank solution (HCl 1.0M) (Bedair et al., 2020; Oguzie et al., 2007; Rbaa et al., 2020a). In the cathodic part, the H^+ proton reduction reaction was reduced by the addition of phosphate materials:

$$2H^+ + 2e^-$$

H_2 Eq.4

This reaction requires two successive stages (Anusuya et al., 2017), the first of which is the discharge reaction known as the VOLMER reaction:

$$H^+_{ads} + e^-$$

H_{ads} Eq.5

The second stage may be purely chemical (TAFEL reaction) or electrochemical (HEYROVSKI reaction):

$$H_{ads} + H_{ads}$$

$$H^+_{ads} + H^+ + 2e^-$$

H_2 Réaction chimique Eq.6

H_2 Réaction électrochimique Eq.7

In the anodic part, the dissolution of iron corresponds to the following oxidation reaction:

$$Fe$$

$Fe^{2+} + 2e^-$ Eq.8

The cathodic and anodic curves had about the same form following the addition of the phosphate inhibitors to the 1.0 M HCl medium, with a small shift in corrosion potential toward the cathodic values. Additionally, compared to the blank, the steel's corrosion mechanism remained unchanged (Rbaa et al., 2020b). When the inhibitors were added, the corrosion current density dropped proportionately and peaked at a concentration of 50 ppm. Table 8 lists the electrochemical parameters for various PTC_a and OCP_a concentrations in 1.0M HCl, including anodic TAFEL slopes (β_a), cathodic TAFEL slopes (β_c), corrosion potential (E_{corr}), corrosion current density (i_{corr}), and inhibitory efficiency (η_{pp} %).

Table 8. Electrochemical parameters and corrosion-inhibiting efficacy of carbon steel in 1.0M HCl without and with addition of inhibitors

Compounds	[ ] ppm	E_{corr} mV/Ag/AgCl	i_{corr} µA.cm^{-2}	β_c mV.dec^{-1}	β_a mV.dec^{-1}	η_{PP} %
HCl 1.0M	--	-498	983	140	150	-
PTCa	25	-626	332	127	135	66.2
	50	-520	58	108	130	94.1
	100	-522	146	128	132	85.1
	200	-639	193	135	136	80.3
	300	-616	208	121	147	78.8
OCP$_a$	25	-523	215	130	134	78.1
	50	-530	126	114	121	87.2
	100	-541	271	126	128	72.4
	200	-531	345	132	141	64.9
	300	-536	437	125	131	55.5

It is currently shown that the addition of various inhibitor concentrations has caused the corrosion current density (i_{corr}) to decrease in comparison to the blank. The lowest corrosion current density value is observed at 50 ppm, when they decrease from 983 µA.cm^{-2} (the Blank value) to 58 µA.cm^{-2} and 126 µA.cm^{-2} in the presence of PTC_a and OCP_a, respectively. The inhibitory effectiveness increases with concentration, reaching a maximum of 94.1% (PTC_a) and 87.2% (OCP_a) at the same concentration. From higher concentrations to the ideal concentration, the corrosion current density increased and the inhibitory effectiveness slightly decreased(Ferraa, Aribou, et al., 2025). Additionally, the addition of inhibitors causes a slight shift in the cathodic potential (β_c). This suggests that the adsorption of this molecule on the carbon steel surface blocks the active sites available for corrosion without altering the cathodic reaction mechanism(Rbaa et al., 2019). The existence of PTC_a and OCP_a in the aggressive environment has altered the corrosion potential values from that of the blank medium; the shift in potential is more than -85 mV, which means phosphate inhibitors are considered as mixed inhibitors with a cathodic predominance (Anyiam et al., 2020; Palaniappan et al., 2019). A good correlation is obtained between the efficiencies calculated from EIS and PDP techniques. In addition, the results clearly show how effective phosphate compounds are in providing good protection against steel corrosion in 1.0 M HCl, which is further reflected in the electrochemical parameters calculated from both techniques.

C. Temperatures Effect And Activation Thermodynamic Parameters

Temperature is a key factor in deciding the corrosion performance of materials exposed to aggressive environments. It affects the interaction between the carbon steel surface and the acidic solution in the absence and presence of an inhibitor. Figure 13 presents the polarization curves of carbon steel in HCl medium recorded at different temperatures, in the absence and presence of 50 ppm of PTC_a and OCP_a, respectively. The corresponding electrochemical parameters are presented in Table 9. Within the temperature range from 298 K to 328 K, an increase in both cathodic and anodic current densities can be observed. In the uninhibited solution, this increase is more pronounced, indicating an acceleration of the process of metal dissolution by temperature. In the presence of inhibitors, however, the increase in corrosion current density is much less significant, showing that the inhibitors remain highly efficient at higher temperatures as well. Moreover, the nearly parallel appearance of the polarization curves suggests that temperature mainly influences the corrosion rate without significantly altering the basic corrosion mechanism(Rbaa et al., 2020).

Figure 13. Polarization curves of carbon steel in 1 M HCl, without and with PTCa and OCPa at various temperatures.

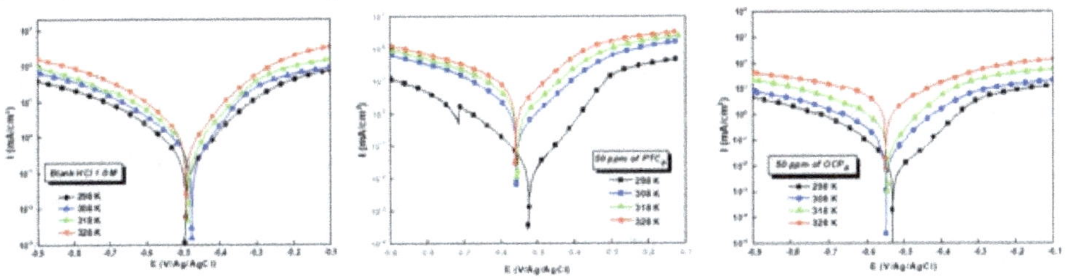

Table 9. Electrochemical parameters and corrosion inhibition efficiency of carbon steel in 1.0 M HCl, with and without inhibitors at various temperatures.

Medium	**Temperature in K**	**E_{corr} mV/Ag/AgCl**	**i_{corr} µA cm^{-2}**	**$-\beta_c$ mV dec^{-1}**	**β_a mV dec^{-1}**	*pp %*
HCl 1.0 M	298	-498	**983**	**140**	**150**	-
	308	-477	1200	184	112	-
	318	-487	1450	171	124	-
	328	-493	2200	161	118	-

continued on following page

Table 9. Continued

Medium	Temperature in K	E_{corr} mV/Ag/AgCl	i_{corr} µA cm^{-2}	$-\beta_c$ mV dec^{-1}	β_a mV dec^{-1}	*pp %*
PTCa	298	-520	58	108	130	94.0
	308	-556	97	115	138	91.9
	318	-557	164	121	145	88.6
	328	-558	327	118	142	85.1
OCP_a	298	-530	126	114	121	87.2
	308	-549	170	122	134	82.7
	318	-541	257	108	116	73.8
	328	-547	374	126	144	61.9

Table 9 shows the dependence of the electrochemical parameters with temperature, through which an increase in corrosion current density and a rise in the corrosion rate of carbon steel in the uninhibited medium can be observed. This indicates an acceleration of the dissolution process(Anupama et al., 2016; Cao et al., 2021). A similar but less pronounced tendency was observed in the presence of phosphate-based inhibitors. The inhibition efficiency decreases slightly with the increase in temperature, achieving 85.1% at 328 K for PTCa and 61.9% for OCPa, which is indicative of partial deterioration of the protective film formed on the steel surface(Essaadaoui et al., 2018).

The parameters extracted through the temperature dependent tests allow the determination of some important thermodynamic parameters, which provide more detailed information on the adsorption behavior of inhibitors in the corrosive medium. In this context, the ARRHENIUS relationship and the variation of ln(i_{corr}) as a function of 1000/T, in both the absence and presence of PTC_a and OCP_a, are presented in Figure 15 (Tiskar et al., 2016).

Figure 14. ARRHENIUS lines of carbon steel in 1.0 M HCl and 0.5 M H_2SO_4 in the absence and presence of PTCa.

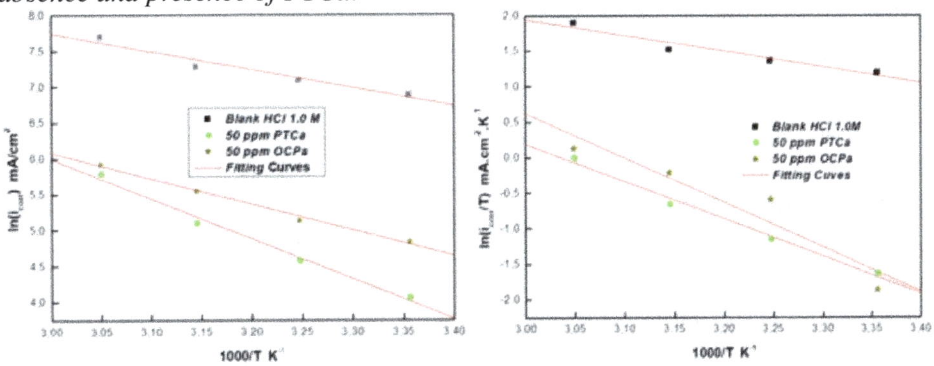

Table 10. Activation parameters for carbon steel in 1.M HCl in the absence and presence of phosphate inhibitors

		E_a (KJ/mol)	ΔH_a (KJ/mol)	ΔS_a (J/mol.K)
HCl 1 M	*Blank*	21,0	18,4	-126,0
	PTCa	45,5	43,7	-64,8
	OCPa	29.9	52.1	-36.1

As indicated in Table 10, the addition of calcium phosphate results in an increase in the activation energy (E_a), suggesting that the energy barrier for the corrosion reactions has been altered in the presence of the inhibitors. This change is also associated with either physical adsorption or weak chemical bonds between the inhibitors species and the steel surface (Boughoues et al., 2020). Additionally, the positive enthalpy values confirm that the adsorption of PTC_a and OCP_a nto the steel surface in the aggressive medium is endothermic. The observed increase in disorder in the presence of inhibitors is explained by the higher ΔS_a values, calculated relative to the blank solution(Boughoues et al., 2020).

4. SURFACE MORPHOLOGY CHARACTERIZATION

The morphology of the surface and elemental composition were studied by scanning electron microscopy coupled with energy-dispersive X-ray spectroscopy (SEM-EDX) to confirm and characterize the protective film formation. This analysis also allowed for the quantification of the elements presents in the carbon steel/3% NaCl interface in the absence and presence of phosphates inhibitors (PTC_a, HAp and OCP_a), thereby supporting the electrochemical study. Figure 15 represent the SEM images and EDX spectra of the carbon steel in 3% NaCl solution, both without and with various inhibitors at their optimal concentrations after 24 hours of immersion. For the uninhibited sample, some distinct scratches on the surface are observed, which are due to the manual polishing, small white dots corresponding to the first corrosion attacks are also present. However, when inhibitors were added, the carbon steel surface is covered by a uniform and compact protective layer. This layer acts as an efficient barrier that strongly reduces surface degradation and inhibits the nucleation and growth of corrosion.

The EDX of blank shows the presence of peaks for oxygen, iron and carbon. The detection of oxygen on the carbon steel surface is due to corrosion products such as iron(II) hydroxide ($Fe(OH)_2$) and iron(III) hydroxide ($Fe(OH)_3$), as confirmed by the SEM observations. On the other hand, the EDX analysis of carbon steel, immersed in NaCl solutions containing the phosphate-based inhibitors, shows additional peaks for calcium (Ca^{2+}) and phosphorus, main constituents of the studied phosphate materials. Indeed, these findings confirm the formation of a protective layer on the carbon steel surface, consisting of the protective ions from the phosphates. On the other hand, by comparing the corroded carbon steel in NaCl 3% to the carbon steel exposed to inhibitor solutions, a decrease in the oxygen mass percentage is observed after inhibitor addition. Thus, this confirms that OCP_a prevents carbon steel corrosion by forming a protective layer, which limits access of the electrolyte to the surface more effectively than HAp and PTC_a. Calcium phosphates present a strong inhibitive action for carbon steel in aggressive marine media. This finding is in excellent agreement with the electrochemical results obtained to evaluate the performance of our inhibitors.

Figure 15. SEM/EDX of the carbon steel surface in the presence and absence of PTCa, HAp and OCPa after 24h immersion in 3% NaCl.

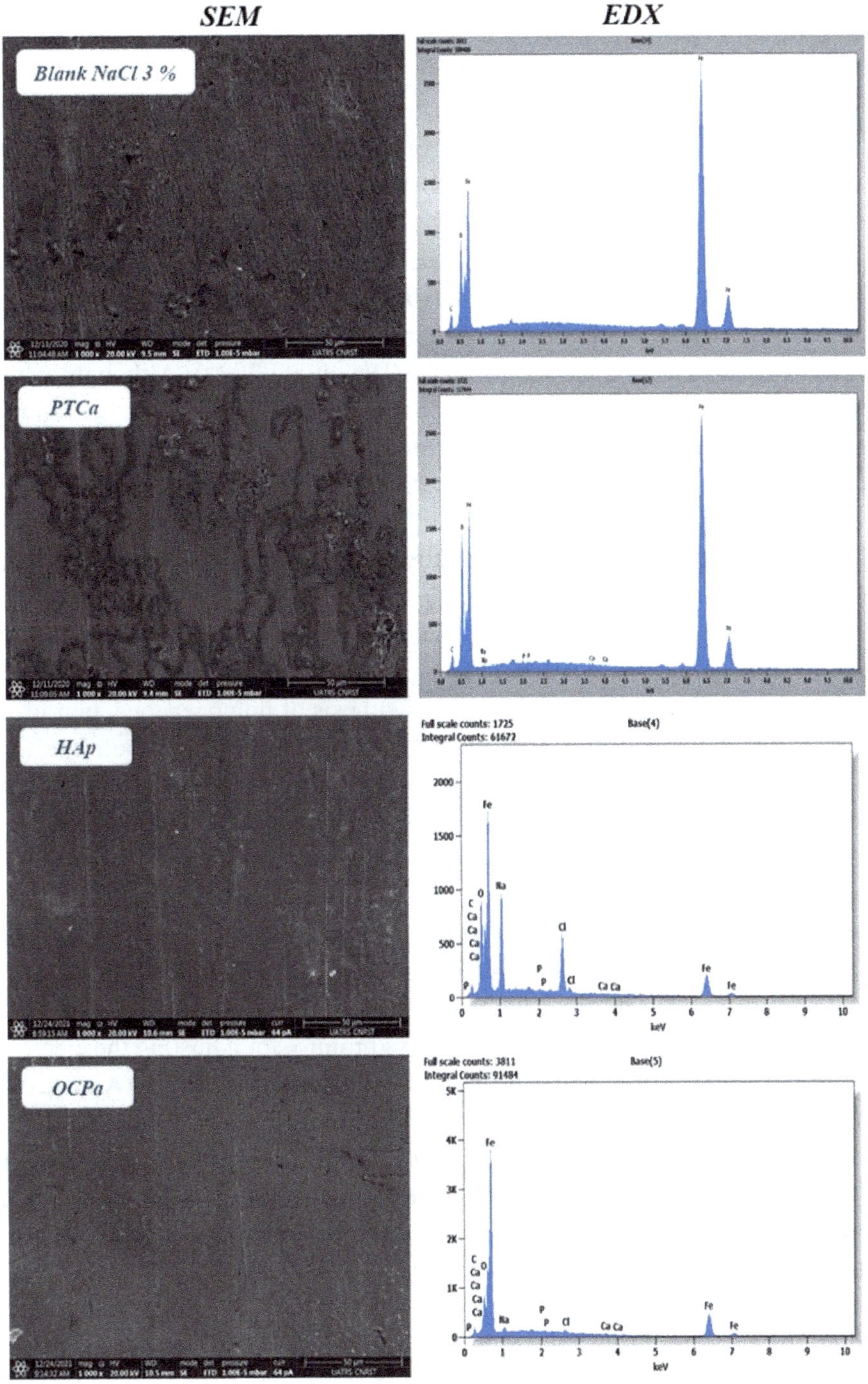

5. CHALLENGES AND FUTURE PERSPECTIVES OF CALCIUM PHOSPHATE-BASED MATERIALS IN ENERGY STORAGE APPLICATIONS

Calcium phosphates recently attracted broad attention for electrochemical energy storage, especially as electrode materials in batteries and supercapacitors(Aldayyat & Yam, 2025; Panneerselvam et al., 2024). For example, various studies have pointed to the potential of hydroxyapatite ($Ca_{10}(PO_4)_6(OH)_2$) for application in calcium-ion batteries, which is an emerging alternative to lithium-ion batteries, by virtue of the abundance, low cost and safety that calcium can offer(Aldayyat & Yam, 2025; Eagambaram et al., 2025b; Omar et al., 2016). In addition, the crystalline structures of calcium phosphates allow for reversible insertion and extraction of ions, making them suitable for use as electrode materials(Dev et al., 2022). Although naturally insulating, calcium phosphates can be turned into effective storage materials through various strategies such as doping, increasing porosity or incorporation with conductive materials. These approaches address their intrinsically low electronic and ionic conductivity and form the main limitation to performance and support the integration of conductive nanomaterials, like carbon and graphene(Deng et al., 2022; Islam et al., 2023).

However, several technical challenges still exist: their specific capacity is still modest compared to transition metal oxides; cycling stability has to be further improved for long-time applications; and the synthesis of homogeneous, porous and doped materials remains challenging and expensive. Current efforts, therefore, focus on engineering these materials to become appropriate components of future energy storage devices(Yan et al., 2023b). Future research on calcium phosphate materials for energy storage devices will probably continue to be developed by advanced material engineering strategies that focus on surmounting their intrinsic limitations. In this respect, the design of nanostructures with high specific surface area and controlled porosity will likely contribute to the improvement of ion exchange kinetics and electrolyte diffusion, hence enhancing the overall electrochemical performance. Multicationic doping with Fe^{2+}, Mn^{2+}, and Co^{2+} shows promise for adjusting the electronic structure and improving charge transport within the phosphate matrix (Ahmed et al., 2024; Ajitha et al., 2025). Additionally, hybrid bio-inspired structures that merge the natural biocompatibility and chemical stability of calcium phosphates with the excellent conductivity of carbon-based materials like graphene, carbon nanotubes or amorphous carbon may pave the way for the development of multifunctional electrodes. These electrodes could offer better mechanical flexibility, lightweight properties and improved electrical conductivity(X. Wang et al., 2023). These developments place the composite materials based on calcium phosphate in an excellent position to develop and be integrated into high-performance, next-generation energy storage systems such as flexible supercapacitors and low-cost sodium-ion batteries that emphasize sustainable and portable electronics(Rudra et al., 2024; Yang et al., 2025). Of course, to reveal their complete potential, deep fundamental insights into the electrochemical mechanisms of ion insertion/extraction dynamics and structural evolution upon cycling are required. Thereby, mastering such challenging tasks takes advantage of advanced in situ/ex situ analytical techniques such as synchrotron X-ray diffraction, transmission electron microscopy, and electrochemical impedance spectroscopy that follow real-time changes, hence helping to navigate the rational design of these materials. Collectively, the various strategies discussed here outline a pathway to the scalable development of CaP-based materials as viable components in next-generation, high-performance energy storage devices(Arroyo-de Dompablo et al., 2020; Rbaa et al., 2019).

6. RELATION BETWEEN CORROSION INHIBITION AND ENERGY STORAGE DEVICES

Recent research and the present analysis point out the crucial relationship between the corrosion prevention performance of calcium phosphate-based materials and their possible electrochemical role in energy storage. The characteristics in terms of structure and composition determining the protective performance, such as crystallinity, ionic substitution, and surface charge, are crucial for defining its charge transfer and ion diffusion capabilities. For example, magnesium or cobalt-substituted apatites exhibit enhanced inhibition efficiency due to their higher surface reactivity and increased capability of adsorption onto metallic substrates.

These substitutions also boost ionic mobility and redox activity, both of which are desirable in electrode materials for supercapacitors or batteries. Additionally, forming a compact, adherent phosphate layer on the metal surface in corrosive conditions is similar to the solid electrolyte interface found in electrochemical storage devices where stable surface films ensure effective charge transport and long-lasting durability. This dual function shows that calcium phosphate compounds can act not only as eco-friendly corrosion inhibitors but also as promising options for energy storage and conversion systems. Their ability to combine surface protection with reversible charge exchange presents new opportunities for developing multifunctional materials that link corrosion science and electrochemical energy technologies.

7. CONCLUSION

The current chapter examines the electrochemical uses of calcium phosphates. It highlights their dual role as corrosion inhibitors and as new materials for energy storage systems. The chapter focuses on recent studies and provides an overview of how calcium phosphate compounds, mainly hydroxyapatite and similar structures, perform well against corrosion in neutral and acidic environments. Their ability to protect carbon steel surfaces was confirmed by a comparative electrochemical study of three types of calcium phosphate, showing differences in effectiveness based on the medium and material structure composition. In addition to corrosion protection, early studies on their energy storage capabilities have shown promising but limited results. Calcium phosphates have some inherent advantages, such as being widely available, environmentally friendly and having stable crystal structures. However, issues like poor electrical conductivity and low specific capacity hinder their use in modern energy devices. Despite these challenges, the outlook is optimistic. Advances in material engineering, particularly in nanostructuring, adding new elements and creating hybrid composites, have opened new possibilities for boosting their electrochemical performance. These advances could translate to multi-functional calcium phosphate-based materials that would successfully marry corrosion resistance with energy storage attributes. This supports the goals of sustainability and green technology.

REFERENCES

Abdallah, M., Al-Tass, H. M., & Jahdaly, A. L. (2016). Inhibition properties and adsorption behavior of 5-arylazothiazole derivatives on 1018 carbon steel in 0.5 M H2SO4 solution. *Journal of Molecular Liquids*, *216*, 590–597. DOI: 10.1016/j.molliq.2016.01.077

Abouricha, S., Ihechmyen, A., Aziam, H., Said, H. A., Ouarga, A., Oueldna, N., Sabi, N., Noukrati, H., Lahcini, M., Saadoune, I., & Ben Youcef, H. (2024). From bone tissue to batteries : Hydroxyapatite as a filler to enhance the mechanical, thermal, and electrochemical properties of electrolytes for solid-state sodium-ion batteries. *Journal of Energy Storage*, *91*, 111967. DOI: 10.1016/j.est.2024.111967

Ahmed, A., Abdul Karim, M. R., & Usmana, M. (2024). Enhancing Charge Storage Mechanism in Mixed Metal Phosphate as Anodic Electrode Material for Energy Storage Systems. *2024 21st International Bhurban Conference on Applied Sciences and Technology (IBCAST)*, 642-651. DOI: 10.1109/IBCAST61650.2024.10876893

Ajitha, M., Selvarani, K., Gopinath, S. C. B., & Theivasanthi, T. (2025). Hydrothermal Production of Multifunctional Zinc Ferrite Nanoparticles as Fertilizer, Supercapacitor Electrode, and NPK Sensor. *BioNanoScience*, *15*(2), 312. DOI: 10.1007/s12668-025-01925-3

Al-Sabagh, A. M., Abd-El-Bary, H. M., El-Ghazawy, R. A., Mishrif, M. R., & Hussein, B. M. (2012). Corrosion inhibition efficiency of heavy alkyl benzene derivatives for carbon steel pipelines in 1M HCl. *Egyptian Journal of Petroleum*, *21*(2), 89–100. DOI: 10.1016/j.ejpe.2012.11.003

Aldayyat, A., & Yam, F. K. (2025). Recent development in metal phosphate-based electrode materials for supercapacitor application : A review. *Journal of Solid State Electrochemistry*, *29*(10), 4249–4280. Advance online publication. DOI: 10.1007/s10008-025-06280-7

Anupama, K. K., Ramya, K., & Joseph, A. (2016). Electrochemical and computational aspects of surface interaction and corrosion inhibition of mild steel in hydrochloric acid by Phyllanthus amarus leaf extract (PAE). *Journal of Molecular Liquids*, *216*, 146–155. DOI: 10.1016/j.molliq.2016.01.019

Anusuya, N., Saranya, J., Sounthari, P., Zarrouk, A., & Chitra, S. (2017). Corrosion inhibition and adsorption behaviour of some bis-pyrimidine derivatives on mild steel in acidic medium. *Journal of Molecular Liquids*, *225*, 406–417. DOI: 10.1016/j.molliq.2016.11.015

Anyiam, C. K., Ogbobe, O., Oguzie, E. E., Madufor, I. C., Nwanonenyi, S. C., Onuegbu, G. C., Obasi, H. C., & Chidiebere, M. A. (2020). Corrosion inhibition of galvanized steel in hydrochloric acid medium by a physically modified starch. *SN Applied Sciences*, *2*(4), 520. DOI: 10.1007/s42452-020-2322-2

Aribou, Z., Mzioud, K., Ferraa, N., Ouakki, M., Khemmou, N., & Touhami, M. E. (2025). Advancements in Electrodeposition Techniques for Enhancing Copper-Based Composite Coatings : A Comprehensive Overview. In El Kacimi, Y., & Alaoui, K. (Eds.), (pp. 89–144). Advances in Chemical and Materials Engineering. IGI Global., DOI: 10.4018/979-8-3693-6341-6.ch004

Arroyo-de Dompablo, M. E., Ponrouch, A., Johansson, P., & Palacín, M. R. (2020). Achievements, Challenges, and Prospects of Calcium Batteries. *Chemical Reviews*, *120*(14), 6331–6357. DOI: 10.1021/acs.chemrev.9b00339 PMID: 31661250

Arul, K. T., Ramesh, M., Chennakesavan, C., Karthikeyan, V., Manikandan, E., Umar, A., Maaza, M., & Henini, M. (2018). Novel multifunctional of magnesium ions (Mg++) incorporated calcium phosphate nanostructures. *Journal of Alloys and Compounds*, *730*, 31–35. DOI: 10.1016/j.jallcom.2017.09.254

Asghar, A., Chen, Z., Al-Harthi, E. A., Hakami, J., Rashid, M. S., Sultana, H., Hussain, S., Javed, Y., Shad, N. A., & Imran, M. (2023). High-Performance Electrode Materials for Electrochemical Energy Storage Devices Based on Microrod-Like Structures of Calcium Phosphate ($Ca_2P_2O_7$). *Physica Status Solidi (RRL) –. Physica Status Solidi. Rapid Research Letters*, *17*(10), 2300178. DOI: 10.1002/pssr.202300178

Attia, M. E. H., Driss, Z., Kabeel, A. E., Alagar, K., Athikesavan, M. M., & Sathyamurthy, R. (2021). Phosphate bags as energy storage materials for enhancement of solar still performance. *Environmental Science and Pollution Research International*, *28*(17), 21540–21552. DOI: 10.1007/s11356-020-12018-x PMID: 33411289

Bakri, A. E., Ferraa, N., Lachheb, A., & Bennani-Ziatni, M. (2025). Adsorption of resorcinol onto synthetic calcium phosphate compounds : Kinetic, isotherm, and thermodynamic studies. *Ecological Engineering*.

Ballouze, R., Marahat, M. H., Mohamad, S., Saidin, N. A., Kasim, S. R., & Ooi, J. P. (2021). Biocompatible magnesium-doped biphasic calcium phosphate for bone regeneration. *Journal of Biomedical Materials Research. Part B, Applied Biomaterials*, *109*(10), 1426–1435. DOI: 10.1002/jbm.b.34802 PMID: 33484103

Bedair, M. A., Soliman, S. A., Bakr, M. F., Gad, E. S., Lgaz, H., Chung, I.-M., Salama, M., & Alqahtany, F. Z. (2020). Benzidine-based Schiff base compounds for employing as corrosion inhibitors for carbon steel in 1.0 M HCl aqueous media by chemical, electrochemical and computational methods. *Journal of Molecular Liquids, 317*, 114015. DOI: 10.1016/j.molliq.2020.114015

Behpour, M., Ghoreishi, S. M., Mohammadi, N., Soltani, N., & Salavati-Niasari, M. (2010). Investigation of some Schiff base compounds containing disulfide bond as HCl corrosion inhibitors for mild steel. *Corrosion Science, 52*(12), 4046–4057. DOI: 10.1016/j.corsci.2010.08.020

Benzidia, B., Barbouchi, M., Hsissou, R., Zouarhi, M., Erramli, H., & Hajjaji, N. (2022). A combined experimental and theoretical study of green corrosion inhibition of bronze B66 in 3% NaCl solution by Aloe saponaria (syn. Aloe maculata) tannin extract. *Current Research in Green and Sustainable Chemistry, 5*, 100299. DOI: 10.1016/j.crgsc.2022.100299

Bouali, I., Rocca, E., Veys-Renaux, D., Rhouta, B., Khalil, A., & Aït Aghzzaf, A. (2017). Ca2+-exchange in layered zirconium orthophosphate, α-ZrP : Chemical study and potential application for zinc corrosion inhibition. *Applied Surface Science, 422*, 778–786. DOI: 10.1016/j.apsusc.2017.06.083

Boughoues, Y., Benamira, M., Messaadia, L., & Ribouh, N. (2020). Adsorption and corrosion inhibition performance of some environmental friendly organic inhibitors for mild steel in HCl solution via experimental and theoretical study. *Colloids and Surfaces. A, Physicochemical and Engineering Aspects, 593*, 124610. DOI: 10.1016/j.colsurfa.2020.124610

Cao, Y., Zou, C., Wang, C., Chen, W., Liang, H., & Lin, S. (2021). Green corrosion inhibitor of β-cyclodextrin modified xanthan gum for X80 steel in 1 M H2SO4 at different temperature. *Journal of Molecular Liquids, 341*, 117391. DOI: 10.1016/j.molliq.2021.117391

Chafki, L., Galai, M., Rifi, E. H., Touir, R., Touhami, M. E., Hatim, Z., & Tazouti, A. (2015). *Corrosion inhibition of mild steel in 1 M HCl Solution by hydroxyapatite*. 10.

Choe, H.-B., Lee, H.-S., Ismail, M. A., & Hussin, M. W. (2015). Evaluation of Electrochemical Impedance Properties of Anti- corrosion Films by Arc Thermal Metal Spraying Method. *International Journal of Electrochemical Science, 10*(11), 16. DOI: 10.1016/S1452-3981(23)11219-3

Corrales-Luna, M., Le Manh, T., Romero-Romo, M., Palomar-Pardavé, M., & Arce-Estrada, E. M. (2019). 1-Ethyl 3-methylimidazolium thiocyanate ionic liquid as corrosion inhibitor of API 5L X52 steel in H2SO4 and HCl media. *Corrosion Science*, *153*, 85–99. DOI: 10.1016/j.corsci.2019.03.041

Crespilho, F. N., Costa, C. M., & Lanceros-Méndez, S. (2025). Sustainable Battery Biomaterials. *ChemElectroChem*, *12*(3), e202400530. DOI: 10.1002/celc.202400530

Dagdag, O., Gouri, M. E., Galai, M., Touhami, M. E., Essamri, A., & Harfi, A. E. (2015). *Influence of a derivative of cyclotriphosphazene as a corrosion inhibitor for carbon steel in 3% NaCl medium*. 9.

Dahmani, K., Galai, M., & Cherkaoui, M. (2017). *Cinnamon essential oil as a novel eco-friendly corrosion inhibitor of copper in 0.5 M Sulfuric Acid medium*. 14.

Daoud, D., Douadi, T., Issaadi, S., & Chafaa, S. (2014). Adsorption and corrosion inhibition of new synthesized thiophene Schiff base on mild steel X52 in HCl and H2SO4 solutions. *Corrosion Science*, *79*, 50–58. DOI: 10.1016/j.corsci.2013.10.025

Deng, X., Li, L., Zhang, G., Zhao, X., Hao, J., Han, C., & Li, B. (2022). Anode chemistry in calcium ion batteries : A review. *Energy Storage Materials*, *53*, 467–481. DOI: 10.1016/j.ensm.2022.09.033

Dev, P. R., Anand, C. P., Michael, D. S., & Wilson, P. (2022). Hydroxyapatite coatings : A critical review on electrodeposition parametric variations influencing crystal facet orientation towards enhanced electrochemical sensing. *Materials Advances*, *3*(21), 7773–7809. DOI: 10.1039/D2MA00620K

Drevet, R., Aaboubi, O., & Benhayoune, H. (2012). In vitro corrosion behavior of electrodeposited calcium phosphate coatings on Ti6Al4V substrates. *Journal of Solid State Electrochemistry*, *16*(9), 3069–3077. DOI: 10.1007/s10008-012-1742-3

Eagambaram, M., Mani, K. S., & Sahoo, S. (2025a). Cobalt intercalated strontium phosphate for cost-effective high-performance supercapacitors. *Journal of Energy Storage*, *119*, 116312. DOI: 10.1016/j.est.2025.116312

Eagambaram, M., Mani, K. S., & Sahoo, S. (2025b). Cobalt intercalated strontium phosphate for cost-effective high-performance supercapacitors. *Journal of Energy Storage*, *119*, 116312. DOI: 10.1016/j.est.2025.116312

El Aggadi, S., Ennouhi, M., Boutakiout, A., & El Hourch, A. (2023). Progress towards efficient phosphate-based materials for sodium-ion batteries in electrochemical energy storage. *Ionics*, *29*(6), 2099–2113. DOI: 10.1007/s11581-023-04936-w

El Bakri, A., El Boujaady, H., Ferraa, N., & Bennani-Ziatni, M. (2024). The Removal of Phenol Through Adsorption onto Synthetic Calcium Phosphates – A Study Encompassing Analyses of Kinetics and Thermodynamics. *Ecological Engineering & Environmental Technology*, *25*(4), 301–315. DOI: 10.12912/27197050/184155

El Hallaoui, A., Ferraa, N., Merroun, Y., Chehab, S., Hassoune, H., Ziatni, M. B., Ghailane, R., & Souizi, A. (2024). Innovative catalytic synthesis of 5-unsubstituted 3,4-dihydropyrimidine-2-ones : Use of CoCa3(HPO4)(PO4)2(OH) as a novel bimetallic catalyst in multicomponent condensation of Biginelli reaction. *Polyhedron*, *253*, 116915. DOI: 10.1016/j.poly.2024.116915

El Harmouchi, H., Alami, S., Moumouche, O., El Amri, A., Ouakki, M., Brouzi, K., Khaoulaf, R., Dkhireche, N., & Harcharras, M. (2024). Synthesis of ZnSrP2O7 and CoSrP2O7 Pyrophosphates : Characterization, vibrational properties and electrochemical performance on mild steel in an acidic environment. *Inorganic Chemistry Communications*, *166*, 112593. DOI: 10.1016/j.inoche.2024.112593

Elaidi, H., Ismi, I., Rifi, E., Galai, M., Touir, R., & Touhami, M. E. (2014). *Study of the Influence of Amorphous Tricalcium Phosphate on Corrosion Inhibition of Mild Steel in Hydrochloric Acidic Medium 1M*. *3*(11).

Essaadaoui, Y., Galai, M., Ouakki, M., Kadiri, L., Ouass, A., Cherkaoui, M., Rifi, H., Lebkiri, A., & Rifi, E. (2018). *STUDY OF THE ANTICORROSIVE ACTION OF EUCALYPTUS CAMALDULENSIS EXTRACT IN CASE OF MILD STEEL IN 1.0 M HCl*. 431-442.

Ferraa, N., Aribou, Z., Mzioud, K., Ouakki, M., Cherkaoui, M., & Ziatni, M. B. (2025). Advancements in Green Inorganic Corrosion Inhibitors : A Comprehensive Review and Comparison. In El Kacimi, Y., & Alaoui, K. (Eds.), (pp. 255–298). Advances in Chemical and Materials Engineering. IGI Global., DOI: 10.4018/979-8-3693-6341-6.ch009

Ferraa, N., Ouakki, M., Alaoui, K., Cherkaoui, M., & Bennani Ziatni, M. (2025). Sustainable enhancement of steel corrosion inhibition in saline environments using Ca2.7Mg6.3 (HPO4)(PO4)5(OH) and Ca3Mg7(PO4)6(OH)2 : Synthesis, characterization, electrochemical and chemometric investigations. *Journal of Molecular Structure*, *1348*, 143480. DOI: 10.1016/j.molstruc.2025.143480

Ferraa, N., Ouakki, M., Barbouchi, M., Cherkaou, M., & Ziatni, B. (2024). *Theoretical Investigation and Electrochemical Study of Stoichiometric Hydroxyapatite as an Ecofriendly Corrosion Inhibitor for Steel in 3% NaCl*. *14*(3), 54. https://doi.org/doi.org/10.33263/BRIAC143.054

Ferraa, N., Ouakki, M., Cherkaoui, M., & Bennani Ziatni, M. (2022). Study of the Inhibitory Action of Apatitic Tricalcium Phosphate on Carbon Steel in Two Acidic Media (HCl 1.0 M and H2SO4 0.5 M). In Vaseashta, A., Achour, M. E., Mabrouki, M., Fasquelle, D., & Tachafine, A. (Éds.), *Proceedings of the Sixth International Symposium on Dielectric Materials and Applications (ISyDMA'6)* (p. 159-176). Springer International Publishing. DOI: 10.1007/978-3-031-11397-0_14

Ferraa, N., Ouakki, M., Cherkaoui, M., & Ziatni, M. B. (2022). Synthesis, Characterization and Evaluation of Apatitic Tricalcium Phosphate as a Corrosion Inhibitor for Carbon Steel in 3 wt% NaCl. *Journal of Bio- and Tribo-Corrosion, 8*(1), 23. DOI: 10.1007/s40735-021-00622-4

Ferraa, N., Ouakki, M., El Harmouchi, H., Cherkaoui, M., & Bennani Ziatni, M. (2023). Investigation of the inhibition behavior of an octacalcium phosphate as a green corrosion inhibitor against carbon steel in 3% NaCl medium. *Inorganic Chemistry Communications, 157*, 111343. DOI: 10.1016/j.inoche.2023.111343

Gerengi, H., & Sahin, H. I. (2012). Schinopsis lorentzii Extract As a Green Corrosion Inhibitor for Low Carbon Steel in 1 M HCl Solution. *Industrial & Engineering Chemistry Research, 51*(2), 780–787. DOI: 10.1021/ie201776q

Ghenimi, G., Ouakki, M., Ferraa, N., Barebita, H., Guedira, T., & Cherkaoui, M. (2025). Elaboration and investigation of corrosion inhibition by bismuth glasses for mild steel in 1.0 M HCl solution. *Moroccan Journal of Chemistry, 13*(1), 440–458. DOI: 10.48317/IMIST.PRSM/MORJCHEM-V13I1.53043

Goyal, M., Kumar, S., Bahadur, I., Verma, C., & Ebenso, E. E. (2018). Organic corrosion inhibitors for industrial cleaning of ferrous and non-ferrous metals in acidic solutions : A review. *Journal of Molecular Liquids, 256*, 565–573. DOI: 10.1016/j.molliq.2018.02.045

Habraken, W., Habibovic, P., Epple, M., & Bohner, M. (2016). Calcium phosphates in biomedical applications : Materials for the future? *Materials Today, 19*(2), 69–87. DOI: 10.1016/j.mattod.2015.10.008

Harmouchi, H. E., Moumouche, O., Alami, S., Hmada, A., Khaoulaf, R., Brouzi, K., Touhami, M. E., & Harcharras, M. (2023). Vibrational Spectroscopy and electrochemical study of MCuP2O7 (M = Ba, Ca and Zn), improvement of corrosion resistance of mild steel in 1.0 M HCl medium. *Journal of Molecular Structure, 1284*, 135452. DOI: 10.1016/j.molstruc.2023.135452

Heakal, F. E.-T., Deyab, M. A., Osman, M. M., Nessim, M. I., & Elkholy, A. E. (2017). Synthesis and assessment of new cationic gemini surfactants as inhibitors for carbon steel corrosion in oilfield water. *RSC Advances*, *7*(75), 47335–47352. DOI: 10.1039/C7RA07176K

Heughebaert, J. C., & Montel, C. (1970). Reception de phosphate pure tricalcic. *Bulletin de la Société Chimique de France*, *8/9*, 2923–2924.

Hossain, N., Chowdhury, M., & Kchaou, M. (2020). An overview of green corrosion inhibitors for sustainable and environment friendly industrial development An overview of green corrosion inhibitors for sustainable and environment friendly industrial development. *Journal of Adhesion Science and Technology*, *35*(7), 673–690. Advance online publication. DOI: 10.1080/01694243.2020.1816793

Huang, Y., Yao, K., Zhang, Q., Huang, X., Chen, Z., Zhou, Y., & Yu, X. (2024). Bioelectronics for electrical stimulation : Materials, devices and biomedical applications. *Chemical Society Reviews*, *53*(17), 8632–8712. DOI: 10.1039/D4CS00413B PMID: 39132912

Hwang, J.-W., Park, J.-S., Lee, J.-S., Jung, U.-W., Kim, C.-S., Cho, K.-S., Lee, Y.-K., & Choi, S.-H. (2012). Comparative evaluation of three calcium phosphate synthetic block bone graft materials for bone regeneration in rabbit calvaria. *Journal of Biomedical Materials Research. Part B, Applied Biomaterials*, *100B*(8), 2044–2052. DOI: 10.1002/jbm.b.32768 PMID: 22865716

Iqbal, M. F., Nasir, F., Shabbir, F., Babar, Z. U. D., Saleem, M. F., Ullah, K., Sun, N., & Ali, F. (2025). Supercapacitors : An Emerging Energy Storage System. *Advanced Energy and Sustainability Research*, *n/a*(n/a), 2400412. DOI: 10.1002/aesr.202400412

Islam, M., Mubarak, M., & Lee, H.-J. (2023). Hybrid Nanostructured Materials as Electrodes in Energy Storage Devices. *Inorganics*, *11*(5), 183. DOI: 10.3390/inorganics11050183

Jain, A., & Payal, R. (2020). Ca-Ion Batteries. In R. Boddula, Inamuddin, R. Pothu, & A. M. Asiri (Éds.), *Rechargeable Batteries* (1re éd., p. 195-215). Wiley. DOI: 10.1002/9781119714774.ch10

Kardas, G. (2005). The Inhibition Effect of 2-Thiobarbituric Acid on the Corrosion Performance of Mild Steel in HCl Solutions. *Materials Science*, *41*(3), 337–343. DOI: 10.1007/s11003-005-0170-2

Laamari, M. R., Derja, A., Benzakour, J., & Berraho, M. (2004). Calcium monofluorophosphate : A new class of corrosion inhibitors in NaCl medium. *Journal of Electroanalytical Chemistry (Lausanne, Switzerland), 569*(1), 1–6. DOI: 10.1016/j.jelechem.2003.12.045

Lai, X., Hu, J., Ruan, T., Zhou, J., & Qu, J. (2021). Chitosan derivative corrosion inhibitor for aluminum alloy in sodium chloride solution : A green organic/inorganic hybrid. *Carbohydrate Polymers, 265*, 118074. DOI: 10.1016/j.carbpol.2021.118074 PMID: 33966838

Li, X., Elshahawy, A. M., Guan, C., & Wang, J. (2017). Metal Phosphides and Phosphates-based Electrodes for Electrochemical Supercapacitors. *Small, 13*(39), 1701530. DOI: 10.1002/smll.201701530 PMID: 28834280

Li, X., Xiao, X., Li, Q., Wei, J., Xue, H., & Pang, H. (2018). Metal (M = Co, Ni) phosphate based materials for high-performance supercapacitors. *Inorganic Chemistry Frontiers, 5*(1), 11–28. DOI: 10.1039/C7QI00434F

Loto, R. T., Loto, C. A., Popoola, A. P., & Fedotova, T. (2016). Electrochemical Studies of the Inhibition Effect of 2-Dimethylaminoethanol on the Corrosion of Austenitic Stainless Steel Type 304 in Dilute Hydrochloric Acid. *Silicon, 8*(1), 145–158. DOI: 10.1007/s12633-015-9344-1

M., P., Ammal, P., & Joseph, A. (2017). Comparative studies on the corrosion inhibition characteristics of three different triazine based Schiff's bases, HMMT, DHMMT and MHMMT, for mild steel exposed in sulfuric acid. *Egyptian Journal of Petroleum, 27*. DOI: 10.1016/j.ejpe.2017.07.011

Ma, W., & Zhang, D. (2019). Multifunctional structural supercapacitor based on graphene and magnesium phosphate cement. *Journal of Composite Materials, 53*(6), 719–730. DOI: 10.1177/0021998318790322

Mansoori, H., Young, D., Brown, B., & Singer, M. (2018). Influence of calcium and magnesium ions on CO2 corrosion of carbon steel in oil and gas production systems—A review. *Journal of Natural Gas Science and Engineering, 59*, 287–296. DOI: 10.1016/j.jngse.2018.08.025

Masese, T., & Kanyolo, G. M. (2024). Electrode materials for calcium batteries : Future directions and perspectives. *EcoEnergy, 2*(3), 339–368. DOI: 10.1002/ece2.53

Matić, T., Zebić, M. L., Miletić, V., Cvijović-Alagić, I., Petrović, R., Janaćković, Dj., & Veljović, Dj. (2022). Sr,Mg co-doping of calcium hydroxyapatite : Hydrothermal synthesis, processing, characterization and possible application as dentin substitutes. *Ceramics International, 48*(8), 11155–11165. DOI: 10.1016/j.ceramint.2021.12.335

Maurya, S., Jaiswal, A., Ji, G., & Prakash, R. (2021). Waste Solanum melongena stem extract for corrosion inhibition of mild steel in 1M NaCl. *Materials Today: Proceedings*, *44*, 2716–2720. DOI: 10.1016/j.matpr.2020.12.691

Meeusen, M., Zardet, L., Homborg, A. M., Lekka, M., Andreatta, F., Fedrizzi, L., Boelen, B., Terryn, H., & Mol, J. M. C. (2019). A Complementary Electrochemical Approach for Time-Resolved Evaluation of Corrosion Inhibitor Performance. *Journal of the Electrochemical Society*, *166*(11), C3220–C3232. DOI: 10.1149/2.0271911jes

Mehta, R. K., Gupta, S. K., & Yadav, M. (2022). Studies on pyrimidine derivative as green corrosion inhibitor in acidic environment : Electrochemical and computational approach. *Journal of Environmental Chemical Engineering*, *108499*(5), 108499. Advance online publication. DOI: 10.1016/j.jece.2022.108499

Mzioud, K., Ferraa, N., Aribou, Z., Rached, S., Ouakki, M., Habsaoui, A., & Touhami, M. E. (2025). Green Corrosion Protection : Plant Extracts as Sustainable Solutions. In El Kacimi, Y., & Alaoui, K. (Eds.), (pp. 215–254). Advances in Chemical and Materials Engineering. IGI Global., DOI: 10.4018/979-8-3693-6341-6.ch008

Mzioud, K., Habsaoui, A., Rached, S., Ech-chihbi, E., Ouakki, M., Salghi, R., & Touhami, M. E. (2024). Experimental investigation and theoretical modeling of copper corrosion inhibition by *Urginea maritima* essential oil. *Materials Today Sustainability*, *27*, 100906. DOI: 10.1016/j.mtsust.2024.100906

Nahali, H., Dhouibi, L., & Idrissi, H. (2015). Effect of Na3PO4 addition in mortar on steel reinforcement corrosion behavior in 3% NaCl solution. *Construction & Building Materials*, *78*, 92–101. DOI: 10.1016/j.conbuildmat.2014.12.099

Öğretir, C., Mihçi, B., & Bereket, G. (1999). Quantum chemical studies of some pyridine derivatives as corrosion inhibitors. *Journal of Molecular Structure THEOCHEM*, *488*(1-3), 223–231. DOI: 10.1016/S0166-1280(99)00033-0

Oguzie, E. E., Li, Y., & Wang, F. H. (2007). Effect of 2-amino-3-mercaptopropanoic acid (cysteine) on the corrosion behaviour of low carbon steel in sulphuric acid. *Electrochimica Acta*, *53*(2), 909–914. DOI: 10.1016/j.electacta.2007.07.076

Omar, F. S., Numan, A., Duraisamy, N., Bashir, S., Ramesh, K., & Ramesh, S. (2016). Ultrahigh capacitance of amorphous nickel phosphate for asymmetric supercapacitor applications. *RSC Advances*, *6*(80), 76298–76306. DOI: 10.1039/C6RA15111F

Ouakki, M., Galai, M., Cherkaoui, M., Rifi, E.-H., & Hatim, Z. (2018). *Inorganic Compound (Apatite doped by Mg and Na) as a Corrosion Inhibitor for Mild Steel in Phosphoric Acidic Medium. 10*(7), 18.

Oyekunle, D. T., Agboola, O., & Ayeni, A. O. (2019). Corrosion Inhibitors as Building Evidence for Mild Steel : A Review. *Journal of Physics: Conference Series, 1378*(3), 032046. DOI: 10.1088/1742-6596/1378/3/032046

Palaniappan, N., Alphonsa, J., Cole, I. S., Balasubramanian, K., & Bosco, I. G. (2019). Rapid investigation expiry drug green corrosion inhibitor on mild steel in NaCl medium. *Materials Science and Engineering B, 249*, 114423. DOI: 10.1016/j.mseb.2019.114423

Panneerselvam, K., Warrier, A. R., Mathiarasu, R. R., Thuy Nga, T. T., J, R. R., Kumaravelu, T. A., Chou, W.-C., Huang, Y.-C., Chen, J.-L., Chen, C.-L., Kandasami, A., & Dong, C.-L. (2024). Enhancement in supercapacitive performance of calcium phosphate nanoneedles through Ni-ion incorporation : Insights into atomic and electronic structures. *Materials Today Sustainability, 26*, 100769. DOI: 10.1016/j.mtsust.2024.100769

Pokhmurs'kyi, V. I., Zin', I. M., Layon, S. B., & Bilyi, L. M. (2003). Synergistic Effect of Phosphate and Calcium-Containing Pigments on the Corrosion Resistance of Galvanized Steel. *Materials Science, 39*(2), 153–160. DOI: 10.1023/B:MASC.0000010264.27310.9e

Przywecka, K., Grzmil, B., & Kowalczyk, K. (2019). Anticorrosive and physico-chemical properties of modified phosphate pigments. *Polish Journal of Chemical Technology, 21*(1), 20–23. DOI: 10.2478/pjct-2019-0004

Rabecca, H. J. T., Priyajanani, Y., Manivannan, S., & Lourduraj, A. J. C. (2025). Supercapacitor electrode with constructive formation of manganese cobalt phosphate structures. *Materials Science and Engineering B, 313*, 117964. DOI: 10.1016/j.mseb.2024.117964

Rahmouni, K., Keddam, M., Srhiri, A., & Takenouti, H. (2005). Corrosion of copper in 3% NaCl solution polluted by sulphide ions. *Corrosion Science, 47*(12), 3249–3266. DOI: 10.1016/j.corsci.2005.06.017

Rbaa, M., Abousalem, A. S., Touhami, M. E., Warad, I., Bentiss, F., Lakhrissi, B., & Zarrouk, A. (2019). Novel Cu (II) and Zn (II) complexes of 8-hydroxyquinoline derivatives as effective corrosion inhibitors for mild steel in 1.0 M HCl solution : Computer modeling supported experimental studies. *Journal of Molecular Liquids, 290*, 111243. DOI: 10.1016/j.molliq.2019.111243

Rbaa, M., Ouakki, M., Galai, M., Berisha, A., Lakhrissi, B., Jama, C., Warad, I., & Zarrouk, A. (2020a). Simple preparation and characterization of novel 8-Hydroxyquinoline derivatives as effective acid corrosion inhibitor for mild steel : Experimental and theoretical studies. *Colloids and Surfaces. A, Physicochemical and Engineering Aspects*, *602*, 125094. DOI: 10.1016/j.colsurfa.2020.125094

Rbaa, M., Ouakki, M., Galai, M., Berisha, A., Lakhrissi, B., Jama, C., Warad, I., & Zarrouk, A. (2020b). Simple preparation and characterization of novel 8-Hydroxyquinoline derivatives as effective acid corrosion inhibitor for mild steel : Experimental and theoretical studies. *Colloids and Surfaces. A, Physicochemical and Engineering Aspects*, *602*, 125094. DOI: 10.1016/j.colsurfa.2020.125094

Rhilassi, A. E., Ferraa, N., & Bennani-Ziatni, M. (2024). Kinetic, isotherm and thermodynamic studies of the adsorption of phenol and tyrosine onto apatitic tricalcium phosphate. *Current Chemistry Letters*, *13*(2), 403–416. DOI: 10.5267/j.ccl.2023.10.005

Rudra, S., Seo, H. W., Sarker, S., & Kim, D. M. (2024). Supercapatteries as Hybrid Electrochemical Energy Storage Devices : Current Status and Future Prospects. *Molecules (Basel, Switzerland)*, *29*(1), 243. DOI: 10.3390/molecules29010243 PMID: 38202828

Safari-Gezaz, M., Parhizkar, M., & Asghari, E. (2025). Effect of cobalt ions doping on morphology and electrochemical properties of hydroxyapatite coatings for biomedical applications. *Scientific Reports*, *15*(1), 149. DOI: 10.1038/s41598-024-84055-2 PMID: 39747275

Saji, V. S. (2010). A Review on Recent Patents in Corrosion Inhibitors. *Recent Patents on Corrosion Science*, *2*(1), 6–12. DOI: 10.2174/1877610801002010006

Saleem, I., Shukrullah, S., Munir, M. A., Kareri, T., Irfan, M., & Jalalah, M. (2025). Performance Analysis of Electrochemical and Dielectric Parameters of CdS–CaP Binary Nanocomposites for Energy Storage Application. *Journal of Electronic Materials*, *54*(11), 9640–9653. DOI: 10.1007/s11664-025-12294-8

Sane, A. R., & Universite de Toulouse, E. D. M.-G. des procedes et de l'Environnement. (2017). *Phosphate-based materials for thermal energy storage* (No. FRNC-TH--15033). Universite de Toulouse, Ecole Doctorale Megep - Genie des procedes et de l'Environnement, ecole Nationale Superieure des Mines d'Albi-Carmaux - IMT, Centre Rapsodee, UMR CNRS 5302 (France). https://inis.iaea.org/records/8q316-d8967

Shadanbaz, S., & Dias, G. J. (2012). Calcium phosphate coatings on magnesium alloys for biomedical applications : A review. *Acta Biomaterialia*, *8*(1), 20–30. DOI: 10.1016/j.actbio.2011.10.016 PMID: 22040686

Shao, H., Yin, X., Zhang, K., Yang, W., Chen, Y., & Liu, Y. (2022). N-[2-(3-indolyl) ethyl]-cinnamamide synthesized from cinnamomum cassia presl and alkaloid tryptamine as green corrosion inhibitor for Q235 steel in acidic medium. *Journal of Materials Research and Technology*, *20*, 916–933. DOI: 10.1016/j.jmrt.2022.07.122

Shrivastava, K., Gill, F. S., Juyal, S., Lal, M., & Jain, A. (2025). Green Carbon for Clean Energy : Biomass-Derived Hierarchical Structures in Energy Storage. *Wiley Interdisciplinary Reviews. Energy and Environment*, *14*(2), e70007. DOI: 10.1002/wene.70007

Šimková, L., Gorodylova, N., & Šulcová, P. (2019). *APPLICATION OF DOPED HYDROXYAPATITE AS A CORROSION INHIBITOR.*

Suzuki, O. (2010). Octacalcium phosphate : Osteoconductivity and crystal chemistry. *Acta Biomaterialia*, *6*(9), 3379–3387. DOI: 10.1016/j.actbio.2010.04.002 PMID: 20371385

Tan, B., Zhang, S., Qiang, Y., Guo, L., Feng, L., Liao, C., Xu, Y., & Chen, S. (2018). A combined experimental and theoretical study of the inhibition effect of three disulfide-based flavouring agents for copper corrosion in 0.5 M sulfuric acid. *Journal of Colloid and Interface Science*, *526*, 268–280. DOI: 10.1016/j.jcis.2018.04.092 PMID: 29747040

Tasić, Ž. Z., Petrović Mihajlović, M. B., Radovanović, M. B., Simonović, A. T., & Antonijević, M. M. (2018). Cephradine as corrosion inhibitor for copper in 0.9% NaCl solution. *Journal of Molecular Structure*, *1159*, 46–54. DOI: 10.1016/j.molstruc.2018.01.031

Tazi, S., Raissouni, I., Chaoukat, F., Bouchta, D., Dahdouh, A., Elkhamlichi, R., & Douhri, H. (2016). *L'effet Inhibiteur d'Eugénol sur la corrosion du Laiton dans NaCl 3% (The Inhibition effect of Brass corrosion in NaCl 3% by Eugenol).* 11.

Tiskar, M., Galai, M., Elhadiri, H., Ebn Touhami, M., Sfaira, M., Satrani, B., Ghanmi, M., Chaouch, A., & Touir, R. (2016). Juniperus Phoenicea essential oil as green corrosion inhibitor for mild steel in molar hydrochloric acid. *Matériaux & Techniques*, *104*(6-7), 609. DOI: 10.1051/mattech/2017003

Tristijanto, H., Ilman, M. N., & Tri Iswanto, P. (2020). Corrosion Inhibition of Welded of X – 52 Steel Pipelines by Sodium Molybdate in 3.5% NaCl Solution. *Egyptian Journal of Petroleum*, *29*(2), 155–162. DOI: 10.1016/j.ejpe.2020.02.001

Tuncer, M., Bakan, F., Gocmez, H., & Erdem, E. (2019). Capacitive behaviour of nanocrystalline octacalcium phosphate (OCP) (Ca8H2(PO4)6·5H2O) as an electrode material for supercapacitors : Biosupercaps. *Nanoscale, 11*(39), 18375–18381. DOI: 10.1039/C9NR07108C PMID: 31573596

Umoren, S. A., Abdullahi, M. T., & Solomon, M. M. (2022). An overview on the use of corrosion inhibitors for the corrosion control of Mg and its alloys in diverse media. *Journal of Materials Research and Technology, 20*, 2060–2093. DOI: 10.1016/j.jmrt.2022.08.021

Velrani, S., Jeyaprabha, B., & Prakash, P. (2014). *Inhibition of mild steel corrosion in 3.5% NaCl medium using 1-butyl-3-methylimidazolium chloride. 1*(10), 13.

Wang, J., Jin, C., Mei, D., Ding, Y., Chang, L., Zhu, S., Wang, L., Feng, Y., & Guan, S. (2021). Synthesis and degradation behaviour of Zn-modified coating on Mg alloy. *Surface Engineering, 37*(8), 963–971. DOI: 10.1080/02670844.2020.1840827

Wang, X., Zhai, X., Lv, P., Jiao, Y., Wang, S., & Chi, J. (2023). Amorphous nickel phosphate as a high performance electrode material for supercapacitor. *Synthetic Metals, 292*, 117217. DOI: 10.1016/j.synthmet.2022.117217

Xu, X., & Huang, H. (2023). The corrosion inhibition mechanism of Ca(H2PO4)2 on AZ91D magnesium alloy in 3.5 wt% NaCl solution. *New Journal of Chemistry, 47*(42), 19726–19736. DOI: 10.1039/D3NJ04280D

Yan, L., Yang, W., Yu, H., Zhang, L., & Shu, J. (2023a). Recent progress in rechargeable calcium-ion batteries for high-efficiency energy storage. *Energy Storage Materials, 60*, 102822. DOI: 10.1016/j.ensm.2023.102822

Yan, L., Yang, W., Yu, H., Zhang, L., & Shu, J. (2023b). Recent progress in rechargeable calcium-ion batteries for high-efficiency energy storage. *Energy Storage Materials, 60*, 102822. DOI: 10.1016/j.ensm.2023.102822

Yang, X., Fang, Y., Miao, G., Wang, Z., & Li, P. (2025). Ion-exchange induced statically activated zinc-containing hydroxyapatite interface toward stable zinc metal anodes. *Journal of Power Sources, 644*, 237049. DOI: 10.1016/j.jpowsour.2025.237049

Yohai, L., Vázquez, M., & Valcarce, M. B. (2013). Phosphate ions as corrosion inhibitors for reinforcement steel in chloride-rich environments. *Electrochimica Acta, 102*, 88–96. DOI: 10.1016/j.electacta.2013.03.180

Zakir Hossain, S. M., Kareem, S. A., Alshater, A. F., Alzubair, H., Razzak, S. A., & Hossain, M. M. (2020). Effects of Cinnamaldehyde as an Eco-Friendly Corrosion Inhibitor on Mild Steel in Aerated NaCl Solutions. *Arabian Journal for Science and Engineering, 45*(1), 229–239. DOI: 10.1007/s13369-019-04236-4

Zhang, B., Wang, X., Wang, S., Li, Y., Chen, L., Jiao, H., Yu, Z., & Tu, J. (2025). High-energy-density lithium manganese iron phosphate for lithium-ion batteries : Progresses, challenges, and prospects. *Journal of Energy Chemistry*, *100*, 1–17. DOI: 10.1016/j.jechem.2024.08.011

Zhou, X., Yang, H., & Wang, F. (2012). Investigation on the inhibition behavior of a pentaerythritol glycoside for carbon steel in 3.5% NaCl saturated Ca(OH)2 solution. *Corrosion Science*, *54*, 193–200. DOI: 10.1016/j.corsci.2011.09.018

Chapter 9
A Comprehensive Review of State-of-the-Art Battery Energy Systems for Green Transportation

G.V. Keerthi Sri
https://orcid.org/0009-0007-3213-2867
Department of EEE, SR University, India

Venkataramana Veeramsetty
https://orcid.org/0000-0002-2512-5761
Department of EEE, SR University, India

D. M. Vinod Kumar
Department of EEE, SR University, India

ABSTRACT

The shift to environmentally friendly mobility is central to slowing down climate change, cutting dependence on fossil fuels, and curbing greenhouse gas emissions, thus speeding up the worldwide use of electric vehicles. Battery systems stand at the centre of this movement as they have direct impacts on EV performance, efficiency, and sustainability. This assessment examines the electrochemical processes, strengths, and weaknesses of different contemporary EV battery types, including lithium-ion, lead-acid, nickel-based, sodium-based, and those employed in hybrid, plug-in hybrid, and fuel cell vehicles. Lithium-ion batteries are notable for their ability to store a lot of energy per unit mass, have long lifetimes, and are efficient, yet they are the most widely used battery type in most EVs. Though cheaper, lead-acid and others are generally employed in HEVs and PHEVs because of lesser

DOI: 10.4018/979-8-3373-3156-0.ch009

performance. A comparison highlights the role of each type in EV evolution. The significance of Battery Management Systems for safety, lifespan, and performance is also highlighted.

1. INTRODUCTION

The automotive industry has been a cornerstone of global transportation, enabling mobility, economic growth, and technological advancements (Hossain et al., 2022). Automobile manufacturers strongly dependent on fossil fuels have caused environmental problems that primarily stem from carbon emissions and climate-altering effects. The automotive industry has transitioned to electric vehicles (EVs) as sustainable options for people who no longer use conventional internal combustion engine (ICE) vehicles. Sustainability has driven increased investments from both developed and developing countries, who want to start making electric vehicles to replace non-renewable energy sources and deliver higher-demand, clean, and efficient transportation options (Ali & Naushad, 2022). Due to this emission, nowadays, renewable energy sources are preferred.

Renewable energy sources are naturally replenished and can be used continuously without depleting the Earth's resources. These include solar, wind, hydro, geothermal, and biomass energy, all of which provide cleaner alternatives to fossil fuels. Unlike conventional energy sources, renewables produce little to no greenhouse gas emissions, making them a key tool in addressing climate change and reducing air pollution (Renewables, 2021). In addition to environmental benefits, renewable energy also promotes energy security and economic development through job creation and local investments (Hillerbrand, 2018). As global demand for sustainable energy grows, renewable sources are becoming a vital part of the modern energy landscape.

At the core of EV technology lies the battery system, which directly influences vehicle performance, range, and longevity (Manzetti & Mariasiu, 2015). Lithium-ion batteries (LIBs) have emerged as the leading energy storage solution for EVs due to their high energy density, long cycle life, and efficiency. Other battery technologies, including lead-acid, nickel-metal hydride (NiMH), and sodium nickel chloride batteries, also play roles in specific applications, particularly in hybrid EVs (HEVs) and plug-in hybrid EVs (PHEVs). Equally important is the Battery Management System (BMS), which ensures battery safety, efficiency, and longevity. The BMS tracks essential battery elements, including temperature and voltage, and state of charge (SOC), to stop overcharging and deep discharging, and runaway thermals (UDHAYAKUMAR et al.). The BMS improves both battery life expectancy along vehicle reliability through optimized battery maintenance. Switches to smart BMS technologies result in better energy efficiency, together with adaptive charging techniques and live performance checks that strengthen EV sustainability.

The comprehensive review paper is structured to provide a holistic understanding of state-of-the-art battery energy systems for green transportation. Section 2 begins by classifying the different types of electric vehicles (EVs), including BEVs, HEVs, PHEVs, FCEVs, and ER-EVs, highlighting their main characteristics and advantages. Part 4 discusses the various battery performance metrics.

Part 5 discusses the charging infrastructure for electric vehicles.

Part 6 provides a comparative analysis of battery parameters, a discussion about future growth trends, an overview of EV sales and research publications, offering insights into the evolving landscape of EV technologies.

Part 7 evaluates the environmental impact of EVs.

Part 8 explores the latest innovations and challenges in EV battery technologies.

Finally, Part 9 concludes the review by summarizing the key findings, highlighting the significance of battery energy systems in enabling green transportation, and providing recommendations for future research and development.

2. ELECTRIC VEHICLES

The automotive industry evolves through electric vehicles (EVs) because they bring sustainable electric drive systems instead of traditional internal combustion engine (ICE) vehicles. Powered by electricity, EVs eliminate tailpipe emissions, significantly reducing greenhouse gases and combating climate change. EVs function with improved energy efficiency while demanding fewer components, which reduces their running expenses and boosts national security against fossil fuel reliance.

2.1. Advantages of EVs:

- Environmentally Friendly: Since EVs emit no tailpipe emissions, the air is cleaner and the environment is healthier.
- Lower Emissions of Greenhouse Gases: As EVs use electricity to power them rather than fossil fuels, they drastically lower carbon dioxide (CO2) emissions, particularly when fueled by renewable sources.
- Lower Operating Costs: Refueling an EV is typically less expensive than filling up a gasoline vehicle, and electricity tends to be less volatile than the price of gasoline.
- Minimal Maintenance: EVs tend to have fewer moving components compared to conventional automobiles, thus minimizing the requirement for frequent oil changes and decreasing total maintenance costs.

- Smooth and Quiet Driving: EVs are driven with minimal noise, and the ride is smooth and comfortable.
- Home Charging Convenience: Owners have the convenience of charging their EVs at home with a standard or fast charger, which reduces the need for constant trips to fuel stations.
- Improved Energy Efficiency: EVs use more of the energy from the power source for motion than internal combustion engine (ICE) cars, so they are more efficient.
- Compatibility with Renewable Energy: EVs can be powered from solar, wind, or other renewable energy sources, which will make transportation more sustainable.
- Enhanced Technology: With advancing battery technology, EVs are equipped with higher ranges, quick charging times, and improved overall performance, rendering them a convenient option for most consumers.

2.2. Types of EVs

Electric vehicles (EVs) are now divided into many categories according to the technology of their engines (Sanguesa et al., 2021). Generally, they can be categorized into five main types, each differing in power source and operational mechanism, as shown in Figure 1.

Figure 1. Types of Electric Vehicles

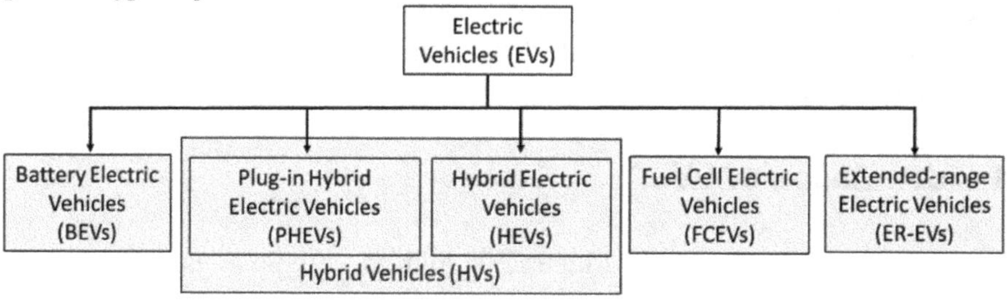

a) **Zero Emission Electric Vehicle (ZEV):** ZEVs operate entirely using electricity, producing no direct emissions during their operation. It uses a battery pack to store electrical energy, which powers an electric motor to drive the wheels. A power electronic converter regulates the energy flow between the battery and motor, ensuring efficient operation as shown in Figure 2. Since it has no internal combustion engine, it eliminates the need for fossil fuels, making it environmentally friendly. Regenerative braking technology increases the overall economy by allowing the car to recharge the battery and collect energy while braking. ZEVs are perfect for cutting greenhouse gas emissions and air pollution (Muratori et al., 2021) . The Nissan Leaf is a real-world example of a ZEV. Air pollution is decreased because it is entirely electric and emits no exhaust emissions (Oki & Sato, 2024).

Figure 2. Zero-Emission Electric Vehicle

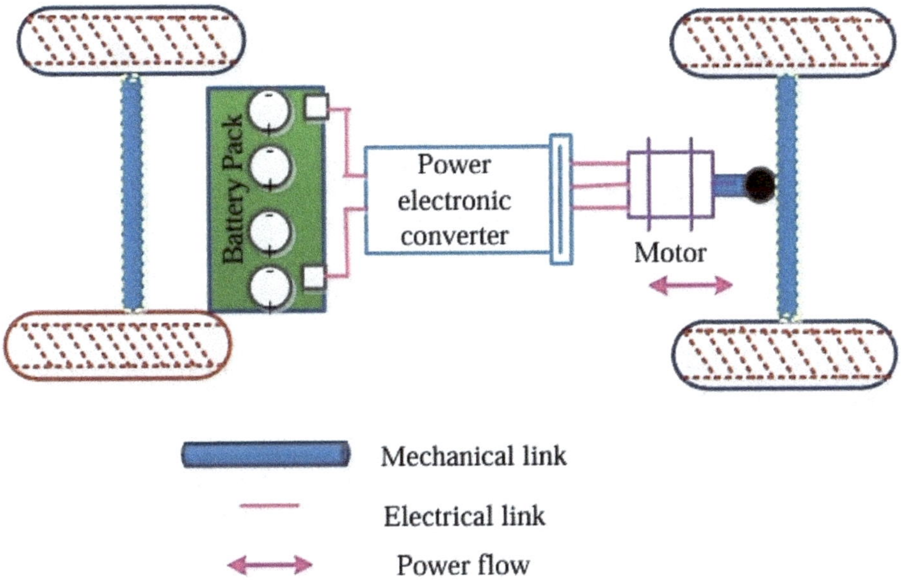

b) **Hybrid Electric Vehicle (HEV):** Hybrid electric vehicles (HEVs) combine an internal combustion (IC) engine with an electric motor and battery system to boost performance, reduce emissions, and improve fuel efficiency. The wheels are driven by the IC engine, which runs on fuel such as petrol or diesel and can recharge the battery. Electrical energy stored in the battery drives the electric motor for propulsion, particularly at low speeds, or it can be used in conjunction with the ICE for increased efficiency.

As seen in Figure 3, a mechanical connection joins the ICE, motor, and drivetrain for seamless power transfer. In contrast, a power electronic converter regulates the power flow between the battery and motor. This vehicle also employs regenerative braking, which converts braking force into electrical energy stored in the battery by using the motor as a generator(Khalid et al., 2022).

Figure 3. Hybrid Electric Vehicle

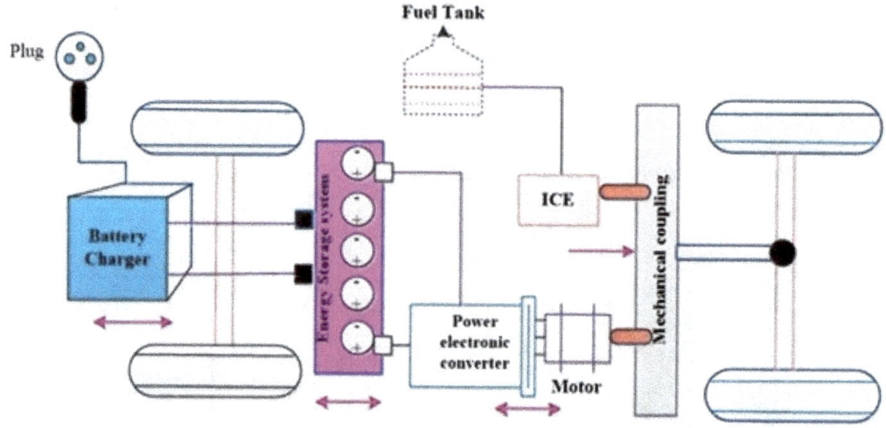

c) **Plug-in Hybrid Electric Vehicle (PHEV):** As seen in Figure4, PHEVs combine a conventional IC engine with an electric motor and a battery pack. The battery may be charged from an external power source through a charging connection. The IC engine and the generator work together to generate power when needed, either to drive the wheels or to recharge the battery. The battery-powered electric motor may either operate alone or in combination with the internal combustion engine to propel the vehicle. A controller manages the power flow between the IC engine, motor, battery, and transmission, ensuring efficient performance.

Figure 4. Plug-in Hybrid Electric Vehicle

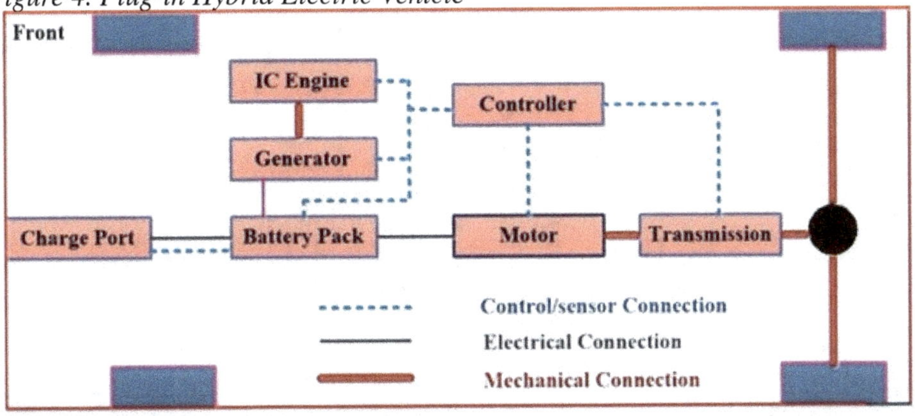

This dual system provides flexibility, as the vehicle can run on electricity for shorter trips and switch to the IC engine for longer distances (Al-Alawi & Bradley, 2013). The real-time example for PHEV is the Chevrolet Volt, which operates primarily as an electric vehicle but includes a gasoline engine to extend its range when the battery is depleted (Duhon et al., 2015)

d) **Fuel Cell Electric Vehicle (FCEV):** Fueled by hydrogen, FCEVs produce energy to run an electric motor. The first step involves feeding hydrogen into the fuel cells from a cylinder. Here, an electrochemical reaction between hydrogen and oxygen (from the air) yields heat, water, and electricity. The electric motor, which propels the vehicle, is powered by the generated electricity after it has been controlled by a power electronic converter (Larminie et al., 2003). Additionally, a battery pack is included to store excess energy and provide extra power during acceleration or high load, ensuring smooth operation as shown in Figure 5.

Figure 5. Fuel Cell Electric Vehicle

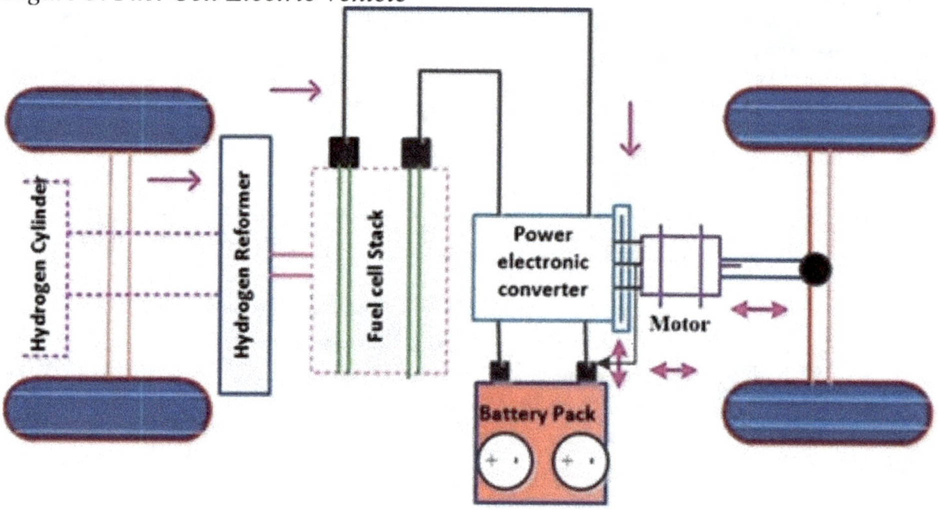

The real-time example for FCEV is the Toyota Mirai utilizes hydrogen fuel cells to generate electricity, emitting only water vapour as a by-product (Gómez & Santos, 2023). A ZEV is powered solely by a battery. It is more efficient than hybrid and plug-in hybrid vehicles, which rely on a combination of gasoline and electric power. In cases where the battery is depleted, HEVs use the engine to power the battery. This makes them less efficient than electric or PHEV vehicles. ZEV and PHEV vehicles can be charged directly. Among them, HEVs are superior to conventional hybrid vehicles but slightly less efficient than ZEVs.

e) **Extended-Range Electric Vehicles (ER-EVs):** Although they include an extra combustion engine, Extended-Range Electric Vehicles (ER-EVs) are quite like Battery Electric Vehicles (BEVs). In contrast to hybrid electric vehicles (HEVs) and plug-in hybrid electric vehicles (PHEVs), this engine does not power the wheels; instead, it just recharges the battery as necessary. One well-known example is the BMW i3, which has a 42.2 kWh battery and can go 260 km on pure electricity and an additional 130 km when using the extended-range functionality. The battery technologies employed in EVs are examined below based on the previously mentioned findings (Chan, 2007).

3. BATTERY SYSTEMS

In an electrochemical cell, during charging, electrons travel from the cathode to the anode through the external circuit as shown in Fig 6, while positive ions move through the electrolyte (Petrovic & Petrovic, 2021). During discharging, this process is reversed, with electrons flowing from the anode to the cathode and positive ions traveling through both the circuit and the electrolyte. The negative electrode donates electrons to the external circuit during discharge, while the positive electrode donates electrons during charging.

Figure 6. Electrochemical cell

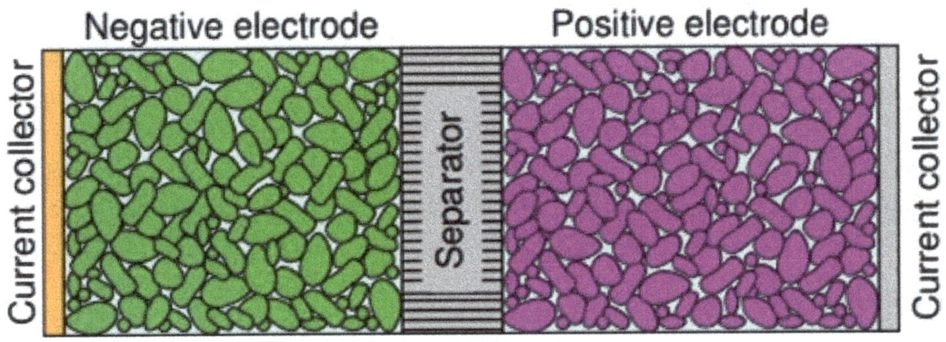

The electrolyte facilitates internal charge transfer by allowing the movement of cations towards the positive electrode and anions towards the negative electrode, a process driven by diffusion, which is shown in Figure 6. The separator prevents short circuits and self-discharge by isolating the electrodes, typically made of materials like glass mat, fiber, polyethylene, or polymer. Current collectors, often in the form of metal foils, supply electrical current to the cell terminals. The various types of batteries in EVs (Muslimin et al., 2022). They are pictorially shown in Figure 7.

3.1 Types of Batteries

Different battery types power electric vehicles, each with unique materials, chemistry, and performance characteristics. This section introduces the major categories lead-acid, nickel-metal hydride, sodium-nickel, and lithium-ion. It explains how each type functions and where it is best applied in EV systems.

Figure 7. Types of Batteries

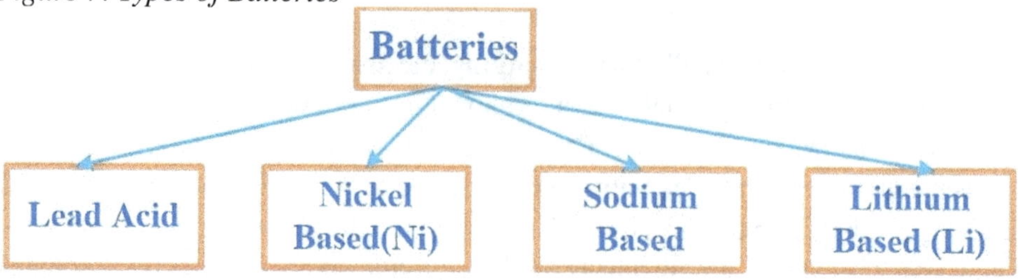

3.1.1. Lead Acid Battery

Lead acid battery vehicles with IC engines, which are more costly to manufacture, are frequently started by batteries. Spongy lead is an active ingredient in lead-acid cells' negative plates, as shown in Figure8. One of the active ingredients in the positive plates is lead dioxide (Rand, 1980). with lead and lead oxide, releasing electrical energy in the process, as shown in Eq. 1 (Larminie & Lowry, 2012).

Figure 8. Lead-Acid Battery

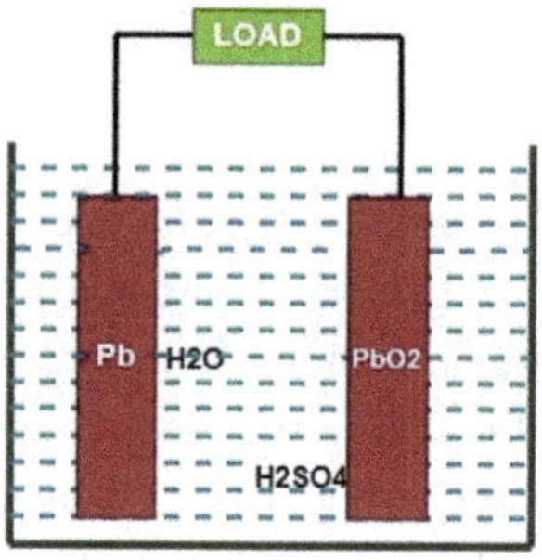

$$Pb + PbO_2 + 2H_2SO_4 \leftrightarrow 2PbSO_4 + 2H_2O \quad (1)$$

During the discharge phase, lead peroxide is oxidized to lead sulfate by the positive electrode. Lead peroxide is switched with lead sulfate as the negative electrode (Rand, 1979). Both electrode reactions during discharge result in the production of lead sulfate. During charging, lead sulfate is converted to lead peroxide on the positive electrode. Lead sulfate is switched with lead peroxide as the negative electrode. When a lead-acid battery is being charged, the amount of sulfuric acid increases. The positive electrode in the discharge reaction is shown in Eq. 2:

$$PbO2 + SO2-4 + 4H+ + 2e- \rightarrow PbSO4 + 2H2O \quad (2)$$

The discharge reaction's negative electrode is shown in Eq. 3:

$$Pb + SO2-4 \rightarrow PbSO4 + 2e- \quad (3)$$

The positive electrode's charge response is represented using Eq. 4:

$$PbSO4 + 2H2O \rightarrow PbO2 + SO2-4 + 4H+ + 2e- \quad (4)$$

The charge reaction at the negative electrode is represented using Eq. 5:

$$PbSO4 + 2e- \rightarrow Pb + SO2-4 \quad (5)$$

Lead-acid batteries are affordable and widely used, these batteries offer high surge currents and reliability. However, they are heavy and have a short lifespan with maintenance. They are used in automotive starters, backup power systems, and uninterruptible power supplies (UPS) (Ruetschi, 1977). Lead-acid batteries are not the best choice for modern electric vehicles as they have a shorter lifespan, slower charging times, and limited range. In HEVs, they can be used for auxiliary power and engine starting but are heavier and less efficient. For PHEVs, lead-acid batteries do not provide enough range and take longer to charge, making them less suitable. Similarly, in FCEVs, lead-acid batteries do not perform well since these vehicles need a lightweight, high-efficiency battery, where lithium-ion is a better fit.

3.1.2. Nickel Metal Hydride Battery (NiMH)

While the NiMH batteries function similarly, the NiMH battery's negative electrode is cadmium-free because it employs hydrogen stored in metal hydride rather than cadmium. The positive electrode of this battery functions similarly to other battery cells, whereas the negative electrode functions similarly to a fuel cell. Hydrogen is liberated from the metal at the negative electrode during discharge, where it combines with nickel oxyhydroxide to generate electrons and water. The reactions are inverted when the NiMH battery is charged. Water is produced at the same rate throughout charging and discharging, and the electrolyte remains unchanged (Kopera, 2004) . The fundamental idea is a reversible process in which hydrogen forms a connection with the metal and then releases itself as free hydrogen when required. The cell must be sealed because the alloys that absorb hydrogen inside the cell can react with the air. If they're exposed, other molecules can take the place of hydrogen in the alloys, affecting the battery's performance. Overall reaction for NiMH battery is computed using Eq. 6 and graphically shown in Figure 9.

$$MH + NiOOH \leftrightarrow M + Ni(OH)_2 \tag{6}$$

Energy is released during discharge when a metal hydride (MH) is at the negative electrode reacts with nickel hydroxide at the positive electrode to form nickel hydroxide. The positive electrode in the discharge reaction is shown in Eq. 7:

$$Ni(OH) + H_2O + 2e^- \rightarrow Ni + 2OH^- \tag{7}$$

Figure 9. Nickel metal-hydride Battery

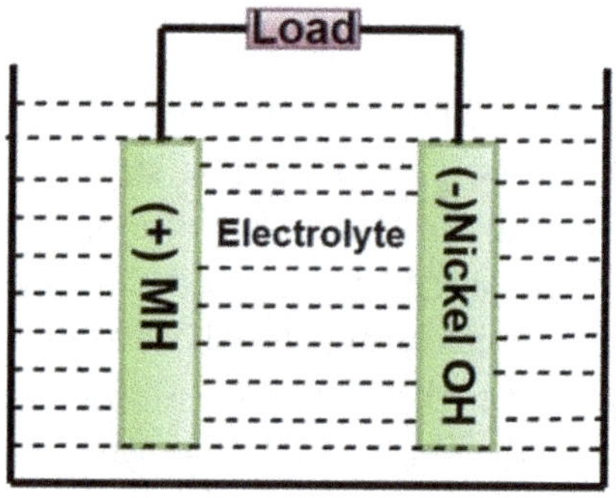

The negative electrode in the discharge reaction is shown in Eq. 8:

$$MH + OH^- \rightarrow M + H_2O + e^- \tag{8}$$

When charging, the reactions go the other way. Metal hydride and nickel are transformed back into their corresponding hydroxides. The positive and negative electrodes are represented using Eqns. 9 and 10.

$$Ni + 2OH^- \rightarrow Ni(OH) + 2e^- \tag{9}$$

$$M + H_2O + e^- \rightarrow MH + OH^- \tag{10}$$

Ni-Mh batteries have a higher energy density than various batteries that are less toxic and have a moderate lifespan. However, they suffer from high self-discharge rates and heat generation during charging. They are used in HEVs, consumer electronics, and medical equipment (Young & Yasuoka, 2016). NiMH batteries are best suited for HEVs due to their reliability, moderate energy density, and ability to handle frequent charge-discharge cycles efficiently. While they were once used in PHEVs, they have largely been replaced by lithium-ion batteries due to energy density limitations (Taniguchi et al., 2001) . ZEVs and FCEVs typically rely on lithium-ion batteries or hydrogen fuel cells, as NiMH lacks the range and efficiency required for full-electric applications. However, NiMH remains a strong choice for conventional hybrids like the Toyota Prius, where long cycle life, safety, and cost-effectiveness outweigh the need for higher energy density (Ovshinsky et al., 1993).

3.1.3. Sodium Nickel Battery (Na-Ni)

Na-Ni battery is a type of high-temperature, molten-salt battery that uses sodium and nickel compounds for energy storage, as shown in Figure10. Similar to sodium-sulfur (Na-S) batteries, Na-Ni batteries operate at elevated temperatures (typically around 300–350°C), which allows the sodium and nickel components to remain in a molten state, facilitating the flow of ions (Sudworth, 2001). Charge Reactions of Sodium- Nickel Battery at the Anode (Sodium Electrode), Cathode (Nickel Electrode), and the overall charge reaction is given in Eqns. 11, 12, and 13.

$$Na_2O + 2e^- \rightarrow 2Na + O^{2-} \tag{11}$$

$$NiO_2 + 2e^- + 2O^{2-} \rightarrow NiO_2 + 2Na^+ + O^{2-} \tag{12}$$

$$Na_2O + NiO_2 \rightarrow 2Na + 2NiO \tag{13}$$

Discharge Reactions of Sodium-Nickel Battery at the Anode (Sodium Electrode), the Cathode (Nickel Electrode), and the Overall Discharge Reaction are given in Eqns. 14, 15, and 16.

$$Na \rightarrow Na^+ + e^- \tag{14}$$

$$NiO_2 + 2e^- + 2Na^+ \rightarrow NiO_2 \tag{15}$$

$$Na + NiO_2 \rightarrow Na_2O + 2NiO_2 \tag{16}$$

Sodium-Nickel Chloride Battery (Na-NiCl2): With high-temperature operation and good cycle life, these batteries are efficient but require thermal management. They are used in grid storage, heavy-duty transportation, and renewable energy backup (X. Lu et al., 2012).

Figure 10. Sodium-Nickel Battery

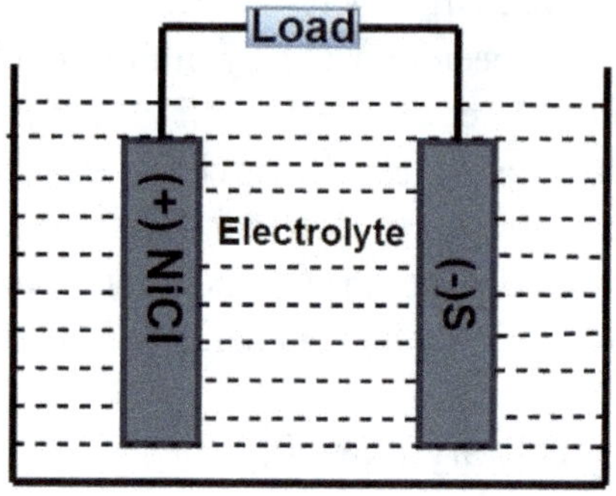

Sodium-Nickel batteries are an emerging technology that could be well-suited for ZEVs, as they offer advantages in terms of energy density and environmental sustainability due to the abundance of sodium. These batteries could also be considered for HEVs and PHEVs, where they can complement traditional power sources with a more sustainable option, enhancing energy efficiency and reducing emissions. While the Sodium-Nickel battery technology may not be the primary choice for FCEVs due to the different energy storage needs, it could still play a role in supporting auxiliary functions or acting as a secondary power source in some configurations, helping to improve the overall energy management system in these vehicles (Accardo et al., 2021).

3.1.4. Lithium-ion battery (Li-ion)

Li-ion batteries were first used in the early 1990s and are now commonly used to power a variety of compatible devices (M. Li et al., 2018). It is well known that Li-ion batteries are the priority option for cell phones, laptops, and PCs. Their specific energy levels were predicted to be about 10 times greater than lead-acid batteries by 2020, surpassing 300 watt-hours per kilogram. Because of their high energy density, specific energy, and ability to provide high power, Li-ion batteries are now widely used in a wide variety of portable devices, such as smartphones, tablets, laptops, digital cameras, toys, and power tools. Li-ion batteries and energy packs for EVs and HEVs are becoming popular in the automotive industry. These batteries have higher energy storage capacity and a wider temperature range for their size. This allows them to last longer between charges and power devices more efficiently in hot or cold conditions than other batteries. They are smaller and lighter as well. However, because they are delicate and require a safety circuit to avoid overcharging, which can be dangerous, they must be handled with care.

Li-ions are drawn out of the cathode, which is typically lithium cobalt oxide, and transferred via the electrolyte to the anode, which is often graphite, where they are intercalated during the charging process, which is shown in Figure 11. The following is a summary of the total response. Anode Reaction and Cathode Reactions are given below as shown in Eqns. 17 and 18:

$$Li^{+} + e^{-} \rightarrow Li_{x}C \tag{17}$$

$$Li_{1-x}CoO_{2} \rightarrow LiCoO_{2} + Li^{+} + e^{-} \tag{18}$$

Lithium ions leave the anode during discharging and return to the cathode, where they re-enter the lithium cobalt oxide. The general reaction between the Anode and Cathode Reactions is shown below in Eqns. 19 and 20:

$$Li_{x}C \rightarrow C + Li^{+} + e^{-} \tag{19}$$

$$LiCoO_{2} + Li^{+} + e^{-} \rightarrow Li_{1-x}CoO_{2} \tag{20}$$

Li-ion batteries are renowned for their high energy density, lightweight design, and low self-discharge, making them the preferred choice for smartphones, laptops, renewable energy storage, and electric vehicles (EVs). Despite their advantages, challenges such as safety risks, gradual degradation, and high costs persist. In ZEVs like battery electric vehicles (BEVs), Li-ion batteries enable fully emission-free operation as the primary power source. HEVs and PHEVs use them to enhance fuel efficiency and extend electric driving ranges, respectively. Even hydrogen FCEVs, which rely on fuel cells, integrate Li-ion batteries to stabilize power fluctuations and boost system efficiency (Kennedy et al., 2000).

Figure 11. Lithium Battery

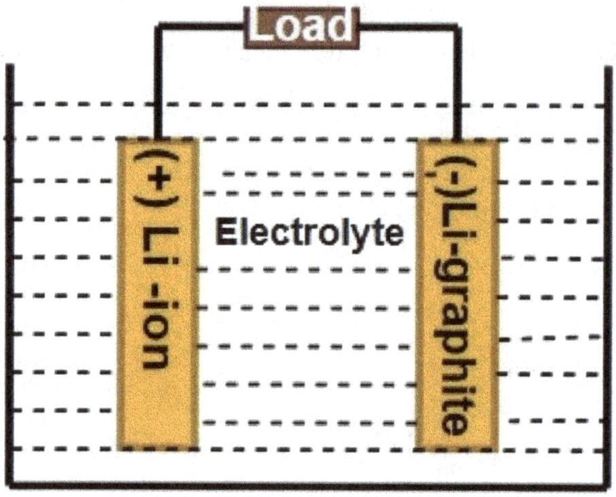

3.2 Key Characteristics of Various Battery Types

The efficiency, self-discharge rate, cost, energy density, and life cycle of Li-ion, NaNiCl, Lead Acid, and NiMH batteries are determined using Indian standard values for these parameters 1.

Efficiency (in percentage):

- Li-ion (Lithium-ion) batteries exhibit the highest efficiency, close to 95%, making them ideal for electric vehicles and portable electronics.
- NaNiCl (Sodium Nickel Chloride) batteries also have high efficiency, around 88%, making them suitable for energy storage applications.
- Lead-acid batteries have moderate efficiency (around 80%), primarily used in automotive starter applications and backup power supplies.
- NiMH (Nickel-Metal Hydride) batteries have the lowest efficiency (65%) due to higher internal resistance and heat generation, often used in hybrid vehicles.
- Comparison of the efficiency of different battery types is graphically shown in Figure 12.

Figure12. Comparison of Battery Technologies Figure13. Self-discharge of Battery Technologies

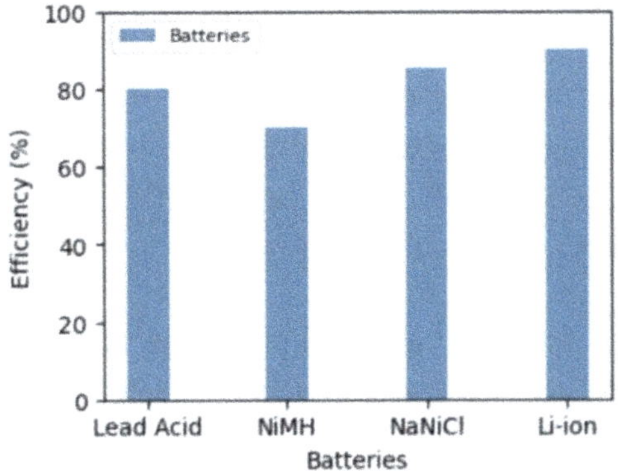

Self -Discharge (percentage per month):

- NiMH batteries have the highest self-discharge rate (25%), requiring frequent recharging.
- Lead-acid batteries are appropriate for standby applications due to their mild self-discharge rate (10%).
- Li-ion batteries are widely used in consumer devices and electric cars because of their much-reduced self-discharge rate (3%).
- NaNiCl batteries are the best for long-term energy storage because they have the lowest self-discharge rate (1%).
- This graph displays the monthly percentage of charge lost when the battery is not in use, as seen in Figure 13.

Cost (/kWh):

- NaNiCl batteries are the most expensive ($35/kWh) due to their advanced chemistry and limited production scale.
- NiMH batteries have a relatively high cost ($18/kWh), making them less favorable for high-energy applications.
- Li-ion batteries have a moderate cost ($15/kWh) while offering superior performance.
- Lead Acid batteries are the cheapest ($8/kWh), making them suitable for budget applications.
- Figure 14 compares the cost per kilowatt-hour of each battery type.

Energy Density (Wh/kg):

- Because Li-ion batteries offer the highest energy density (200 Wh/kg), they are perfect for applications requiring high capacity and low weight.
- NaNiCl batteries are frequently utilized in large-scale storage applications due to their high energy density (110 Wh/kg).
- NiMH batteries are a good option for hybrid cars because of their modest energy density (80 Wh/kg).

Lead-acid batteries are hefty and bulky because they have the lowest energy density (40 Wh/kg). Figure 15 shows the energy stored per unit weight of the battery.

Figure 14. Cost Comparison of Battery Technologies

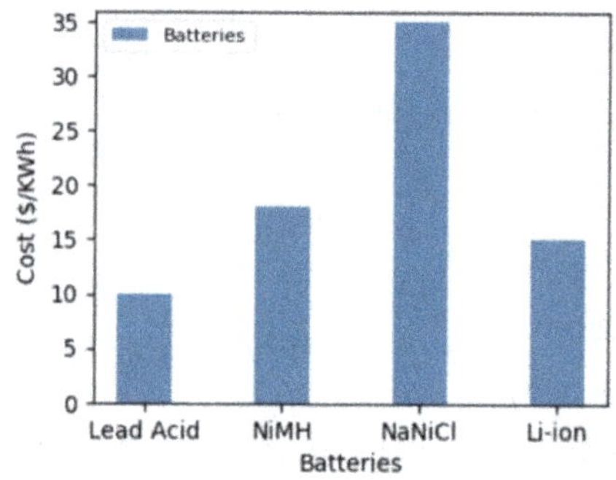

Figure 15. Energy Density of Battery Technologies

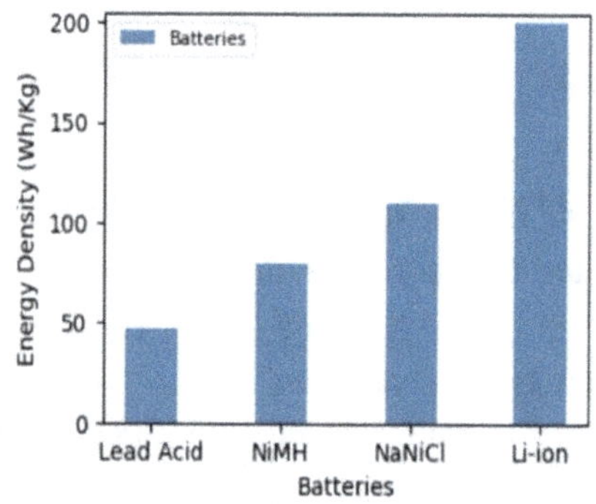

Life Cycle (cycles):

- NaNiCl batteries have the longest lifespan (2500 cycles), making them durable for industrial applications.
- Li-ion batteries also have a long lifespan (2000 cycles), supporting their dominance in electric vehicles and portable devices.
- NiMH batteries have a moderate life cycle (800 cycles), making them suitable for hybrid vehicles.
- Lead-acid batteries have the shortest lifespan (400 cycles), requiring frequent replacement.
- Figure 16 compares the number of charge-discharge cycles before the battery significantly degrades.

Figure 16. Life Cycle of Battery Technologies

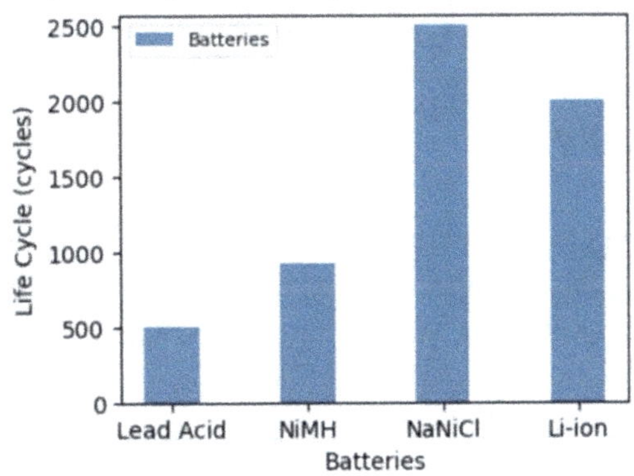

Table 1. Comparison of Different Battery Types

Key Characteristics	Lead Acid	NiMH	NaNiCl	Li-ion
Efficiency (%)	70-85%	60-70%	85-90%	90-98%
Self-Discharge (per month)	5-15%	20-30%	<1%	2-5%
Cost	6,000-10,000 ($70-$120)	15,000-20,000 ($180-$240)	25,000-40,000 ($300-$480)	10,000-20,000 ($120-$240)
Energy Density	30-50 Wh/kg	60-100 Wh/kg	100-120 Wh/kg	150-250 Wh/kg
Life Cycle	300-500 cycles	500-1000 cycles	2000-3000 cycles	1000-3000 cycles

Table 1 compares four battery types, such as Lead Acid, Nickel Metal Hydride (NiMH), Sodium Nickel Chloride (NaNiCl), and Li-ion, based on key parameters. In terms of efficiency, Lead Acid and NiMH batteries have lower efficiency (70-85% and 60-70%, respectively), while NaNiCl and Li-ion offer higher efficiency (85-90% and 90-98%). Self-discharge is highest in NiMH (20-30%), meaning it loses charge quickly when unused, whereas NaNiCl has the lowest (<1%), making it ideal for long-term storage. Cost-wise, Lead Acid is the cheapest (6,000-10,000 per kWh), NaNiCl is the most expensive (25,000-40,000 per kWh), and Li-ion falls in the moderate range (10,000-20,000 per kWh). Energy density, which indicates how much energy a battery can store per kg, is lowest in Lead Acid (30-50 Wh/kg) and highest in Li-ion (150-250 Wh/kg), making Li-ion more compact and lightweight. Lastly, the lifespan, measured in charge-discharge cycles, is shortest for Lead Acid (300-500 cycles) and longest for NaNiCl and Li-ion (up to 3000 cycles), making them more durable options.

4. BATTERY PERFORMANCE METRICS

4.1. Energy Density

Energy density acts as a fundamental measurement parameter for batteries because it describes the energy capacity that batteries store within a standard volume or weight measurement. The measurement unit for energy density includes Wh/L for volumetric analysis and Wh/kg for gravimetric assessment, and it strongly affects battery performance as shown in Figure17. High energy density allows manufacturers to design compact, lightweight batteries that benefit portable electronic devices at present, such as laptops and smartphones. With this technology, devices maintain longer operational periods between recharging sessions while manufacturers produce smaller portable products, including drone systems and electric aircraft systems (Wu et al., 2020).

Energy Density: The energy density of a system is defined as the total energy stored per unit volume, as given in Eq. 21.

$$\text{Energy Density} = \frac{\text{Total Energy Stored }(W_h)}{\text{Volume (L)}} \tag{21}$$

Gravimetric Energy Density: Gravimetric energy density represents the total energy stored per unit mass and is expressed as Eq. 22.

$$\text{Energy Density} = \frac{\text{Total Energy Stored }(W_h)}{\text{Mass (kg)}} \tag{22}$$

4.2. Energy Density Comparison of Size and Weight

The volumetric and specific energy densities are displayed in the battery comparison table below, which also shows lighter and smaller cells.

Figure 17. Battery Types and Their Energy Densities

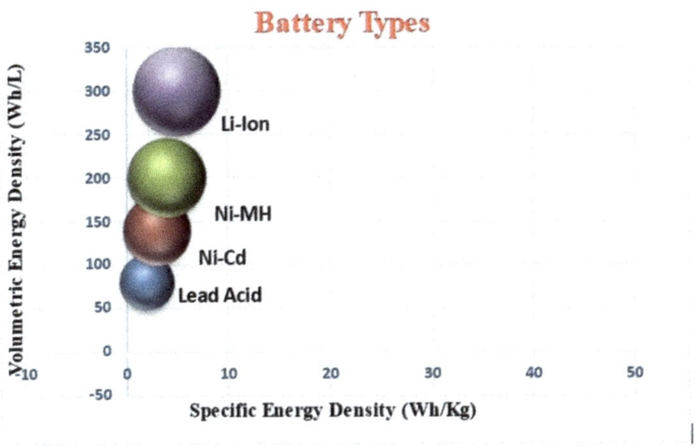

4.3. Power Density

The essential battery characteristic is power density which describes the rate of electrical energy transfer which is measured in W/L (watts per liter) or W/kg (watts per kilogram). Power specifications exceeding 450 W/L and 500 W/Kg are crucial for speed-to-charge and peak power supply in vehicles and tools and integrated energy systems. The infrastructure affects performance characteristics in applications that need different levels of power. Electric vehicles need strong power for fast acceleration but must also maintain steady power for cruising (Lin et al., 2018).

Volumetric Power Density: Volumetric power density defines the power output per unit volume and is given by Eq. 23.

$$\text{Power Density} = \frac{\text{Power Output (W)}}{\text{Volume (L)}} \tag{23}$$

Gravimetric Power Density Gravimetric power density represents the power output per unit mass and is expressed as Eq. 24.

$$\text{Power Density} = \frac{\text{Power Output (W)}}{\text{Mass (kg)}} \quad (24)$$

4.4. Cycle Life

The vital battery measurement cycle life establishes the maximum number of full charge-discharge operations that lead to a substantial capacity reduction in the storage unit. The performance of batteries over the long term, along with their cost efficiency, depends on this parameter. Battery life duration works as a cost-saving factor since it reduces the need for battery replacement during ownership. The length of cycles is determined mainly from battery chemistry characteristics and the way users operate them. The battery cycle life becomes shorter because of deep discharging and exposing it to high temperatures, combined with fast charging rates. The cycle capability rate matters most for electric vehicles and grid storage systems as well as portable electronics, which undergo many charge-discharge cycles (Porzio & Scown, 2022).

4.5. State of Charge

The remaining energy capacity of a battery is presented through SOC, which functions as a percentage. BMS functions because of the SOC implementation. The precise determination of remaining operating time happens when SOC correctly estimates battery capacity. The system enhances battery cycle operations by preventing deterioration due to excessive charging or emptying of the power storage system. Battery capacity gets optimized through efficient energy management strategies because of SOC monitoring. Users receive essential information from the SOC, which enables them to check their batteries' remaining power duration precisely. The actual amount of stored energy that exists within the battery at present stands as the current battery capacity. The maximum energy capacity that a battery can store during normal operation is known as nominal battery capacity (Hickey & Jahns, 2019).

$$\text{SOC (\%)} = \frac{\text{Current Battery Capacity}}{\text{Nominal Battery Capacity}} * 100 \quad (25)$$

4.6. State of Health

A battery's overall health condition, together with its predicted service duration, becomes apparent through its State of Health measure. Professional evaluation of capacity fades along with changes in voltage behavior, together with charge-discharge cycles, determines the State of Health. EIS stands as one of several advanced techniques that deliver a detailed examination of battery systems.

$$SOC\ (\%) = \frac{\text{Current Capacity}}{\text{Initial Capacity}} * 100 \qquad (26)$$

Where the maximum battery capacity currently possible represents the current maximum charge, while initial capacity denotes the maximum potential of a fresh battery. SOH estimation provides essential knowledge for forecasting battery lifespan and developing optimal usage guidelines and proactive maintenance practices to optimize performance during all stages of battery life (Shen et al., 2021).

5. CHARGING OF ELECTRIC VEHICLES

One of the key factors influencing the success of electric vehicles (EVs) is not just their range but also the efficiency and convenience of the charging process. For widespread EV adoption, users must have access to fast and simple charging solutions. This requires a well-developed charging infrastructure, including home charging options and strategically placed fast-charging stations for long-distance travel.

Various charging standards exist worldwide, primarily defined by regional requirements. In North America and parts of the Pacific region, the SAE-J1772 standard is commonly used (Coyner & Bittner, 2023). Europe has embraced the IEC-62196 standard, whereas China adheres to the GB/T 20234 standard. The main difference between these standards is how they are classified: European standards classify charging according to charging power levels, whereas North American and Chinese standards classify charging according to power type (AC or DC). The Bharat EV Charging Standard, which includes AC-001 and DC-001 for low-power charging, is followed for charging electric vehicles in India. International standards for rapid charging are also being established, including CHAdeMO, CCS (Combined Charging System), and GB/T. These rules are monitored by the Ministry of Power and the Bureau of Indian Standards (BIS) to guarantee a uniform and expandable charging infrastructure throughout the nation(Farhadi & Moghaddas Tafreshi, 2022).

The SAE-J1772 standard, introduced in 1996 by SAE International, outlines four charging modes:

- AC Level 1: Delivers up to 1.9 kW of electricity using a 120V AC standard outlet with a maximum current of 16A.
- AC Level 2: Provides up to 19.2 kW of power and operates at 240V AC with a maximum current of 80A.
- DC Level 1: Makes use of an external charger that can supply up to 40 kW of power with a maximum voltage of 500V DC and 80A current.
- DC Level 2 utilizes an external charger as well, but it can give upto 100 kW of power and supports up to 500V DC and 200A current.

These charging standards play a crucial role in ensuring compatibility, efficiency, and safety for EV users across different regions.

6. COMPARATIVE ANALYSIS AND DISCUSSIONS

6.1. Discussions

The bar chart illustrates a substantial increase in values from 2023 to 2030, reflecting strong and continuous growth. Initially, there is a steady rise, followed by rapid expansion from 2027 onward, culminating in the highest value by 2030. This trend suggests that factors such as rising demand, increased investments, and technological advancements may be driving growth. A study by the European Commission (2018) predicts that the number of utilized batteries in electric vehicles within the European Union will surge by over 1000 percent during this period. As shown in Figure 18, this implies that by 2030, more than 1.1 million electric vehicle battery systems (EVBS) from end-of-life electric vehicles will be in circulation. This sharp increase highlights the growing significance of battery utilization and the need for sustainable management strategies (Kvalnes & Pripp, 2023).

The bar chart in Figure19 depicts variations in the number of publications per year, highlighting phases of both growth and stagnation. From 2012 to 2017, publication numbers remained consistently low, with some years showing minimal activity. However, from 2018 onwards, there is a steady rise, culminating in a sharp surge between 2020 and 2022, reaching peak levels. Although 2023 sees a slight decline, the numbers remain significantly higher than in earlier years.

Figure 18. Future Growth Trend Analysis (2023-2030)

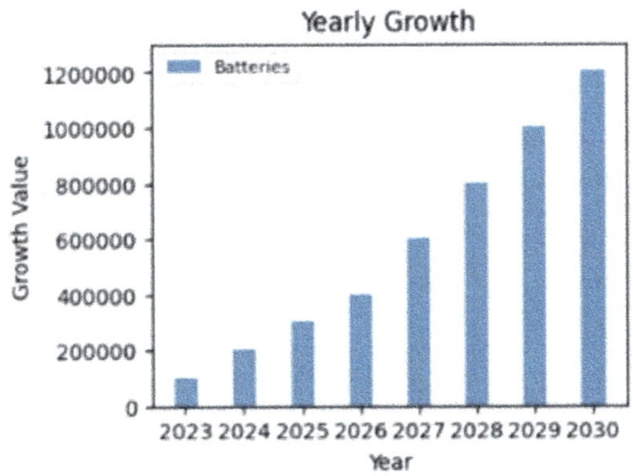

This pattern indicates a gradual buildup followed by rapid expansion, likely driven by growing interest, technological advancements, or market shifts. Despite this upward trend, the apparent stagnation in recent years is unexpected given the importance of the topic.

This suggests a potential lack of clear direction, underscoring the necessity of further research to guide future developments.

Figure 19. Yearly Variation in Quantity (2012-2023)

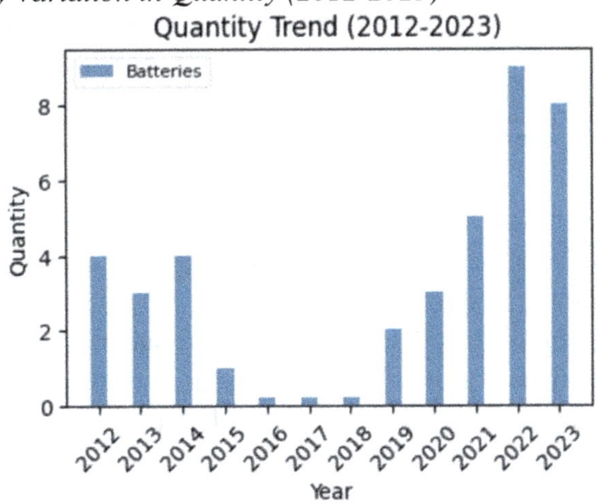

6.2. Comparative Study of Parameters of Batteries

Table 2. Comparison of Different Battery Types

Type of battery	Cost	Nominal Voltage	Self-Discharge-rate	Overcharge Tolerance	Life Cycle	Energy Density (Wh/kg)	Thermal Stability	Efficiency (%)	Power Density
Lead-Acid	Low	2V	Low	High	200–300	30–50	Low	85-90	180
Nickel-Cadmium (Ni-Cd)	Low	1.25V	Medium	Moderate	1500	45–80	Low	95	150
Lithium-Ion (Li-Ion)	High	3.6V	Very Low	Very Low	500–1000	110–160	High	98-100	1800
Sodium-Ion	Moderate	2.3–3.6V	Low	Moderate	1000–3000	90–150	Moderate	90-95	5001-000

Table 2 provides a comparative analysis of various battery types used in electric vehicles, focusing on key parameters such as cost, voltage, self-discharge rate, overcharge tolerance, lifespan, operating temperature range, energy density, toxicity, thermal stability, efficiency, and power density. Lead-acid batteries are cost-effective but have low energy density, limited lifespan, and stability concerns due to their hazardous sulphuric acid content. Nickel-Cadmium (Ni-Cd) batteries offer a longer cycle life and moderate energy density but are environmentally harmful due to toxic cadmium. Lithium-ion batteries, though expensive, provide high energy density, excellent efficiency, and better stability, making them a preferred choice despite potential electrolyte toxicity. Sodium-ion batteries emerge as a promising alternative with moderate cost, decent energy density, eco-friendliness, and a long cycle life, making them suitable for sustainable energy applications. Despite their high cost, lithium batteries are excellent in terms of energy density, cycle life, and low maintenance. Although lead-acid batteries are less expensive, their energy density, cycle life, and environmental effects are all limited. Batteries based on sodium and nickel provide balanced performance; sodium-based batteries stand out for their minimal environmental effect and moderate cycle life (Y. Li & Dai, 2014).

Figure 20. Battery EV vs Plug-in EV Sales in the US

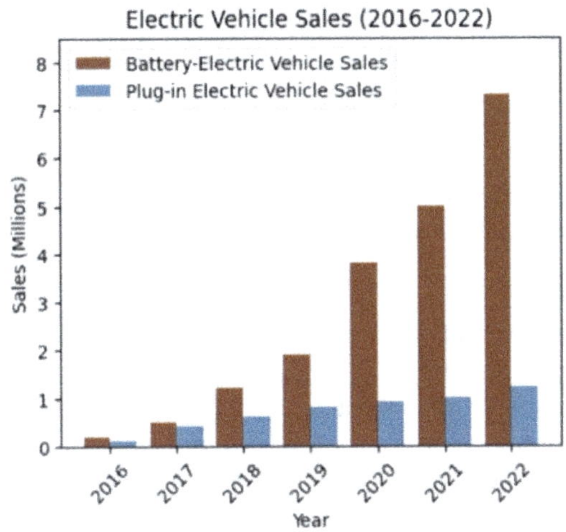

This chart shows the growth of Battery Electric Vehicles (BEVs) and Plug-in Hybrid Electric Vehicles (PHEVs) in the U.S. from 2016 to 2022 as shown in Fig 20. BEVs, which run entirely on electricity, saw a significant rise in sales, especially from 2020 onwards, reaching 7.3 million units in 2022. In contrast, PHEVs, which use both an electric motor and a gasoline engine, experienced slower growth, with sales at 0.92 million in 2022. Overall, the data highlights the increasing preference for fully electric vehicles over hybrids.

Table 3, a detailed comparison of various studies on Battery Management Systems (BMS) for electric vehicles, highlighting their methodologies and monitored parameters. It categorizes each study by author, publication year, and algorithm type, emphasizing common approaches like neural networks, extended Kalman filters, and machine learning techniques. Most studies prioritize monitoring the State of Charge (SOC) and State of Health (SOH), while fewer focus on parameters such as current, voltage, and temperature. Advanced methods, such as deep learning, transfer learning, convolutional layers, and multi-head attention mechanisms, are used to enhance accuracy and robustness in estimation. The table underscores the evolving landscape of BMS technologies and the diverse strategies adopted to improve battery performance and reliability.

continued on following page

Table 3. Continued

Table 3. Comparison Of BMS Parameters(Suganya et al., 2024).

S. No	Author and Year	Algorithm	Current	Voltage	SOC	SOH	Temperature
1	(Chemali et al., 2018)	Deep neural network	NO	NO	YES	YES	NO
2	(Tudoroiu et al., 2018)	Unscented Kalman Filter, Particle filter, and a nonlinear observer	NO	NO	YES	YES	NO
3	(C. Chen et al., 2019)	Feed-forward neural network and extended Kalman filter algorithm	NO	NO	YES	YES	NO
4	(Sheikh et al., 2020)	Machine learning algorithm	NO	NO	YES	YES	NO
5	(Ren et al., 2019)	Extended Kalman filter	NO	NO	YES	YES	NO
6	(Datta et al., 2020)	PID controller and Raspberry Pi 3 Model B+	YES	YES	YES	YES	NO
7	(How et al., 2020)	Deep neural networks	NO	NO	YES	YES	NO
8	(He et al., 2020)	Adaptive extended Kalman filter algorithm and recursive least square method	NO	NO	YES	YES	NO
9	(Jenkins et al., 2019)	Linear ODE Model	NO	NO	YES	YES	NO
10	(Hannan et al., 2020)	Recurrent nonlinear autoregressive with exogenous inputs (RNARX) neural network. algorithm and lightning search algorithm (LSA)	NO	NO	YES	YES	NO
11	(Gade, 2021)	Simulation mode	YES	YES	YES	NO	NO
12	(Alsabari et al., 2021)	Coulomb counting	NO	NO	YES	YES	NO
13	(Bhovi et al., 2021)	Kalman filtering	NO	NO	YES	YES	NO
14	(Roman et al., 2021)	Coulomb counting method, Extended Kalman filter, Unscented Kalman filter	NO	NO	YES	YES	YES
15	(Olarte et al., 2021)	Electrochemical impedance spectroscopy lab-testing equipment	NO	NO	YES	YES	NO
16	(Mohammed & Saif, 2021)	Coulomb counting and open circuit voltage method	NO	NO	YES	YES	NO
17	(Ravi, 2021)	Improved extended Kalman filter	NO	NO	YES	YES	NO

S. No	Author and Year	Algorithm	Current	Voltage	SOC	SOH	Temperature
18	(Sahoo et al., 2022)	Base Artificial Neural Network Using Transfer Learning	NO	NO	YES	YES	NO
19	(Cui et al., 2022)	Neural networks	NO	NO	YES	YES	NO
20	(Tran et al., 2022)	Linear Regressor, KNNR, RFR, DTR	NO	NO	YES	YES	NO
21	(Lipu et al., 2022)	Deep learning	NO	NO	YES	YES	NO
22	(Vishnu & Saleem, 2022)	An integral correction-based SOC estimation	NO	NO	YES	YES	NO
23	(Tang et al., 2023)	Multi-head attention mechanism, Convolutional layers, GRU networks	NO	NO	YES	YES	NO
24	(Saday et al., 2023)	FPGA-based BMS	NO	NO	YES	YES	NO
25	(J. Lu et al., 2023)	Deep learning	NO	NO	YES	YES	NO
26	(Yun et al., 2023)	Extended Kalman filter	NO	NO	YES	YES	NO
27	(Y. Chen et al., 2023)	H-infinity particle filter (HIPF)	NO	NO	YES	YES	NO

7. IMPACT OF EVS ON THE ENVIRONMENT

- Greenhouse Gas Emissions Reduction: You don't get tailpipe emissions from EVs, which are different from traditional gasoline and diesel-powered vehicles. They reduce dangerous emissions like carbon dioxide ($CO2$) and nitrogen oxides NOx that lead to air pollution and global warming.
- Greater Energy Efficiency: EVs make better use of their battery's energy to power the wheels than internal combustion engine vehicles, with less energy being lost on the way. They help achieve energy conservation overall and greater transport efficiency.
- Integration with Renewable Energy: EVs can be powered by renewable energy sources such as solar and wind power, reducing their environmental footprint even further. Clean energy charging reduces fossil fuel dependency and carbon footprint.
- Ecological Issues in Battery Manufacturing: The production of lithium-ion batteries demands raw materials such as lithium, cobalt, and nickel, which

necessitate a large amount of mining. The mining process may destroy ecosystems and demand huge amounts of water.

- Challenges in Battery Disposal and Recycling: Used batteries, if not disposed of properly, can cause environmental pollution. The key to minimizing waste and recovering valuable materials is the development of efficient recycling techniques.
- Long-Term Sustainability of EVs: Even with these challenges, studies show that EVs have a lower environmental footprint throughout their life cycle than traditional vehicles.

Continuous improvements in battery technology, recycling processes, and cleaner electricity generation can further enhance their sustainability.

7.1. Safety Concerns

- Thermal Runaway: The battery cells can experience an escalating reaction of fire and explosion due to additional heat-production sequences activated by an excessive temperature increase.
- Short Circuits: The battery system faces the risk of short internal circuits from multiple existing production problems and physical damage, and deteriorating components.
- Other Safety Concerns: Long periods of charging battery cells will result in overheating injuries that damage the device's battery systems. The battery undergoes permanent damage when over-discharged, and its operational lifespan gets reduced significantly. Therefore, when electrolytes leak out, the resulting corrosion harms both the environment and exposed equipment (Y. Chen et al., 2021).

8. INNOVATIONS AND CHALLENGES IN EV BATTERY TECHNOLOGIES

Batteries play a crucial role in electric vehicles, as they constitute a major portion of the vehicle's cost and directly influence performance. While lithium-ion batteries currently dominate the market, there is an ongoing pursuit of alternative technologies that offer better durability, energy density, and efficiency in charging and discharging processes. Some of the promising battery technologies under investigation include:

- Lithium Iron Phosphate ($LiFePO4$): Known for its energy density of approximately 220 Wh/L and extended lifespan ranging between 2,000 and 10,000 cycles, this battery type demonstrates high thermal stability. Although it is being tested in EV applications, it remains in the early stages of development. Research conducted at MIT has led to weight reduction and the development of prototype cells capable of charging within 10–20 seconds, a significant improvement over standard charging times of six minutes (Millner, 2010).

- Magnesium-Ion (Mg-Ion) Batteries: By utilizing magnesium instead of lithium, these batteries can store a significantly larger charge, potentially achieving an energy density of 6.2 kWh/L, 8.5 times higher than current lithium-ion batteries. Research institutions such as ARPA-E, Toyota, and NASA are actively exploring the viability of this technology (Zhao-Karger & Fichtner, 2017).
- Lithium-Metal Batteries: These batteries replace the conventional graphite anode with a thin lithium-metal layer, effectively doubling energy storage capacity compared to standard lithium-ion counterparts. Startups like Solid Energy Systems have initiated their deployment of drones, with future applications anticipated in electric vehicles. Lithium-metal batteries exhibit exceptional Coulombic efficiency (above 99.1%), maintaining a high efficiency of 98.4% even after 1,000 cycles (Qian et al., 2015).
- Lithium-Air (Li-Air) Batteries: Requiring a continuous supply of oxygen to facilitate reactions with lithium, this battery type was first conceptualized in the 1970s but has gained research traction only recently. With a projected specific energy of 12 kWh/kg—nearly 45 times that of conventional lithium batteries—Li-Air technology holds the potential to rival fuel-based energy systems (Kosivi et al., 2014).
- Aluminum-Air Batteries: This technology generates electricity by reacting aluminum with oxygen, delivering high energy densities of approximately 6.2 kWh/L. These batteries could enable EVs to achieve driving ranges of up to 1,600 km. Additionally, they offer a cost advantage, with current pricing at approximately €300/kWh, and they are fully recyclable (Gelman et al., 2014).
- Sodium-Air (Na2O2) Batteries: Developed by BASF, sodium-air batteries boast an energy density of 4.5 kWh/L, significantly increasing the driving range compared to lithium-based batteries. Since sodium is one of the most abundant elements on Earth, this battery type presents a cost-effective and sustainable alternative for EV applications (Adelhelm et al., 2015).
- Graphene-Based Batteries: Composed of ultra-light pure carbon sheets, graphene batteries exhibit excellent thermal conductivity and minimal heat generation, allowing for rapid charging. A prototype graphene battery developed by Graphenano enabled a GTA Spano vehicle (900 hp) to travel 800 km on a single charge. Under high-power charging conditions, these batteries can be replenished in as little as five minutes. Though still in early development,

some prototypes have already achieved a specific power of 1 kWh/kg, with projections indicating potential increases to 6.4 kWh/kg (Kim et al., 2014).

While lithium-ion batteries continue to be the dominant choice for EVs, ongoing research into alternative battery technologies could lead to substantial breakthroughs in energy storage, efficiency, and sustainability. Advancements in these areas will not only enhance vehicle performance but also contribute to the global transition toward cleaner and more sustainable transportation solutions.

9. CONCLUSION

Electric vehicles (EVs) have gained widespread attention as a sustainable alternative to conventional fuel-powered vehicles. Battery energy systems play a crucial role in advancing green transportation by enabling the development of efficient and sustainable EVs. This review provides a comprehensive analysis of EV classifications, battery technologies, performance metrics, charging infrastructure, environmental impacts, and recent innovations in the field. The discussion on charging infrastructure highlights the significance of fast-charging and wireless technologies in improving user convenience and addressing range anxiety. A comparative analysis of battery parameters, combined with an examination of future growth trends and market dynamics, underscores the rapid global adoption of EVs.

Furthermore, the environmental impact of EVs has been critically assessed, demonstrating their potential to significantly reduce greenhouse gas emissions while also addressing challenges related to battery recycling and resource sustainability. While EV adoption has contributed substantially to lowering carbon emissions, challenges such as battery degradation, charging infrastructure limitations, and environmental concerns regarding battery disposal remain pressing issues. Innovations in solid-state batteries, fast-charging solutions, and advanced battery recycling techniques present promising avenues for improvement, but further advancements are required to enhance scalability and sustainability.

This review highlights the vast potential of battery energy systems while emphasizing the need for continuous innovation and strategic development to overcome existing challenges. Addressing these key areas will optimize the future of electric mobility, paving the way for widespread adoption and long-term environmental benefits. Future research should focus on developing cost-effective, high-performance battery materials, improving recycling processes, and integrating renewable energy sources into EV charging networks. Additionally, strong policy support, industry collaboration, and technological breakthroughs will be essential in accelerating the transition toward a greener and more sustainable transportation ecosystem.

REFERENCES

Accardo, A., Dotelli, G., Musa, M. L., & Spessa, E. (2021). Life cycle assessment of an NMC battery for application to electric light-duty commercial vehicles and comparison with a sodium-nickel-chloride battery. *Applied Sciences (Basel, Switzerland)*, *11*(3), 1160. DOI: 10.3390/app11031160

Adelhelm, P., Hartmann, P., Bender, C. L., Busche, M., Eufinger, C., & Janek, J. (2015). From lithium to sodium: Cell chemistry of room temperature sodium–air and sodium–sulfur batteries. *Beilstein Journal of Nanotechnology*, *6*(1), 1016–1055. DOI: 10.3762/bjnano.6.105 PMID: 25977873

Al-Alawi, B. M., & Bradley, T. H. (2013). Review of hybrid, plug-in hybrid, and electric vehicle market modeling studies. *Renewable & Sustainable Energy Reviews*, *21*, 190–203. DOI: 10.1016/j.rser.2012.12.048

Ali, I., & Naushad, M. (2022). A study to investigate what tempts consumers to adopt electric vehicles. *World Electric Vehicle Journal*, *13*(2), 26. DOI: 10.3390/wevj13020026

Alsabari, A. M., Hassan, M. K., & Zafira, R. (2021). Modeling and validation of lithium-ion battery with initial state of charge estimation. *Indonesian Journal of Electrical Engineering and Computer Science*, *21*(3), 1317–1331. DOI: 10.11591/ijeecs.v21.i3.pp1317-1331

Bhovi, R., Ranjith, K., Sachin, B., & Kariyappa, S. (2021). Modeling and Simulation of Battery Management System (BMS) for Electric Vehicles. *Shanghai Ligong Daxue Xuebao*, *23*(06), 805–815. DOI: 10.51201/JUSST/21/05347

Chan, C. C. (2007). The state of the art of electric, hybrid, and fuel cell vehicles. *Proceedings of the IEEE*, *95*(4), 704–718. DOI: 10.1109/JPROC.2007.892489

Chemali, E., Kollmeyer, P. J., Preindl, M., & Emadi, A. (2018). State-of-charge estimation of Li-ion batteries using deep neural networks: A machine learning approach. *Journal of Power Sources*, *400*, 242–255. DOI: 10.1016/j.jpowsour.2018.06.104

Chen, C., Xiong, R., Yang, R., Shen, W., & Sun, F. (2019). State-of-charge estimation of lithium-ion battery using an improved neural network model and extended Kalman filter. *Journal of Cleaner Production*, *234*, 1153–1164. DOI: 10.1016/j.jclepro.2019.06.273

Chen, Y., Kang, Y., Zhao, Y., Wang, L., Liu, J., Li, Y., Liang, Z., He, X., Li, X., Tavajohi, N., & Li, B. (2021). A review of lithium-ion battery safety concerns: The issues, strategies, and testing standards. *Journal of Energy Chemistry*, *59*, 83–99. DOI: 10.1016/j.jechem.2020.10.017

Chen, Y., Li, R., Sun, Z., Zhao, L., & Guo, X. (2023). SOC estimation of retired lithium-ion batteries for electric vehicle with improved particle filter by H-infinity filter. *Energy Reports*, *9*, 1937–1947. DOI: 10.1016/j.egyr.2023.01.018

Coyner, K., & Bittner, J. (2023). *Automated Vehicles and Infrastructure Enablers: Logistics and Delivery*. SAE International.

Cui, Z., Wang, L., Li, Q., & Wang, K. (2022). A comprehensive review on the state of charge estimation for lithium-ion battery based on neural network. *International Journal of Energy Research*, *46*(5), 5423–5440. DOI: 10.1002/er.7545

Datta, M. L., Shwetha, D. S., Kumar, Y., & Raghavendra, M. (2020). Smart Monitoring and Simulation of E-vehicle using Simulink. [IJERT]. *International Journal of Engineering Research & Technology (Ahmedabad)*, *8*(13), 212–214.

Duhon, A. N., Sevel, K. S., Tarnowsky, S. A., & Savagian, P. J. (2015). Chevrolet volt electric utilization. *SaE International Journal of Alternative Powertrains*, *4*(2), 269–276. DOI: 10.4271/2015-01-1164

Farhadi, P., & Moghaddas Tafreshi, S. M. (2022). Charging stations for electric vehicles; A comprehensive review on planning, operation, configurations, codes and standards, challenges and future research directions. *Smart Science*, *10*(3), 213–245. DOI: 10.1080/23080477.2021.2003947

Gade, A. R. (2021). The new battery management system in electric vehicle. IJERT-International Journal of Engineering Research Technology, 7.

Gelman, D., Shvartsev, B., & Ein-Eli, Y. (2014). Aluminum–air battery based on an ionic liquid electrolyte. *Journal of Materials Chemistry. A, Materials for Energy and Sustainability*, *2*(47), 20237–20242. DOI: 10.1039/C4TA04721D

Gómez, J. A., & Santos, D. M. F. (2023). The status of on-board hydrogen storage in fuel cell electric vehicles. *Designs*, *7*(4), 97. DOI: 10.3390/designs7040097

Hannan, M. A., Lipu, M. S. H., Hussain, A., Ker, P. J., Mahlia, T. M. I., Mansor, M., Ayob, A., Saad, M. H., & Dong, Z. Y. (2020). Toward enhanced state of charge estimation of lithium-ion batteries using optimized machine learning techniques. *Scientific Reports*, *10*(1), 4687. DOI: 10.1038/s41598-020-61464-7 PMID: 32170100

He, Z., Yang, Z., Cui, X., & Li, E. (2020). A method of state-of-charge estimation for EV power lithium-ion battery using a novel adaptive extended Kalman filter. *IEEE Transactions on Vehicular Technology*, *69*(12), 14618–14630. DOI: 10.1109/TVT.2020.3032201

Hickey, R., & Jahns, T. M. (2019). Direct comparison of state-of-charge and state-of-energy metrics for li-ion battery energy storage. 2019 IEEE Energy Conversion Congress and Exposition (ECCE), 2466–2470.

Hillerbrand, R. (2018). Why affordable clean energy is not enough. A capability perspective on the sustainable development goals. *Sustainability (Basel)*, *10*(7), 2485. DOI: 10.3390/su10072485

Hossain, M. S., Kumar, L., Islam, M. M., & Selvaraj, J. (2022). A comprehensive review on the integration of electric vehicles for sustainable development. *Journal of Advanced Transportation*, *2022*(1), 3868388. DOI: 10.1155/2022/3868388

How, D. N. T., Hannan, M. A., Lipu, M. S. H., Sahari, K. S. M., Ker, P. J., & Muttaqi, K. M. (2020). State-of-charge estimation of li-ion battery in electric vehicles: A deep neural network approach. *IEEE Transactions on Industry Applications*, *56*(5), 5565–5574. DOI: 10.1109/TIA.2020.3004294

Jenkins, B., Krupadanam, A., & Annaswamy, A. M. (2019). Fast adaptive observers for battery management systems. *IEEE Transactions on Control Systems Technology*, *28*(3), 776–789. DOI: 10.1109/TCST.2019.2891234

Kennedy, B., Patterson, D., & Camilleri, S. (2000). Use of lithium-ion batteries in electric vehicles. *Journal of Power Sources*, *90*(2), 156–162. DOI: 10.1016/S0378-7753(00)00402-X

Khalid, M., Ahmad, F., Panigrahi, B. K., & Al-Fagih, L. (2022). A comprehensive review on advanced charging topologies and methodologies for electric vehicle battery. *Journal of Energy Storage*, *53*, 105084. DOI: 10.1016/j.est.2022.105084

Kim, H., Park, K.-Y., Hong, J., & Kang, K. (2014). All-graphene-battery: Bridging the gap between supercapacitors and lithium ion batteries. *Scientific Reports*, *4*(1), 5278. DOI: 10.1038/srep05278 PMID: 24923290

Kopera, J. J. C. (2004). Inside the Nickel metal hydride battery. Cobasys, MI, USA Http://Www. Cobasys. Com/Pdf/Tutorial/Inside_nimh_battery_technology. Pdf (Accessed 08/11).

Kosivi, J., Gomez, J., Nelson, R., Kalu, E. E., & Weatherspoon, M. H. (2014). Non-paste based composite cathode electrode for lithium air battery. Electrochemical Society Meeting Abstracts 225, 2, 206. DOI: 10.1149/MA2014-01/2/206

Kvalnes, F., & Pripp, S. W. (2023). *Techno-economic assessment of an automated lithium-ion battery module disassembly process: Investigating the optimal level of automated disassembly*. University of Agder.

Larminie, J., Dicks, A., & McDonald, M. S. (2003). Fuel cell systems explained: Vol. 2. *J*. Wiley Chichester.

Larminie, J., & Lowry, J. (2012). *Electric vehicle technology explained*. John Wiley & Sons. DOI: 10.1002/9781118361146

Li, M., Lu, J., Chen, Z., & Amine, K. (2018). 30 years of lithium-ion batteries. *Advanced Materials*, *30*(33), 1800561. DOI: 10.1002/adma.201800561 PMID: 29904941

Li, Y., & Dai, H. (2014). Recent advances in zinc–air batteries. *Chemical Society Reviews*, *43*(15), 5257–5275. DOI: 10.1039/C4CS00015C PMID: 24926965

Lin, Z., Liu, T., Ai, X., & Liang, C. (2018). Aligning academia and industry for unified battery performance metrics. *Nature Communications*, *9*(1), 5262. DOI: 10.1038/s41467-018-07599-8 PMID: 30531912

Lipu, M. S. H., Ansari, S., Miah, M. S., Meraj, S. T., Hasan, K., Shihavuddin, A. S. M., Hannan, M. A., Muttaqi, K. M., & Hussain, A. (2022). Deep learning enabled state of charge, state of health and remaining useful life estimation for smart battery management system: Methods, implementations, issues and prospects. *Journal of Energy Storage*, *55*, 105752. DOI: 10.1016/j.est.2022.105752

Lu, J., Xiong, R., Tian, J., Wang, C., & Sun, F. (2023). Deep learning to estimate lithium-ion battery state of health without additional degradation experiments. *Nature Communications*, *14*(1), 2760. DOI: 10.1038/s41467-023-38458-w PMID: 37179411

Lu, X., Li, G., Kim, J. Y., Lemmon, J. P., Sprenkle, V. L., & Yang, Z. (2012). The effects of temperature on the electrochemical performance of sodium–nickel chloride batteries. *Journal of Power Sources*, *215*, 288–295. DOI: 10.1016/j.jpowsour.2012.05.020

Manzetti, S., & Mariasiu, F. (2015). Electric vehicle battery technologies: From present state to future systems. *Renewable & Sustainable Energy Reviews*, *51*, 1004–1012. DOI: 10.1016/j.rser.2015.07.010

Millner, A. (2010). Modeling lithium ion battery degradation in electric vehicles. *2010 IEEE Conference on Innovative Technologies for an Efficient and Reliable Electricity Supply*, 349–356. DOI: 10.1109/CITRES.2010.5619782

Mohammed, N., & Saif, A. M. (2021). Programmable logic controller based lithium-ion battery management system for accurate state of charge estimation. *Computers & Electrical Engineering*, *93*, 107306. DOI: 10.1016/j.compeleceng.2021.107306

Muratori, M., Alexander, M., Arent, D., Bazilian, M., Cazzola, P., Dede, E. M., Farrell, J., Gearhart, C., Greene, D., Jenn, A., Keyser, M., Lipman, T., Narumanchi, S., Pesaran, A., Sioshansi, R., Suomalainen, E., Tal, G., Walkowicz, K., & Ward, J. (2021). The rise of electric vehicles—2020 status and future expectations. *Progress in Energy*, *3*(2), 22002. DOI: 10.1088/2516-1083/abe0ad

Muslimin, S., Nawawi, Z., Suprapto, B. Y., & Dewi, T. (2022). Comparison of batteries used in electrical vehicles. 5th FIRST T1 T2 2021 International Conference (FIRST-T1-T2 2021), 421–425.

Oki, S., & Sato, Y. (2024). Nissan leaf and e-power: Evolution of motors and inverters. *IEEJ Journal of Industry Applications*, *13*(1), 8–16. DOI: 10.1541/ieejjia.23006552

Olarte, J., Martínez de Ilarduya, J., Zulueta, E., Ferret, R., Fernández-Gámiz, U., & Lopez-Guede, J. M. (2021). A battery management system with EIS monitoring of life expectancy for lead–acid batteries. *Electronics (Basel)*, *10*(11), 1228. DOI: 10.3390/electronics10111228

Ovshinsky, S. R., Fetcenko, M. A., & Ross, J. (1993). A nickel metal hydride battery for electric vehicles. *Science*, *260*(5105), 176–181. DOI: 10.1126/science.260.5105.176 PMID: 17807176

Petrovic, S., & Petrovic, S. (2021). Basic electrochemistry concepts. Electrochemistry Crash Course for Engineers, 3–10.

Porzio, J., & Scown, C. D. (2022). Life-Cycle Assessment Considerations for Batteries and Battery Materials. *Advanced Energy Materials*, *12*(14), 2200477. DOI: 10.1002/aenm.202200477

Qian, J., Henderson, W. A., Xu, W., Bhattacharya, P., Engelhard, M., Borodin, O., & Zhang, J.-G. (2015). High rate and stable cycling of lithium metal anode. *Nature Communications*, *6*(1), 6362. DOI: 10.1038/ncomms7362 PMID: 25698340

Rand, D. A. J. (1979). Battery systems for electric vehicles—A state-of-the-art review. *Journal of Power Sources*, *4*(2), 101–143. DOI: 10.1016/0378-7753(79)85001-6

Rand, D. A. J. (1980). Lead—acid traction batteries for electric road vehicle propulsion—directions for research and development. *Journal of Power Sources*, *5*(3), 221–234. DOI: 10.1016/0378-7753(80)80001-2

Ravi, R. (2021). Battery management system in electric vehicles. SGS-Engineering & Sciences, 1(01).

Ren, H., Zhao, Y., Chen, S., & Wang, T. (2019). Design and implementation of a battery management system with active charge balance based on the SOC and SOH online estimation. *Energy*, *166*, 908–917. DOI: 10.1016/j.energy.2018.10.133

Renewables, I. E. A. (2021). *Analysis and Forecast to 2026*. International Energy Agency.

Roman, D., Saxena, S., Robu, V., Pecht, M., & Flynn, D. (2021). Machine learning pipeline for battery state-of-health estimation. *Nature Machine Intelligence*, *3*(5), 447–456. DOI: 10.1038/s42256-021-00312-3

Ruetschi, P. (1977). Review on the lead—Acid battery science and technology. *Journal of Power Sources*, *2*(1), 3–120. DOI: 10.1016/0378-7753(77)85003-9

Saday, A., Ozkan, İ. A., & Sarıtas, I. (2023). FPGA-Based battery management system for real-time monitoring and instantaneous SOC prediction. *International Journal of Applied Mathematics Electronics and Computers*, *11*(1), 55–61. DOI: 10.18100/ijamec.1233451

Sahoo, S., Hariharan, K. S., Agarwal, S., Swernath, S. B., Bharti, R., Han, S., & Lee, S. (2022). Transfer learning based generalized framework for state of health estimation of Li-ion cells. *Scientific Reports*, *12*(1), 13173. DOI: 10.1038/s41598-022-16692-4 PMID: 35915128

Sanguesa, J. A., Torres-Sanz, V., Garrido, P., Martinez, F. J., & Marquez-Barja, J. M. (2021). A review on electric vehicles: Technologies and challenges. *Smart Cities*, *4*(1), 372–404. DOI: 10.3390/smartcities4010022

Sheikh, S. S., Anjum, M., Khan, M. A., Hassan, S. A., Khalid, H. A., Gastli, A., & Ben-Brahim, L. (2020). A battery health monitoring method using machine learning: A data-driven approach. *Energies*, *13*(14), 3658. DOI: 10.3390/en13143658

Shen, S., Liu, B., Zhang, K., & Ci, S. (2021). Toward fast and accurate SOH prediction for lithium-ion batteries. *IEEE Transactions on Energy Conversion*, *36*(3), 2036–2046. DOI: 10.1109/TEC.2021.3052504

Sudworth, J. L. (2001). The sodium/nickel chloride (ZEBRA) battery. *Journal of Power Sources*, *100*(1–2), 149–163. DOI: 10.1016/S0378-7753(01)00891-6

Suganya, R., Joseph, L., & Kollem, S. (2024). Optimizing Energy Storage Systems: Interplay of Current, Voltage, and Temperature Parameters in Batteries. 2024 3rd International Conference on Computational Modelling, Simulation and Optimization (ICCMSO), 18–21.

Tang, P., Hua, J., Wang, P., Qu, Z., & Jiang, M. (2023). Prediction of lithium-ion battery SOC based on the fusion of MHA and ConvolGRU. *Scientific Reports*, *13*(1), 16543. DOI: 10.1038/s41598-023-43858-5 PMID: 37783740

Taniguchi, A., Fujioka, N., Ikoma, M., & Ohta, A. (2001). Development of nickel/metal-hydride batteries for EVs and HEVs. *Journal of Power Sources*, *100*(1–2), 117–124. DOI: 10.1016/S0378-7753(01)00889-8

Tran, M., Panchal, S., Chauhan, V., Brahmbhatt, N., Mevawalla, A., Fraser, R., & Fowler, M. (2022). Python-based scikit-learn machine learning models for thermal and electrical performance prediction of high-capacity lithium-ion battery. *International Journal of Energy Research*, *46*(2), 786–794. DOI: 10.1002/er.7202

Tudoroiu, R.-E., Zaheeruddin, M., Radu, S.-M., & Tudoroiu, N. (2018). Estimation techniques for state of charge in battery management systems on board of hybrid electric vehicles implemented in a real-time MATLAB/SIMULINK environment. *New Trends in Electrical Vehicle Powertrains*, *1*, 13.

UDHAYAKUMAR. (n.d.). *K., DEVAKIRUBAKARAN, S., & SEKAR, K. S.* Advances in Batteries, Battery Modeling, Battery Management System, Battery Thermal Management, SOC, SOH, and Charge/Discharge Characteristics in EV Applications.

Vishnu, C., & Saleem, A. (2022). Adaptive integral correction-based state of charge estimation strategy for lithium-ion cells. *IEEE Access : Practical Innovations, Open Solutions*, *10*, 69499–69510. DOI: 10.1109/ACCESS.2022.3187193

Wu, Y., Xie, L., Ming, H., Guo, Y., Hwang, J.-Y., Wang, W., He, X., Wang, L., Alshareef, H. N., Sun, Y.-K., & Ming, J. (2020). An empirical model for the design of batteries with high energy density. *ACS Energy Letters*, *5*(3), 807–816. DOI: 10.1021/acsenergylett.0c00211

Young, K., & Yasuoka, S. (2016). Capacity degradation mechanisms in nickel/metal hydride batteries. *Batteries*, *2*(1), 3. DOI: 10.3390/batteries2010003

Yun, J., Choi, Y., Lee, J., Choi, S., & Shin, C. (2023). State-of-charge estimation method for lithium-ion batteries using extended kalman filter with adaptive battery parameters. *IEEE Access : Practical Innovations, Open Solutions*, *11*, 90901–90915. DOI: 10.1109/ACCESS.2023.3305950

Zhao-Karger, Z., & Fichtner, M. (2017). Magnesium–sulfur battery: Its beginning and recent progress. *MRS Communications*, *7*(4), 770–784. DOI: 10.1557/mrc.2017.101

Chapter 10
Breakthroughs in Electrochemical Innovations Into Scalable Applications

J. Mangaiyarkkarasi
https://orcid.org/0000-0003-1431-9584
Nadar Mahajana Sangam S. Vellaichamy Nadar College, India

J. Shanthalakshmi Revathy
https://orcid.org/0000-0003-1724-7117
Velammal College of Engineering and Technology, India

ABSTRACT

Electrochemical technologies are emerging as pivotal solutions for the global transition to sustainable and scalable energy systems. This chapter bridges recent laboratory breakthroughs with pathways for industrial deployment, emphasizing innovations that extend beyond conventional proof-of-concept studies. Advances in nanostructured electrodes, solid-state electrolytes, and high-performance catalysts have redefined benchmarks of efficiency, stability, and safety. Novel manufacturing strategies including additive manufacturing and roll-to-roll processing are presented as transformative enablers of cost-effective scale-up, while AI-driven optimization accelerates the design-to-application cycle. The chapter also highlights electrochemical routes for green hydrogen production, CO_2 reduction, and nitrogen fixation, which are reshaping sustainable chemical manufacturing. Distinctive case studies in battery storage and hydrogen transport illustrate how scientific progress translates into real-world applications.

DOI: 10.4018/979-8-3373-3156-0.ch010

1. INTRODUCTION

Electrochemical technologies have emerged as key enablers in advancing sustainable energy systems, green chemistry, and modern manufacturing. Spanning applications from energy storage and conversion to chemical synthesis and environmental remediation, these processes provide distinctive advantages, including high efficiency, modularity, and seamless integration with renewable energy sources. However, despite significant advancements in laboratory-scale research, translating electrochemical innovations into scalable, real-world applications remains a complex and multifaceted challenge (Ferretti et al., 2025). The journey from breakthrough discovery to large-scale implementation is often hindered by a variety of technical, economic, and regulatory barriers. Materials that perform exceptionally well under controlled lab conditions frequently encounter stability, cost, or manufacturability issues when scaled up. Similarly, reactor designs optimized for fundamental research may not translate seamlessly into commercial systems due to limitations in mass transfer, energy efficiency, or system integration. As a result, the transition from bench to market demands not only scientific excellence but also interdisciplinary collaboration and strategic innovation across engineering, materials science, and systems design. In recent years, however, a new wave of breakthroughs has started to reshape the landscape of electrochemical applications (Buglioni et al., 2022). Advances in nanomaterials, electrode architecture, and electrolyte formulation have led to remarkable improvements in performance, selectivity, and durability. At the same time, innovations in process intensification, additive manufacturing, and artificial intelligence have opened new pathways for the scale-up of electrochemical systems. These developments are enabling researchers and industries to address long-standing limitations and accelerate the deployment of electrochemical technologies in sectors such as energy storage (e.g., batteries and supercapacitors), hydrogen production, carbon dioxide conversion, wastewater treatment, and more (Zheng et al., 2024).

This chapter explores the latest breakthroughs that are propelling electrochemical technologies toward scalable and impactful applications. It highlights key areas of progress, such as the development of robust and earth-abundant catalysts, the design of scalable electrochemical reactors, and the integration of electrochemical systems with digital tools for process optimization. The chapter also examines case studies where innovations have successfully bridged the gap between research and commercialization, offering insights into the critical success factors for scale-up (Olabi, Abbas, et al., 2023). Moreover, the chapter delves into the broader ecosystem that supports electrochemical innovation—policy frameworks, market drivers, and collaborative initiatives that foster technology transfer and industrial adoption. As global tasks like climate change, resource scarcity, and energy insecurity intensify, the demand for scalable, sustainable technologies continues to grow. Electrochemical processes, with their inherent versatility and environmental compatibility, are uniquely positioned to meet this demand provided that innovation is matched by practical scalability. By synthesizing insights from recent research, engineering advances, and industrial experience, this chapter aims to offer a comprehensive outline of the state-of-the-art in scalable electrochemical innovations. It sets the stage for a deeper exploration of the technical strategies, design principles, and real-world implementations that are shaping the future of electrochemistry in scalable applications (Olabi et al., 2022).

2. LITERATURE SURVEY

The field of electrochemical systems has seen remarkable improvements over the past few decades, transitioning from laboratory-scale innovations to scalable applications across various industries. These breakthroughs have emerged across diverse applications, such as energy storage, environmental improvement, and green chemical production. This literature survey explores the key innovations, challenges, and progress made in scaling electrochemical technologies.

2.1 Electrochemical Energy Storage and Conversion

A major focus of electrochemical innovation has been energy storage, especially in the development of batteries and supercapacitors. Lithium-ion batteries (LIBs), in particular, have achieved broad commercial use in portable electronics, electric vehicles, and grid energy storage systems. Recent breakthroughs focus on improving energy density, lifespan, and safety, often by exploring alternative anode and cathode materials, such as silicon anodes and lithium-sulfur cathodes (Han et al., 2024). Similarly, redox flow batteries (RFBs), particularly vanadium-based systems, have garnered attention for their potential in large-scale energy storage, benefiting from innovations in electrolyte composition and membrane materials to improve efficiency and cycle life (Amir et al., 2023). In addition, fuel cell technologies—especially proton exchange membrane fuel cells (PEMFCs)—have progressed significantly through advancements in catalyst development. The move from platinum to more affordable materials like palladium and nickel-based alloys is a significant step toward scalable hydrogen energy solutions (Huang et al., 2025). Electrochemical CO_2 reduction, which converts CO_2 to valuable chemicals, is another exciting field. Researchers have made progress in catalyst development, increasing the efficiency and selectivity of CO_2 conversion while addressing challenges in scaling up reaction rates for industrial applications.

2.2 Electrochemical Water Treatment and Environmental Applications

Electrochemical methods for wastewater treatment have emerged as viable alternatives to conventional physical and chemical treatment processes. Electrocoagulation and electrodialysis are some of the techniques explored for large-scale water purification. Innovations in electrode materials, such as conducting polymers and composite materials, have improved the performance of these systems by enhancing ion selectivity and reducing energy consumption. These technologies are being scaled in pilot plants and industrial applications to treat heavy metal-contaminated water and remove organic pollutants efficiently.

2.3 Materials and Reactor Design

A significant challenge in scaling electrochemical applications lies in material selection and reactor design. Recent literature highlights the importance of designing scalable, high-performance electrodes with large surface areas. Innovations in 3D porous electrodes, nanostructured catalysts, and flow field designs have been critical in improving the performance of electrochemical systems for both energy and environmental applications (Amir et al., 2023). In particular, the development of progressive manufacturing techniques like 3D printing and roll-to-roll processing has enabled the production of large-scale, cost-effective electrodes and membranes for commercial electrochemical systems.

2.4 Economic and Commercial Viability

While scientific progress in electrochemical systems has been promising, the transition to large-scale commercialization requires overcoming economic barriers. Recent studies emphasize the need for innovations that reduce the cost of key materials, enhance system longevity, and improve manufacturing processes. Techniques such as modular system designs and hybrid energy solutions (combining electrochemical systems with renewable energy sources) are key strategies being explored to reduce the costs and boost the scalability of electrochemical systems (Luo et al., 2024).

3. BACKGROUND STUDY

Electrochemical technologies, which convert chemical energy into electrical energy and vice versa, have long been recognized for their potential in addressing critical global challenges, including energy storage, environmental remediation, and sustainable chemical production. These systems, based on fundamental principles of electrochemistry, are seen as essential for advancing a cleaner, more sustainable future. However, the changeover from small-scale, laboratory-based experiments to large-scale, commercially viable applications remains one of the most significant challenges in the field (Abdelkareem et al., 2022). The concept of electrochemical processes dates back to the early 19th century, with the invention of the battery by Alessandro Volta. Since then, electrochemical systems have evolved significantly, becoming integral to technologies like fuel cells, batteries, electrolysis, and electrochemical synthesis. In particular, advancements in electrochemical energy storage, like lithium-ion batteries (LIBs) and more recently, solid-state batteries, have revolutionized portable electronics and electric vehicles. Similarly, the development of fuel cells for clean power generation and electrolysis for hydrogen production is gaining momentum as a key solution to decarbonize industries and lessen reliance on fossil fuels. Despite their promise, scaling up electrochemical technologies for industrial applications presents numerous challenges. At the laboratory scale, electrochemical systems can achieve high efficiency, selectivity, and performance under ideal conditions. However, when transitioning to larger scales, issues such as cost, material limitations, long-term stability, and mass transport limitations often hinder performance. For instance, the use of noble metals like platinum as catalysts in fuel cells or electrolyzers is costly and unsustainable for large-scale production. Similarly, scaling up electrode and electrolyte systems while maintaining optimal performance requires innovations in material science and reactor design (Abo-Khalil et al., 2022).

Nowadays, several breakthroughs have been made to address these tasks. Advances in materials science, such as the development of earth-abundant catalysts, 3D electrode architectures, and novel membrane technologies, have shown promising results in improving the effectiveness and cost-efficiency of electrochemical systems. The incorporation of artificial intelligence (AI) and machine learning (ML) has extra accelerated optimization processes, enabling real-time observation, predictive maintenance, and dynamic system control, all of which contribute to enhancing the scalability of these technologies. Moreover, new design strategies, such as modular and flexible system architectures, allow for easier scale-up without significant losses in performance. Innovations in manufacturing techniques have also enabled cost-effective production of large-scale electrochemical systems, such as batteries and fuel cells. Despite these advancements, a major hurdle remains: the economic feasibility of large-scale deployment (Olabi et al., 2021). Although the technical performance of many electrochemical technologies has improved, achieving commercial viability on a global scale requires addressing issues such as material costs, energy efficiency, and integration with existing infrastructure. The success of electrochemical systems at scale will ultimately depend on ongoing research, innovative design strategies, and collaborative efforts between academia, industry, and policymakers to foster cost-effective, large-scale solutions for global sustainability challenges. These advancements and challenges set the stage for continued breakthroughs in electrochemical innovations, with the goal of realizing their full potential in large-scale applications across energy, environmental, and industrial sectors.

4. FROM LAB TO MARKET: BRIDGING THE GAP IN ELECTROCHEMICAL TECHNOLOGIES

Electrochemical technologies harness chemical reactions involving electron transfer to generate, store, or utilize electrical energy. They underpin diverse applications, including energy storage, water treatment, corrosion prevention, and chemical manufacturing. At their core are electrochemical cells, composed of two electrodes immersed in an electrolyte, where redox (reduction–oxidation) reactions drive electron transfer. In galvanic (voltaic) cells, this process generates electrical energy, while in electrolytic cells, electrical input drives non-spontaneous reactions.

Batteries represent the most widely recognized electrochemical technology. Lithium-ion batteries, dominant in electric vehicles (EVs), illustrate the challenge of scaling: while laboratory cells achieve high energy density, commercial-scale modules face issues of thermal management, safety, and limited cycle life, as seen in the Tesla Model S battery recalls due to overheating risks. Lead–acid batteries, despite their robustness, struggle with recyclability and energy density, while emerging solid-state batteries face difficulties in fabricating defect-free solid electrolytes at scale.

Supercapacitors offer rapid charging and long lifetimes but have low energy densities compared with batteries. Scaling up their use in grid storage requires breakthroughs in electrode materials with higher surface area and improved energy-per-cost ratios, as demonstrated in trials by Skeleton Technologies, which reported challenges balancing cost with industrial-grade performance. Fuel cells highlight additional barriers. Proton exchange membrane (PEM) fuel cells have been successfully deployed in vehicles such as the Toyota Mirai, yet rely on platinum-group catalysts, raising cost and supply chain concerns. Solid oxide fuel cells (SOFCs), while efficient, face durability issues at high operating temperatures, limiting widespread commercialization.

In environmental applications, electrochemical water treatment systems, such as electrocoagulation units, show laboratory-scale success in removing dyes and heavy metals, but their energy consumption remains prohibitively high for municipal deployment. Similarly, electrochemical CO_2 reduction, though promising in producing value-added chemicals, suffers from poor selectivity and high overpotentials, as reported in recent pilot studies by Siemens Energy. Electrolysis remains central to hydrogen production. While proton exchange membrane electrolyzers powered by wind or solar energy are advancing, their large-scale deployment is limited by the high cost of iridium-based catalysts and the durability of membranes under fluctuating renewable inputs.

These examples highlight that bridging the gap from laboratory research to market requires addressing concrete challenges: cost-effective material synthesis, long-term stability, safe operation at scale, and integration with existing infrastructure. Overcoming these obstacles is essential for electrochemical technologies to realize their transformative potential in energy and environmental systems.

4.1 The Innovation Bottleneck

Electrochemical research has yielded an impressive array of discoveries over the past two decades, from high-performance catalysts and novel electrode materials to efficient reactor designs and unique electrolytes. Figure 1 explains Electrochemical energy storage conversion.

Figure 1. Electrochemical energy storage conversion.

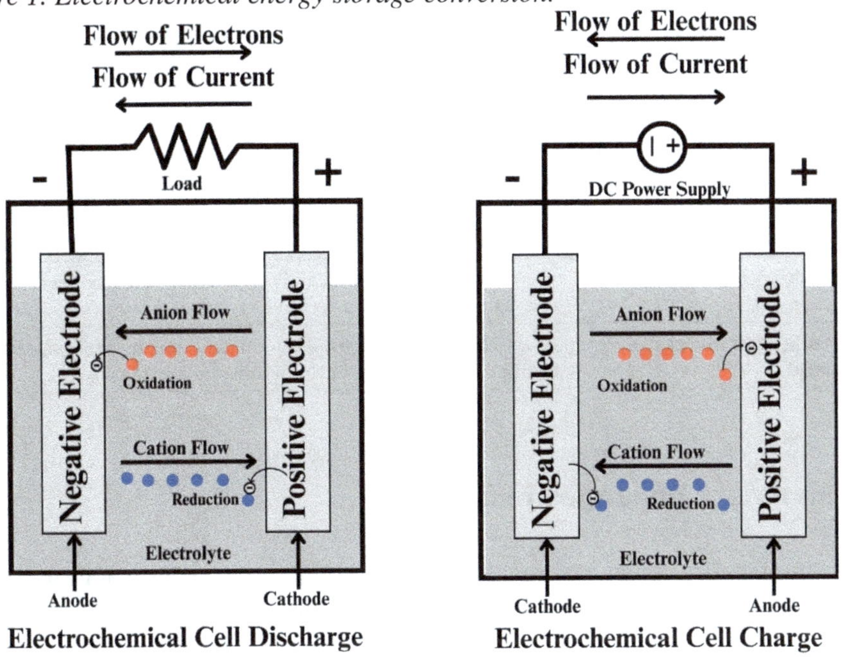

These innovations hold immense promise for clean energy storage, green hydrogen production, and carbon-neutral chemical manufacturing. However, despite this progress, a persistent bottleneck exists between laboratory success and commercial adoption. Many technologies that demonstrate high efficiency or selectivity at the bench scale fail to meet the requirements for real-world deployment, largely due to issues with scalability, cost, and long-term durability (Xu et al., 2025).

Despite rapid advances, electrochemical technologies face distinct bottlenecks that hinder large-scale deployment. These challenges vary by technology but generally fall into four categories: materials, cost, durability/safety, and scalability/integration. A comparative synthesis is provided in Table 1 to highlight the most critical barriers for each technology.

Table 1. Key Innovation Bottlenecks in Electrochemical Technologies

Technology	Main Bottlenecks	Details / Examples
Lithium-ion batteries	Safety, material supply, recycling	Thermal runaway risks in large packs; dependence on cobalt/nickel; limited recycling infrastructure.
Solid-state batteries	Manufacturing defects, cost	Difficulty producing uniform solid electrolytes at scale; high interfacial resistance increases costs.
Supercapacitors	Low energy density, cost-performance balance	High power but insufficient energy for grid applications; electrode materials remain expensive.
PEM Fuel cells	Catalyst cost, durability	Reliance on platinum catalysts; membrane degradation under cycling limits lifetime.
SOFCs	High operating temperature, stability	Materials degrade at >700 °C; scaling down for portable use is challenging.
Electrolyzers (PEM/ Alkaline)	Catalyst cost, intermittent stability	Precious-metal catalysts (iridium, ruthenium) drive up cost; renewable intermittency reduces efficiency.
Electrochemical CO_2 reduction	Selectivity, overpotential, scale-up	Poor selectivity for desired products; high energy input; pilot plants not yet cost-competitive.
Electrochemical water treatment	Energy demand, electrode fouling	High operational costs; electrodes degrade in real wastewater conditions.

4.2 Material Challenges in Scale-Up

At the heart of many electrochemical systems are materials—catalysts, membranes, electrolytes, and electrode substrates—that determine performance and stability. In laboratory studies, researchers often rely on rare, high-purity, or costly materials to demonstrate proof-of-concept results. Scaling these materials for industrial use, however, demands a shift toward abundant, cost-effective alternatives that maintain performance under operational stress. Additionally, ensuring consistent material quality at large production volumes is crucial for reproducibility and reliability, often requiring collaboration with materials manufacturers and suppliers (Ryabicheva et al., 2025).

4.3 Engineering and System Integration

Translating electrochemical processes from beaker-scale experiments to industrial-scale systems introduces numerous engineering complexities. Heat management, mass transport, current distribution, and fluid dynamics all behave differently at larger scales. Designing reactors that can operate continuously, safely, and efficiently under industrial conditions is a major engineering hurdle. Furthermore, electrochemical systems must be integrated into broader process flows, often requiring compatibility with existing infrastructure and hybridization with other energy or chemical technologies. These factors make it essential to engage chemical engineers and system designers early in the development cycle.

4.4 Techno-Economic and Lifecycle Considerations

Innovative electrochemical technologies must not only work effectively they must be economically viable and environmentally sustainable. Techno-economic analysis (TEA) plays a critical role in identifying cost drivers and informing design decisions that affect scalability. Lifecycle assessment (LCA), meanwhile, helps quantify eco-friendly impacts from raw material sourcing to end-of-life disposal (Agrawal et al., 2024). By incorporating TEA and LCA early in the R&D process, developers can prioritize the most promising pathways and avoid investing in technically sound but commercially unfeasible solutions.

4.5 Bridging Through Collaboration and Commercialization Pathways

Bridging the lab-to-market gap requires more than technical ingenuity—it demands robust collaboration between academia, industry, investors, and policy-makers. Startup incubators, government-funded pilot programs, and public-private partnerships provide critical support for maturing technologies. Intellectual property management, regulatory navigation, and access to capital are also vital components of the commercialization journey. By fostering an innovation ecosystem that supports both deep science and practical deployment, we can accelerate the transition of electrochemical technologies from academic novelty to industrial norm.

5. SCALING UP: TRANSFORMING ELECTROCHEMICAL CONCEPTS INTO INDUSTRIAL SOLUTIONS

Turning an electrochemical concept into a viable industrial solution is one of the most critical and challenging phases in the innovation pipeline. While laboratory-scale experiments often prove the fundamental feasibility of new reactions, catalysts, or cell designs, scaling these technologies to an industrially relevant level involves navigating a complex web of technical, economic, and practical hurdles. The process demands a multidisciplinary approach that brings together electrochemistry, chemical engineering, materials science, manufacturing, and systems integration. At the core of scaling up is the transition from milligram- or milliliter-scale prototypes to kilogram- or kiloliter-scale systems. This requires a thorough understanding of how electrochemical behavior changes with scale. For example, current density, heat management, fluid dynamics, and gas evolution all of which are relatively straightforward to control in a lab can become limiting factors in large-scale operations. If not addressed early, these factors can significantly impact efficiency, selectivity, and safety at scale.

One of the most pressing challenges in scale-up is maintaining performance and stability under continuous, real-world operating conditions. Laboratory setups are often optimized for short-term tests, frequently using purified reactants, controlled temperatures, and intermittent operation(Shi et al., 2024). Industrial systems, on the other hand, must operate for thousands of hours, tolerate impurities, and deliver consistent output under fluctuating demand. These differences mean that performance metrics obtained in the lab rarely translate directly to industrial relevance. Another key consideration in scale-up is the design of reactors and systems. Electrochemical cells must be re-engineered to ensure even current distribution, optimal flow paths, and effective removal of heat and by-products. The geometry of electrodes, the choice of membrane materials, and the configuration of flow channels all have a profound impact on large-scale system performance. Innovative designs such as modular stacks, flow batteries, and gas-diffusion electrodes are being developed to address these needs, offering flexible, scalable architectures that can be tailored to different applications. Cost and manufacturability are also central to the scale-up process. Materials used in early-stage research such as platinum-based catalysts or high-purity solvents are often too expensive for commercial use. Therefore, identifying alternatives that offer similar performance at a fraction of the cost is crucial. Moreover, the ability to mass-produce key components like electrodes, membranes, and catalysts with tight quality control and reproducibility is essential for industrial adoption. Techniques such as roll-to-roll processing, 3D printing, and automated assembly are increasingly being employed to make scale-up more efficient and reliable(Qadeer et al., 2024).

Finally, the scale-up journey must be guided by real-world application needs. This involves engaging with stakeholders across the value chain from raw material suppliers to end users to understand system requirements, safety standards, and market dynamics. Field trials and pilot plants are often necessary to validate system performance under realistic conditions and build confidence among investors and partners. Transforming electrochemical concepts into industrial solutions is a multifaceted process that extends far beyond lab success. It requires engineering ingenuity, material innovation, economic foresight, and a systems-level mindset to create technologies that are not only ground breaking, but also practical, scalable, and impactful.

6. KEY INNOVATIONS DRIVING COMMERCIAL ELECTROCHEMICAL SYSTEMS

6.1 Advanced Catalysts for Selectivity and Stability

Catalysts lie at the heart of most electrochemical reactions, from water splitting to carbon dioxide reduction. Traditional catalysts such as platinum and iridium offer excellent performance but are prohibitively expensive and scarce. Recent innovations have focused on developing earth-abundant alternatives including transition metal oxides, nitrides, and single-atom catalysts that can match or even exceed the activity and selectivity of noble metals (Varhade et al., 2025). Moreover, catalyst engineering now includes tailoring surface structures at the atomic level, enabling enhanced reaction kinetics and long-term operational stability under industrial conditions. These advances are making it feasible to deploy high-performance electrochemical systems at scale and at lower cost.

6.2 Scalable Electrode Architectures

The structure and design of electrodes play a vital role in dictating reaction efficiency and mass transport characteristics. Modern commercial systems have moved beyond flat electrodes to incorporate porous, 3D-structured materials that maximize surface area while enabling efficient flow of reactants and products. Innovations such as gas diffusion electrodes, foam-based electrodes, and nanostructured coatings have significantly improved current densities and minimized energy losses (Yammine et al., 2024). These scalable electrode architectures also facilitate integration into modular stack systems, supporting industrial applications from electrolysis to electrochemical synthesis.

6.3 Solid-State and Flow-Based Electrolyte Systems

Electrolytes serve as the medium for ion transport and are essential for maintaining reaction stability. Recent progress in solid-state and flow-based electrolyte systems has enhanced the safety, flexibility, and scalability of electrochemical devices. Solid electrolytes eliminate the risks of leakage and improve system durability, while flow electrolytes offer tunable performance and easier heat management in large systems.

6.3.1 Lithium-Ion Battery Systems

Lithium-ion (Li-ion) battery systems are currently the dominant rechargeable energy storage technology in consumer electronics and electric vehicles, owing to their high gravimetric and volumetric energy densities, low mass, and extended cycle life. These batteries operate via reversible intercalation and deintercalation of lithium ions between a graphite anode and a metal oxide cathode commonly lithium cobalt oxide ($LiCoO_2$), lithium iron phosphate ($LiFePO_4$), or other layered, spinel, or olivine-type materials—facilitated by a liquid organic electrolyte containing a lithium salt (e.g., $LiPF_6$) in a carbonate solvent. During charge, lithium ions migrate from the cathode to the anode, embedding into the graphite structure; during discharge, they return to the cathode, generating electrical energy. One of the principal advantages of Li-ion technology is its high specific energy, making it especially suitable for applications requiring lightweight, compact power sources, such as smartphones, laptops, and electric drivetrains. Additional benefits include low self-discharge rates and rapid charge capability. Nevertheless, Li-ion cells are susceptible to thermal runaway, fire, or explosion when subjected to mechanical damage, overcharging, or internal short circuits. As a result, robust battery management systems (BMS) are essential for monitoring cell voltage, temperature, current, and state-of-charge to ensure safe operation. Challenges facing Li-ion technology include resource constraints for critical materials (e.g., cobalt, lithium, and nickel), as well as environmental and economic concerns related to extraction, processing, and end-of-life recycling. Research continues to focus on advancing cell chemistry, electrode architecture, and electrolyte formulations to improve performance metrics such as energy density, power density, safety, and cycle durability. Figure 2 presents schematic representations of both lithium-ion and solid-state battery architectures.

Figure 2. Lithium-ion battery and Solid state battery.

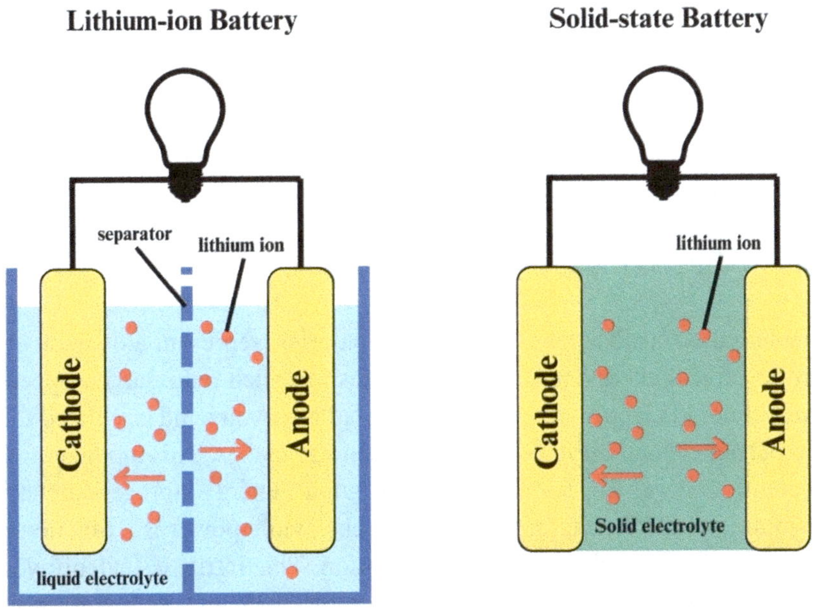

6.3.2 Solid-State Electrolyte Systems

Solid-state electrolyte (SSE) systems represent a major innovation in electrochemical energy storage, replacing conventional liquid electrolytes with solid ionic conductors to facilitate lithium-ion transport between electrodes. These systems are central to the development of solid-state batteries (SSBs), which are being engineered as inherently safer, higher energy-density alternatives to conventional lithium-ion cells. SSE materials include a range of ceramics (e.g., garnet-type $Li_7La_3Zr_2O_{12}$, NASICON-type conductors), glassy electrolytes, and polymer-based systems (e.g., PEO-based composites), each offering distinct advantages in terms of ionic conductivity, mechanical strength, and electrochemical stability. One of the most promising aspects of SSBs is the compatibility with lithium metal anodes, which significantly boosts specific energy and volumetric efficiency. A primary benefit of solid-state configurations is improved safety, as solid electrolytes are non-flammable and eliminate the risk of electrolyte leakage, thereby mitigating hazards such as thermal runaway. Additional benefits include extended cycle life and improved thermal stability across a wider operating temperature range. Despite these advantages, several critical challenges remain, including limited ionic conductivity at ambient temperatures (especially in polymer-based systems), poor interfacial contact and stability between solid electrolytes and electrode materials, and complex fabrication techniques that hinder large-scale manufacturability. Current research efforts are focused on optimizing electrolyte composition, engineering stable interfaces, and developing scalable processing methods to enable the commercial viability of SSBs. As these obstacles are addressed, solid-state electrolyte systems are poised to play a pivotal role in next-generation energy storage, particularly for electric mobility and high-performance portable electronics.

6.3.3 Flow-Based Electrolyte Systems

Flow-based electrolyte systems, or flow batteries, represent a distinctive class of electrochemical energy storage technologies in which redox-active species are dissolved in liquid electrolytes stored in external reservoirs and circulated through an electrochemical cell during charge and discharge cycles. This modular architecture decouples energy capacity from power output: energy storage is determined by the volume of electrolyte in the storage tanks, while power is a function of the electrochemical cell stack size and configuration. This intrinsic scalability makes flow batteries highly attractive for stationary, large-scale applications such as grid energy storage, load leveling, and renewable energy integration. A key advantage of flow batteries is their exceptional cycle stability, often exceeding tens of thousands of cycles with minimal performance degradation due to the physical separation of electroactive materials from the electrodes. They also offer design flexibility, ease of maintenance, and enhanced safety profiles due to their aqueous, non-flammable electrolytes. However, flow batteries typically exhibit lower energy densities compared to lithium-ion and solid-state systems, rendering them unsuitable for compact or mobile applications. The most mature and commercially developed variant is the vanadium redox flow battery (VRFB), which employs vanadium ions in multiple oxidation states (V^{2+}/V^{3+} and VO^{2+}/VO_2^{+}) in both half-cells, thereby minimizing cross-contamination and enhancing coulombic efficiency. Alternative chemistries under investigation or development include zinc-bromine, iron-chromium, and organic-based redox couples. As the demand for long-duration, grid-scale energy storage grows in support of intermittent renewable energy sources, flow battery systems are emerging as robust and sustainable candidates for future deployment. *Figure 3 shows Flow battery system*

Figure 3. Flow battery system

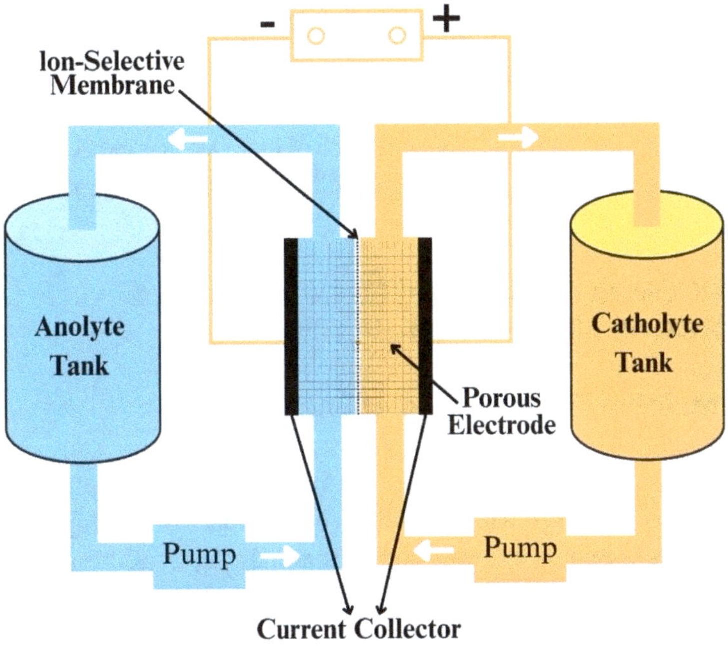

These innovations are especially relevant in applications like redox flow batteries, where long-duration energy storage and operational flexibility are critical.

6.4 Smart System Monitoring and AI Integration

As electrochemical systems become more complex, the need for intelligent control and diagnostics has grown. The integration of sensors, machine learning algorithms, and digital twin technologies has enabled real-time monitoring and optimization of electrochemical processes. These tools can predict system failures, adapt to changing inputs, and maximize performance through automated feedback loops. By reducing downtime and improving process efficiency, smart systems contribute significantly to the commercial viability of electrochemical technologies.

6.5 Modular and Flexible Reactor Designs

Traditional chemical plants rely on large, centralized infrastructures. In contrast, many electrochemical systems now leverage modular reactor designs that can be scaled incrementally based on demand. This approach reduces capital expenditure, accelerates deployment, and allows for distributed manufacturing or energy generation. Modular electrochemical reactors are already being used in applications like on-site hydrogen generation and small-scale CO_2 conversion, offering a more adaptable model for industrial adoption (Brinkert & Mandin, 2022). These innovations collectively form the foundation of next-generation electrochemical systems that are not only technically impressive but also commercially viable. As research and industry continue to collaborate, the pace of innovation is expected to accelerate further opening the door to widespread deployment across multiple sectors.

7. OVERCOMING CHALLENGES IN ELECTROCHEMICAL SCALABILITY

7.1 Maintaining Performance at Scale

Laboratory-scale electrochemical cells often achieve high efficiencies under ideal conditions. However, when scaled up, performance often drops due to uneven current distribution, poor mass transport, or thermal effects. Addressing this requires optimized cell geometries, improved flow designs, and advanced modeling to ensure uniform conditions across large electrodes.

7.2 Electrode Material Cost and Availability

High-performance electrodes often rely on scarce or expensive materials (e.g., platinum, iridium). To scale affordably, research is shifting toward earth-abundant alternatives such as nickel, iron, cobalt, and carbon-based materials, along with strategies like doping or alloying to enhance activity and durability(Sohrabi et al., 2022).

7.3 Durability and Long-Term Stability

Electrochemical devices must operate continuously for thousands of hours in harsh chemical environments. Stability testing under real-world conditions, development of corrosion-resistant materials, and protective coatings are key to enhancing longevity and reducing system maintenance.

7.4 Mass and Heat Transfer Limitations

At scale, inefficient heat dissipation or poor reactant delivery can limit performance and cause degradation. Engineers are using computational fluid dynamics (CFD) to design optimized flow fields and cooling systems that enhance thermal and mass transfer efficiency.

7.5 Electrolyte Management

Electrolytes must be stable, conductive, and compatible with all components. At scale, maintaining pH balance, avoiding cross-contamination, and managing ionic species buildup become more complex. Innovations include using solid or gel electrolytes, flow systems, and ion-exchange membranes to manage this challenge.

7.6 System Integration and Compatibility

Electrochemical systems must be integrated into larger industrial processes or grids. This includes synchronization with renewable power sources, compatibility with chemical feedstocks, and safety management. Hybrid system designs and flexible operating modes are being developed to improve integration(Boumya et al., 2021).

7.7 Manufacturing and Assembly

Producing electrochemical systems at scale requires consistent and cost-effective manufacturing methods. Roll-to-roll processing, additive manufacturing (3D printing), and automated assembly lines are helping scale up production while maintaining quality and reducing costs.

7.8 Reactor and Stack Design

Single-cell designs are not enough; full-scale applications need modular stacks that can be scaled linearly. Design innovations include better sealing methods, gas-liquid separation techniques, and pressure/temperature control systems to improve efficiency and scalability.

7.9 Monitoring and Process Control

Large-scale systems require robust real-time monitoring and fault detection. Integration of sensors, remote diagnostics, and AI-based control systems helps in predictive maintenance and dynamic optimization, increasing system uptime and performance.

7.10 Economic and Regulatory Barriers

Even technically sound systems may struggle to scale without a supportive business case. Lifecycle cost analysis, carbon pricing, and regulatory incentives (like green energy credits) play a crucial role. Policy support and clear commercialization pathways are essential for de-risking large-scale investment.

Each of these points represents a crucial step in the journey toward scalable, commercially viable electrochemical technologies(Patra et al., 2023). Tackling them collectively through interdisciplinary research, strategic investment, and industry partnerships is key to unlocking the full potential of electrochemical systems in the global transition to sustainable technologies.

8. MATERIALS AND DESIGN STRATEGIES FOR SCALABLE ELECTROCHEMICAL DEVICES

8.1 Modular Cell and Stack Architecture

Designing electrochemical cells in a modular format allows for easy scale-up by replicating individual units (cells) into stacks or arrays. Modular designs support gradual capacity expansion without redesigning the entire system. They enable flexibility in manufacturing, transport, and maintenance. Common in applications like fuel cells, electrolyzers, and redox flow batteries, modular stacks help maintain uniform current distribution and heat management at scale(Rahman et al., 2025). Key benefits are, Scalability without compromising performance, simplified system integration, easier fault isolation and maintenance

8.2 High Surface Area and 3D Electrode Architectures

Incorporating 3D or porous electrode structures increases the effective surface area, enhancing reaction rates and improving mass transport. High surface area allows for higher current densities and improved efficiency in reactions like water splitting, CO_2 reduction, or energy storage. Porous and foam-based electrodes improve access of reactants and removal of products, especially in gas-evolving or multi-phase reactions. Key benefits are, higher throughput per unit volume, reduced ohmic losses and better catalyst utilization, Improved heat and mass transfer at large scale.

8.3 Optimized Flow Field and Mass Transport Design

Engineering the flow paths of reactants and electrolytes within the electrochemical cell to ensure uniform distribution, minimize concentration gradients, and manage by-product removal. Poor flow design can lead to hotspots, uneven reactions, and catalyst degradation. Optimized flow fields (e.g., serpentine, interdigitated, or parallel) ensure consistent performance across the entire electrode area. Key benefits are, Enhanced efficiency and system lifetime, better control of local chemical environments (e.g., pH, temperature), scalability without performance loss due to non-uniform operation.

9. ELECTROCHEMICAL ENGINEERING: THE PATHWAY TO REAL-WORLD IMPACT

Electrochemical engineering integrates fundamental science with system-level design to translate laboratory advances into scalable, reliable, and efficient technologies. A multiscale perspective is essential, beginning with materials development and extending to cell and stack design, manufacturing strategies, and digital control systems. Each layer introduces unique opportunities and bottlenecks that must be addressed in a coordinated manner. (Xiao et al., 2024). Table 2 explains Comparative synthesis of electrochemical engineering elements

Table 2 Comparative synthesis of electrochemical engineering elements

Aspect	Role / Focus	Current Advances	Challenges / Bottlenecks
Materials (electrodes, electrolytes, membranes, catalysts)	Define efficiency, stability, conductivity, and reaction selectivity	Nanostructured electrodes, solid-state electrolytes, non-precious metal catalysts	Limited lifetime under cycling, high cost of rare metals, scalability of advanced materials
Cell Design	Optimize reaction interfaces, ion/electron transport, thermal management	3D electrode architectures, bipolar plates, advanced separators	Balancing compactness with performance; safety issues (e.g., dendrite growth in Li batteries)
Stacks / Modules	Scale up single cells to useful power/energy levels	Modular fuel cell stacks, high-capacity battery packs, hybrid systems	Uniform gas/liquid distribution, heat management, cumulative degradation across cells
Manufacturing Processes	Enable reproducible, cost-effective production	Additive manufacturing, roll-to-roll electrode coating, automated assembly	High capital cost, quality control at scale, defect-free production of solid electrolytes/membranes
Digital Control & System Integration	Enhance reliability, efficiency, and safety at scale	AI-driven optimization, digital twins, predictive maintenance, smart battery management systems (BMS)	Integration with grids/renewables, cyber-physical security, lack of standardized protocols

A clearer appreciation of these interconnected layers highlights that breakthroughs at the materials level are insufficient without parallel progress in manufacturing and digital control. By aligning innovations across all scales, electrochemical engineering can accelerate the transition from laboratory prototypes to robust, industrial-scale systems capable of addressing global energy and sustainability demands (Silveri et al., 2025).

10. CASE STUDIES IN SCALABLE **ELECTROCHEMICAL** *APPLICATIONS ACROSS INDUSTRIES*

Electrochemical technologies are increasingly bridging the gap between laboratory-scale experiments and industrial applications. The scalable integration of electrochemical systems into various industries offers real-world solutions for energy, chemicals, and environmental management. Below are seven case studies highlighting how electrochemical innovations are being successfully scaled to meet industry needs.

10.1 Lithium-Ion Batteries for Electric Vehicles (EVs)

One of the most significant success stories in electrochemical scaling is the development of lithium-ion batteries for electric vehicles (EVs).

Figure 4. Lithium ion battery models and applications.

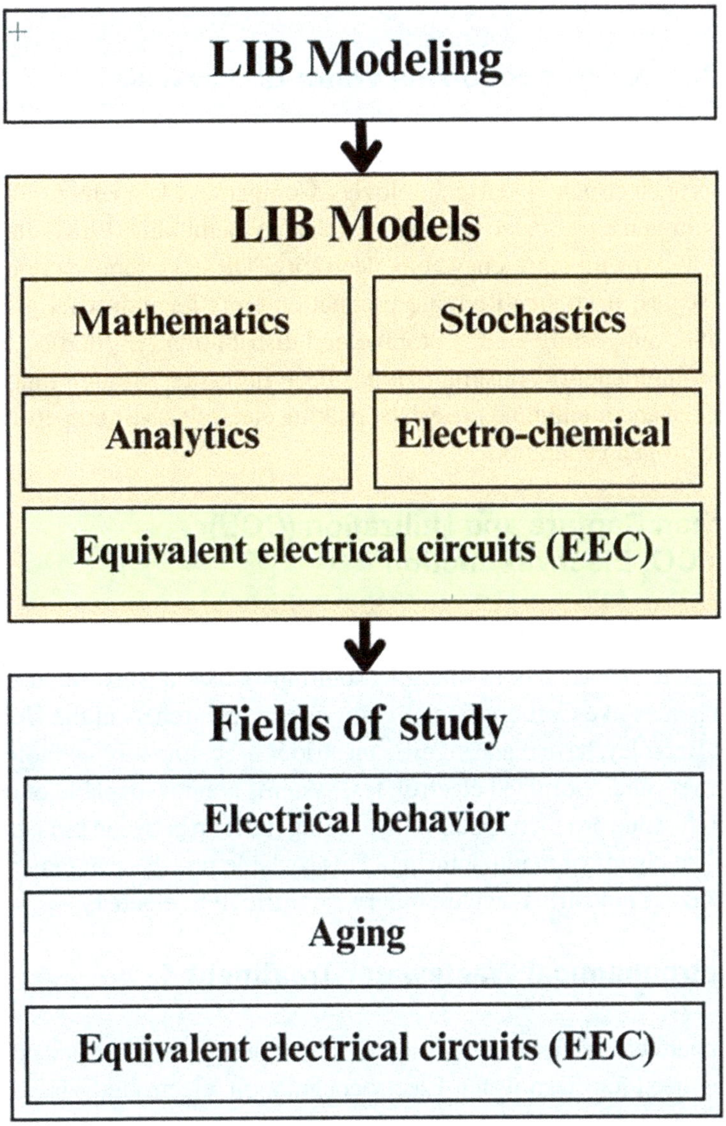

Initial laboratory research into high-energy-density electrodes and electrolytes has been scaled up by companies like Tesla, Panasonic, and CATL to produce large quantities of batteries with exceptional energy storage capacities. These systems power EVs, making them more sustainable and viable for mass adoption (Koech et al., 2024). The success of these batteries can be attributed to advancements in materials (e.g., lithium iron phosphate and nickel-cobalt-manganese cathodes) and manufacturing techniques that enable high-volume, cost-effective production. Figure 4. explains lithium ion battery models and their applications.

10.2 Hydrogen Production via Water Electrolysis

Green hydrogen production through water electrolysis is one of the most promising applications of electrochemical technologies. Companies like Siemens, ITM Power, and Nel Hydrogen have scaled up proton exchange membrane (PEM) electrolyzers to produce hydrogen using renewable electricity. These systems can be deployed at various scales, from small on-site production units for industrial use to large-scale facilities supporting energy storage and distribution (El-Shafie, 2023). The challenge of high energy consumption has been mitigated through innovations in catalyst design and membrane materials, making electrolysis a competitive method for green hydrogen generation.

10.3 Carbon Capture and Utilization (CCU) Through CO_2 Electroreduction

The electrochemical reduction of CO_2 into useful chemicals, such as ethylene, methanol, and formic acid, is an emerging solution for tackling greenhouse gas emissions. Companies like Carbon Clean Solutions and researchers at the University of California, Berkeley, have made significant strides in scaling this technology. Using electrocatalysts and optimized electrolytes, systems are now capable of converting CO_2 into high-value products, potentially closing the loop between industrial carbon emissions and chemical manufacturing. This scalable process can play a vital role in decarbonizing industries that are otherwise difficult to electrify.

10.4 Electrochemical Wastewater Treatment

Electrochemical processes for wastewater treatment are being scaled up to address the growing need for clean water. Electrocoagulation, electrodialysis, and electrochemical oxidation have been successfully implemented in industrial wastewater management. Companies like Aqualia and Electrochemical Solutions have developed large-scale systems to treat water efficiently, removing contaminants such as heavy metals, organics, and pathogens. These electrochemical systems are gaining favor due to their lower energy consumption, ease of automation, and minimal chemical use compared to traditional methods like flocculation(Yoshimura et al., 2022).

10.5 Redox Flow Batteries for Large-Scale Energy Storage

Redox flow batteries (RFBs) are a promising electrochemical technology for large-scale energy storage, providing grid stability and balancing intermittent renewable energy sources like solar and wind. Companies such as Vanadium Flow and ESS Inc. have scaled RFB technology, focusing on enlightening the efficiency, cycle life, and cost-effectiveness of these batteries. RFBs store energy in liquid electrolytes and offer unique advantages like scalable capacity, long cycle life, and safety. Their ability to provide grid-scale storage solutions makes them critical in the transition to renewable energy infrastructure.

10.6 Electrochemical Synthesis of Fine Chemicals

The electrochemical synthesis of fine chemicals, such as pharmaceuticals, agrochemicals, and specialty chemicals, is rapidly scaling up as an alternative to traditional catalytic processes. Companies like BASF and Dow Chemical are exploring electrochemical methods for efficient and sustainable chemical production. Innovations in electrochemical reactors and flow chemistry systems enable the direct conversion of low-cost feedstocks into high-value chemicals with minimal waste and energy consumption (Olabi, Allam, et al., 2023). These electrochemical processes offer a cleaner, more sustainable approach to industrial chemical production.

10.7 Electrorefining in Metal Production

Electrorefining, the process of purifying metals like copper, gold, and nickel through electrochemical means, has been successfully scaled in industries like mining and metallurgy. Companies such as Glencore and Aurubis have adopted large-scale electrorefining systems, replacing traditional pyrometallurgical processes with more energy-efficient electrochemical methods. Electrorefining offers several advantages, including better metal purity, reduced environmental impact, and the ability to recycle materials efficiently. With growing demand for sustainable mining practices, electrorefining is controlled to play a essential role in the future of the metals industry.

CONCLUSION

The journey from laboratory breakthroughs to scalable electrochemical applications remains one of the most promising frontiers in modern engineering and sustainability. This chapter has highlighted how innovations in materials, device design, and system integration are steadily addressing core challenges of efficiency, durability, and cost. Yet, for these technologies to move beyond proof-of-concept and achieve widespread commercialization, critical issues such as material availability, long-term stability, and manufacturing efficiency must be systematically addressed. Importantly, scalability is not driven by scientific advances alone but also by economic viability. Reductions in capital and operational costs through the use of earth-abundant materials, automated manufacturing, and process intensification will play a decisive role in accelerating adoption. Coupled with supportive industrial policies, market incentives, and collaborative partnerships, these cost-driven strategies can create favorable conditions for rapid deployment at scale. Encouragingly, progress across sectors such as energy storage, carbon capture, water treatment, and green chemical production already demonstrates the transformative potential of electrochemical systems. These technologies are uniquely positioned to provide low-carbon, resource-efficient solutions to some of the world's most pressing challenges. In conclusion, the transition of electrochemical innovations into scalable, real-world applications will require a convergence of scientific breakthroughs, engineering optimization, and economic strategies. With sustained innovation, investment in infrastructure, and alignment with global sustainability goals, electrochemical technologies are poised to become central pillars of a cleaner, more resilient, and economically viable future.

REFERENCES

Abdelkareem, M. A., Maghrabie, H. M., Abo-Khalil, A. G., Adhari, O. H. K., Sayed, E. T., Radwan, A., Rezk, H., Jouhara, H., & Olabi, A. G. (2022). Thermal management systems based on heat pipes for batteries in EVs/HEVs. *Journal of Energy Storage*, *51*, 104384. DOI: 10.1016/j.est.2022.104384

Abo-Khalil, A. G., Abdelkareem, M. A., Sayed, E. T., Maghrabie, H. M., Radwan, A., Rezk, H., & Olabi, A. G. (2022). Electric vehicle impact on energy industry, policy, technical barriers, and power systems. *International Journal of Thermofluids*, *13*, 100134. DOI: 10.1016/j.ijft.2022.100134

Agrawal, D., Mahajan, N., Singh, S. A., & Sreedhar, I. (2024). Green hydrogen production pathways for sustainable future with net zero emissions. *Fuel*, *359*, 130131. DOI: 10.1016/j.fuel.2023.130131

Amir, M., Deshmukh, R. G., Khalid, H. M., Said, Z., Raza, A., Muyeen, S. M., Nizami, A.-S., Elavarasan, R. M., Saidur, R., & Sopian, K. (2023). Energy storage technologies: An integrated survey of developments, global economical/environmental effects, optimal scheduling model, and sustainable adaption policies. *Journal of Energy Storage*, *72*, 108694. DOI: 10.1016/j.est.2023.108694

Boumya, W., Taoufik, N., Achak, M., Bessbousse, H., Elhalil, A., & Barka, N. (2021). Electrochemical sensors and biosensors for the determination of diclofenac in pharmaceutical, biological and water samples. *Talanta Open*, *3*, 100026. DOI: 10.1016/j.talo.2020.100026

Brinkert, K., & Mandin, P. (2022). Fundamentals and future applications of electrochemical energy conversion in space. *NPJ Microgravity*, *8*(1), 52. DOI: 10.1038/s41526-022-00242-3 PMID: 36424420

Buglioni, L., Raymenants, F., Slattery, A., Zondag, S. D. A., & Noël, T. (2022). Technological Innovations in Photochemistry for Organic Synthesis: Flow Chemistry, High-Throughput Experimentation, Scale-up, and Photoelectrochemistry. *Chemical Reviews*, *122*(2), 2752–2906. DOI: 10.1021/acs.chemrev.1c00332 PMID: 34375082

El-Shafie, M. (2023). Hydrogen production by water electrolysis technologies: A review. *Results in Engineering*, *20*, 101426. DOI: 10.1016/j.rineng.2023.101426

Ferretti, A. C., Cohen, B., Deng, L., Diwan, M., Frederick, M. O., & Lehnherr, D. (2025). Adoption of Electrochemistry within the Pharmaceutical Industry: Insights from an Industry-Wide Survey. *Organic Process Research & Development*, *29*(2), 322–332. DOI: 10.1021/acs.oprd.4c00353

Han, J., Bai, X., Xu, X., Bai, X., Husile, A., Zhang, S., Qi, L., & Guan, J. (2024). Advances and challenges in the electrochemical reduction of carbon dioxide. *Chemical Science (Cambridge)*, *15*(21), 7870–7907. DOI: 10.1039/D4SC01931H PMID: 38817558

Huang, Q., Zeng, L., Lashari, N., Huang, Z., & Xie, X. (2025). A comprehensive review on the techno-economic analysis of electrochemical energy storage systems: Technologies, applications, benefits and trends. *Future Batteries*, *5*, 100031. DOI: 10.1016/j.fub.2025.100031

Koech, A. K., Mwandila, G., & Mulolani, F. (2024). A review of improvements on electric vehicle battery. *Heliyon*, *10*(15), e34806. DOI: 10.1016/j.heliyon.2024.e34806 PMID: 39170484

Luo, Y., Rao, Z., Yang, X., Wang, C., Sun, X., & Li, X. (2024). Safety concerns in solid-state lithium batteries: From materials to devices. *Energy & Environmental Science*, *17*(20), 7543–7565. DOI: 10.1039/D4EE02358G

Olabi, A. G., Abbas, Q., Shinde, P. A., & Abdelkareem, M. A. (2023). Rechargeable batteries: Technological advancement, challenges, current and emerging applications. *Energy*, *266*, 126408. DOI: 10.1016/j.energy.2022.126408

Olabi, A. G., Abdelkareem, M. A., Wilberforce, T., Alkhalidi, A., Salameh, T., Abo-Khalil, A. G., Hassan, M. M., & Sayed, E. T. (2022). Battery electric vehicles: Progress, power electronic converters, strength (S), weakness (W), opportunity (O), and threats (T). *International Journal of Thermofluids*, *16*, 100212. DOI: 10.1016/j.ijft.2022.100212

Olabi, A. G., Allam, M. A., Abdelkareem, M. A., Deepa, T. D., Alami, A. H., Abbas, Q., Alkhalidi, A., & Sayed, E. T. (2023). Redox Flow Batteries: Recent Development in Main Components, Emerging Technologies, Diagnostic Techniques, Large-Scale Applications, and Challenges and Barriers. *Batteries*, *9*(8), 409. DOI: 10.3390/batteries9080409

Olabi, A. G., Bahri, A., Abdelghafar, A. A., Baroutaji, A., Sayed, E. T., Alami, A. H., Rezk, H., & Abdelkareem, M. A. (2021). Large-vscale hydrogen production and storage technologies: Current status and future directions. *International Journal of Hydrogen Energy*, *46*(45), 23498–23528. DOI: 10.1016/j.ijhydene.2020.10.110

Patra, S., Verma, J., Mishra, Y. K., Kurinec, S., Wang, Q., Syväjärvi, M., & Tiwari, A. (2023). The positioning of biofuel cells-based biobatteries for net-zero energy future. *Journal of Energy Storage*, *72*, 107919. DOI: 10.1016/j.est.2023.107919

Qadeer, M. A., Zhang, X., Farid, M. A., Tanveer, M., Yan, Y., Du, S., Huang, Z.-F., Tahir, M., & Zou, J.-J. (2024). A review on fundamentals for designing hydrogen evolution electrocatalyst. *Journal of Power Sources, 613*, 234856. DOI: 10.1016/j.jpowsour.2024.234856

Rahman, M. M., Imani, S., Anjum, N., Sijuade, A. A., & Okoli, O. (2025). Materials and design strategies for next-generation energy storage: A review. *Renewable & Sustainable Energy Reviews, 212*, 115368. DOI: 10.1016/j.rser.2025.115368

Ryabicheva, M., Zhigalenok, Y., Abdimomyn, S., Skakov, M., Miniyazov, A., Zhanbolatova, G., Mukhamedova, N., Ospanova, Z., Djenizian, T., & Malchik, F. (2025). From lab to market: Economic viability of modern hydrogen evolution reaction catalysts. *Fuel, 395*, 135227. DOI: 10.1016/j.fuel.2025.135227

Shi, Y., Xiao, S., Jiao, S., Wang, Y., Jiang, F., Wang, R., Zhang, Y., Liu, J., Qiao, J., Hu, Z., Yang, S., & Zhu, X. (2024). Hydrogen evolution electrodes: Materials and mechanisms in alkaline electrolysis. *Desalination, 586*, 117887. DOI: 10.1016/j.desal.2024.117887

Silveri, F., Della Pelle, F., & Compagnone, D. (2025). Recent advances in sustainable strategies for the integration of nanostructured sensing surfaces in electroanalytical devices. *Trends in Analytical Chemistry, 185*, 118175. DOI: 10.1016/j.trac.2025.118175

Sohrabi, H., Dezhakam, E., Khataee, A., Nozohouri, E., Majidi, M. R., Mohseni, N., Trofimov, E., & Yoon, Y. (2022). Recent trends in layered double hydroxides based electrochemical and optical (bio)sensors for screening of emerging pharmaceutical compounds. *Environmental Research, 211*, 113068. DOI: 10.1016/j.envres.2022.113068 PMID: 35283073

Varhade, S., Guruji, A., Singh, C., Cicero, G., García-Melchor, M., Helsen, J., & Pant, D. (2025). Electrochemical CO 2 Reduction: Commercial Innovations and Prospects. *ChemElectroChem, 12*(2), e202400512. Advance online publication. DOI: 10.1002/celc.202400512

Xiao, B.-H., Xiao, K., Li, J.-X., Xiao, C.-F., Cao, S., & Liu, Z.-Q. (2024). Flexible electrochemical energy storage devices and related applications: Recent progress and challenges. *Chemical Science (Cambridge), 15*(29), 11229–11266. DOI: 10.1039/D4SC02139H PMID: 39055032

Xu, J., Yang, Y., Jin, H., Zheng, Y., & Qiao, S.-Z. (2025). Bridging gaps between lab- and fab-oriented anode design for proton exchange membrane water electrolyzers. *Chem, 11*(1), 102305. DOI: 10.1016/j.chempr.2024.09.004

Yammine, P., El-Nakat, H., Kassab, R., Mansour, A., El Khoury, B., Koumeir, D., Matar, Z., & Chmayssem, A. (2024). Recent Advances in Applied Electrochemistry: A Review. *Chemistry (Basel, Switzerland)*, *6*(3), 407–434. DOI: 10.3390/chemistry6030024

Yoshimura, R., Wai, S., Ota, Y., Nishioka, K., & Suzuki, Y. (2022). Effects of Artificial River Water on PEM Water Electrolysis Performance. *Catalysts*, *12*(9), 934. DOI: 10.3390/catal12090934

Zheng, L., Cao, M., Du, Y., Liu, Q., Emran, M. Y., Kotb, A., Sun, M., Ma, C.-B., & Zhou, M. (2024). Artificial enzyme innovations in electrochemical devices: Advancing wearable and portable sensing technologies. *Nanoscale*, *16*(1), 44–60. DOI: 10.1039/D3NR05728C PMID: 38053393

About the Contributors

Khaoula Alaoui received his PhD. in Chemistry (2024) and she was awarded his M.Sc. degree in Organic chemistry and Environnemental Sciences (2015) from Ibn Tofail University. Then she started to PhD thesis. She had published more than 20 papers in international journals. His work has been cited 300 total citations by 204 documents with h-index=10 (since 2016 in SCOPUS) and presented several communications in symposia and national/international meetings. She has supervised more than 30 Master theses in corrosion sciences. She is a reviewer at several national and international journals. She is the author of a 4 book schapters. Her main research interests are in the field of corrosion sciences and molecular modeling simulation using new tools and approaches such as Artificial intelligence, Machine and Deep Learning.

Moussa Ouakki received his PhD thesis in Physical Chemistry of Materials from Ibn Tofail University, Kenitra in 2019. He is an expert in corrosion sciences, especially in corrosion of metals, alloys, and their inhibition in acidic medium using organic and inorganic compounds and plant extracts. His work deals with corrosion inhibition by using chemical, electrochemical techniques and computational calculations. He is currently working at the National Superior School of Chemistry, Ibn Tofail University, Kenitra. He has published more than 105 papers including 5 book chapters cited by 2449 documents and an inventor of 1 patent, with a focus on materials and their applications. He has participated in more than 76 conferences. His H-index is 31 on Scopus.

Younes El Kacimi is currently working as Quality Director in Automotive Industry. He has more than 15 years of industry core experience in Quality and Project Management and academic levels. Younes has built an international profile over the years through research, teaching, and professional experience. He holds expertise in corrosion and corrosion prevention by studying and developing advanced engineering tools and technology. Younes has authored 1 Patent, 8 Books and Books chapters and many other publications in corrosion sciences, nanomaterials and analytical chemistry. He is also an editorial board member and reviewer in different well-renowned international journals.

***

A. V. V. Sudhakar is a Professor in the Department of Electrical and Electronics Engineering at SR University, Warangal. He received his Ph.D. in Power Systems from JNTU Hyderabad, his M.Tech. in High Voltage Engineering from JNTU Kakinada, and his B.E. in Electrical Engineering from Andhra University, Visakhapatnam. With over 23 years of academic and research experience, his areas of expertise include Power System Operation and Control, Multi-area Economic Operation, Grid Integration of Renewable Energy Sources, Smart and Microgrids, Electric Vehicle Charging Technologies, and Battery Management Systems. His current research focuses on Intelligent Algorithms for Multi-area Operation of Power Systems under Deregulated Environments. Dr. Sudhakar has held various academic positions since 1999 and has contributed extensively to the field through teaching, research, and professional development. He can be contacted at email: avv.sudhakar@sru.edu.in

Said Bakkali is a researcher specializing in electrochemistry and materials science, with a strong focus on surface modification techniques, including electrodeposition and electroless deposition. Within the field of electroplating, his research particularly examines the role of organic additives in the deposition mechanism and their impact on the properties of metal coatings

Mounia Bennani Ziatni is a Professor at the Faculty of Science, Ibn Tofail University, Morocco. She specializes in the synthesis of biomaterials based on phosphates, particularly calcium phosphates, focusing on their innovative applications. Her research has extensively explored the adsorption of amino acids on these materials, and more recently, she has shifted towards using phosphates as green corrosion inhibitors, offering eco- friendly solutions. With an h- index of 12 on Scopus and 927 citations, her contributions are widely recognized through numerous publications, cementing her role in the development of sustainable materials for environmental applications.

Mohammed Cherkaoui is a Professor at the Faculty of Science, Ibn Tofail University, Morocco. Cherkaoui specializes in corrosion science, materials chemistry, and electrochemistry, focusing on the development of innovative corrosion inhibitors. His research incorporates electrochemical, computational, and surface characterization techniques, with recent publications highlighting the effectiveness of synthesized compounds and natural extracts in corrosion protection across various environments. He has made significant contributions to the development of eco- friendly corrosion inhibitors and authored numerous publications, with an h- index of 26 on Scopus and 1,979 citations. His work employs advanced methods to enhance the understanding of corrosion mechanisms and inhibitor performance.

Ram Raghotham Rao Deshmukh is a Professor in the Department of Electrical and Electronics Engineering at SR University, Warangal. He holds a Ph.D. in Electrical Engineering from Cardiff University, UK, and a Postgraduate Certificate in Higher Education from the University of Sunderland, UK. He completed his B.Tech. from Nagpur University. Dr Deshmukh brings over 20 years of diverse academic and industry experience, including significant roles in the UK electrical engineering sector. His research interests span Electrical Machines, Electric Vehicles, and Artificial Intelligence. He has held various academic and industrial positions, including Senior Design Engineer at Brush Electrical Machines and Chief Electrical Engineer at ATB Laurence-Scott Ltd., UK. He is currently serving as the Director of Training and Placements at SR University. He can be contacted by email: ramdeshmukh@sru.edu.in

Mohamed Ebn Touhami is a research professor in Ibn Tofail University, specialist in materials sciences engineering and environment. He was born in 1970 in Kenitra City, Morocco. He received his Ph.D. in Physics Chemistry, from Ibn Tofail University, Morocco, 1996. He had published more than 160 papers in international journals. His work has been cited 2744 total citations by 1494 documents with h-index = 27 (since 1996 in SCOPUS) and presented several communications in symposia and national/international meetings. He has developed 7 patents and he has participated in 6 nationalresearch projects. He has supervised 100 Doctoral Theses and 300 Master theses in science. He is a reviewer at several national and international journals and has three books. His main research interests are in the field of chemistry especially for electroless plating of many alloys and corrosion inhibition in several industrial media. He wasresponsible forseveral Master formations. He was director of University Center for Analysis, Expertise, Technology Transfer and Incubator. Currently, he is a Dean of Faculty of Science, Ibn Tofail University, Morocco.

Mohamed Ebn Touhami is a research professor in Ibn Tofail University, specialist in materials sciences engineering and environment. He was born in 1970 in Kenitra City, Morocco. He received his Ph.D. in Physics Chemistry, from Ibn Tofail University, Morocco, 1996. He had published more than 160 papers in international journals. His work has been cited 2744 total citations by 1494 documents with h-index = 27 (since 1996 in SCOPUS) and presented several communications in symposia and national/international meetings. He has developed 7 patents and he has participated in 6 nationalresearch projects. He has supervised 100 Doctoral Theses and 300 Master theses in science. He is a reviewer at several national and international journals and has three books. His main research interests are in the field of chemistry especially for electroless plating of many alloys and corrosion inhibition in several industrial media. He wasresponsible forseveral Master formations. He was director of University Center for Analysis, Expertise, Technology Transfer and Incubator. Currently, he is a Dean of Faculty of Science, Ibn Tofail University, Morocco.

Nouhaila Ferraa completed her Ph.D. in Physical Chemistry of Materials at Ibn Tofail University, Kenitra, in 2023. She specializes in the synthesis and characterization of inorganic compounds, focusing on their use as corrosion inhibitors through electrochemical techniques. Her research involves investigating corrosion inhibition through a blend of chemical and electrochemical methods, along with computational analysis. She has published over seven articles, which have been cited in 20 other works, and has an h- index of 3 on Scopus.

Nouhaila Ferraa completed her Ph.D. in Physical Chemistry of Materials at Ibn Tofail University, Kenitra, in 2023. She specializes in the synthesis and characterization of inorganic compounds, focusing on their use as corrosion inhibitors through electrochemical techniques. Her research involves investigating corrosion inhibition through a blend of chemical and electrochemical methods, along with computational analysis. She has published over seven articles, which have been cited in 20 other works, and has an h- index of 3 on Scopus.

Gaddala Anusha is a Research Scholar in the Department of Electrical and Electronics Engineering at SR University, Warangal. She is currently pursuing her PhD in Power Electronics with a research focus on power electronic converter applications in hydrogen and electric vehicles. She holds a Master of Technology (MTech) in Power Electronics from Balaji Institute of Technology, affiliated with JNTU Hyderabad, and a Bachelor of Technology (B. Tech) in Electrical and Electronics Engineering from Sudheer Reddy College of Engineering, Nizamabad, also under JNTU Hyderabad. Her research interests include high-gain DC-DC converters, intelligent MPPT control techniques, fuel cell systems, and renewable energy integration for electric vehicle applications. She is actively involved in the simulation and modelling of advanced power converters using MATLAB/Simulink. She can be contacted by email: at 2305c20004sru.edu. in.

Amar Habsaoui is a Professor at the Faculty of Science, Ibn Tofail University, Morocco. His research encompasses corrosion inhibition, environmental chemistry, and innovative materials. His significant contributions include studies on the corrosion inhibition properties of natural extracts, such as Mentha pulegium L. and Citrullus colocynthis seed oil, emphasizing eco- friendly solutions. He has authored over a hundred publications, which have been cited in 256 articles, and has an h-index of 12 on Scopus.

John Henao is currently professor at the SECIHTI-CIATEQ A.C. in Querétaro, Mexico, and is member of the mexican system of researchers (SNI). He is materials engineer from the University of Antioquia, Colombia, and obtained his master degree in high performance materials engineering from the University of Limoges, France. He also obtained his PhD. Degree in Engineering and Advanced Technologies from the University of Barcelona, Spain. During the last years, He has been working on the field of thermal spray, biomaterials, and materials for energy and corrosion applications.

Khaoula Mzioud earned her Ph.D. in Physical Chemistry from Ibn Tofail University, Kenitra, in 2023. She specializes in the valorization of aromatic and medicinal plants for sustainable environmental applications. Her research focuses on the use of plant-derived compounds as antioxidants, antifungals, and corrosion inhibitors, as well as on sustainability and materials science. She has published more than 19 scientific articles, which have been cited 196 times, and she holds an h-index of 8 on Scopus.

Carlos Poblano is a metallurgical engineer graduated from Universidad Nacional Autonoma de Mexico where he also obtained a MSc degree in metallurgy. He earned a PhD degree in materials science at the University of Sheffield, UK. Currently, he is director of the advanced materials department at CIATEQ A.C. His research lines include the fabrication of different metallic, ceramic, and composite materials by additive manufacturing and thermal spraying; he also works in powder metallurgy and corrosion science projects.

Ouigua Rochdi A third-year doctoral student in the Department of Chemistry, she is currently investigating the applications of natural and industrial products in various fields, including pharmaceuticals and cosmetics, with a particular focus on metal corrosion and degradation. Her research emphasizes their use as corrosion inhibitors. She has extensive experience in separation and characterization techniques, as well as electrochemical studies.

Madgula Satyanrayana completed his under graduation degree in Electrical and Electronics Engineering from Jatipitha College of Engineering Adilabad, Telangana in 2003 and post-graduation from JNTU Hyderabad in 2011. Currently as a Research Scholar at SR University, Warangal, Telangana. His area of interests are power quality issues classification and mitigation methods using AI techniques, smart grid, power system stability, Integration of renewable energy into the power-grid, power system deregulation. Venkataramana Veeramsetty completed his under graduation from G.M.R Institute of Technology, Rajam, Andhra Pradesh in 2006 and post-graduation from Osmania university in 2011, and Ph.D. from NIT Warangal in 2018. Currently he is working as assistant professor in the department of electrical and electronics engineering in S R engineering college Warangal, Telangana. His area of interests are applications of artificial intelligence in electrical engineering, Integration of renewable energy into the power-grid, power system deregulation, power system state estimation, image classifications and object detection using deep learning techniques, and sequential data processing using recurrent neural networks. He published his research works on these domains in various international journal and conferences. He is also acting as a head of the center for artificial intelligence and deep learning, and he is an official instructor for NVIDIA deep learning institute. Durgam Rajababu received his B.Tech degree in Electrical and Electronics Engineering from JNTUCE, Ananthapuram, India in 1999. He received his M.Tech degree in electrical Power Systems from the same University in 2002. He received his Ph.D degree in 2022 from JNTU, Hyderabad, Telangana, India. He is working as Associate Professor in Electrical and Electronics Engineering department, SR University, Warangal, Telangana, India. His research area of interest is Applications of Power Electronic in Power System and Renewable Energy systems

Victor Manuel Zezatti Flores completed his undergraduate studies at the Faculty of Chemical Sciences and Engineering (FCQeI) of the Autonomous University of the State of Morelos (UAEM), earning a degree in Mechanical Engineering. He continued his academic training at the Center for Research in Engineering and Applied Sciences (CIICAp) at the same university, where he earned his Master's and Doctorate degrees in Applied Sciences, with a specialization in Mechanics. He is an active member of the State and National System of Researchers of CONAHCYT, participating in high-impact collaborative projects aimed at generating knowledge, publishing scientific articles, and writing chapters in specialized books. He currently works as a full-time Research Professor, part of the Academic Faculty "Engineering and Technology Applied to Sustainable Processes." His research focuses on Advanced Technologies in Thermal Engineering and Systems Optimization using the Internet of Things (IoT). In addition to his academic career, he has extensive professional experience as an auditor and technical expert in the metalworking and education sectors. He has worked under the guidelines of the Mexican Accreditation Entity (EMA) and various standards such as ISO 9001, ISO 45001, ISO 19011, and ISO 21001, contributing to the development and certification of industrial processes and educational placements.

Index

A

B

C

D

E

F

G

H

I

L

M

N

P

S

www.ingramcontent.com/pod-product-compliance
Lightning Source LLC
LaVergne TN
LVHW062336300926
849161LV00007B/1710